Hyperthermic Oncology from Bench to Bedside

Satoshi Kokura • Toshikazu Yoshikawa •
Takeo Ohnishi
Editors

Hyperthermic Oncology from Bench to Bedside

Springer

Editors
Satoshi Kokura
Kokura Lab, Health and Medical Sciences
Kyoto Gakuen University
Kyoto, Japan

Toshikazu Yoshikawa
Department of Inflammation and Immunology
Kyoto Prefectural University of Medicine
Kyoto, Japan

Takeo Ohnishi
Department of Radiation Oncology
Nara Medical University
Kashihara, Japan

ISBN 978-981-10-0717-0   ISBN 978-981-10-0719-4 (eBook)
DOI 10.1007/978-981-10-0719-4

Library of Congress Control Number: 2016941936

Printed on acid-free paper

This Springer imprint is published by Springer Nature
The registered company is Springer Science+Business Media Singapore Pte Ltd.

Preface

Heat has known to be effective in treating various diseases including cancer for many centuries. Since the 1960s, scientific hyperthermia research has made remarkable progress and many excellent outcomes have been achieved from various basic research. Clinical hyperthermia research on the other hand has resulted in limited success. There has been some progress however; in recent years, meta-analysis and phase 3 clinical trials for hyperthermia combined with radiotherapy or chemoradiotherapy has shown promising efficacy in various malignancies. Thus, I believe that the *Hyperthermic Oncology from Bench to Bedside* is timely in that it gives a comprehensive overview of the significant findings of basic and clinical hyperthermia research.

Hyperthermia alone has the capability of killing tumor cells, but it has limited potency in clinical practice. The effects of combining hyperthermia with chemotherapeutic agents and radiation have been investigated extensively in clinical settings. As a result, the mechanisms of the synergistic effects have already been clarified by basic research. Recently, many reports have demonstrated that hyperthermia can boost anti-tumor immunity. The efficacy of cancer immunotherapy such as immune checkpoint inhibitors is becoming increasingly evident and it will undoubtedly be more widely utilized in various cancer treatments. Therefore, we believe it necessary to examine the effects of hyperthermia on immune functions at the molecular level and verify the synergystic effects when combined with cancer immunotherapy in clinical practice. This book includes five chapters that cover the latest information concerning the effects of hyperthermia on the immune system from aspects of both basic and clinical science. In addition, we believe it is vital to improve the heating device, heating technology and methods to measure temperature for clinical hyperthermia. As such, our final chapter focuses on, and gives suggestions for such technical improvements.

The *Hyperthermic Oncology from Bench to Bedside* represents the most up-to-date overview of current research in Hyperthermic Oncology and provides readers with valuable information and useful research tools to facilitate their work in this challenging field.

We are delighted to have been involved with this project and we thank all contributors to this book for their outstanding efforts.

Kyoto, Japan Toshikazu Yoshikawa

Contents

Part I
Basic Science in Cultured Cells

Chapter 1
Introduction

Satoshi Kokura

Abstract This book is expected to serve as a source of the latest information on hyperthermia, a type of treatment that utilizes various heating methods, such as electromagnetic therapy, to increase temperature in cells and tissues, which is thought to result in an antitumor effect. The contributors range from basic researchers to clinical specialists, each of whom presents the most recent observations in their respective field. The information contained herein should provide medical staff with useful information in relation to cancer treatment.

Keywords Hyperthermia • Local hyperthermia • Regional hyperthermia • Whole body hyperthermia

1.1 Hyperthermia

Hyperthermia is rather unique in that it is an old yet novel cancer therapy. The earliest known use of hyperthermia to treat cancerous tissue was by the Ancient Greeks, Romans, and Egyptians around 2000 BCE. That treatment method was not entirely unlike present-day thermal ablation. Controlled clinical trials on hyperthermia began in the 1960s in Western countries, and based on the results, continued into the 1970s, expanding to include heating methods such as microwaves and radio-frequency waves for treating cancer. Although results from clinical trials on the effects of hyperthermia with concomitant radiotherapy showed a curative effect, heating the cancer at the correct depth remained extremely difficult. Therefore, to allow more precise heating, capacitive heating devices using 8 MHz radio waves were developed. These devices made hyperthermia acceptable for clinical use, and soon after, hyperthermia began being carried out much more frequently in Japan.

The use of hyperthermia in cancer treatment has numerous benefits: It is relatively inexpensive compared with other treatments; it has not been associated

S. Kokura (✉)
Kokura Lab, Faculty of Health and Medical Sciences, Kyoto Gakuen University, 18, Gotanda-cho, Yamanouchi, Ukyo-ku, Kyoto-shi, Kyoto 615-8577, Japan
e-mail: kokura@kyotogakuen.ac.jp

S. Kokura et al. (eds.), *Hyperthermic Oncology from Bench to Bedside*,
DOI 10.1007/978-981-10-0719-4_1

with any adverse effects, despite the fact that it is often repeated several times for each patient; and, unlike treatment using chemotherapy, it does not induce sickness.

Recently, hyperthermia using relatively low temperatures showed a better anticancer effect when hyperthermia was combined with both radiation therapy and anticancer drugs. The anticancer effect of hyperthermia combined with immunotherapy has also recently been reported, as have similar findings in other novel fields.

In this book, the most recent observations on hyperthermia from basic researchers and clinical specialists have been compiled to serve as a textbook or manual for anyone interested in the use of hyperthermia for cancer treatment.

Presently, hyperthermia is divided into the following three categories: local, regional/part-body, and whole body. When local hyperthermia is used for cancer treatment, heat is applied to a solid tumor. In local hyperthermia, the heating temperature can be as high as 80 °C because the purpose is to completely ablate the tumor [1]. In regional/part-body hyperthermia, the goal is to achieve certain physiological effects such as cell death without causing serious injury to adjoining normal tissue, and thus the heating temperature is typically in the range of 41–45 °C [2]. Alternatively, a heating temperature of 39–41 °C can be used to mimic fever-range effects that do not damage any tissues. This milder temperature range is typically used in whole-body hyperthermia, which is used in conjunction with other therapies to treat metastatic cancer that has spread throughout the body [3].

Currently, there are several methods for inducing hyperthermia. One more traditional method is to expose tumors to external stimuli such as an infrared light, or to submerge the tumor site in a hot water bath. Although these strategies are suitable for tumors that are easily accessible, such as skin tumors, they are not appropriate for tumors of internal organs; in fact, even for skin tumors, precise temperature and spatial control is difficult. Another widely used method is to raise the tumor temperature by inserting metallic probes into the tumor and delivering precise amounts of energy using microwaves or radio waves. This method enables rapid and substantial increases in temperature and is therefore well suited for ablation purposes; however, this method tends to form a significant temperature gradient across the tumor [4], which makes it less attractive for treatments sensitive to small temperature differences, such as those that aim for immunostimulatory responses [5, 6]. Recent progress in techniques such as high-intensity focused ultrasound has enabled non-invasive heating of internal organs and shown promise for local hyperthermia of tumors not near the surface [7]. Imaging techniques are often used to confirm that the heating is occurring at the intended location and temperature range within the tumor [8]. Recently, nanomaterials activated by external energy sources have started being used for local hyperthermia [9]. When nanoparticles are distributed throughout the tumor reasonably uniformly, this method enables more precise temperature control and more uniform temperature distribution [10]. This improved temperature control and distribution is facilitating studies regarding how local hyperthermia stimulates systemic antitumor immune responses.

Cancer stem cells have been proposed as the driving force behind tumorigenesis, recurrence, and metastases, which contribute to the failure of some cancer treatments. Water bath hyperthermia, a novel treatment for cancer, can be highly effective when combined with chemotherapy, radiotherapy, or immunotherapy. Side population (SP) cells are part of a subpopulation of cells that have stem cell-like characteristics. Preliminary experiments have reported a decrease in SP cells in heat-treated colon cancer cells [11], and this result encouraged us to investigate whether hyperthermia had in fact played a role in regulating the stemness of colon cancer cells.

Hyperthermia treatment has broad clinical applications and can be appropriate for a wide variety of situations. This book includes discussions, fundamental descriptions, and clinical observations by a number of clinical specialists on the antitumor effects of hyperthermia treatment, with a primary focus on regional/part-body hyperthermia.

References

1. den Brok MH, Sutmuller RP, van der Voort R, Bennink EJ, Figdor CG, Ruers TJ, et al. *In situ* tumor ablation creates an antigen source for the generation of antitumor immunity. Cancer Res. 2004;64:4024–9.
2. van der Zee J. Heating the patient: a promising approach? Ann Oncol. 2002;13:1173–84.
3. Wust P, Hildebrandt B, Sreenivasa G, Rau B, Gellermann J, Riess H, et al. Hyperthermia in combined treatment of cancer. Lancet Oncol. 2002;3:487–97.
4. Curley SA. Radiofrequency ablation of malignant liver tumors. Oncologist. 2001;6:14–23.
5. Takada T, Yamashita T, Sato M, Sato A, Ono I, Tamura Y, et al. Growth inhibition of re-challenge B16 melanoma transplant by conjugates of melanogenesis substrate and magnetite nanoparticles as the basis for developing melanoma-targeted chemo-thermo-immunotherapy. J Biomed Biotechnol. 2009;2009:457936.
6. Toraya-Brown S, Sheen MR, Zhang P, Chen L, Baird JR, Demidenko E, et al. Local hyperthermia treatment of tumors induces CD8 (+) T cell-mediated resistance against distal and secondary tumors. Nanomed Nanotechnol Biol Med. 2014;10:1273–85.
7. Xiaoping L, Leizhen Z. Advances of high intensity focused ultrasound (HIFU) for pancreatic cancer. Int J Hyperthermia. 2013;29:678–82.
8. Partanen A, Tillander M, Yarmolenko PS, Wood BJ, Dreher MR, Kohler MO. Reduction of peak acoustic pressure and shaping of heated region by use of multifoci sonications in MR-guided high-intensity focused ultrasound mediated mild hyperthermia. Med Phys. 2013;40:013301.
9. Krishnan S, Diagaradjane P, Cho SH. Nanoparticle-mediated thermal therapy: evolving strategies for prostate cancer therapy. Int J Hyperthermia. 2010;26:775–89.
10. Wust P, Gneveckow U, Johannsen M, Bohmer D, Henkel T, Kahmann F, et al. Magnetic nanoparticles for interstitial thermotherapy – Feasibility, tolerance and achieved temperatures. Int J Hyperthermia. 2006;22:673–85.
11. Gao F, Ye YF, Zhang YQ, Yang JQ. Water bath hyperthermia reduces stemness of colon cancer cells. Clin Biochem. 2013;46:1747–50.

Chapter 2
Induction of Oxidative Stress by Hyperthermia and Enhancement of Hyperthermia-Induced Apoptosis by Oxidative Stress Modification

Yoshiaki Tabuchi, Kanwal Ahmed, and Takashi Kondo

Abstract Hyperthermia (HT) is considered to be a possible treatment modality for various cancers, and its pleiotropic effects support its combined use with radiotherapy and/or chemotherapy. However, clinical results by HT alone have not always been satisfactory. In mammalian cells, HT elicits a wide spectrum of alterations in cellular morphology, biochemistry and function. One of these HT-induced alterations, oxidative stress, has been attributed to the increased production of reactive oxygen spaces (ROS), and is known to play an important role as an intracellular mediator of HT-induced cell death, including apoptosis. Indeed, it has been well established that increases in intracellular oxidative stress significantly enhance HT-induced apoptosis. Attention has therefore been focused on the development of heat sensitizers to modulate the intracellular ROS. Interestingly, the modification of oxidative stress via addition of ROS-generating compounds significantly enhanced the apoptosis elicited by HT. In this chapter, we describe the induction of oxidative stress by HT and enhancement of HT-induced apoptosis by oxidative stress modification, and we discuss the possible mechanisms underlying this enhancement.

Keywords Hyperthermia • Oxidative stress • ROS • Apoptosis

Y. Tabuchi (✉)
Division of Molecular Genetics Research, Life Science Research Center, University of Toyama, 2630 Sugitani, Toyama 930-0194, Japan
e-mail: ytabu@cts.u-toyama.ac.jp

K. Ahmed • T. Kondo
Department of Radiological Sciences, Graduate School of Medicine and Pharmaceutical Sciences, University of Toyama, 2630 Sugitani, Toyama 930-0194, Japan

S. Kokura et al. (eds.), *Hyperthermic Oncology from Bench to Bedside*,
DOI 10.1007/978-981-10-0719-4_2

2.1 Introduction

Hyperthermia (HT) induced by heat stress elicits a wide spectrum of alterations in cellular morphology, biochemistry and function in mammalian cells [1]. The most striking cellular changes in response to HT are the disruptions of cytoskeletal structures such as microtubules and microfilaments. Such disruptions lead to loss of the correct organelle localization and the breakdown of intracellular transport processes. Cellular susceptibility to HT depends on the phase of the cell cycle. It has been recognized that the S or M phase of the cell cycle is more sensitive to HT than the G_1 or G_2 phase [2]. On the other hand, recent findings indicate that heat stress induces DNA damage that is detected as DNA double-strand breaks in G_1 and G_2 phase cells but not in S phase cells [3]. Exposure of cells to heat stress leads to protein denaturation and aggregation in the cells, assuming that the accumulation of unfolding proteins is the signal which initiates the heat-shock response. Heat activates the expression of heat responsive genes, including heat shock proteins (HSPs), coincident with global decreases in mRNA and protein synthesis [4]. It is well known that HT induce apoptosis [10]. Heat stress has also been shown to induce oxidative stresses because of the increased production of reactive oxygen spaces (ROS) [5, 6], and oxidative stresses play important roles as intracellular mediators of heat-induced cell death [7–10].

HT has shown promising anticancer effects for various cancers with the advantages of relatively low side effects and minimal damage to normal tissue. Combined treatments with HT and either chemotherapy, radiotherapy or both have been used for patients with cancer in various organs, and the anticancer effects of these combination therapies have been verified in many clinical trials [11]. However, the clinical results by HT alone have not always been sufficient due to biological and technical difficulties. One of the strategies to overcome these difficulties is the development of thermo-sensitizers without affecting normal conditions. HT-induced cell killing, including apoptosis, has been shown to be enhanced by excess oxidative stress via the generation of various compounds [12–25]. Here, the induction of oxidative stress by HT and enhancement of HT-induced apoptosis by the modification of oxidative stress are summarized, and the mechanisms of the enhancement are discussed.

2.2 Induction of Oxidative Stress by HT

ROS, such as superoxide anion (O_2^-), hydrogen peroxide (H_2O_2), hydroxyl radical ($OH^\cdot$), nitric oxide ($^\cdot NO$), peroxynitrite ($ONOO^-$), and lipid peroxides (LPO), are continuously generated in electron transfer reactions in aerobic cells. Sources of ROS have been suggested to include mitochondria, peroxisome, nicotinamide adenine dinucleotide phosphate (NADPH) oxidases, xanthine oxidase and an uncoupled nitric oxide synthase. The levels of ROS are strictly regulated by a

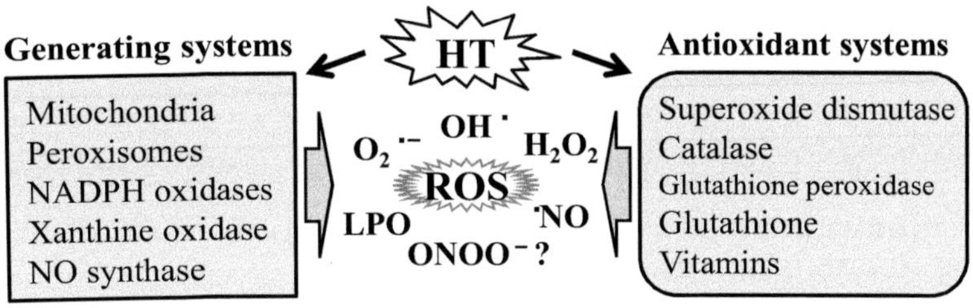

Fig. 2.1 The generating and antioxidant systems for reactive oxygen spaces (ROS). Levels of ROS strictly maintain a balance between generating and antioxidant systems. Hyperthermia (HT) may increase in the levels of ROS to affect one or both systems

system of antioxidants including superoxide dismutase (SOD), catalase, glutathione peroxidase, glutathione and vitamins (Fig. 2.1). Oxidative stress occurs as a consequence of an imbalance between the production of ROS and antioxidant defenses. Relatively low levels of ROS are essential to maintain physiological homeostasis of the cell. On the other hand, increased levels of ROS may be detrimental and lead to cell damage and cell death. HT is one of the factors responsive for production of ROS in cells [5, 6] (Fig. 2.1).

Exposure of cells to HT increases in levels of intracellular O_2^- [19, 20]. The HT-induced elevation of O_2^- is thought to be due to the activation of xanthine oxidase and/or mitochondria respiratory reaction chain [7]. In addition, HT has been shown to decrease the expression level of SOD1 (CuZnSOD) and its enzyme activity, leading to an increase in ROS production [9]. Wan et al. [10] recently reported that a decrease in SOD2 (MnSOD) activity accounted for the enhanced ROS levels in heat-treated platelets. Furthermore, overexpression of SOD2 [26] or addition of SOD/catalase-mimicking nanoparticles [27] significantly decreased the levels of ROS and protected cells against heat stress, suggesting that ROS, mainly O_2^-, play important roles in the cell injury elicited by heat stress. Intracellular SODs converts O_2^- to H_2O_2, and the reaction of these ROS with low-molecular-weight iron complexes can induce LPO [28]. Increases in levels of H_2O_2, $OH^{\cdot}$ and/or LPO were observed in a variety of samples exposed to HT [24, 26]. It has been shown that O_2^- reacts with other radicals such as NO, and one product of the interaction between O_2^- and NO, $ONOO^-$, is a powerful oxidant [29]. The accumulation of inducible NO synthase (iNOS) and NO was caused by heat treatment of the human glioblastoma cell lines bearing a mutant *p53* gene [30]. It is postulated that NO or $ONOO^-$ may play a role under HT conditions. NADPH oxidase is another source of ROS in addition to mitochondria or peroxisomes, and seems to play an important role in the ROS production induced by heat treatment [31]. Moon et al. [31] demonstrated that NADPH-mediated ROS production activates hypoxia-inducible factor-1 via the ERK pathway after HT treatment in human tumor cell lines. On the other hand, either intracellular or mitochondrial ROS generation was significantly blocked by the mitochondria-targeted ROS scavenger Mito-TEMPO in HT-induced apoptosis of human platelets, but not by inhibitors of NADPH oxidase and iNOS, indicating that mitochondrial ROS play a critical role in platelet apoptosis induced by HT [10] (Fig. 2.1).

2.3 Enhancement of HT-Induced Apoptosis by Oxidative Stress Modification

A combination of HT with radio and/or chemotherapy has been shown to be effective for various cancers [11]. However, eradication of cancer cells is not always sufficient due to biological and technical difficulties. To overcome these difficulties, chemical thermos-sensitizers that promote apoptosis may be of benefit. Previous reports [7–10] indicated that the elevations of intracellular ROS induced by HT are one of the major mechanisms involved in HT-induced cell damage. It was also proposed that oxidative stress could be a general mediator of apoptosis [32]. For this reason, modulation of the intracellular generation of ROS is one approach to enhance HT-induced apoptosis. In addition, an ideal sensitizer would be nontoxic at normal temperatures but would became toxic at HT temperatures. Various types of thermo-sensitizers that modify oxidative stress to potentiate apoptosis have been demonstrated [12–25].

The water soluble azo compound, 2,2'-azobis (2-amidinopropane) dehydrochloride (AAPH), has been used as a free radical to generate compounds activated by heat or light [33]. HT combined with AAPH generates alkyl radicals and alkyl peroxy radicals. This compound has been reported to increase the thermo-sensitivity in Chinese hamster V79 cells [12] or human colon cancer HT-29 cells [13]. We found that HT (44 °C, 10 min) increased the intracellular calcium ion concentration ($[Ca^{2+}]_i$) arising from increased expression of inositol 1,4,5-trisphosphate receptor 1 (IP_3R1) and LPO in human lymphoma U937 cells, while an additional increase in $[Ca^{2+}]_i$ due to further increase in LPO and the activation of a mitochondria-caspase-dependent pathway played a major role in thermosensitization by AAPH [14] (Fig. 2.2). On the other hand, the ability of this compound to promote HT was attributed to elevations of LPO and $[Ca^{2+}]_i$ but not to activation of the mitochondria-caspase-dependent pathway in human cervical cancer cells, HeLa and CaSki, suggesting that the thermosensitizing mechanism of AAPH is cell-type specific [16]. In addition, the thermosensitizing effect of the more lipophilic azo compound 2,2'-azobis (2,4-dimethylvaleronitrile) (AMVN), a free radical initiator, was found to be approximately 2-fold more potent than that of the water soluble AAPH on the apoptosis in U937 cells [15].

A member of pterin family, 6-formylpterin (6-FP; 2-amino-4-oxo-3,4-dihydropteridine-6-carbaldehyde), is a metabolite of folic acid in cells, and was found to be more abundant in some cancer cells [34]. This compound has unique effects, such as the inhibition of xanthine oxidase, the modulation of iNOS and the generation of intracellular H_2O_2 by transforming electrons from NAD(P)H to oxygens [35]. When U937 cells were treated by the combination of HT (44 °C,

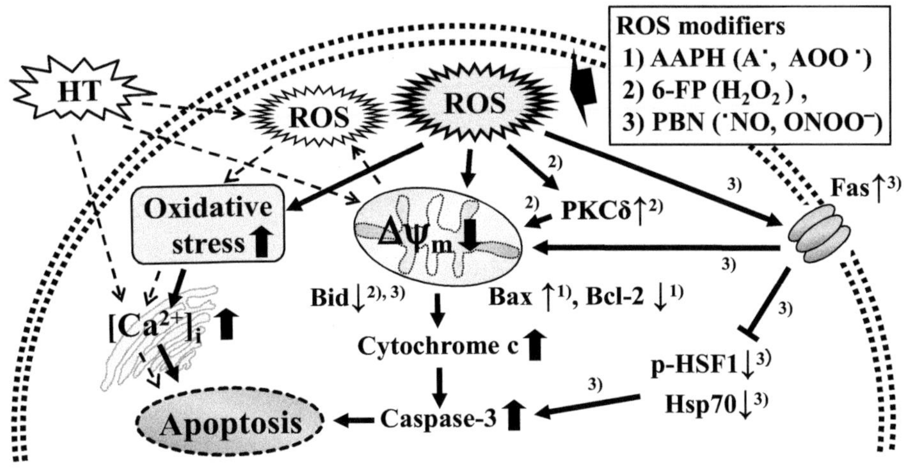

Fig. 2.2 A schematic illustration of the mechanism of enhancement of hyperthermia (HT)-induced apoptosis by reactive oxygen spaces (ROS) modifiers in human lymphoma U937 cells. HT (44 °C, 10 or 20 min) combined with AAPH, 6-formylpterin (6-FP) or PBN generated alkyl radicals (A) and alkyl peroxy radicals (AOO), hydrogen peroxide (H_2O_2) or nitric oxide (NO) and peroxynitrite ($ONOO^-$), respectively, and increased intracellular oxidative stress. Both elevation of the intracellular calcium ion concentration ($[Ca^{2+}]_i$) and loss of mitochondrial membrane potential (ΔΨm) followed by cytochrome c release and caspase-3 activation were commonly detected in each combination, and al combinations were found to significantly enhance apoptosis. In addition, compound-specific mechanisms were described: (1, AAPH) A significant increase in Bax expression and decrease in Bcl-2 expression were observed. (2, 6-FP) A significant decrease in Bid expression and increase in protein kinase δ (PKCδ) expression in mitochondria were observed. (3) A significant decrease in Bid expression and increase in Fas expression on the cell surface was observed. Both the level of phosphorylated heat shock transcription factor 1 (p-HSF1) and the level of Hsp70 declined

20 min) and 6-FP, the levels of both intracellular H_2O_2 and apoptosis were significantly higher than those of treatment by HT alone. Several functional analyses have revealed that the increase in $[Ca^{2+}]_i$, activation of the mitochondria-caspase-dependent pathway and the translocation of PKCδ to mitochondria play important roles in the enhancement of HT-induced apoptosis by 6-FP [17] (Fig. 2.2). Previous findings suggest that PKCδ, a member of the protein kinase C family, is activated in apoptotic cells treated with oxidative stress or HT [36].

Although the spin-trapping reagent, N-tert-butyl-alpha-phenylnitrone (PBN), has been known as antioxidant, it was shown that this reagent can be decomposed with hydroxyl radicals by the Fenton reaction under oxidative conditions, thereby generating NO [37]. We hypothesized that PBN could be a sensitizer of HT-induced apoptosis, because HT causes intracellular oxidative stress [5, 6]. Treatment of U937 cells with HT (44 °C, 10 min) plus PBN increased the levels of nitrite in the culture medium, whereas this treatment decreased the production of O_2^- [20]. Enhancement of HT-induced apoptosis by PBN was observed in U937 cells [18–20]. Under the conditions employed, scavengers of NO and $ONOO^-$ markedly prevented the enhancement of apoptosis, suggesting that the NO released from PBN by HT and the $ONOO^-$ generated by reaction between NO with O_2^- are involved in the enhancement of apoptosis. Furthermore, we found that the mitochondria caspase and $[Ca^{2+}]_i$-dependent pathways also contributed to HT-induced apoptosis [20] (Fig. 2.2). In addition, an elevation of Fas and reduction of Hsp70 and phosphorylated heat shock transcription factor 1 (HSF1) were observed following the combined treatment. Much evidence has been obtained supporting the activation of Fas in NO-induced apoptosis and $ONOO^-$-induced apoptosis [38]. Fas, a death receptor, is able to trigger apoptosis via direct activation of the caspase cascade. The activation of Fas has been known to reduce the levels of upregulated Hsp70 and phosphorylated HSF1 in U937 cells exposed to heat stress [39]. The roles of Fas in the enhancement of HT-induced apoptosis by PBN are thought to be related to the downregulation of both HSF1 and Hsp70 [20] (Fig. 2.2).

The nitroxide Tempo (2,2,6,6-tetramethylpiperidine-N-oxyl) exerts the SOD-mimic antioxidant action, but produces H_2O_2 upon O_2^- dismutation in cells [40]. Matsuya et al. [41] newly synthesized benzocycloalkene and furan-fused tetracyclic compounds. From the viewpoint of the bioactivity, such as the antiviral and anti-apoptotic properties, a TIPS (triisopropylsilyl) substituent plays an important role. Interestingly, benzocycloalkene and furan-fused tetracyclic compounds with TIPS were found to produce ROS including O_2^- and H_2O_2 [21, 22]. Sanazole (AK-2123; N-(2-Methoxy- ethyl)-3-nitro-1H-1,2,4-triazole-1-acetamide) was a perspective hypoxic cell rediosensitizer due to its lower toxicity and hyper-radiosensitising effects [42]. Under aerobic conditions, sanazole may generate O_2^- in the presence of the enzyme system NADPH/cytochrome P450 reductase [43]. As we expected, the combination of each compound and HT treatment significantly enhanced the apoptosis via ROS generation in U937 cells [21–25].

2.4 Induction of Genes Involved in Apoptosis by the Combination of a ROS Modifier and HT

HT effects many aspects of the cellular functions in mammalian cells [1]. However, the overall responses to HT in the cells remain unknown. Transcript profiling technologies including DNA microarray have made it possible to profile global gene expression patterns associated with different biological responses. Moreover, bioinformatics analysis now allows for the mapping of gene expression data into relevant pathway maps based on their functional annotation and known molecular interactions. Using either microarray or bioinformatics technology, many genes and gene networks responsive to HT have been identified [18, 19, 44, 45]. Moreover, proteomics techniques have revealed valuable information about the HT response in cancer cells [46, 47].

When U937 cells were exposed to HT (44 °C, 10 min; 15 % apoptosis level), 39 upregulated and 3 downregulated genes were identified. From these genes and our previous findings [18, 19, 45], we found a gene network containing many HSPs,

BCL2-associated athanogene 3 (BAG3), protein kinases c-JUN N-terminal kinase (JNK; MAPK8), nuclear factor, erythroid 2-like 2 (NFE2L2; NRF2) and caspase-3 (CASP3) [18] (Fig. 2.3a). In mammals, the expression of HSPs, HSPA (heat shock 70 kDa proteins) family members, such as HSPA1A/B and HSPA6, DNAJ (DNAJ (Hsp40) heat shock proteins) family members, such as DNAJA1 and DNAJB1, or BAG3 is mainly regulated by HSF1 [48, 49], and these proteins function as cytoprotective and anti-apoptotic molecules against various stressors, particularly HT [1, 48, 49]. The induction of HSPs appears to be a common response to HT as described in many previous microarray studies [18, 44, 45]. Sequential activation of caspases plays a central role in the execution phase of apoptosis. It is already well known that the JNK pathway is involved in apoptosis evoked by HT [50]. In U937 cells, activations of CASP3 and JNK have been observed under apoptotic conditions [18, 19, 45]. Interestingly, functional category analysis predicted activation of an NRF2-mediated antioxidant response in HT-treated U937 cells [45]. NRF2 is an emerging regulator of cellular resistance to oxidants, including ROS. Heme oxygenase 1 (HMOX1) functions as an adaptive mechanism to protect cells from

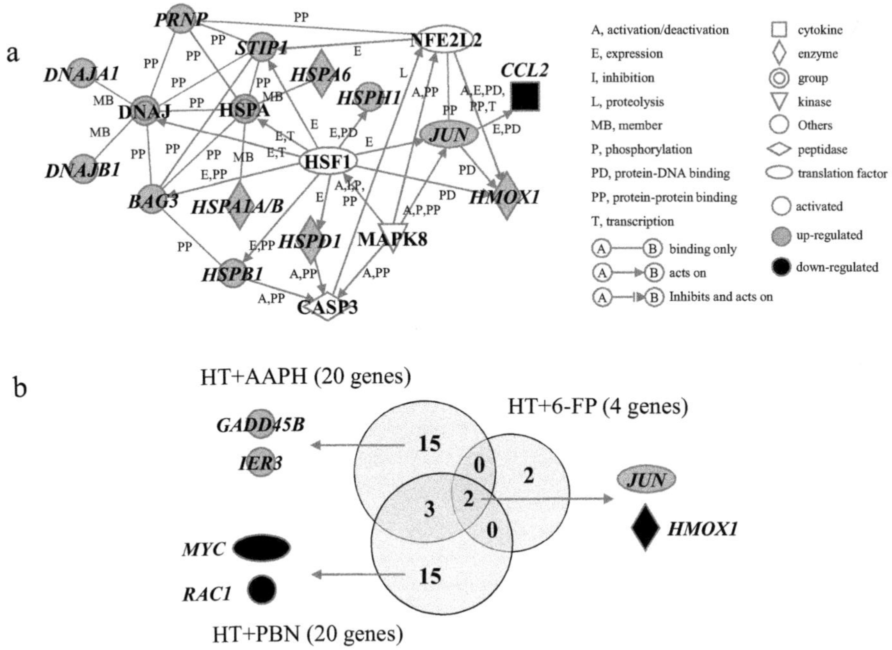

Fig. 2.3 (**a**) A gene network containing genes that were differentially expressed in U937 cells exposed to HT (44 °C, 10 min) only. The network is shown graphically as nodes (genes) and edges (the biological associations between the nodes). Nodes and edges are displayed using various shapes and labels that present the functional class of genes and the nature of the relationship between the nodes, respectively. (**b**) Venn diagram of genes that were differentially expressed in the cells treated with the combination of HT plus AAPH, 6-FP or PBN. The numbers of gene are shown. Genes that are associated with apoptosis are also demonstrated

oxidative damage during stress, and is induced by HT in U937 cells [18]. HT has been reported to induce oxidative stress accompanied by an elevation of ROS [8] and an increase in the expression level of NRF2 [51]. Although heat is a principal activator of HSF1, previous findings suggested that this transcription factor can be activated directly by oxidative stress [52]. In general, the heat shock response and oxidative stress response are thought to be closely related. In fact, HSF1 has been shown to be directly involved in the regulation of HMOX1 via its function as an anti-oxidative [53]. The expressions of HMOX1 is also regulated by NRF2. Wang et al. [54] also demonstrated that HSF1 and NRF2 are prominent mediators of oxidative signaling in oxidative stress-mediated apoptosis in cancer cells. It is noteworthy that HSF1 and NRF2 engage in crosstalk for cytoprotection [55].

In U937 cells, the combined treatment with HT (44 °C, 10 min) and either a ROS modifier, AAPH (20 mM), 6-FP (0.3 mM) or PBN (2 mM), enhanced the apoptosis level (approximately 30 %). The numbers of genes that were differentially expressed in cells treated with each combination compared with the number in cells treated by HT alone are demonstrated in Fig. 2.3b. These results indicated the chemical-specific gene expression pattern [18]. However, unexpectedly, only modification of change of gene expression was observed. On the other hand, interestingly, a decrease of HMOX1 expression and an increase of Jun proto-oncogene (JUN) were commonly observed in HT-treated cells under 3 different oxidative stress conditions. HMOX1 [56] and JUN [57] are known to inhibit and promote apoptosis induced by stresses including HT. The expression JUN, a major component of AP-1 transcription factor, is known to be regulated by JNK. JNK/c-JUN signaling pathway has been reported to play a role in the regulation of HMOX1 expression and HMOX1-mediated cytoprotection [58]. Recently, Sawai et al. [59] showed that HSF1 bound to the Jun promoter and was necessary for its efficient response to heat shock. Under the combination of HT and AAPH conditions, the gene expressions of growth arrest and DNA-damage-inducible, beta (GADD45B) and immediate early response 3 (IER3) were upregulated; both genes have been reported to participate in the induction of apoptosis [60, 61]. On the other hand, the downregulation of gene expressions of v-myc avian myelocytomatosis viral oncogene homolog (MYC) [62] and ras-related C3 botulinum toxin substrate 1 (rho family, small GTP binding protein Rac1) (RAC1) [63], which have cytoprotective activity, was observed in the HT-treated cells with PBN. In a future study, it will be important to elucidate the functions of these candidate genes in order to better understand the mechanism of enhancement of HT-induced apoptosis by ROS modulators.

2.5 Conclusion

HT is considered to be a promising approach in cancer therapy. The combination of a ROS modulator plus HT may come to be a promising approach for the treatment of cancer. Moreover, further studies elucidating the detail molecular mechanism responsible for the enhancement of HT-induced apoptosis will lead to the development of novel strategies for use in HT therapy.

References

1. Richter K, Haslbeck M, Buchner J. The heat shock response: life on the verge of death. Mol Cell. 2010;40:253–66. doi:10.1016/j.molcel.2010.10.006.
2. Westra A, Dewey WC. Variation in sensitivity to heat shock during the cell-cycle of Chinese hamster cells in vitro. Int J Radiat Biol Relat Stud Phys Chem Med. 1971;19:467–77.
3. Velichko AK, Petrova NV, Kantidze OL, Razin SV. Dual effect of heat shock on DNA replication and genome integrity. Mol Biol Cell. 2012;23:3450–60. doi:10.1091/mbc.E11-12-1009.
4. Mariner PD, Walters RD, Espinoza CA, Drullinger LF, Wagner SD, Kugel JF, et al. Human Alu RNA is a modular transacting repressor of mRNA transcription during heat shock. Mol Cell. 2008;29:499–509. doi:10.1016/j.molcel.2007.12.013.
5. Finkel T, Holbrook NJ. Oxidants, oxidative stress and the biology of ageing. Nature. 2000;408:239–47.
6. Slimen IB, Najar T, Ghram A, Dabbebi H, Ben Mrad M, Abdrabbah M. Reactive oxygen species, heat stress and oxidative-induced mitochondrial damage. A review. Int J Hyperthermia. 2014;30:513–23. doi:10.3109/02656736.2014.971446.
7. Skibba JL, Quebbeman EJ, Kalbfleisch JH. Nitrogen metabolism and lipid peroxidation during hyperthermic perfusion of human livers with cancer. Cancer Res. 1986;46:6000–3.
8. Katschinski DM, Boos K, Schindler SG, Fandrey J. Pivotal role of reactive oxygen species as intracellular mediators of hyperthermia-induced apoptosis. J Biol Chem. 2000;275:21094–8.
9. El-Orabi NF, Rogers C, Edwards HG, Schwartz DD. Heat-induced inhibition of superoxide dismutase and accumulation of reactive oxygen species leads to HT-22 neuronal cell death. J Therm Biol. 2011;36:49–56. doi:10.1016/j.jtherbio.2010.11.002.
10. Wang Z, Cai F, Chen X, Luo M, Hu L, Lu Y. The role of mitochondria-derived reactive oxygen species in hyperthermia-induced platelet apoptosis. PLoS One. 2013;8:e75044. doi:10.1371/journal.pone.0075044.
11. Westermann A, Mella O, Van Der Zee J, Jones EL, Van Der Steen-Banasik E, Koper P, et al. Long-term survival data of triple modality treatment of stage IIB-III-IVA cervical cancer with the combination of radiotherapy, chemotherapy and hyperthermia – an update. Int J Hyperthermia. 2012;28:549–53. doi:10.3109/02656736.2012.673047.
12. Krishna MC, Dewhirst MW, Friedman HS, Cook JA, DeGraff W, Samuni A, Russo A, Mitchell JB. Hyperthermic sensitization by the radical initiator 2,2'-azobis (2-amidinopropane) dihydrochloride (AAPH). I. In vitro studies. Int J Hyperthermia. 1994;10:271–81.
13. Wang CC, Chen F, Kim E, Harrison LE. Thermal sensitization through ROS modulation: a strategy to improve the efficacy of hyperthermic intraperitoneal chemotherapy. Surgery. 2007;142:384–92.
14. Li FJ, Kondo T, Zhao QL, Tanabe K, Ogawa R, Li M, et al. Enhancement of hyperthermia-induced apoptosis by a free radical initiator, 2,2'-azobis (2-amidinopropane) dihydrochloride, in human histiocytic lymphoma U937 cells. Free Radic Res. 2001;35:281–99.
15. Li FJ, Kondo T, Zhao QL, Hayashi Y, Ogawa R, Cui ZG, et al. A lipophilic free radical initiator, 2,2'-azobis (2,4-dimethylvaleronitrile) (AMVN) enhances caspase-dependent apoptosis induced by hyperthermia. Int J Hyperthermia. 2003;19:165–77.
16. Yuki H, Kondo T, Zhao QL, Fujiwara Y, Tanabe K, Ogawa R, et al. A free radical initiator, 2,2'-azobis (2-aminopropane) dihydrochloride enhances hyperthermia-induced apoptosis in human uterine cervical cancer cell lines. Free Radic Res. 2003;37:631–43.
17. Wada S, Cui ZG, Kondo T, Zhao QL, Ogawa R, Shoji M, et al. A hydrogen peroxide-generating agent, 6-formylpterin, enhances heat-induced apoptosis. Int J Hyperthermia. 2005;21:231–46.
18. Wada S, Tabuchi Y, Kondo T, Cui ZG, Zhao QL, Takasaki I, et al. Gene expression in enhanced apoptosis of human lymphoma U937 cells treated with the combination of different free radical generators and hyperthermia. Free Radic Res. 2007;41:73–81.

19. Hirano H, Tabuchi Y, Kondo T, Zhao QL, Ogawa R, Cui ZG, et al. Analysis of gene expression in apoptosis of human lymphoma U937 cells induced by heat shock and the effects of alpha-phenyl N-tert-butylnitrone (PBN) and its derivatives. Apoptosis. 2005;10:331–40.
20. Cui ZG, Kondo T, Matsumoto H. Enhancement of apoptosis by nitric oxide released from alpha-phenyl-tert-butyl nitrone under hyperthermic conditions. J Cell Physiol. 2006;206:468–76.
21. Yu DY, Matsuya Y, Zhao QL, Ahmed K, Wei ZL, Nemoto H, et al. Enhancement of hyperthermia-induced apoptosis by a new synthesized class of furan-fused tetracyclic compounds. Apoptosis. 2007;12:1523–32.
22. Yu DY, Matsuya Y, Zhao QL, Ahmed K, Wei ZL, Hori T, et al. Enhancement of hyperthermia-induced apoptosis by a new synthesized class of benzocycloalkene compounds. Apoptosis. 2008;13:448–61. doi:10.1007/s10495-008-0178-9.
23. Yu DY, Zhao QL, Wei ZL, Shehata M, Kondo T. Enhancement of hyperthermia-induced apoptosis by sanazole in human lymphoma U937 cells. Int J Hyperthermia. 2009;25:364–73. doi:10.1080/02656730902967418.
24. Zhao QL, Fujiwara Y, Kondo T. Mechanism of cell death induction by nitroxide and hyperthermia. Free Radic Biol Med. 2006;40:1131–43.
25. Zhao QL, Fujiwara Y, Kondo T. Synergistic induction of apoptosis and caspase-independent autophagic cell death by a combination of nitroxide Tempo and heat shock in human leukemia U937 cells. Apoptosis. 2010;15:1270–83. doi:10.1007/s10495-010-0522-8.
26. Venkataraman S, Wagner BA, Jiang X, Wang HP, Schafer FQ, Ritchie JM, et al. Overexpression of manganese superoxide dismutase promotes the survival of prostate cancer cells exposed to hyperthermia. Free Radic Res. 2004;38:1119–32.
27. Yoshihisa Y, Zhao QL, Hassan MA, Wei ZL, Furuichi M, Miyamoto Y, et al. SOD/catalase mimetic platinum nanoparticles inhibit heat-induced apoptosis in human lymphoma U937 and HH cells. Free Radic Res. 2011;45:326–35. doi:10.3109/10715762.2010.532494.
28. Minotti G. Sources and role of iron in lipid peroxidation. Chem Res Toxicol. 1993;6:134–46.
29. Radi R, Cassina A, Hodara R, Quijano C, Castro L. Peroxynitrite reactions and formation in mitochondria. Free Radic Biol Med. 2002;33:1451–64.
30. Matsumoto H, Hayashi S, Hatashita M, Ohnishi K, Ohtsubo T, Kitai R, et al. Nitric oxide is an initiator of intercellular signal transduction for stress response after hyperthermia in mutant *p53* cells of human glioblastoma. Cancer Res. 1999;59:3239–44.
31. Moon EJ, Sonveaux P, Porporato PE, Danhier P, Gallez B, Batinic-Haberle I, et al. NADPH oxidase-mediated reactive oxygen species production activates hypoxia-inducible factor-1 (HIF-1) via the ERK pathway after hyperthermia treatment. Proc Natl Acad Sci U S A. 2010;107:20477–82. doi:10.1073/pnas.1006646107.
32. Matés JM, Sánchez-Jiménez FM. Role of reactive oxygen species in apoptosis: implications for cancer therapy. Int J Biochem Cell Biol. 2000;32:157–70.
33. Terao K, Niki E. Damage to biological tissues induced by radical initiator 2,2'-azobis (2-amidinopropane) dihydrochloride and its inhibition by chain-breaking antioxidants. J Free Radic Biol Med. 1986;2:193–201.
34. Ziegler I. Pteridine formation during lectin-induced lymphocyte activation. J Cell Biochem. 1985;28:197–206.
35. Arai T, Endo N, Yamashita K, Sasada M, Mori H, Ishii H, et al. 6-formylpterin, a xanthine oxidase inhibitor, intracellularly generates reactive oxygen species involved in apoptosis and cell proliferation. Free Radic Biol Med. 2001;30:248–59.
36. Kanthasamy AG, Kitazawa M, Kanthasamy A, Anantharam V. Role of proteolytic activation of protein kinase C delta in oxidative stress-induced apoptosis. Antioxid Redox Signal. 2003;5:609–20.
37. Saito K, Yoshioka H, Kazama S, Cutler RG. Release of nitric oxide from a spin trap, N-tert-butyl-alpha-phenylnitrone, under various oxidative conditions. Biol Pharm Bull. 1998;21:401–4.

38. Shrivastava P, Pantano C, Watkin R, McElhinney B, Guala A, Poynter ML, et al. Reactive nitrogen species-induced cell death requires Fas-dependent activation of c-Jun N-terminal kinase. Mol Cell Biol. 2004;24:6763–72.
39. Schett G, Steiner CW, Gröger M, Winkler S, Graninger W, Smolen J, et al. Activation of Fas inhibits heat-induced activation of HSF1 and up-regulation of hsp70. FASEB J. 1999;13:833–42.
40. Krishna MC, Russo A, Mitchell JB, Goldstein S, Dafni H, Samuni A. Do nitroxide antioxidants act as scavengers of $O2^{-\cdot}$ or as SOD mimics? J Biol Chem. 1996;271:26026–31.
41. Matsuya Y, Sasaki K, Nagaoka M, Kakuda H, Toyooka N, Imanishi N, et al. Synthesis of a new class of furan-fused tetracyclic compounds using o-quinodimethane chemistry and investigation of their antiviral activity. J Org Chem. 2004;69:7989–93.
42. Shibamoto Y, Sakano K, Kimura R, Nishidai T, Nishimoto S, Ono K, et al. Radiosensitization in vitro and in vivo by 3-nitrotriazoles. Int J Radiat Oncol Biol Phys. 1986;12:1063–6.
43. Schepetkin IA, Cherdyntseva NV, Kagiya VT. Sanazole as substrate of xanthine oxidase and microsomal NADPH/cytochrome P450 reductase. Pathophysiology. 2001;8:119–27.
44. Murray JI, Whitfield ML, Trinklein ND, Myers RM, Brown PO, Botstein D. Diverse and specific gene expression responses to stresses in cultured human cells. Mol Biol Cell. 2004;15:2361–74.
45. Furusawa Y, Tabuchi Y, Wada S, Takasaki I, Ohtsuka K, Kondo T. Identification of biological functions and gene networks regulated by heat stress in U937 human lymphoma cells. Int J Mol Med. 2011;28:143–51. doi:10.3892/ijmm.2011.702.
46. Kajihara A, Takahashi A, Ohnishi K, Imai Y, Yamakawa N, Yasumoto J, et al. Protein microarray analysis of apoptosis-related protein expression following heat shock in human tongue squamous cell carcinomas containing different *p53* phenotypes. Int J Hyperthermia. 2008;24:605–12. doi:10.1080/02656730802348339.
47. Jia M, Souchelnytskyi S. Proteome profiling of heat shock of human primary breast epithelial cells, a dataset report. Cell Stress Chaperones. 2011;16:459–67. doi:10.1007/s12192-010-0253-3.
48. Akerfelt M, Morimoto RI, Sistonen L. Heat shock factors: integrators of cell stress, development and lifespan. Nat Rev Mol Cell Biol. 2010;11:545–55. doi:10.1038/nrm2938.
49. Franceschelli S, Rosati A, Lerose R, De Nicola S, Turco MC, Pascale M. Bag3 gene expression is regulated by heat shock factor 1. J Cell Physiol. 2008;215:575–7. doi:10.1002/jcp.21397.
50. Verheij M, Bose R, Lin XH, Yao B, Jarvis WD, Grant S, et al. Requirement for ceramide-initiated SAPK/JNK signalling in stress-induced apoptosis. Nature. 1996;380:75–9.
51. Chang SW, Lee SI, Bae WJ, Min KS, Shin ES, Oh GS, et al. Heat stress activates interleukin-8 and the antioxidant system via Nrf2 pathways in human dental pulp cells. J Endod. 2009;35:1222–8. doi:10.1016/j.joen.2009.06.005.
52. Ahn SG, Thiele DJ. Redox regulation of mammalian heat shock factor 1 is essential for Hsp gene activation and protection from stress. Genes Dev. 2003;17:516–28.
53. Koizumi S, Gong P, Suzuki K, Murata M. Cadmium-responsive element of the human heme oxygenase-1 gene mediates heat shock factor 1-dependent transcriptional activation. J Biol Chem. 2007;282:8715–23.
54. Wang K, Fang H, Xiao D, Zhu X, He M, Pan X, et al. Converting redox signaling to apoptotic activities by stress-responsive regulators HSF1 and NRF2 in fenretinide treated cancer cells. PLoS One. 2009;4:e7538. doi:10.1371/journal.pone.0007538.
55. Dayalan Naidu S, Kostov RV, Dinkova-Kostova AT. Transcription factors Hsf1 and Nrf2 engage in crosstalk for cytoprotection. Trends Pharmacol Sci. 2015;36:6–14. doi:10.1016/j.tips.2014.10.011.
56. Mayerhofer M, Gleixner KV, Mayerhofer J, Hoermann G, Jaeger E, Aichberger KJ, et al. Targeting of heat shock protein 32 (Hsp32)/heme oxygenase-1 (HO-1) in leukemic cells in chronic myeloid leukemia: a novel approach to overcome resistance against imatinib. Blood. 2008;111:2200–10.

57. Kondo T, Matsuda T, Tashima M, Umehara H, Domae N, Yokoyama K, et al. Suppression of heat shock protein-70 by ceramide in heat shock-induced HL-60 cell apoptosis. J Biol Chem. 2000;275:8872–9.
58. Lin X, Fang Q, Chen S, Zhe N, Chai Q, Yu M, et al. Heme oxygenase-1 suppresses the apoptosis of acute myeloid leukemia cells *via* the JNK/c-JUN signaling pathway. Leuk Res. 2015;39:544–52. doi:10.1016/j.leukres.2015.02.009.
59. Sawai M, Ishikawa Y, Ota A, Sakurai H. The proto-oncogene JUN is a target of the heat shock transcription factor HSF1. FEBS J. 2013;280:6672–80. doi:10.1111/febs.12570.
60. Takekawa M, Saito H. A family of stress-inducible GADD45-like proteins mediate activation of the stress-responsive MTK1/MEKK4 MAPKKK. Cell. 1998;95:521–30.
61. Jin H, Suh DS, Kim TH, Yeom JH, Lee K, Bae J. IER3 is a crucial mediator of TAp73-β-induced apoptosis in cervical cancer and confers etoposide sensitivity. Sci Rep. 2015;5:8367. doi:10.1038/srep08367.
62. Dong J, Naito M, Tsuruo T. c-Myc plays a role in cellular susceptibility to death receptor-mediated and chemotherapy-induced apoptosis in human monocytic leukemia U937 cells. Oncogene. 1997;15:639–47.
63. Jayachandran G, Sazaki J, Nishizaki M, Xu K, Girard L, Minna JD, et al. Fragile histidine triad-mediated tumor suppression of lung cancer by targeting multiple components of the Ras/Rho GTPase molecular switch. Cancer Res. 2007;67:10379–88.

Chapter 3
Molecular Damage: Hyperthermia Alone

Akihisa Takahashi

Abstract The goal of cancer therapy using hyperthermia is to kill cancer cells and shrink tumors directly. Understanding the molecular targets for heat-induced cell killing, and how such targets are protected or repaired in heat-resistant cells, might provide clues to improve cancer therapy with hyperthermia. However, the molecular mechanisms involved in heat-induced cell killing are not yet fully understood. This chapter is intended to provide an overview of hyperthermia-induced molecular damage to important cellular components such as proteins, lipids, and DNA. We will focus on heat-induced DNA damage, review recent literature, and discuss the potential issues and problems that remain to be solved to understand this process fully.

Keywords Heat • Protein denaturation • Lipid peroxidation • DNA double strand breaks (DSBs) • Histone H2AX phosphorylated at serine 139 (γH2AX)

3.1 Introduction

Hyperthermia is defined as a moderate elevation in the 40–47 °C temperature range [1]. Hyperthermia is widely used to treat patients with various types of cancers, and is performed in combination with radiation and/or other anti-cancer drugs [2]. While numerous effects can contribute to clinical hyperthermia at the tissue, organ and whole body level, this chapter focuses on effects at the cellular and molecular levels. Hyperthermia causes many changes of the biomolecules in cells and leads to a loss of cellular homeostasis. The aim of this chapter is to review the mechanisms involved in heat-induced molecular damage, of which there are three primary types: protein denaturation, lipid peroxidation, and DNA damage (Fig. 3.1) [3].

A. Takahashi (✉)
Gunma University Heavy Ion Medical Center, 3-39-22 Showa-machi, Maebashi, Gunma 371-8511, Japan
e-mail: a-takahashi@gunma-u.ac.jp

S. Kokura et al. (eds.), *Hyperthermic Oncology from Bench to Bedside*,
DOI 10.1007/978-981-10-0719-4_3

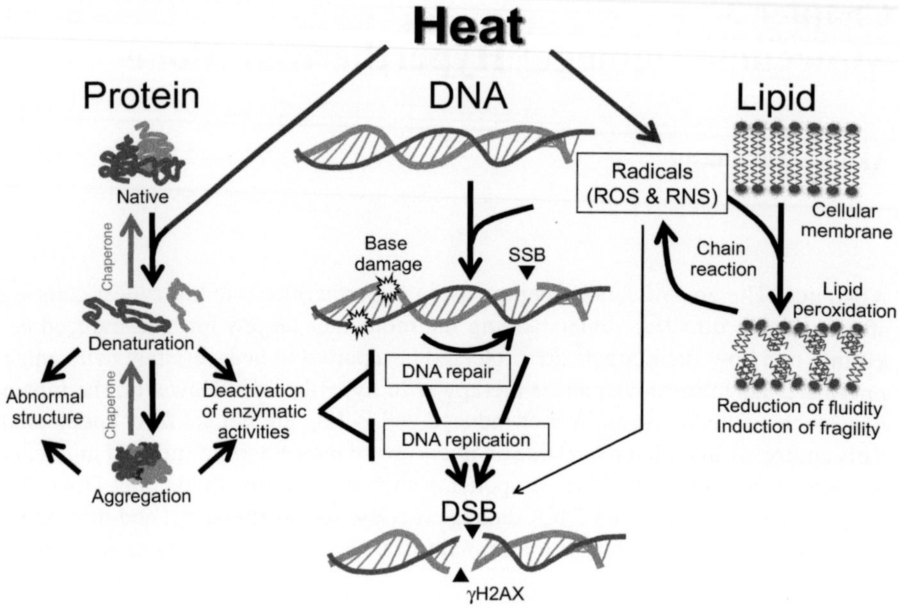

Fig. 3.1 Heat induces molecular damage to three cellular components of proteins, lipids, and DNA. ↑, promotion; ⊤, depression; *ROS* reactive oxygen species, *RNS* reactive nitrogen species, *SSB* single-strand break, *DSB* double-strand break, *γH2AX* histone H2AX phosphorylated at serine 139. Based on [3]

3.2 Heat Induces Protein Denaturation

Proteins are large molecules composed of folded amino acid chains, and have 3-dimensional structures, which reflect their diverse roles in nearly all cellular functions. Therefore, proteins are "the chief actors" in almost every biological process within living cells. Proteins can be classified by two characteristics: (i) structural or mechanical functions and (ii) metabolic functions.

Heat disrupts the secondary and tertiary structure of proteins, although the primary structure (amino acid sequence) remains the same after denaturation. Specifically, heat disrupts hydrogen bonds and non-polar hydrophobic interactions by increasing kinetic energy and causing the molecules to vibrate rapidly and violently. Denaturation of proteins disrupts their normal α-helix and β-sheet structures, allowing the protein to uncoil into a random shape leading to a disruption of its function. Problems occur when proteins become denatured and aggregate non-specifically.

Heat stress leads to damage of structural functions including the reorganization of actin filaments into stress fibers and the aggregation of other filaments. Consequently, heat shock has deleterious effects on the internal organization of the cell as follows [4]. Organelles such as the Golgi and endoplasmic reticulum become fragmented and disassemble. The number and integrity of mitochondria and lysosomes decrease, the nucleoli (sites of ribosome assembly) swell, and large granular depositions consisting of ribosomal proteins become visible. Large depositions, the stress granule, resulting from assemblies of proteins and RNA, are found in the cytosol. Heat shock also induces centrosomal dysfunction causing mitotic catastrophe [5], and damage to the metabolic functions of enzymes, which catalyze biochemical reactions. As a result, heat shock leads to the inhibition of multiple nuclear matrix-dependent functions (DNA replication, DNA transcription, mRNA processing, and DNA repair [6]), and decreased levels of ATP [7]. Taken together, all these effects stop cell growth and lead to cell-cycle arrest, as indicated by noncondensed chromosomes in the nucleus.

To prevent protein denaturation and aggregation, glycerol (a chemical chaperone that increases hydrophobic interactions between solvent and solute macromolecules such as globular proteins) is produced, which protects against thermal killing [8]. Moreover, cells express several types of heat shock proteins (HSPs) (also called stress proteins) that also function as molecular chaperones for the denaturation of heat-labile proteins [9]. Downregulation of HSPs increases heat sensitivity [10] and upregulation of HSPs induced thermo-tolerance [11, 12]. These stress-induced genes encode not only molecular chaperones, but also factors that participate in protein degradation and transport and RNA repair, among other processes [2]. Hyperthermia-induced up-regulation of HSP transcription is mediated by heat shock transcription factors (HSFs), which are constitutively expressed in higher eukaryotes, and become activated under heat shock conditions [13]. From these results, it was generally thought that proteins are the major targets for heat-induced cell killing.

3.3 Heat Induces Lipid Peroxidation

Lipids are a group of naturally occurring molecules and classified into these eight well-defined categories such as fatty acyls, glycerolipids, glycerophospholipids, sphingolipids, sterol lipids, prenol lipids, saccharolipids, and polyketides [14]. The main biological functions of lipids include storing energy, signaling, and acting. The lipid bilayer has been firmly established as the universal basis for cell membrane structure. Cell membranes are selective barriers that separate individual cells and cellular compartments.

Heat shock induces the production of reactive oxygen species (ROS) [15] and reactive nitrogen species (RNS) [16] that can attack all biological molecules. ROS plays an important role in the antitumor effect of hyperthermia [17]. Furthermore, ROS can react with the polyunsaturated fatty acids of lipid cellular membranes and induce lipid peroxidation [18]. This process proceeds through a free radical chain reaction mechanism, and reduces membrane fluidity [19] and increases membrane fragility [20]. Although hyperthermia can cause changes in lipids these appear to be reversible as the viscosity of the plasma membrane decreases with increasing temperature [21], and this may be associated with altered transport functions of the membrane. Changes in membrane viscosity were associated with an elevation in the activity of the ATP-dependent sodium-potassium pump [21], which maintains Na^+ and K^+ levels across the plasma membrane against a concentration gradient. During heat treatment, membrane permeability towards several compounds is altered, including anticancer drugs [22–24].

3.4 Heat Induces DNA Damage

DNA is the important molecule that carries the genetic information in all cellular forms of life. It is generally accepted that hyperthermia inhibits DNA repair [25]. However, whether hyperthermia indirectly and directly induces DNA damage is controversial. Using a sucrose density-gradient centrifugation assay, several authors reported that no DNA strand breaks were detected after heat treatment [26].

On the other hand, early studies reported that increased levels of base damage such as apurinic/apyrimidinic (AP) DNA sites [27], oxidative base lesions [15], deamination of cytosine [28] and other types of damage were detected after heat treatment and were mediated through the production of ROS [15] and RNS [16]. In addition, it was reported that hyperthermia may induce chromosomal aberrations by causing protein denaturation or by interfering with replication [29] and cellular DNA strand breaks (Table 3.1) [30–37].

Using an alkaline elution assay, a development of DNA strand breaks was observed when the cell were incubated at 37 °C for 4–24 h after heat treatment [30] or at temperatures above 45 °C [31] because of its low detection sensitivity. Using a DNA unwinding assay, the number of DNA strand breaks was observed to increase rapidly and quantitatively in a temperature and heating time dependent manner after 41–45 °C heat treatment [32, 33]. Using an *in situ* nick translation (ISNT) assay, DNA strand breaks appeared as early as 15 min following 43 °C heat treatment, and the number of breaks was determined semi-quantitatively in each cellular nucleus [34]. Interestingly, the number of heat-induced DNA DSBs was predicted from the decrease in grain density over 45 min of 43 °C heat treatment. It was suggested many DSBs may form if SSBs or nicks are created near each another in the opposite DNA strands [34]. Using pulse-field gel electrophoresis (PFGE) methods, it was reported that heat treatment for 30 min at 45.5 °C induced a number of DSBs equivalent to those formed after an exposure of approximately 10 Gy in replicating DNA or from an exposure of 3–5 Gy in total or bulk DNA [35]. This report was ignored because heat alone (10–80 min, 45.5 °C) induced very few DSBs, and reached a plateau when compared to ionizing radiation for a given extent of cell killing. In addition to using PFGE, heat-induced DSBs were detected with a neutral comet assay (neutral single cell electrophoresis assay), which is a more direct method of detecting DSBs in a single cell [36, 37].

At the beginning of this century, our deeper understanding of DNA repair networks facilitated the interpretation of results using advanced fluorescence imaging and molecular biology techniques. Especially, the measurement of γH2AX

Table 3.1 Heat induces DNA strand breaks

Assays	DNA damage	Temperature, time	References
Alkaline elution assay	Strand breaks	45 °C, 10–60 min	[30, 31]
DNA unwinding assay	Strand breaks	45 °C, 60–180 min	[32, 33]
		44 °C, 60–240 min	
		43 °C, 120–300 min	
		45 °C, 30–90 min	
		41 °C, 30–90 min	
In situ nick translation (ISNT) assay	SSBs or nick	43 °C, 15–45 min	[34]
Neutral pulse-field gel electrophoresis (PFGE)	DSBs	45.5 °C, 20–120 min	[35]
Neutral comet assay	DSBs	45.5 °C, 30–180 min	[36, 37]
Immunostaining of γH2AX foci	DSBs	45.5 °C, 10–30 min	[36]
		45 °C, 10–40 min	
		44.5 °C, 10–40 min	
		44 °C, 20–80 min	
		43.5 °C, 20–80 min	
		43 °C, 30–120 min	
		42.5 °C, 40–120 min	
		42 °C, 60–240 min	
		41.5 °C, 120–360 min	

SSB single strand breaks, *DSB* double strand breaks

(histone H2AX phosphorylated at serine 139) foci, which is a variant of the core histone H2A, attracted considerable attention. The phosphorylation of H2AX by ataxia telangiectasia mutated (ATM), ATM Rad3-related (ATR), or DNA-dependent protein kinase (DNA-PK) is believed to be one of the first events in the DNA damage response (DDR) [38, 39]. Since its discovery, γH2AX, has been regarded as a universal marker of DNA DSBs. γH2AX can provide a very sensitive and specific indicator for the existence of a single DSB [40–42]. One γH2AX focus can be correlated with one cellular DSB.

Using this method, it was reported that heat triggers the focal phosphorylation of histone H2AX, similar to the formation of ionizing radiation induced foci (IRIF) [35]. Although reports suggested that the formation of heat-induced γH2AX foci is cell type dependent [43, 44], this has been contradicted. There is no question that heat-induced γH2AX focus formation behavior is a widespread phenomenon in mammals [45]. It has been shown by many groups that heat stimulates the phosphorylation of histone H2AX [46–52]. The number of γH2AX foci increased rapidly and reached a maximum in the cell nucleus 30 min after heat treatment [36]. In addition, γH2AX foci were detected quantitatively associated with the temperature and length of hyperthermia exposure at wide temperature range (41.5 °C–45.5 °C) [36]. Moreover, the number of heat-induced γH2AX foci was also in good agreement with previous reports showing that a 30-min heat treatment at 45.5 °C induced a number of DSBs equivalent to exposure of approximately 10 Gy in S-phase cells or 5 Gy in asynchronous cells [35, 36]. As for heat-induced γH2AX foci formation, the published data and their interpretation are contradictory; γH2AX does mark DSBs induced by heat [36, 37, 46, 47], and does not occur due to DNA damage but is a by-product of other cellular processes perturbed by heat [50–53]. It might be premature to judge the absence of heat-induced DSBs, because the temperature and length of the hyperthermia exposure was not sufficient in the study [50]. Arguments for this are that hyperthermia-induced foci did not recapitulate all the characteristics of IRIF in that they failed to co-localize with the phosphorylation of ATM [50, 52], the Mre11-Rad50-Nbs1 (MRN) complex [54–57], p53 binding protein (53BP1) [50, 52] or structural maintenance of chromosomes protein 1 (SMC1) [50, 52], although the induction of γH2AX and mediator of DNA-damage checkpoint 1 (MDC1) foci by hyperthermia was dependent on the DSB signaling factor ATM [36, 48–51, 58]. Different DDRs between heat-treated cells and irradiated cells was predicted because these factors are proteins that are denatured by heat. However, these DDR factors subsequently re-formed foci at damaged sites over time, and the behavior of these factors exhibited a pattern related to the temperature and length of hyperthermia exposure [49, 55, 59]. It was reported that cellular chromosome aberrations were observed after heat treatment in plateau phase cells using premature chromosome condensation and fluorescence *in situ* hybridization methods [60]. Therefore, it currently cannot be claimed that γH2AX foci do not represent DSBs following heat treatment [61].

Adding to this debate, it was recently reported that two different patterns of γH2AX foci can be detected in heat-treated cells: the larger IRIF-like foci in G_1- and G_2-phase cells, and the smaller but more numerous foci in S-phase cells [37]. Surprisingly, heat-induced DSBs were detected in heated G_1/G_2 cells but not in S-phase cells, while the inverse was true for SSBs. Furthermore, the authors demonstrated inhibitory effects of heat on replication fork progression. The absence of DSBs in cells heated in the S-phase can be caused by the suppression of replication fork progression that might, in turn, prevent DSB formation [62]. The S-phase specific 'protective' foci may thus mark sites of stalled replication forks that are not yet converted to DSBs. On the other hand, the foci in non S-phase cells could mark DSBs that were directly induced by heat or, alternatively, persistent DSBs [63] that were unmasked by heat-related chromatin changes. This latter explanation may be difficult to prove because only a limited number of persistent DSBs have been observed while hyperthermia often induces a large number of foci. However, the results from this study do not match a previous report where the formation of heat-induced DSBs and chromosomal aberrations were observed in S-phase cells [35]. Consequently, the different mechanisms involved in heat-induced DSB formation in different cell cycle phases are required to identify the exact mechanisms involved.

3.5 Relationship Between Heat-Sensitivity and DNA Damage

It has been accepted that heat-induced cell killing is not dependent on the formation of DSBs, although X-ray-induced cell killing is dependent on DSBs. This conclusion is based on results showing no correlation between thermosensitivity and radiosensitivity [35] or between cell cycle-dependent heat- and X-ray-induced cell killing [64]. Heat-induced cell killing was more efficient than X-ray-induced cell killing on that condition, although heat- and X-ray-induced cell killing would be expected to show the same kinetics when heat and X-rays induced the same numbers of DSBs. It is a simple and crude notion that these observations were considered to be evidence to indicate that heat-induced cell killing was caused by mechanisms other than DSBs, the mechanism by which X-rays induce cell killing [35]. It was also proposed that small numbers of heat-induced DSBs may not be a major determinant of cell survival, but could modify the effects of other stimuli *via* DSB-related signaling [46]. Therefore, DSBs had been not regarded as an important factor in heat-induced cell killing. However, it is possible that this difference in cell killing kinetics might have been caused by the cellular repair of DSBs because repair enzymes for DSBs are deactivated by heat treatment [25].

Furthermore, using Arrhenius plots analysis, cell killing and γH2AX foci formation were same temperature (42.5 °C) of inflection points, and were similar to the thermal activation energies [36]. In addition, the cell cycle-dependent pattern of heat sensitivity was the same as the cell cycle pattern observed for heat-induced γH2AX foci formation and thermotolerance was caused by a depression in the number of γH2AX foci formed after heat treatment [36]. This depression occurred when cells were pre-treated before being exposed to heat exposure [36]. DNA polymerase β (POLβ) contributed to thermotolerance through its reactivation and stimulation by HSPs [12]. These findings suggested that heat-sensitivity might be correlated with DSB formation.

Most reports that confirmed the induction of DSBs by heat used indirect assays such as γH2AX foci or the accumulation of repair-related proteins. Other studies confirmed DSB induction by direct methods and showed that γH2AX foci correlated with cell killing and thermotolerance. More sensitive and specific methods to detect DNA DSBs and SSBs directly may be required to settle this long-standing dispute [25].

3.6 Possible Mechanisms of Heat-Induced DSB Formation

A putative model for the formation of DSBs by heat treatment is shown in Fig. 3.2 [12, 36, 65, 66]. Heat increases the occurrence of base damage through the production of radicals [15, 27, 28]. Base incision enzymes such as glycosylases and AP-endonucleases may cause DNA base damage, but DNA synthesis enzymes

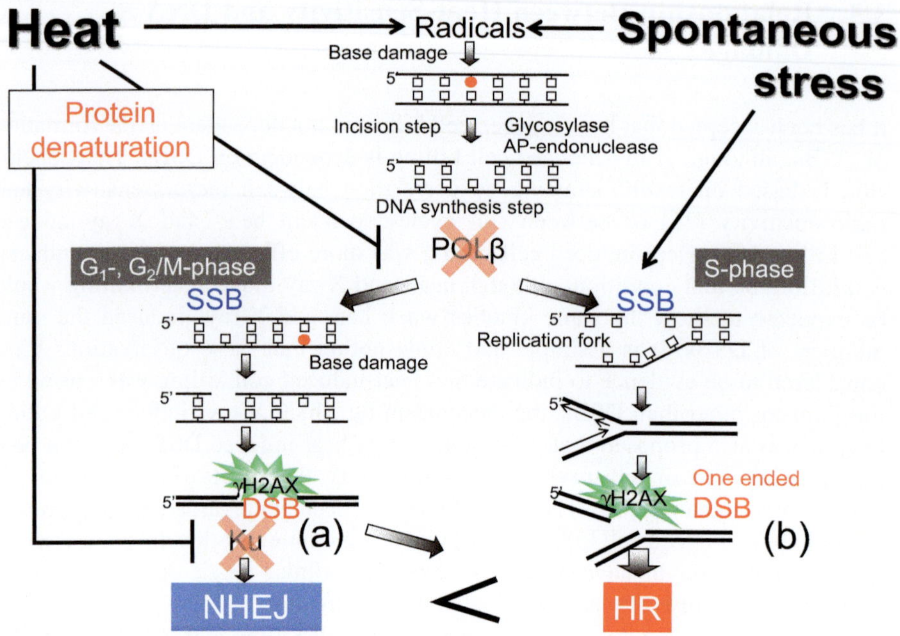

Fig. 3.2 Putative model for the formation of DSBs by heat.↑, promotion; ⊤, depression; *SSB* single-strand break, *DSB* double-strand break, *POLβ* DNA polymerase-β, *Ku* Ku70/Ku80, *γH2AX* histone H2AX phosphorylated at serine 139, *NHEJ* non-homologous end-joining, *HR* homologous recombination. Based on [65, 66]

may not operate at these nick sites. It was also reported that heat induces the activation of the AP-endonuclease [67] and the deactivation of POLβ [68–70]. Furthermore, the deactivation energies of DNA POLβ [68], the formation energies of γH2AX foci [36] and cell killing [36] were similar according to Arrhenius plot analysis. In addition, there is an inflection point at 42.5 °C in the Arrhenius plot [36, 68]. Therefore, these experimental results suggest that heat might induce DSBs *via* the deactivation of DNA POLβ, a key enzyme that protects the genome from DNA damage by base excision repair (BER) in mammalian cells [71, 72]. At the same time, heat induced inhibition of BER might increase the number of inappropriately repaired base damage sites, leading to an elevation in the number of nicks generated during repair attempts. Therefore, it appears that there is a possible mechanism that might explain how heat induces nick formation through enzymatic repair processes at the cellular level. This might explain why few DSBs were detected even at high temperatures in *in vitro* cell-free systems such as polymerase chain reaction (PCR) method (over 60 °C). DSBs might then be generated where nicks form in close proximity to each other on opposite DNA strands. This theoretically provides a mechanism that might account for the increased numbers of DSBs observed in heat treated cells. Indeed, the inhibition of poly (ADP) ribose polymerase (PARP), which is involved in BER and SSB repair, induces γH2AX foci [73]. It has also been hypothesized that a nick is converted to a DSB at a DNA replication fork. This was supported by recent findings that more γH2AX foci were formed when cells were heated in the S-phase than in the G_1- or G_2-phase [36].

There are two major pathways for the repair of DSBs in eukaryotes: non-homologous end-joining (NHEJ) repair and homologous recombination (HR) repair [74, 75]. NHEJ repair is the dominant repair pathway throughout the cell cycle [76, 77], although HR results in the error-free repair of lesions in late S-G_2 phases of the cell cycle, because HR repairs DSBs using homologous undamaged DNA as the repair template [74, 75]. The first NHEJ step involves the Ku70/Ku80 heterodimer that binds to and protects DNA ends from degradation by exonucleases. Ku70/Ku80 is heat labile [59, 78–82] (Fig. 3.2a). In addition, it was reported that heat might inhibit HR repair because HR-related proteins such as Rad51 [59, 83] and breast cancer susceptibility gene 2 (BRCA2) [83] do not translocate into the nucleus and form foci after heat treatment. Delayed recognition of γH2AX foci by HR-related proteins also might contribute to slower repair kinetics for heat-induced DSBs. However, factors important in the HR process [59, 83] are the same as for DDR [49, 55, 59] but NHEJ subsequently re-forms foci at damaged sites over time, and the behavior of these factors exhibits a pattern related to the temperature and length of the hyperthermia exposure. Moreover, it was proposed that HR is the predominant pathway for repairing one-ended DSBs arising in the S-phase when the replication fork encounters SSBs or base damage [84] (Fig. 3.2b). These phenomena suggest that HR repairs heat-induced DSBs rather than NHEJ.

Another possible mechanism is that spontaneous DNA damage could contribute significantly to the production of endogenous DSBs at physiological temperatures [85, 86]. In contrast to DSBs, SSBs are frequently occurring endogenous DNA lesions in cells (10^4/cell/day) [85]. SSBs can be induced directly by free radicals such as ROS and RNS, or more commonly as a consequence of the repair of AP sites generated by the depurination or repair of deaminated cytosine or other damaged bases [85]. These radicals also directly induce DSBs. In normal human cells, it is estimated that approximately 1 % of cellular single-strand lesions are converted to about 50 endogenous DSBs *per* cell *per* cell cycle [86]. This number is similar to the estimate for the number of exogenously generated DSBs produced by 1.5–2.0 Gy of ionizing radiation. Heat-induced DSBs may reflect the depression of cellular repair processes intended to cope with the number of DSBs generated spontaneously during the processes of transcription and replication. Other potential mechanisms leading to DSBs could involve inhibiting the access of repair enzymes to DNA damage sites due to the denaturation of nuclear matrix proteins [1], and/or the translocation and removal of repair related proteins from the nucleus [49, 55, 59].

3.7 Conclusion

A more extensive knowledge of the exact mechanisms involved in molecular damage caused by hyperthermia that kills cells is important to allow the refinement and development of current and future tumor therapies. The use of new techniques has demonstrated DNA damage, as well as protein denaturation and lipid peroxidation, is caused by heat treatment. Of note, heat-induced DSBs are an important finding that requires further attention. Such observations provide support for the concept that heat-induced DSBs contribute to heat-induced cell killing. Future investigations should define the precise mechanisms that lead to heat-induced DSB formation, and will hopefully lead to an enhanced understanding of hyperthermia and improved efficacy of hyperthermia for cancer therapy.

References

1. Roti Roti JL. Cellular responses to hyperthermia (40–46 degrees C): cell killing and molecular events. Int J Hyperthermia. 2008;24:3–15.
2. Falk MH, Issels RD. Hyperthermia in oncology. Int J Hyperthermia. 2001;17:1–18.
3. Takahashi A, Mori E, Ohnishi T. Heat-induced DNA damage. In: Shimizu T, Kondo T, editors. Cellular response to physical stress and therapeutic application. New York: Nova Science Publishers Inc.; 2013. p. 135–47.
4. Richter K, Haslbeck M, Buchner J. The heat shock response: life on the verge of death. Mol Cell. 2010;40:253–66.
5. Nakahata K, Miyakoda M, Suzuki K, Kodama S, Watanabe M. Heat shock induces centrosomal dysfunction, and causes non-apoptotic mitotic catastrophe in human tumour cells. Int J Hyperthermia. 2002;18:332–43.
6. Roti Roti JL, Kampinga HH, Malyapa RS, Wright WD, van der Waal RP, Xu M. Nuclear matrix as a target for hyperthermic killing of cancer cells. Cell Stress Chaperones. 1998;3:245–55.
7. Sonna LA, Fujita J, Gaffin SL, Lilly CM. Invited review: effects of heat and cold stress on mammalian gene expression. J Appl Physiol. 2002;92:1725–42.
8. Henle KJ, Warters RL. Heat protection by glycerol *in vitro*. Cancer Res. 1982;42:2171–6.
9. Morimoto RI, Kline MP, Bimston DN, Cotto JJ. The heat-shock response: regulation and function of heat-shock proteins and molecular chaperones. Essays Biochem. 1997;32:17–29.
10. Rossi A, Ciafre S, Balsamo M, Pierimarchi P, Santoro MG. Targeting the heat shock factor 1 by RNA interference: a potent tool to enhance hyperthermochemotherapy efficacy in cervical cancer. Cancer Res. 2006;66:7678–85.
11. Burdon RH. Thermotolerance and the heat shock proteins. Symp Soc Exp Biol. 1987;41:269–83.
12. Takahashi A, Yamakawa N, Mori E, Ohnishi K, Yokota S, Sugo N, et al. Development of thermotolerance requires interaction between polymerase-beta and heat shock proteins. Cancer Sci. 2008;99:973–8.
13. Anckar J, Sistonen L. Regulation of HSF1 function in the heat stress response: implications in aging and disease. Annu Rev Biochem. 2011;80:1089–115.
14. Fahy E, Subramaniam S, Murphy RC, Nishijima M, Raetz CR, Shimizu T, et al. Update of the LIPID MAPS comprehensive classification system for lipids. J Lipid Res. 2009;50:S9–14.

15. Bruskov VI, Malakhova LV, Masalimov ZK, Chernikov AV. Heat-induced formation of reactive oxygen species and 8-oxoguanine, a biomarker of damage to DNA. Nucleic Acids Res. 2002;30:1354–63.
16. Hall DM, Buettner GR, Matthes RD, Gisolfi CV. Hyperthermia stimulates nitric oxide formation: electron paramagnetic resonance detection of.NO-heme in blood. J Appl Physiol. 1994;77:548–53.
17. Yoshikawa T, Kokura S, Tainaka K, Itani K, Oyamada H, Kaneko T, et al. The role of active oxygen species and lipid peroxidation in the antitumor effect of hyperthermia. Cancer Res. 1993;53:S2326–9.
18. Halliwell B, Chirico S. Lipid peroxidation: its mechanism, measurement, and significance. Am J Clin Nutr. 1993;57:S715–24.
19. Chen JJ, Yu BP. Alterations in mitochondrial membrane fluidity by lipid peroxidation products. Free Radic Biol Med. 1994;117:411–18.
20. Spickett CM, Reglinski J, Smith WE, Wilson R, Walker JJ, McKillop J. Erythrocyte glutathione balance and membrane stability during preeclampsia. Free Radic Biol Med. 1998;24:1049–55.
21. Bates DA, Le Grimellec C, Bates JH, Loutfi A, Mackillop WJ. Effects of thermal adaptation at 40 degrees C on membrane viscosity and the sodium-potassium pump in Chinese hamster ovary cells. Cancer Res. 1985;45:4895–9.
22. Gerner EW, Cress AE, Stickney DG, Holmes DK, Culver PS. Factors regulating membrane permeability alter thermal resistance. Ann N Y Acad Sci. 1980;335:215–33.
23. Lecavalier D, Mackillop WJ. The effect of hyperthermia on glucose transport in normal and thermal-tolerant Chinese hamster ovary cells. Cancer Lett. 1985;29:223–31.
24. Bates DA, Mackillop WJ. Hyperthermia, adriamycin transport, and cytotoxicity in drug-sensitive and -resistant Chinese hamster ovary cells. Cancer Res. 1986;46:5477–81.
25. Oei AL, Vriend LE, Crezee J, Franken NA, Krawczyk PM. Effects of hyperthermia on DNA repair pathways: one treatment to inhibit them all. Radiat Oncol. 2015;10:165. doi:10.1186/s13014-015-0462-0.
26. Corry PM, Robinson S, Getz S. Hyperthermic effects on DNA repair mechanisms. Radiology. 1977;123:475–82.
27. Warters RL, Brizgys LM. Apurinic site induction in the DNA of cells heated at hyperthermic temperatures. J Cell Physiol. 1987;133:144–50.
28. Lindahl T, Nyberg B. Heat-induced deamination of cytosine residues in deoxyribonucleic acid. Biochemistry. 1974;13:3405–10.
29. Wong RS, Kapp LN, Krishnaswamy G, Dewey WC. Critical steps for induction of chromosomal aberrations in CHO cells heated in S phase. Radiat Res. 1993;133:52–9.
30. Warters RL, Henle KJ. DNA degradation in Chinese hamster ovary cells after exposure to hyperthermia. Cancer Res. 1982;42:4427–32.
31. Dikomey E. Effect of hyperthermia at 42 and 45 degrees C on repair of radiation-induced DNA strand breaks in CHO cells. Int J Radiat Biol Relat Stud Phys Chem Med. 1982;41:603–14.
32. Jorritsma JB, Konings AW. The occurrence of DNA strand breaks after hyperthermic treatments of mammalian cells with and without radiation. Radiat Res. 1984;98:198–208.
33. Mitchel RE, Birnboim HC. Triggering of DNA strand breaks by 45 degrees C hyperthermia and its influence on the repair of gamma-radiation damage in human white blood cells. Cancer Res. 1985;45:2940–5.
34. Anai H, Maehara Y, Sugimachi K. *In situ* nick translation method reveals DNA strand scission in HeLa cells following heat treatment. Cancer Lett. 1988;40:33–8.
35. Wong RS, Dynlacht JR, Cedervall B, Dewey WC. Analysis by pulsed-field gel electrophoresis of DNA double-strand breaks induced by heat and/or X-irradiation in bulk and replicating DNA of CHO cells. Int J Radiat Biol. 1995;68:141–52.
36. Takahashi A, Matsumoto H, Nagayama K, Kitano M, Hirose S, Tanaka H, et al. Evidence for the involvement of double-strand breaks in heat-induced cell killing. Cancer Res. 2004;64:8839–45.

37. Velichko AK, Petrova NV, Kantidze OL, Razin SV. Dual effect of heat shock on DNA replication and genome integrity. Mol Biol Cell. 2012;23:3450–60.
38. Rogakou EP, Pilch DR, Orr AH, Ivanova VS, Bonner WM. DNA double-stranded breaks induce histone H2AX phosphorylation on serine 139. J Biol Chem. 1998;273:5858–68.
39. Paull TT, Rogakou EP, Yamazaki V, Kirchgessner CU, Gellert M, Bonner WM. A critical role for histone H2AX in recruitment of repair factors to nuclear foci after DNA damage. Curr Biol. 2000;10:886–95.
40. Sedelnikova OA, Rogakou EP, Panyutin IG, Bonner WM. Quantitative detection of ^{125}IdU-induced DNA double-strand breaks with gamma-H2AX antibody. Radiat Res. 2002;158:486–92.
41. Rothkamm K, Lobrich M. Evidence for a lack of DNA double-strand break repair in human cells exposed to very low x-ray doses. Proc Natl Acad Sci U S A. 2003;100:5057–62.
42. Takahashi A, Ohnishi T. Does γH2AX foci formation depend on the presence of DNA double strand breaks? Cancer Lett. 2005;229:171–9.
43. Zhou C, Li Z, Diao H, Yu Y, Zhu W, Dai Y, et al. DNA damage evaluated by γH2AX foci formation by a selective group of chemical/physical stressors. Mutat Res. 2006;604:8–18.
44. Dong Z, Hu H, Chen W, Li Z, Liu G, Yang J. Heat shock does not induce γH2AX foci formation but protects cells from *N*-methyl-*N*'-nitro-*N*-nitrosoguanidine-induced genotoxicity. Mutat Res. 2007;629:40–8.
45. Takahashi A, Mori E, Somakos GI, Ohnishi K, Ohnishi T. Heat induces γH2AX foci formation in mammalian cells. Mutat Res. 2008;656:88–92.
46. Kaneko H, Igarashi K, Kataoka K, Miura M. Heat shock induces phosphorylation of histone H2AX in mammalian cells. Biochem Biophys Res Commun. 2005;328:1101–6.
47. Paul C, Murray AA, Spears N, Saunders PT. A single, mild, transient scrotal heat stress causes DNA damage, subfertility and impairs formation of blastocysts in mice. Reproduction. 2008;136:73–84.
48. Takahashi A, Mori E, Su X, Nakagawa Y, Okamoto N, Uemura H, et al. ATM is the predominant kinase involved in the phosphorylation of histone H2AX after heating. J Radiat Res. 2008;51:417–22.
49. Takahashi A, Mori E, Ohnishi T. The foci of DNA double strand break-recognition proteins localize with γH2AX after heat treatment. J Radiat Res. 2010;51:91–5.
50. Hunt CR, Pandita RK, Laszlo A, Higashikubo R, Agarwal M, Kitamura T, et al. Hyperthermia activates a subset of ataxia-telangiectasia mutated effectors independent of DNA strand breaks and heat shock protein 70 status. Cancer Res. 2007;67:3010–17.
51. Laszlo A, Fleischer I. The heat-induced γ-H2AX response does not play a role in hyperthermic cell killing. Int J Hyperthermia. 2009;25:199–209.
52. Laszlo A, Fleischer I. Heat-induced perturbations of DNA damage signaling pathways are modulated by molecular chaperones. Cancer Res. 2009;69:2042–9.
53. Kampinga HH, Laszlo A. DNA double strand breaks do not play a role in heat-induced cell killing. Cancer Res. 2005;65:10632–3.
54. Zhu WG, Seno JD, Beck BD, Dynlacht JR. Translocation of MRE11 from the nucleus to the cytoplasm as a mechanism of radiosensitization by heat. Radiat Res. 2001;156:95–102.
55. Seno JD, Dynlacht JR. Intracellular redistribution and modification of proteins of the Mre11/Rad50/Nbs1 DNA repair complex following irradiation and heat-shock. J Cell Physiol. 2004;199:157–70.
56. Xu M, Myerson RJ, Xia Y, Whitehead T, Moros EG, Straube WL, et al. The effects of 41 degrees C hyperthermia on the DNA repair protein, MRE11, correlate with radiosensitization in four human tumor cell lines. Int J Hyperthermia. 2007;23:343–51.
57. Gerashchenko BI, Gooding G, Dynlacht JR. Hyperthermia alters the interaction of proteins of the Mre11 complex in irradiated cells. Cytometry A. 2010;77:940–52.
58. Miyakoda M, Suzuki K, Kodama S, Watanabe M. Activation of ATM and phosphorylation of p53 by heat shock. Oncogene. 2002;21:1090–6.

59. Genet SC, Fujii Y, Maeda J, Kaneko M, Genet MD, Miyagawa K, et al. Hyperthermia inhibits homologous recombination repair and sensitizes cells to ionizing radiation in a time- and temperature-dependent manner. J Cell Physiol. 2013;228:1473–81.
60. Inoue K, Kawata T, Saito M, Liu C, Uno T, Isobe K, et al. Effect of an ATM kinase inhibitor on thermo- and/or radio-sensitization in non-proliferating normal human fibroblasts and osteosarcoma cells. Thermal Med. 2010;26:97–107.
61. Takahashi A, Mori E, Ohnishi T. A possible role of DNA double strand breaks in heat-induced cell killing. Cancer Res. 2005;65:10633.
62. van der Waal RP, Griffith CL, Wright WD, Borrelli MJ, Roti Roti JL. Delaying S-phase progression rescues cells from heat-induced S-phase hypertoxicity. J Cell Physiol. 2001;187:236–43.
63. Lukas C, Savic V, Bekker-Jensen S, Doil C, Neumann B, Pedersen RS, et al. 53BP1 nuclear bodies form around DNA lesions generated by mitotic transmission of chromosomes under replication stress. Nat Cell Biol. 2011;13:243–53.
64. Westra A, Dewey WC. Variation in sensitivity to heat shock during the cell-cycle of Chinese hamster cells *in vitro*. Int J Radiat Biol. 1971;19:467–77.
65. Takahashi A, Ohnishi T. What is the critical hyperthermia target in cancer cells? Thermal Med. 2006;22:229–37.
66. Mori E, Takahashi A, Ohnishi T. The biology of heat-induced DNA double-strand breaks. Thermal Med. 2008;24:39–50.
67. Kenny MK, Mendez F, Sandigursky M, Kureekattil RP, Goldman JD, Franklin WA, et al. Heat shock protein 70 binds to human apurinic/apyrimidinic endonuclease and stimulates endonuclease activity at abasic sites. J Biol Chem. 2001;276:9532–6.
68. Spiro IJ, Denman DL, Dewey WC. Effect of hyperthermia on CHO DNA polymerases alpha and beta. Radiat Res. 1982;89:134–49.
69. Jorritsma JB, Kampinga HH, Scaf AH, Konings AW. Strand break repair, DNA polymerase activity and heat radiosensitization in thermotolerant cells. Int J Hyperthermia. 1985;1:131–45.
70. Mivechi NF, Miyachi H, Scanlon KJ. Heat radiosensitization and the level of DNA polymerases alpha and beta of human colony-forming unit-granulocyte-macrophage and myeloid leukemias sensitive and resistant to chemotherapeutic agents. Cancer Res. 1990;50:2044–8.
71. Sobol RW, Horton JK, Kühn R, Gu H, Singhal RK, Prasad R, et al. Requirement of mammalian DNA polymerase-βin base-excision repair. Nature. 1996;379:183–6.
72. Sobol RW, Prasad R, Evenski A, Baker A, Yang XP, Horton JK, et al. The lyase activity of the DNA repair protein β-polymerase protects from DNA-damage-induced cytotoxicity. Nature. 2000;405:807–10.
73. Bryant HE, Schultz N, Thomas HD, Parker KM, Flower D, Lopez E, et al. Specific killing of BRCA2-deficient tumours with inhibitors of poly(ADP-ribose) polymerase. Nature. 2005;434:913–17.
74. Hall EJ, Giaccia AJ. Radiobiology for the radiologist. 7th ed. Philadelphia: Lippincott Williams & Wilkins; 2012.
75. Wouters BG, Begg AC. Irradiation-induced damage and the DNA damage response. In: Joiner M, van der Kogel A, editors. Basic clinical radiobiology. 4th ed. London: Hodder Arnold Publication; 2009. p. 11–26.
76. Shibata A, Conrad S, Birraux J, Geuting V, Barton O, Ismail A, et al. Factors determining DNA double-strand break repair pathway choice in G_2 phase. EMBO J. 2011;30:1079–92.
77. Terato H, Tanaka R, Nakaarai Y, Nohara T, Doi Y, Iwai S, et al. Quantitative analysis of isolated and clustered DNA damage induced by gamma-rays, carbon ion beams, and iron ion beams. J Radiat Res. 2008;49:133–46.
78. Burgman P, Ouyang H, Peterson S, Chen DJ, Li GC. Heat inactivation of Ku autoantigen: possible role in hyperthermic radiosensitization. Cancer Res. 1997;57:2847–50.

79. Matsumoto Y, Suzuki N, Sakai K, Morimatsu A, Hirano K, Murofushi H. A possible mechanism for hyperthermic radiosensitization mediated through hyperthermic lability of Ku subunits in DNA-dependent protein kinase. Biochem Biophys Res Commun. 1997;234:568–72.
80. Ihara M, Suwa A, Komatsu K, Shimasaki T, Okaichi K, Hendrickson EA, et al. Heat sensitivity of double-stranded DNA-dependent protein kinase (DNA-PK) activity. Int J Radiat Biol. 1999;75:253–8.
81. Beck BD, Dynlacht JR. Heat-induced aggregation of XRCC5 (Ku80) in nontolerant and thermotolerant cells. Radiat Res. 2001;156:767–74.
82. Ihara M, Takeshita S, Okaichi K, Okumura Y, Ohnishi T. Heat exposure enhances radiosensitivity by depressing DNA-PK kinase activity during double strand break repair. Int J Hyperthermia. 2014;30:102–9.
83. Krawczyk PM, Eppink B, Essers J, Stap J, Rodermond H, Odijk H, et al. Mild hyperthermia inhibits homologous recombination, induces BRCA2 degradation, and sensitizes cancer cells to poly(ADP-ribose) polymerase-1 inhibition. Proc Natl Acad Sci U S A. 2011;108:9851–6.
84. Jeggo PA, Geuting V, Löbrich M. The role of homologous recombination in radiation-induced double-strand break repair. Radiother Oncol. 2011;101:7–12.
85. Lindahl T. Instability and decay of the primary structure of DNA. Nature. 1993;362:709–15.
86. Vilenchik MM, Knudson AG. Endogenous DNA double-strand breaks: production, fidelity of repair, and induction of cancer. Proc Natl Acad Sci U S A. 2003;100:12871–6.

Chapter 4
Roles of HSF1 and Heat Shock Proteins in Cancer

Kenzo Ohtsuka

Abstract Heat shock proteins (HSPs) that are induced by various stresses, such as heat, work to protect cells from detrimental environmental stresses as molecular chaperones. The expression of HSPs is regulated by a specific transcription factor HSF1 (heat shock factor 1). The HSF1-HSPs system has long been considered to have an endogenous cytoprotective function. It has recently been shown, however, that HSF1 and HSPs are essential for cancer cell development and progression, and the HSF1-HSPs system seems to be co-opted or hijacked by cancer cells for their own survival. In this review, recent progress in understanding the roles of HSF1 and HSPs in cancer is discussed.

Keywords HSF1 • Heat shock proteins • Molecular chaperones • Cancer • Tumorigenesis

4.1 Introduction

The heat shock response in *Drosophila* was discovered by Ritossa in 1962 [1]. Since then, many researchers throughout the world have been studying the biological functions of heat shock proteins (HSPs) and their associated transcription factor, heat shock factor 1 (HSF1). HSPs are classified into six main families according to their approximate molecular mass: HSP100, HSP90, HSP70, HSP60, HSP40, and small HSPs (sHSP) [2]. It is known that HSPs are not only induced by various environmental (physical, chemical, and biological) stresses, but also synthesized constitutively at normal growth temperatures. It is generally accepted that HSPs have fundamental and indispensable functions as molecular chaperones in protein biogenesis and protein homeostasis, such as folding of nascent polypeptides, assembly/disassembly of multi-subunit oligomers, translocation of proteins across intracellular membranes, and degradation of proteins by lysosomes and proteasomes [3, 4]. When HSPs are induced by environmental stresses, they work

K. Ohtsuka (✉)
Laboratory of Cell & Stress Biology, College of Bioscience & Biotechnology, Chubu University, 1200 Matsumoto-cho, Kasugai, Aichi 487-8501, Japan
e-mail: kohtsuka@isc.chubu.ac.jp

S. Kokura et al. (eds.), *Hyperthermic Oncology from Bench to Bedside*,
DOI 10.1007/978-981-10-0719-4_4

to repair (refold) partially denatured proteins, inhibit protein aggregation, and facilitate the degradation of irreversibly denatured proteins, thus protecting cells from detrimental environmental stresses. Thus, HSPs can be considered to be endogenous cytoprotective factors, or guardians of the proteome [5].

HSF1 and HSPs are also known to have beneficial roles in protein misfolding diseases, especially neurodegenerative diseases, in which characteristic amyloid fibrils or inclusion bodies (large aggregates) of specific mutant proteins are formed inside and/or outside of cells [2]. For example, overexpression of Hsp70 and Hsp40 suppressed the aggregate formation of mutant protein in cultured cells [6], and overexpression of Hsp70 or administration of an HSP inducer (such as paeoniflorin) in a model mouse could ameliorate phenotypes of neurodegenerative disease [7]. Also, lentiviral-mediated delivery of HSF1 into the brain of neurodegenerative model mice topically suppresses the pathogenic mutant protein accumulation and neuronal atrophy [8]. Thus, the HSF1-HSPs system has long been recognized as a provider of cytoprotective and beneficial factors for living organisms [9].

In 2007, however, a shocking paper was reported in which HSF1 have a crucial role in tumor initiation and development [10]. Moreover, each HSP, such as Hsp90 and Hsp70, has been demonstrated to enhance tumor development by regulating the functions of various factors involving tumor initiation, proliferation, invasion, and metastasis [11, 12].

Why does the beneficial HSF1-HSPs system, which works to protect the organisms, help the growth of cancer cells that have developed within the organisms and torment them? In this review, recent progress in understanding the role of HSF1 and HSPs in cancer development, and the therapeutic strategy that is targeting the HSF1-HSPs system will be discussed.

4.2 Overexpression of HSF1 and HSPs in Cancer Cells

As basic research on cancer thermotherapy, the relationship between HSPs and thermo-tolerance have been well studied, and increased levels of HSPs are known to render cells resistant to various environmental stresses such as heat [13]. In the process of carrying out these studies, it has been reported that HSF1 and HSPs are overexpressed in a wide range of cancer cells and that such overexpression is associated with a poor prognosis for many types of cancer [14, 15]. For example, nuclear HSF1 levels were elevated in approximately 80 % of more than 1800 breast cancer patients, and increased HSF1 was associated with a reduced over all survival rate [16]. Why are HSF1 and HSPs activated or overexpressed in cancer cells? Several mechanisms have been proposed as follows: (i) since cancer cells are growing extensively and proteins are synthesized in large quantities, HSPs as molecular chaperones are in demand for protein biogenesis. (ii) Many oncogenes and tumor suppressor genes are mutated and their products are structurally unstable in cancer cells. Molecular chaperones (such as Hsp90 and Hsp70) bind to these mutant proteins, which releases HSF1 from these chaperones and results in its activation. (iii) As the microenvironment of tumor tissues is usually in a stressful condition (low glucose, low pH, low oxygen) that tends to induce HSPs, cancer cells seem to co-opt the HSF1-HSPs system for their own survival. (iv) Signal transduction pathways that are activated in cancer cells, such as EGFR-Her2, Ras-MAPK, and PI3K-mTOR, could phosphorylate and activate HSF1, which in turn induces HSPs [12, 17].

4.3 HSF1 and Hsp70 Are Essential for Cancer Initiation and Development

In 2007, Dai et al. reported the important role of HSF1 in chemical carcinogen (dimethylbenzanthracene, DMBA)-induced skin cancer in mice [10]. They showed that HSF1 knockout mice exhibited extremely reduced tumor incidence as well as increased survival as compared with wild-type mice. Also, $HSF1^{+/-}$ or $HSF1^{-/-}$ mice had increased disease-free survival in a mutant p53 or activated Ras oncogene background as compared with mice expressing wild-type HSF1. A similar result was observed by another group in which deletion of HSF1 inhibited the occurrence and progression of hepatocellular carcinoma induced by a carcinogen (diethylnitrosamine, DEN) [18]. From these results, it is suggested that HSF1 is involved not only in cancer initiation but also in the proliferation of cancer cells. More recently, similar results were obtained in the case of inducible Hsp70 in which mammary tumor initiation was remarkably delayed in Hsp70 knockout mice expressing the polyomavirus middle T oncogene. Furthermore, Hsp70 knockout significantly reduced the ability of mammary tumor cells to migrate and invade [19].

HSF1 is known to be basically a transcription factor for HSPs. Studies in recent years, however, have demonstrated that HSF1 could bind not only to the heat shock element (HSE) present at the promoter region of HSP genes, but also to several hundred other regions of chromosomes [20]. Recently, Mendillo et al. identified more than 500 genes to which HSF1 specifically binds in highly malignant breast cancer cells at normal temperature [21]. In these cells, the expression of genes associated with protein folding/stress response, translation, cell cycle/signaling, and transcription was increased, and that associated with apoptosis, DNA repair, cell adhesion, energy metabolism, and immune process was decreased. All of these alterations in gene expression enhance cancer initiation and progression. Thus, HSF1 is activated in malignant cancer cells even in the absence of heat shock, and acts as a transcription factor or transcriptional repressor to regulate the expression of genes favorable for cancer cells. These authors suggested that HSF1 in cancer cells rewires or changes the transcriptome from the normal cell type to a malignant one [21].

HSF1 and HSPs are expressed at higher levels in many types of cancer, implying that cancer cells are highly dependent on HSF1 and HSPs for their own survival.

For example, it is reported that knock down of HSF1 or some HSP resulted in reduced cell survival in cancer cells but not in normal cells. As HSF1 and HSPs are not themselves oncogenes, cancer cells are often said to be addicted to non-oncogenes [22]. In other word, cancer cells seem to ‘hijack’ the HSF1-HSPs system for their own proliferation and development.

4.4 Association of HSF1 with Factors Involved in Cancer

HSF1 could induce not only HSPs but also various factors that enhance cancer cell proliferation. In some cases, HSF1 binds to a factor to regulate its function. In this section, these typical factors are summarized.

4.4.1 HSF1 Regulates LDHA Transcription and Stimulates Glycolysis of Cancer Cells

As the oxygen concentration of solid tumors is usually low, cancer cells have to obtain energy (ATP) by glycolysis, this feature is well known as the ‘Warburg effect’. It has been reported that overexpression of the HER2/ErbB2 oncogene increased the expression of the glycolysis-regulating enzyme lactate dehydrogenase A (LDHA) and HSF1 [23]. In this report, HER2/ErbB2 was shown to promote HSF1 protein synthesis by enhancing translation and to induce HSF1 trimer formation implying its activation. Also, a putative HSF1 biding site (heat shock element, HSE) was found at the LDHA promoter region and HSF1 was shown to bind to that site. These results suggested that HSF1 regulates LDHA at the transcriptional level, and upregulation of LDHA mediates enhanced glycolysis to support the proliferation of HER2/ErbB2 overexpressing cancer cells. HER2/ErbB2 is a receptor tyrosine kinase and also activates HIF1α (hypoxia-inducible factor 1α) via the PI3K-AKT signaling pathway [24], which in turn induces LDHA transcription (Fig. 4.1).

Moreover, it was recently demonstrated in HER2-overexpressing breast cancer cells that HSF1 is constitutively phosphorylated and activated by the HER2-PI3K-AKT-mTOR signaling pathway and controls Hsp90 client proteins, including MIF (macrophage migration inhibitory factor, a tumor promoting protein) [25].

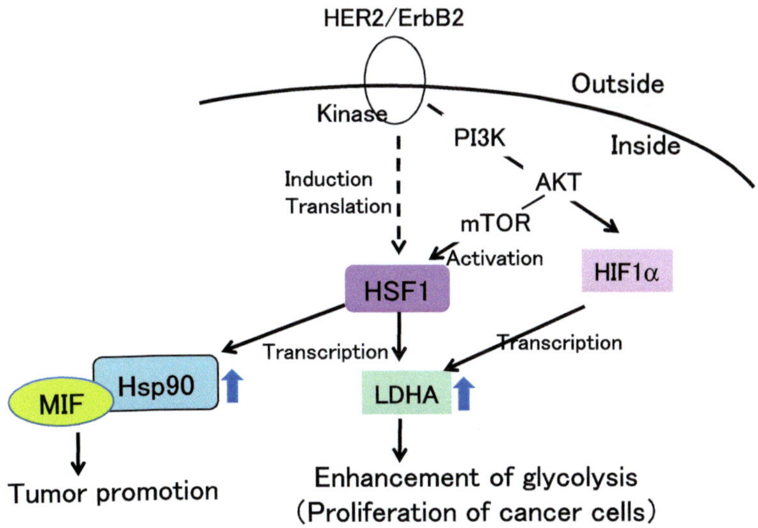

Fig. 4.1 Promotion of LDHA expression by HSF1. HER2/ErbB2 is an EGF receptor (receptor tyrosine kinase) and one of the proto-oncogenes. The HSF1 protein level is elevated in HER2/ErbB2 overexpressing breast cancer cells. LDHA expression is promoted through HSF1-HSE binding. The signal from HER2/ErbB2 also results in HIF1α activation *via* the PI3K-AKT pathway, which in turn causes LDHA transcription. HSF1 is also activated by the PI3K-AKT-mTOR pathway and stabilizes the Hsp90 client protein MIF. *LDHA* lactate dehydrogenase A, *HIF1α* hypoxia-inducible factor 1α, *MIF* macrophage migration inhibitory factor [23, 25]

4.4.2 HSF1 Enhances FoxM1 Transcription and Supports Cancer Cell Progression

Forkhead box M1 (FoxM1) is a pivotal transcription factor for cell cycle progression, which is overexpressed in many types of human cancers, and is shown to be associated with cancer development, invasion, angiogenesis, and metastasis. Thus, FoxM1 is considered to be one of the oncogenes [26]. Recently, it was demonstrated that HSF1 directly binds to the HSE sequence at the FoxM1 promoter region and that FoxM1 is up-regulated under heat-shock stress conditions in an HSF1 dependent manner [27]. Thus, FoxM1 seems to be one of the heat-shock responsive proteins. As a transcription factor, FoxM1 is able to induce VEGF (vascular endothelial growth factor) involved in angiogenesis, MMP2 (matrix metalloproteinase 2) essential for invasion and metastasis, and Cdc2 and Cdc20 necessary for cell cycle progression and proliferation. Therefore it is suggested that HSF1 promotes cancer cell progression via transcription of FoxM1 (Fig. 4.2).

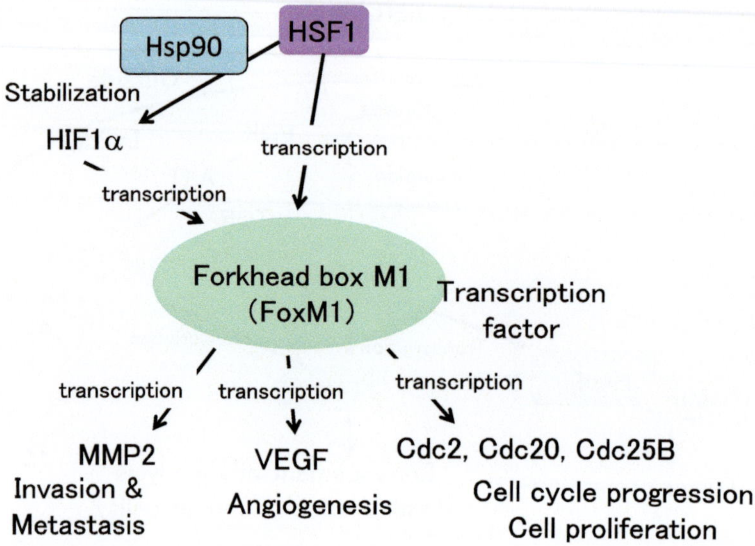

Fig. 4.2 Enhancement of FoxM1 expression by HSF1. As a transcription factor, FoxM1 could induce various genes that are involved in proliferation, angiogenesis, and invasion and metastasis of cancer cells. *FOXM1* forkhead box M1, *HIF1α* hypoxia-inducible factor 1α, *MMP2* matrix metalloproteinase 2, *VEGF* vascular endothelial growth factor, *Cdc* Kinases and phosphatases associated with cell cycle progression [27, 35]

4.4.3 HSF1 Regulates the Level of HuR That Controls Translation of HIF1α and β-Catenin

HuR is an mRNA binding protein that is often overexpressed in various types of cancer, and its enhanced expression correlates with cancer progression [28]. HuR is reported to control mRNA stability and the translation of various proteins associated with cancer including HIF1α and β-catenin.

When HSF1 was knocked out in HER2 oncogene-expressing mammary epithelial cells, HIF1α protein was downregulated, and angiogenesis was suppressed in a xenograft mouse model [29]. The suppression of HIF1α induced by the HSF1 knockout was found to be at the translational level due to downregulation of HuR. HIF1α is a master transcription factor of the genes that control tumor angiogenesis, glycolysis, invasion, and metastasis. Thus, HSF1 could regulate angiogenesis of tumors via the HuR-HIF1α pathway.

Recently, translation of β-catenin was also shown to be controlled by HSF1 through regulating the HuR protein levels in mammary cancer cells [30]. It was previously demonstrated that mTOR serine/threonine kinase is essential for phosphorylation at serine326 and activation of HSF1 [31]. It was found that mTOR is constitutively activated in cancer stem-like cells (CSC) in mammary cancer cells, and phosphorylates and activates HSF1. The protein level of β-catenin was also up-regulated in the CSC, and is well correlated with the activated HSF1 level;

however, HSF1 is not directly involved in transcription and translation of β-catenin. Instead, HSF1 could regulate HuR protein levels, which in turn enhance the translation of β-catenin [30]. Beta-catenin has critical roles in development and carcinogensis as an adaptor protein linking cadherin and actin filaments and as a transcription factor in the Wnt signaling pathway. In breast cancer cells, β-catenin is considered to control the renewal of CSC fraction via regulation of cyclin D1 expression [32].

4.4.4 HSF1 Binds to MTA1 and Represses Gene Expression

MTA1 (metastasis-associated protein 1) is a component of the NuRD (nucleosome remodeling and histone deacetylation) complex. This complex contains histone deacetylases (HDAC1 and 2), chromodomain proteins CHD3 and 4, and ATP-dependent chromaten-remodeling protein Mi2, and acts as a repressor of estrogen-inducible transcription [33]. In a pull-down analysis, HSF1 was demonstrated to bind to MTA1 in human breast carcinoma cells [34]. When the cells were treated with the highly malignant factor heregulin β1, both HSF1 and MTA1 were up-regulated, and HSF1-MTA1-NuRD complexes were formed at the promoters of estrogen-regulated genes, which resulted in the repression of gene expression such as that of c-Myc. Since c-Myc is a pro-apoptotic factor, it is suggested that repression of c-Myc by HSF1-MTA1-NuRD complexes inhibits apoptosis and promotes cancer cell proliferation (Fig. 4.3). Thus, HSF1 is able to work not only as a transcription factor but also as a transcriptional repressor.

4.5 Regulation of Factors Involved in Cancer by HSPs

Each HSP, such as Hsp90 and Hsp70, interacts with and regulates the function of various cancer-associated factors.

4.5.1 Stabilization of HIF1α by Hsp90

HIF1α is very labile under normal oxygen concentrations and is maintained at a low level by the ubiquitin-proteasome degradation system, but stabilized and activated under hypoxia typically in tumor tissues. HIF1α is acknowledged to be a critically important cancer cell survival factor that is required for tumorigenesis, such as angiogenesis and glucose metabolism. The degradation of HIF1α in normoxic cells is mediated by an E3 ubiquitin ligase called VHL (von Hippel-Lindau). Hsp90 was shown to interact with and stabilize HIF1α in renal carcinoma cells [35], suggesting that HIF1α is stabilized by Hsp90 and then activates the transcription of its down-

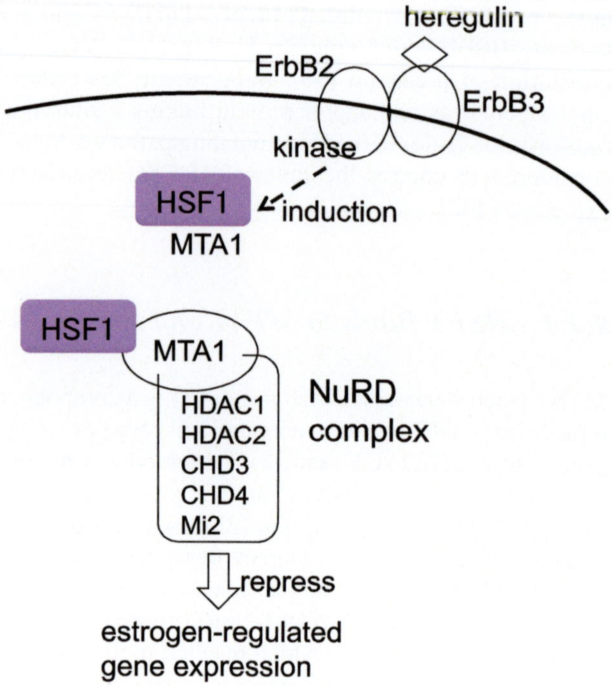

Fig. 4.3 HSF1 works as a transcriptional repressor. Heregulin induces HSF1 and MTA1 through the ErbB2 kinase signal pathway. MTA1 is one of the subunits of the NuRD repressor complex. HSF1 binds to MTA1, which in turn represses estrogen-dependent gene expression such as c-*myc*, resulting in the inhibition of apoptosis. *MTA1* metastasis-associated protein 1 [34]

stream genes and contributes to cancer cell proliferation even under hypoxic conditions (Fig. 4.2).

4.5.2 Hsp90/Hsp70 Chaperone Complex Inactivates LKB1 Tumor Suppressor

One of the tumor suppressor proteins, LKB1 (liver kinase B1), is a serine/threonine kinase that is responsible for Peutz-Jeghers syndrome, which is characterized by multiple hamartomatous polyps of the digestive tract and by an increased incidence of various types of cancer [36]. LKB is maintained in an active state by binding with STRAD (STE-20-related pseudokinase) and scaffolding protein MO25. Active LKB1 could phosphorylate 14 downstream kinases, including AMPK (adenosine monophosphate dependent kinase), and work as a tumor suppressor regulating cell polarity, metabolism, differentiation, and proliferation. LKB1 also interacts with Hsp90 together with its cochaperone Cdc37 and is kept in a catalytically inactive state. When the LKB1-Hsp90-Cdc37 complex was disrupted by an Hsp90 inhibitor geldanamycin, LKB1 was released from the complex and recruited by Hsp70/CHIP for degradation by the ubiquitin/proteasome system [37]. Thus, the Hsp90/Hsp70 chaperone system abrogates the function of the LKB1 tumor suppressor and contributes to tumor cell development (Fig. 4.4).

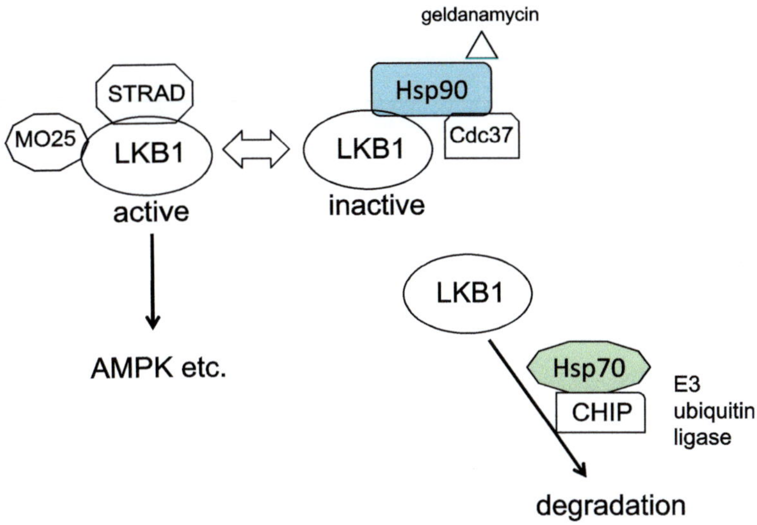

Fig. 4.4 Regulation of LKB1 function by Hsp90 and Hsp70. LKB1 is maintained as an active form by its interaction with STRAD and MO25. Active LKB1 phosphorylates downstream kinases such as AMPK, and regulates cellular energy metabolism and proliferation and works as a tumor suppressor. When Hsp90 and co-chaperone Cdc37 bind to LKB1, it becomes inactive. If the Hsp90 inhibitor geldanamycin is added, LKB1 is released from the Hsp90-Cdc37 complex, and is subjected to degradation by the proteasome through the Hsp70-CHIP pathway [37]

4.5.3 Hsp90/Hsp70 Stabilize and Activate a Metastasis Promoting Protein WASF3

The WASF3 protein was identified as a member of the Wiskott-Aldridge syndrome family of proteins that is involved in actin polymerization and cell movement through interaction with actin binding protein Arp2/3. WASF3 forms a complex with the p85 component of PI3K as well as ABL kinase and several other proteins (Fig. 4.5) and is maintained in an inactive form. The function of WASF3 is regulated via its phosphorylation by PI3K and ABL kinase [38]. As the knockdown of WASF3 by RNAi prevented cancer cell migration and invasion, WASF3 is considered to be a metastasis promoting protein [39]. Recently, it has been demonstrated that WASF3 binds to HSP90 and HSP70 in renal cancer cells, and that treatment of cells with 17-AAG, an Hsp90 inhibitor, resulted in the inhibition of phosphorylation of WASF3 by ABL kinase, suggesting that Hsp90 stabilizes ABL kinase which in turn phosphorylates WASF3 and keeps its active form. Also, Hsp70 directly binds to WASF3 and suppresses its degradation by the proteasome [40]. Therefore, Hsp90 and Hsp70 appear to promote cancer cell invasion and metastasis through stabilization and activation of WASF3 protein.

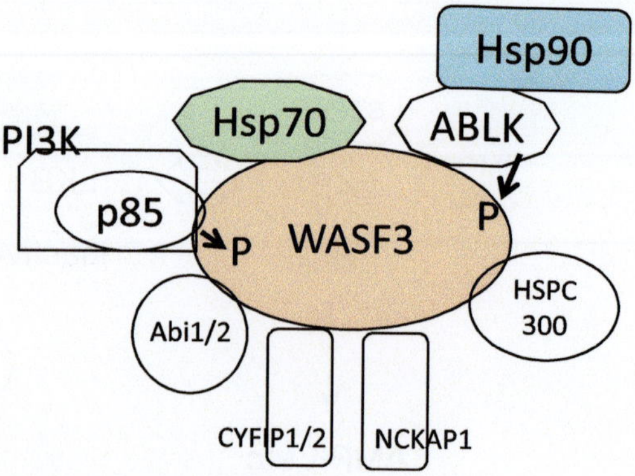

Fig. 4.5 Interaction of metastasis promoting protein WASF3 with Hsp90 and Hsp70. Hsp90 and Hsp70 stabilize and activate the WASF3 protein, which in turn promotes cancer cell invasion and metastasis [40]

4.5.4 Extracellular Hsp90 and Hsp70 Activate MMP2 and Enhance Cancer Cell Migration and Invasion

Hsp90 and Hsp70 are known to be secreted slightly outside of the cells and to function in activation of the immune system. MMP2 (matrix metalloproteinase-2) is a zinc-dependent endopeptidase that degrades various components of the extracellular matrix, and has an essential role in cancer cell invasion and metastasis. Previously, extracellular Hsp90 is reported to associate with MMP2 and to assist the activation of its function [41]. Depletion of extracellular Hsp90 and Hsp70 or the addition of their inhibitors into culture media was shown to lead to the suppression of migration and invasion of cancer cells [42].

4.5.5 Hsp70-Bag3 Complex Regulates Signaling Molecules Associated with Cancer

A growing body of evidence indicates that not only Hsp90 but also Hsp70 could regulate multiple signaling factors, such as NFκB, Src, HuR, HIF1α, and FoxM1. Recently, this chaperoning function of Hsp70 was found to be mediated by the Bag3 co-chaperone [43]. The Bag3 protein is a nucleotide exchange factor of Hsp70 and contains a BAG domain, which binds to the ATPase domain of Hsp70, as well as PxxP and WW domains, which link it, respectively, to the SH3 domain and PPxY motifs of signaling proteins. Notably, Bag3 is up-regulated in a number of cancers and has been implicated in tumor development. When Hsp70 was depleted by RNAi or the Hsp70-Bag3 interaction was disrupted by a small-molecule inhibitor, YM-1, the activities of transcription factors NF-κB, FoxM1, HIF1α, the translation regulator HuR, and the cell cycle regulators p21 and survivin, were suppressed, suggesting that the activities of these cancer cell signaling factors were regulated by Hsp70 through its binding to Bag3 [43].

4.5.6 Nuclear Localized Mortalin Promotes Carcinogensis

Mortalin is a mitochondrial Hsp70 family member that functions in mitochondrial import, oxidative stress response, energy metabolism, and p53 function. Also, mortalin is frequently overexpressed in cancer cells and localizes in various subcellular compartments, including the nuclei. Recently, it was demonstrated that nuclear-localized mortalin could cause strong inactivation of tumor suppressor protein p53 and activation of telomerase and heterogenous ribonucleoprotein K (hnRNP-K), which result in the promotion of carcinogensis and cancer cell migration [44].

4.6 HSF1 and HSPs as Targets for Cancer Therapy

As mentioned above, HSF1 and HSPs are activated and overexpressed in malignant cells to support many cancer-associated signaling kinases and transcription factors. Therefore, the HSF-HSPs system is considered to be an ideal therapeutic target to control malignancy.

4.6.1 Hsp90 Inhibitors

Hsp90 functions in the maturation, stability, and activity of more than 200 proteins, called "client proteins," many of which are oncogenic signaling kinases [45]. When the function of Hsp90 is inhibited, these client proteins are released from the Hsp90 chaperone complex and degraded by the ubiquitin-proteasome system. The first identified inhibitor of Hsp90 is geldanamycin, but it was not used for clinical trials because of its side effect of hepatotoxicity. A geldanamycin derivative, 17-AAG (17-allylamino-17-demethoxygeldanamycin), was the first Hsp90 inhibitor to be evaluated in a clinical study [46]. There are many Hsp90 inhibitors including natural products geldanamycin and radicicol, as well as their synthetic analogs, such as 17-AAG and 17-DMAG, and synthetic small molecule inhibitors, such as AUY922A, BIIB021, and SNX2112. Unfortunately, no phase III clinical trials of these inhibitors have been conducted and no agent has been approved for clinical use. One of the disadvantages of Hsp90 inhibitors, however, is that they induce the cytoprotective chaperone Hsp70. So, induction of Hsp70 is a hallmark of Hsp90 inhibitors. Therefore, combination therapy of Hsp90 and an Hsp70 inhibitor has been conducted by both in vitro and in vivo studies [47].

4.6.2 Hsp70 Inhibitors

Hsp70 has two distinct domains: an amino-terminal ATPase (ATP binding) domain (ABD) and a C-terminal peptide-binding domain (PBD) (Fig. 4.6). The first molecule to be identified as an Hsp70 inhibitor is 2-phenylethynesulfonamid (PES, or

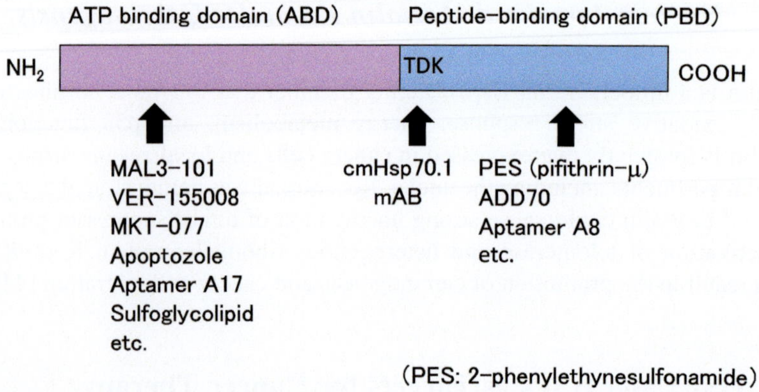

Fig. 4.6 Hsp70 domain structure and its inhibitors. The binding sites of various inhibitors are indicated by arrows. TDK is the extracellular epitope of 14-mer peptide of Hsp70 that targets the monoclonal antibody cmHsp70.1 (Modified from [51])

pifithrin-μ) that binds to the PBD domain. PES was shown to impair autophagy and inhibit lysosomal function, and to have potent anticancer effects in both cell culture and a mouse model of *myc*-induced lymphomagenesis [48]. Recently, a PES-derived compound (PES-Cl) with higher affinity was developed and demonstrated increased cytotoxicity and potency in the same mouse model [49]. Interestingly, a mouse monoclonal antibody (cmHsp70.1) generated against the 14-mer peptide TKDNNLLGRFELSG (TKD) comprising amino acids 450–461 (aa450-461) in the C terminus of inducible Hsp70, which recognizes cell surface Hsp70, could selectively induce antibody-dependent cellular cytotoxicity of membrane Hsp70-positive mouse tumor cells. Also, injections of the cmHsp70.1 mAb into mice bearing membrane Hsp70-positive tumors significantly inhibited tumor growth and enhanced the overall survival [50]. Several other Hsp70 inhibitors are shown in Fig. 4.6 [51]. These compounds, however, are not yet in the stage of clinical trials.

4.6.3 Inhibitors of Heat Shock Response

Small molecule compounds that affect HSF1 expression or activity have been recently developed. One of the triazole nucleoside analogs (named compound 5) could down-regulate mRNA and protein of HSF1 and lead to consequential down-regulation of multiple HSPs (Hsp90, Hsp70, and Hsp27). Down-regulation of these proteins caused the shut-down of several oncogenic signaling pathways and apoptosis resulting in an anticancer effect both in vitro and in vivo [52]. The mechanism by which this compound inhibited HSF1 expression is not yet known, because

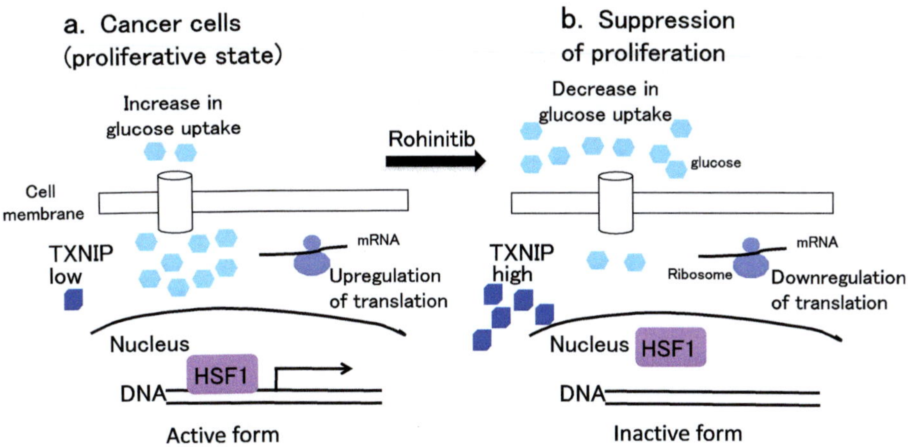

Fig. 4.7 Role of HSF1 in cancer cells and mechanism of Rohinitib action. (**a**): HSF1 is constitutively activated and translation is accelerated in vigorously proliferating cancer cells. Since the expression level of TXNIP (a powerful negative regulator of glucose uptake) is low, glucose uptake is enhanced. (**b**): Addition of Rohinitib causes a decrease in translation on ribosomes and concomitant inactivation of HSF1. Also, Rohinitib induces TXNIP, which in turn downregulates glucose uptake. As a result, proliferation of cancer cells is suppressed. Rohinitib, rocaglate heat shock, initiation of translation inhibitor; *TXNIP* thioredoxin–interacting protein [55]

it could not inhibit RNA and DNA synthesis. Another compound NZ28 has been shown to inhibit the induction of HSPs by heat shock, proteasome inhibitor, and Hsp90 inhibitor, and to sensitize cancer cells to proteasome and Hsp90 inhibitors, and radiation [53, 54]. Although the molecular target of NZ28 is not yet known, it seems to have an inhibitory effect on the translation of major HSPs.

Recently, it has been reported that translational activity on ribosomes is correlated with HSF1 transcriptional activity [55]. When cells were treated with a translational inhibitor such as cycloheximide, the binding of transcriptional factors such as HSF1 to DNA was suppressed by an unknown mechanism. Santagata et al. identified a compound that specifically inhibits HSF1 transcriptional activity by screening more than 300,000 chemical compounds. The compound named rohinitib (RHT) had no effect on the expression of housekeeping genes or on HSF1 mRNA and protein expression. In contrast, RHT specifically inhibited the HSF1-dependent expression of HSPs and other genes, suggesting that HSF1 transcriptional activity was suppressed by RHT (Fig. 4.7). Also, RHT has a cell killing effect on only cancer cells but not normal cells, and an antitumor effect on mouse xenograft tumor model. The mechanism of the inhibitory effect of RHT on HSF1 transcriptional activity, however, remains to be elucidated.

4.7 Summary-Which One Do You Prefer Neurodegenerative Disease or Cancer?

From a biological point of view, the HSF1-HSP system works well when the organism is young, and protects the living body at the cellular level, so that life is inherited successfully to the next generation. In contrast, the function of the HSF1-HSP system is known to gradually decline with age [56]. Therefore, as the organism ages, various cellular functions decline and enter a pathological state, eventually resulting in death. This is the principle of life. Most of the diseases that develop in older age are lifestyle-related diseases, including cancer, and many of the neurodegenerative diseases also develop in later adulthood. The human life span in the developed countries became extraordinarily extended in the twentieth to twenty-first centuries. At the same time, it is obvious that specific diseases of old age are increasing. From the point of view of prevention and treatment of age-related diseases, one of the treatment protocols for neurodegenerative diseases is the activation of the HSF1-HSP system, which results in the prevention of aggregate formation of mutant proteins and the promotion of degradation of abnormal proteins in neuronal cells [57]. On the other hand, down-regulation or inactivation of the HSF1-HSP system by some method is required for the treatment of cancer [58]. It is annoying that we should need to choose either neurodegenerative disease or cancer in older age. Which one would you prefer?

References

1. Ritossa F. A new puffing pattern induced by temperature shock and DNP in Drosophila. Experientia. 1962;18:571–3.
2. Muchowski PJ, Wacker JL. Modulation of neurodegeneration by molecular chaperones. Nat Rev Neurosci. 2005;6:11–22.
3. Hartl FU, Bracher A, Hayer-Hartl M. Molecular chaperones in protein folding and proteostasis. Nature. 2011;475:324–32.
4. Buchberger A, Bukau B, Sommer T. Protein quality control in the cytosol and the endoplasmic reticulum: brothers in arms. Mol Cell. 2010;40:238–52.
5. Ohtsuka K, Kawashima D, Gu Y, et al. Inducers and co-inducers of molecular chaperones. Int J Hyperthermia. 2005;21:703–11.
6. Kobayashi Y, Kume A, Li M, et al. Chaperones Hsp70 and Hsp40 suppress aggregate formation and apoptosis in cultured neuronal cells expressing truncated androgen receptor protein with expanded polyglutamine tract. J Biol Chem. 2000;275:8772–8.
7. Tohnai G, Adachi H, Katsuno M, et al. Paeoniflorin eliminates a mutant AR via NF-YA-dependent proteolysis in spinal and bulbar muscular atrophy. Hum Mol Genet. 2014;23:3552–65.
8. Kondo N, Katsuno M, Adachi H, et al. Heat shock factor-1 influences pathological lesion distribution of polyglutamine-induced neurodegeneration. Nat Commun. 2013;4:1405.
9. Bukau B, Weissman J, Horwich A. Molecular chaperones and protein quality control. Cell. 2006;125:443–51.
10. Dai C, Whitesell L, Rogers AB, et al. Heat shock factor 1 is a powerful multifaceted modifier of carcinogenesis. Cell. 2007;130:1005–18.

11. Sherman MY, Gabai VL. Hsp70 in cancer: back to the future. Oncogene. 2014;34:4153–61.
12. Ciocca DR, Arrigo AP, Calderwood SK. Heat shock proteins and heat shock factor 1 in carcinogenesis and tumor development: an update. Arch Toxicol. 2013;87:19–48.
13. Parsell DA, Lindquist S. The function of heat-shock proteins in stress tolerance: degradation and reactivation of damaged proteins. Annu Rev Genet. 1993;27:437–96.
14. Ciocca DR, Calderwood SK. Heat shock proteins in cancer: diagnostic, prognostic, predictive, and treatment implications. Cell Stress Chaperones. 2005;10:86–103.
15. Mjahed H, Girodon F, Fontenay M, et al. Heat shock proteins in hematopoietic malignancies. Exp Cell Res. 2012;318:1946–58.
16. Santagata S, Hu R, Lin NU, et al. High levels of nuclear heat-shock factor 1 (HSF1) are associated with poor prognosis in breast cancer. Proc Natl Acad Sci U S A. 2011;108:18378–83.
17. Calderwood SK. HSF1, a versatile factor in tumorogenesis. Curr Mol Med. 2012;12:1102–7.
18. Jin X, Moskophidis D, Mivechi NF. Heat shock transcription factor 1 is a key determinant of HCC development by regulating hepatic steatosis and metabolic syndrome. Cell Metab. 2011;14:91–103.
19. Gong J, Weng D, Eguchi T, et al. Targeting the hsp70 gene delays mammary tumor initiation and inhibits tumor cell metastasis. Oncogene. 2015. doi:10.1038/onc.2015.1.
20. Page TJ, Sikder D, Yang L, et al. Genome-wide analysis of human HSF1 signaling reveals a transcriptional program linked to cellular adaptation and survival. Mol Biosyst. 2006;2:627–39.
21. Mendillo ML, Santagata S, Koeva M, et al. HSF1 drives a transcriptional program distinct from heat shock to support highly malignant human cancers. Cell. 2012;150:549–62.
22. Solimini NL, Luo J, Elledge SJ. Non-oncogene addiction and the stress phenotype of cancer cells. Cell. 2007;130:986–8.
23. Zhao YH, Zhou M, Liu H, et al. Upregulation of lactate dehydrogenase A by ErbB2 through heat shock factor 1 promotes breast cancer cell glycolysis and growth. Oncogene. 2009;28:3689–701.
24. Li YM, Zhou BP, Deng J, et al. A hypoxia-independent hypoxia-inducible factor-1 activation pathway induced by phosphatidylinositol-3 kinase/Akt in HER2 overexpressing cells. Cancer Res. 2005;65:3257–63.
25. Schulz R, Streller F, Scheel AH, et al. HER2/ErbB2 activates HSF1 and thereby controls HSP90 clients including MIF in HER2-overexpressing breast cancer. Cell Death Dis. 2014;5: e980. doi:10.1038/cddis.2013.508.
26. Kalin TV, Ustiyan V, Kalinichenko VV. Multiple faces of FoxM1 transcription factor: lessons from transgenic mouse models. Cell Cycle. 2011;10:396–405.
27. Dai B, Gong A, Jing Z, et al. Forkhead box M1 is regulated by heat shock factor 1 and promotes glioma cells survival under heat shock stress. J Biol Chem. 2013;288:1634–42.
28. López de Silanes I, Fan J, Yang X, et al. Role of the RNA-binding protein HuR in colon carcinogenesis. Oncogene. 2003;22:7146–54.
29. Gabai VL, Meng L, Kim G, et al. Heat shock transcription factor Hsf1 is involved in tumor progression via regulation of hypoxia-inducible factor 1 and RNA-binding protein HuR. Mol Cell Biol. 2012;32:929–40.
30. Chou SD, Murshid A, Eguchi T, et al. HSF1 regulation of β-catenin in mammary cancer cells through control of HuR/elavL1 expression. Oncogene. 2015;34:2178–88.
31. Chou SD, Prince T, Gong J, et al. mTOR is essential for the proteotoxic stress response, HSF1 activation and heat shock protein synthesis. PLoS One. 2012;7(6):e39679.
32. Gotoh J, Obata M, Yoshie M, et al. Cyclin D1 over-expression correlates with beta-catenin activation, but not with H-ras mutations, and phosphorylation of Akt, GSK3 beta and ERK1/2 in mouse hepatic carcinogenesis. Carcinogenesis. 2003;24:435–42.
33. Mazumdar A, Wang RA, Mishra SK, et al. Transcriptional repression of oestrogen receptor by metastasis-associated protein 1 corepressor. Nat Cell Biol. 2001;3:30–7.
34. Khaleque MA, Bharti A, Gong J, et al. Heat shock factor 1 represses estrogen-dependent transcription through association with MTA1. Oncogene. 2008;27:1886–93.

35. Isaacs JS, Jung YJ, Mimnaugh EG, et al. Hsp90 regulates a von Hippel Lindau-independent hypoxia-inducible factor-1 alpha-degradative pathway. J Biol Chem. 2002;277:29936–44.
36. Vaahtomeri K, Mäkelä TP. Molecular mechanisms of tumor suppression by LKB1. FEBS Lett. 2011;585:944–51.
37. Gaude H, Aznar N, Delay A, et al. Molecular chaperone complexes with antagonizing activities regulate stability and activity of the tumor suppressor LKB1. Oncogene. 2012;31:1582–91.
38. Sossey-Alaoui K, Li X, Ranalli TA, et al. WAVE3-mediated cell migration and lamellipodia formation are regulated downstream of phosphatidylinositol 3-kinase. J Biol Chem. 2005;280:21748–55.
39. Sossey-Alaoui K, Safina A, Li X, et al. Down-regulation of WAVE3, a metastasis promoter gene, inhibits invasion and metastasis of breast cancer cells. Am J Pathol. 2007;170:2112–21.
40. Teng Y, Ngoka L, Mei Y, et al. HSP90 and HSP70 proteins are essential for stabilization and activation of WASF3 metastasis-promoting protein. J Biol Chem. 2012;287:10051–9.
41. Eustace BK, Sakurai T, Stewart JK, et al. Functional proteomic screens reveal an essential extracellular role for hsp90 alpha in cancer cell invasiveness. Nat Cell Biol. 2004;6:507–14.
42. Sims JD, McCready J, Jay DG. Extracellular heat shock protein (Hsp)70 and Hsp90α assist in matrix metalloproteinase-2 activation and breast cancer cell migration and invasion. PLoS One. 2011;6:e18848.
43. Colvin TA, Gabai VL, Gong J, et al. Hsp70-Bag3 interactions regulate cancer-related signaling networks. Cancer Res. 2014;74:4731–40.
44. Ryu J, Kaul Z, Yoon AR, et al. Identification and functional characterization of nuclear mortalin in human carcinogenesis. J Biol Chem. 2014;289:24832–44.
45. Hong DS, Banerji U, Tavana B, et al. Targeting the molecular chaperone heat shock protein 90 (HSP90): lessons learned and future directions. Cancer Treat Rev. 2013;39:375–87.
46. Neckers L, Workman P. Hsp90 molecular chaperone inhibitors: are we there yet? Clin Cancer Res. 2012;18:64–76.
47. Massey AJ, Williamson DS, Browne H. A novel, small molecule inhibitor of Hsc70/Hsp70 potentiates Hsp90 inhibitor induced apoptosis in HCT116 colon carcinoma cells. Cancer Chemother Pharmacol. 2010;66:535–45.
48. Leu JI, Pimkina J, Frank A, et al. A small molecule inhibitor of inducible heat shock protein 70. Mol Cell. 2009;36:15–27.
49. Balaburski GM, Leu JI, Beeharry N, et al. A modified HSP70 inhibitor shows broad activity as an anticancer agent. Mol Cancer Res. 2013;11:219–29.
50. Stangl S, Gehrmann M, Riegger J, et al. Targeting membrane heat-shock protein 70 (Hsp70) on tumors by cmHsp70.1 antibody. Proc Natl Acad Sci U S A. 2011;108:733–8.
51. Goloudina AR, Demidov ON, Garrido C. Inhibition of HSP70: a challenging anti-cancer strategy. Cancer Lett. 2012;325:117–24.
52. Xia Y, Liu Y, Rocchi P, et al. Targeting heat shock factor 1 with a triazole nucleoside analog to elicit potent anticancer activity on drug-resistant pancreatic cancer. Cancer Lett. 2012;318:145–53.
53. Zaarur N, Gabai VL, Porco Jr JA, et al. Targeting heat shock response to sensitize cancer cells to proteasome and Hsp90 inhibitors. Cancer Res. 2006;66:1783–91.
54. Schilling D, Kühnel A, Konrad S, et al. Sensitizing tumor cells to radiation by targeting the heat shock response. Cancer Lett. 2015;360:294–301.
55. Santagata S, Mendillo ML, Tang YC, et al. Tight coordination of protein translation and HSF1 activation supports the anabolic malignant state. Science. 2013;341:1238303.
56. Kregel KC. Heat shock proteins: modifying factors in physiological stress responses and acquired thermotolerance. J Appl Physiol. 2002;92:2177–86.
57. Neef DW, Jaeger AM, Thiele DJ. Heat shock transcription factor 1 as a therapeutic target in neurodegenerative diseases. Nat Rev Drug Discov. 2011;10:930–44.
58. Whitesell L, Lindquist S. Inhibiting the transcription factor HSF1 as an anticancer strategy. Expert Opin Ther Targets. 2009;13:469–78.

Chapter 5
Cellular Responses in Signal Transduction Pathways Induced by Hyperthermia

Takeo Ohnishi

Abstract This chapter focuses on cell death, the cell cycle, and cell survival signal transduction pathways in mammalian cells. Environmental stresses induce many kinds of signal transduction pathways such as cell death signals and survival signals. Fate of the cells is dependent on those cellular responses. The gene products of tumor suppressor genes such as the *p53* proto-oncogene and the products of other proto-oncogenes play important roles in these signaling pathways. The diverse pathways through which these elements interact with each other, and their close links to other nuclear and extranuclear signaling networks compel us to reconsider the concept of stress responses to cancer therapies such as hyperthermia, radiation and anti-tumor drugs.

Keywords Apoptosis • Caspase • HSP • Bax • Bcl • Signal transduction • Gene expression

5.1 Introduction

Organisms can be exposed to an environment containing genotoxic and non-genotoxic stresses such as hyperthermia, radiation, anti-tumor drugs, pH extremes, hypoxia, and other stresses, and cellular receptors can recognize these stresses, and lead cells to respond and adapt to these stresses at various levels. These responses can involve molecules, cells, organs and entire organisms. To obtain optimal and maximal effects from hyperthermic cancer therapy, it is important to define the cellular responses and mechanisms involved after being exposed to a heat treatment. Obvious questions are which molecules are targeted in the cellular membrane and how many types of proteins and/or DNA molecules are involved? The early events after a heat exposure lead to radical formation and the denaturation of proteins. Some recent advances have also provided additional understanding of enzyme/substrate relationships and of phosphorylation cascades which have been demonstrated in several distinct pathways in membrane-cytoplasm signaling. The next events are believed to involve the activation

T. Ohnishi (✉)
Department of Radiation Oncology, Nara Medical University, 840 Shijo-cho, Kashihara, Nara 634-8521, Japan
e-mail: tohnishi@ares.eonet.ne.jp

S. Kokura et al. (eds.), *Hyperthermic Oncology from Bench to Bedside*,
DOI 10.1007/978-981-10-0719-4_5

of protein kinases which targets proteins such as heat shock factors, ATM, ATR and others. These phenomenon then lead to a successive round of chemical chain reactions. In these reactions, the focus is on cell death, the cell cycle and cell survival signal transduction pathways. Signaling networks are also emerging from responses to different types of stress between the different pathways which respond to these stresses. In an effort to use a defined system to study these mechanisms, we established different cell lines only in the status of their *p53* gene. The molecular dissection of p53 activity and the products of other proto-oncogenes have greatly advanced the studies of nuclear signaling and expanded the concept of checkpoints to mammalian cells. This approach focuses on studies of the molecular targets of hyperthermic therapy, and the effects of combining hyperthermic therapy with other cancer therapies. Many mutations have been reported in the *p53* tumor suppressor gene in advanced human cancers [1]. The function of the tumor suppressor gene *p53* focuses on cellular signaling pathways because p53 modulates many functions in death signaling, survival signaling and cell proliferation pathways [2]. Consequently, mutations or deletions in the *p53* gene can lead to resistance to hyperthermic cancer therapies as well resistance to as radiotherapy and chemotherapy [3].

5.2 Gene Expression Responses after Hyperthermia

To understand the complex cellular responses induced by heat stress, microarrays and bioinformatic systems are the most powerful and widely applied tools used for these studies. Gene expression profiling data and gene networks are expected to provide the molecular basis for a better understanding of the mechanisms involved in biological responses to heat stresses.

5.2.1 Gene Expression Patterns Observed with DNA Chips

Analysis of gene expression has commonly been obtained from cDNA arrays using mRNA, and also with protein arrays. To clarify the function of p53 in signal transduction pathways after hyperthermic cancer therapy, these two methods were used with a mutated *p53* cell line (m*p53*) and a parental wild-type *p53* cell line (wt*p53*) derived from a human squamous cell carcinoma (SAS) cell line. These two cell lines have identical genetic backgrounds except for their *p53* gene status. Using these two cell lines, it is also possible to analyze cell behavior in cultured cell systems (*in vitro*) and in nude mice (*in vivo*).

After applying hyperthermia (44 °C for 40 min) to cultured cells with mutated and wild type *p53* genes, the cells were cultured for different periods after heat exposure, and total RNA was then extracted for cDNA synthesis. Gene expression was compared in heat treated and non-heat treated control cells [4]. Responses to hyperthermia were standardized by comparing the results to the behavior of house-keeping genes. Responses in gene expression after hyperthermia were classified into five groups depending on the degree of gene expression in wt*p53* and m*p53* cells. Among the apoptosis inducing genes, *p53* and *CD40* (cluster of differentiation 40) were induced, and *Bim EL* (Bcl-2 interacting mediator of cell death EL) was up-regulated. *β-NGF* (beta-nerve growth factor) which is considered to be an apoptosis suppressor and an apoptosis promoter, was up-regulated in wt*p53* cells and down-regulated in m*p53* cells. *Caspase-9*, *CD30* (cluster of differentiation 30) and *CD40*, considered to be apoptosis promoting genes were induced in both wt*p53* and m*p53* cells. The expression of apoptosis suppressing genes of *FLIPL* (Fas-associated death domain protein-like interleukin-1 beta-converting enzyme-inhibitory proteins L), *IL-12 R β1* (interleukin-12 receptor beta 1) and VEGI (vascular endothelial growth inhibitor) increased in m*p53* cells. A salient point to notice is that IL-12 R showed a decreased expression in wt*p53* and increased expression in m*p53* cells. On the other hand, IL-12 p35 was down-regulated in heat-treated wt*p53* cells.

In heat-treated wt*p53* cells, the expression of the apoptosis-suppressor genes *DFF 45* (DNA fragmentation factor 45), *Bcl-2, MDM2* (murine double minute 2), *IKK-α* (I-kappa-B kinase-alpha) and *IKK-β* (I-kappa-B kinase-beta) decreased. Eighteen genes were down-regulated in heat treated m*p53* cells. These included *p75 NGFR* (*p75* nerve growth factor receptor), *CRADD* (Caspase and RIP adaptor with death domain), *APRIL* (a proliferation-inducing ligand), *TRAF 5* (tumor necrosis factor receptor-associated factors 5), *TRAIL R1* (tumor necrosis factor-related apoptosis inducing ligand receptor 1), *TWEAK* (tumor necrosis factor-like weak inducer of apoptosis), and others which are considered to be apoptosis promoting genes. In looking at IL-12-related gene expression in wt*p53* cells, both *IL-12 p35* and *IL-12 p40* appeared at only one point after 6 h, but expression was not seen at all except at that one time point. In SAS/m*p53* cells, *IL-12 p35* expression appeared at numerous points, but the expression of *IL-12 p40* occurred at two time points, at 6 and, 12 h. Because it occurred at only at 2 specific two points, increased expression over the entire 5 time points spanning the experiment was not detected. The receptor levels of IL-12 increased in m*p53* cells. IL-12 induces NF-κB, and NF-κB2 levels were increased in m*p53* cells.

5.2.2 Gene Expression Patterns Obtained from DNA Chips and Network Analysis

Differences between gene expression patterns in cells subjected to apoptosis inducing conditions (42 °C, 90 min, and 44 °C, 15 min) and to non-apoptosis inducing conditions (42 °C, 15 min) were examined using DNA chip analysis in a human leukemia U937 cell line [5]. The 1009 and 220 probe sets revealed alterations found under common apoptotic conditions, as well as other common alterations induced after being subjected to the three sets of heating conditions. In two significant gene networks (Fig. 5.1), C (common) and A (apoptosis-related) levels indicate differentially expressed transcripts with the C or common chip set (220 probes) and the A or apoptotic chip set (1009 probes), respectively. In gene network C, the commonly expressed transcripts containing many heat shock proteins (HSPs) were associated

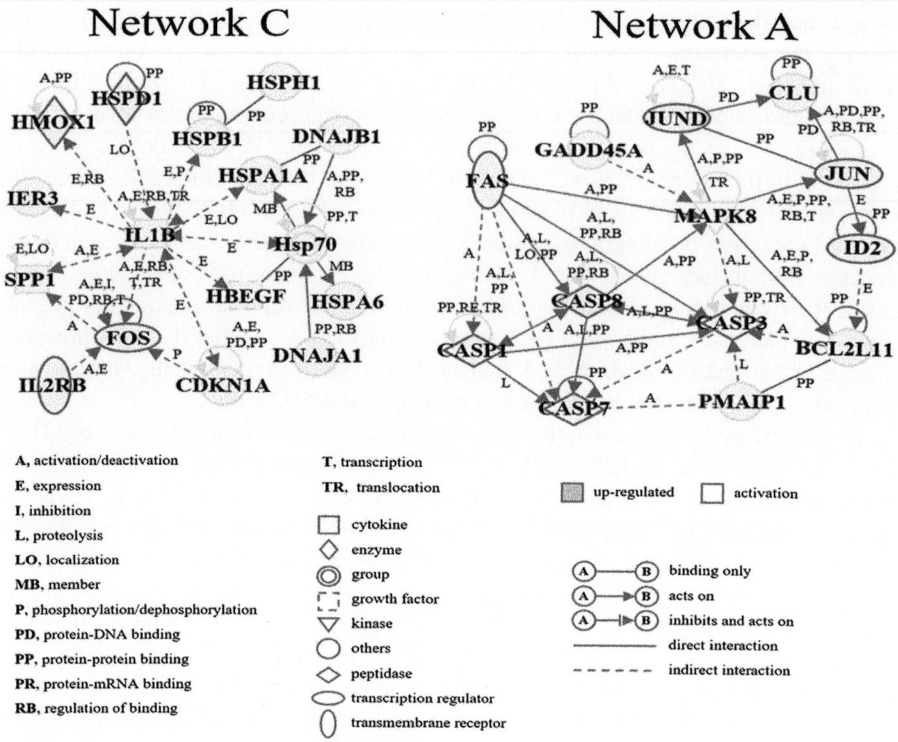

Fig. 5.1 Relationship among many genes expressed in the apoptosis induction network A and the non-apoptosis network C after a heat treatment

with biological functions such as cellular function and maintenance. The induction of HSPs appeared to be a common response to heat, regardless of apoptosis status. On the other hand, Gene network A derived from apoptosis-specific transcripts contained Caspases (CASP1, CASP3, CASP7 and CASP8) and MAPK8 (JNK) which are primarily associated with the molecular functions leading to cell death which includes apoptosis. Sequential activation of Caspases, a member of the cysteine-aspartic acid protease (Caspase) family, plays a central role in the progression of cell apoptosis. It is known that the JNK pathway is involved in apoptosis evoked by heat stress.

5.2.3 Gene Expression Examined with Protein Chips

The analysis of protein microarrays in w*tp53* and m*p53* cells showed that the levels of Fas-associated death domain (FADD)/Mort-1, Ku70, protein kinase C (PKC)δ, PKCε, E2F- and DP-like subunits-1 (E2F-1) and poly(ADP-ribose)polymerase (PARP) increased after exposure to heat in both w*tp53* and m*p53* cells. FADD was induced by heat, and overexpression of FADD resulted in the induction of apoptosis. X-ray cross-complementing gene 6 (XRCC6, Ku70), PKCδ, PKCε and

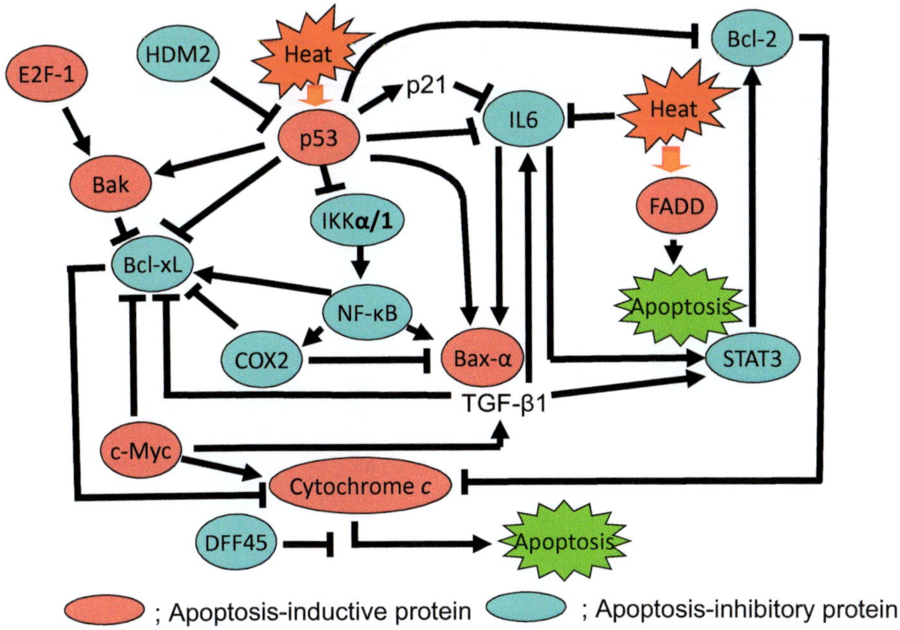

Fig. 5.2 Relationship among apoptosis-related gene products after a heat treatment as reported by protein chip analysis. The induction of protein synthesis was analyzed from observing apoptosis-relating pathways after a heat treatment. *Arrows*, promotion; "**T**" symbols and lines indicate depression

E2F-1 are also involved in apoptosis. Based on these reports, it appears that FADD/Mort-1, Ku70, PKCδ, PKCε, and the transcription factor family including E2F-1 and PARP are involved in heat-induced *p53*-independent apoptosis in wt*p53* and m*p53* cells. It was found that apoptosis inhibitory proteins such as nuclear factor κB (NF-κB), B-cell leukemia (Bcl)-2, (an apoptosis inhibitory protein (Bcl-xL), cytochrome oxidase2 (COX2), signal transducer and activator of transcription (STAT3) and Interleukin (IL)-6 were up-regulated in m*p53* cells after heating. IL-6 has been reported to be down-regulated by heat to a greater extent in wt*p53* cells. Bcl-2, Bcl-xL and IκB kinase (IKK)α (an apoptosis inhibitory protein) were also down-regulated preferentially in wt*p53* cells. From these results, it appears that in m*p53* cells, it would be hard to induce apoptosis after heating when compared to wt*p53* cells. Considering apoptosis-inhibitory proteins, it appears that heat-induced apoptosis may be suppressed by NF-κB, Bcl-2, Bcl-xL, COX2, STAT3, IL-6 and IKKα in m*p53* cells. In view of this, a possible heat-induced *p53*-dependent signal transduction pathway for apoptosis can be proposed: it is suggested that a heat shock can activate a subset of apoptosis-related proteins. The finding from this present study is that *p53* status is closely related to heat-induced apoptosis, and regulates the expression of apoptosis-inhibitory proteins such as Bcl-2, Bcl-xL and IL-6. This could mean that heat-resistance in m*p53* cells may result from the up-regulation of these apoptosis-inhibitory proteins. The proposed signal transduction pathways are summarized in Fig. 5.2 [6].

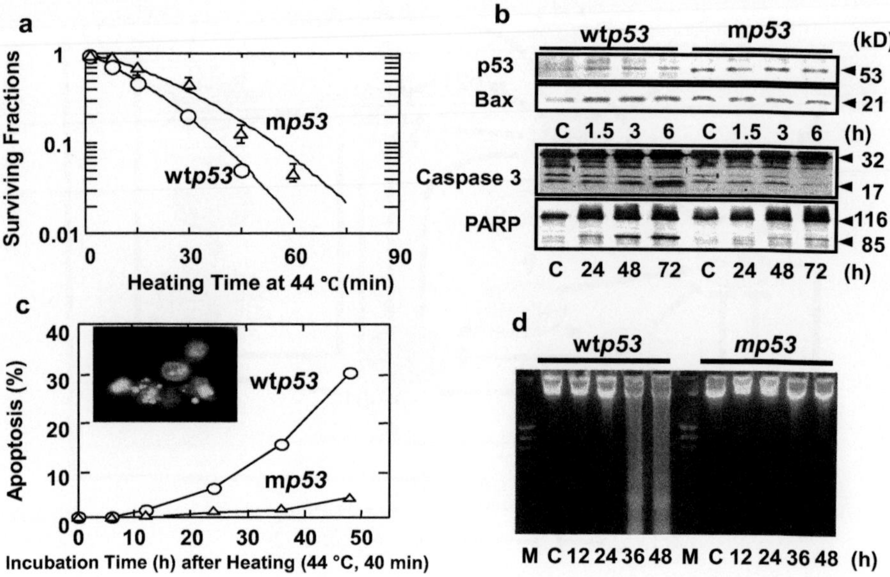

Fig. 5.3 Heat induced apoptosis. (**a**) Survival. (**b**) Western blot analysis. *Arrows* indicate molecular weight (kD). h, incubation periods after heat treatment. c, non-heated cells. (**c**) Apoptosis (inserted figure, typical photographs of apoptic cells). (**d**) DNA degradation. DNA ladder was found in wt*p53* cells, but not in m*p53* cells. M, Size markers for DNA fragments

5.3 *p53*-Dependent Cell Death Mechanism Induced by Heat

The tumor suppressor gene product p53 is activated by genotoxic stresses such as hyperthermia, radiation and DNA damaging drugs. It was found that wt*p53* cells were more sensitive to heat than m*p53* cells (Fig. 5.3a). This suggested that *p53*-dependent apoptosis is at least partially involved in heat sensitivity. Heat treatments activate p53 which regulates the expression of downstream genes. The gene product of *p53* was found to contribute to heat sensitivity. After heat treatment, p53 and Bax were induced (Fig. 5.3b). Bax-mediated apoptosis was promoted *via* the activation of Caspase-3 which degrades Caspase-3 itself and PARP. This was confirmed using Western blot analysis (Fig. 5.3c) in human cancer cells. The frequency of heat induced apoptosis in wt*p53* cells was higher than in m*p53* cells. In fact, DNA degradation (a DNA ladder) which is observed after a heat treatment appeared in the wt*p53* cells but not in m*p53* cells (Fig. 5.3d). These studies indicate heat-induced *p53*-dependent signal transduction and heat sensitivity *via* a *p53*-regulated pathway for apoptosis function in human cancer cells [7]. This supports the use of cancer therapies using not only hyperthermia, but also radiation and anti-tumor drugs such as cisplatin which might be dependent on *p53* gene status. To produce a higher efficacy in hyperthermic cancer therapy, it is proposed to manipulate m*p53* with additional new cancer therapies such as chemical chaperones and gene therapy in patients with m*p53*-tumors.

5.4 Heat-induced *p53* Signal Transduction Pathways

In response to a heat treatment, many chemical chain reactions can be initiated. Among the early events in such responses, p53 molecules are modified through phosphorylation, acetylation, and poly(ADP-ribosyl)ation, and attention has been focused on phosphorylation events after a heat treatment. There are many reports concerning the phosphorylation of proteins such as phosphoinositide 3-kinase (PI3-K), DNA-dependent protein kinase (DNA-PK), ataxia telangiectasia mutated (ATM), and the ATM- and radiation sensitive mutant 3-related protein kinase (ATR) [8, 9]. Although it has been reported that there are substrate specificities for these protein kinases, their functions overlap with each other in cultured cells *in vitro* and in transplanted cells in mice *in vivo*.

Phosphorylation activity has been thought to be dependent on DSB formation because many reports have been published concerning the phosphorylation of histone H2AX which is an indicator of DSB formation. ATM is required for this response to ionizing radiation in human cultured cells. On the other hand, it was reported that ATM or DNA-PK can promote ionizing radiation-induced H2AX phosphorylation redundantly. Histone H2AX was reported to be phosphorylated in an ATR-dependent manner in response to replication stresses such as treatment with hydroxyurea and ultraviolet light. Heat shock as well as X-irradiation induces histone H2AX phosphorylated at serine 139 (γH2AX) in numerous mammalian cell lines. To clarify the details and mechanisms involved in the phosphorylation of H2AX after a heat treatment [9], ATM-knockout cell lines were used as well as DNA-PKcs-knockout cells and their parental cell lines. An ATR-kinase inhibitor (CGK733), and a DNA-PK-inhibitor (NU7026) were also examined, and the intensity of the γH2AX signals was assayed with flow cytometry. The time for γH2AX foci to appear after heating was similar to what was seen after heat treatment in DNA-PKcs−/− cells and in the corresponding parental cells. On the other hand, the slope of the curve showing the extent of phosophorylation of H2AX *vs* heating time in ATM−/− cells was 40 % of the slope in ATM+/+ cells. Phosphorylation of H2AX after heat-treatment was suppressed by a combination of CGK733 and NU7026 in the culture medium with either cell line. Although the phosphorylation of H2AX after heat treatment was not suppressed by NU7026 in the parental cells, such phosphorylation was suppressed by CGK733 in the parental cells. These results suggest that ATM is the predominant protein active in the phosphorylation of histone H2AX after heat treatment.

The activation of p53 is controlled by human double minute 2 (HDM2) with feedback systems (Fig. 5.4). This system is reversible, and p53 activation is depressed in normal conditions. Activated p53 introduces *p53*-regulated apoptosis induced-protein 1 (53AIP1), and then activates the bcl-2 associated x protein (Bax) although bcl-2 inhibits Bax. The next chemical reactions lead to Caspase-9 and Caspase-3 activation through cytochrome c. Finally heat-induced apoptosis is detected (Fig. 5.4). However, other apoptosis pathways can lead to another process involving death ligands, death receptors (FADD, and FS-7-associated-surface

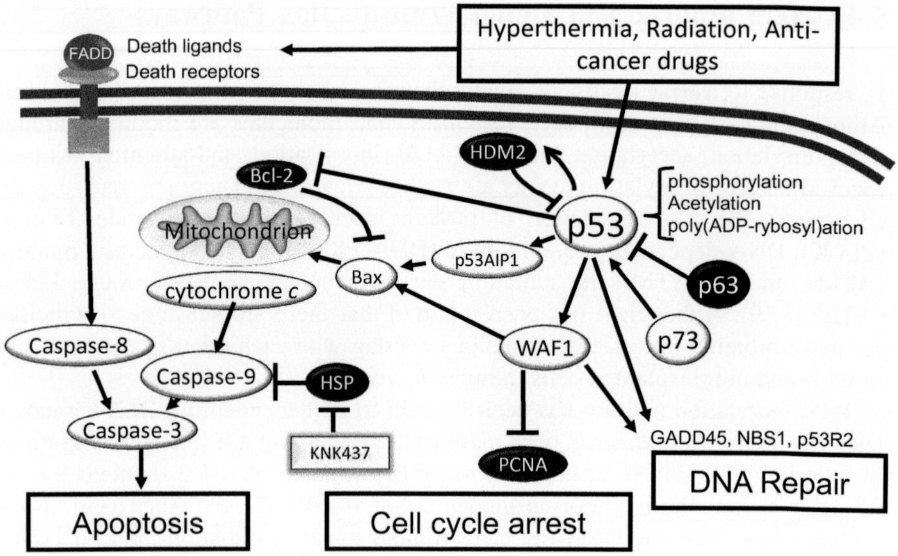

Fig. 5.4 *p53*-Dependent signal transduction pathways after hyperthermia, radiation, and anti-tumor drugs. *Arrows* indicate promotion; "**T**" symbols and lines indicate inhibition

antigen (Fas)) through Caspase-8. This process is also induced by heat treatment and was analyzed using protein chips [6, 10].

Figure 5.4 shows the *p53*-dependent signal transduction pathways after hyperthermia, radiation, and anti-tumor drugs [10–12]. There are two other p53 family members (p63 and p73) which interfere with p53 functions in different systems. A gene product of *p21/WAF1* is a major gene which responds downstream of *p53* and is the principle mediator of cell cycle arrest in response to DNA damage. The gene product of *p21* depresses the proliferating cell nuclear antigen's (PCNA) function. It primarily mediates G_1 cell cycle arrest by inactivating G_1-associated cyclin A- and cyclin E-containing cyclin/cdk complexes. The gene product of *p21* was reported to be induced by heat as well as by other genotoxic stresses (ionizing radiation and DNA-damaging drugs). The possibility that heat can also induce DNA DSB (double-strand break) formation has been supported by previous studies of γ-H2AX (histone H2AX phosphorylated at serine 139) foci formation. On the other hand, p53 activation enhanced DNA repair through growth arrest and the DNA-damage-inducible gene 45 (GADD45), Nijmegen breakage syndrome 1 (NBS1), and p53R2 (Ribonucleotide reductase subunit M2B, also known as RRM2B or p53R2).

5.5 Cell Survival Pathways

Heat treatment as well as radiation induces not only cell death signals but also cell survival signals (Fig. 5.5) [13]. One of the signal cascades which promotes cell survival utilizes a serine/threonine kinase, protein kinase B (Akt) which is activated *via* phosphoinositide 3 kinase (PI(3)K) and 3-phosphoinositide-dependent kinase-1 (PDK1), the *Akt*-mediated survival pathways in the presence of various growth factors. In normal cells, the phosphatase and tensin homologue deleted on chromosome 10 (PTEN) inhibits the activity of Akt. Many cancer cells have a PTEN deficiency, and show a tendency to have a highly activated level of Akt. Because excessive activation of Akt can lead to tumor formation and abnormal cell proliferation, a cancer cell specific cytotoxic reaction could result from the specific inhibition of Akt. Phosphorylation stabilizes Akt, which activates NF-κB (nuclear factor-κB) and induces the phosphorylation of Bad (Bcl-2-antagonist of cell death protein), which results in the latter's dissociation from Bcl-xL. Phosphorylated Bad is sequestered by the cytosolic 14-3-3 protein, and this prevents its translocation into the mitochondria. Consequently, downstream apoptotic events are not triggered.

Under normal conditions, NF-κB remains bound and sequestered in the cytosol by its inhibitor IκB kinase. However, upon exposure to stimuli, including the tumor

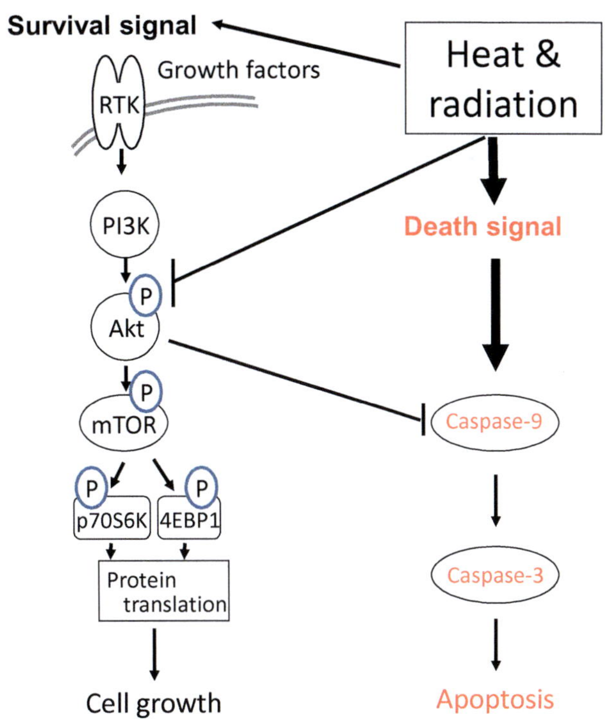

Fig. 5.5 Cell survival pathways. Survival signal pathways interfere with death signal pathways and vice versa. *Arrows*, promotion; "T" symbols and lines indicate inhibition

necUnder normal conditions, NF-κB remains bound and sequestered in the cytosol by its inhibitor IκB kinase. However, upon exposure to stimuli, including the tumor necrosis factor (TNF), IκB is degraded, resulting in the release of NF-κB, which can then translocate into the nucleus and activate the transcription of cell survival genes. Although TNF is a crucial inducer of apoptosis, it may also stimulate cell survival through NF-κB, an important link between various biological processes which include stress responses, cell growth, and cell death. Phosphorylation and inactivation are mediated by a protein kinase complex, the IκB kinase (IKK) complex. TNF induces the activation of IKK through its association with signal transducing molecules like the receptor interacting protein (RIP), and the TNF receptor-associated factor 2 (TRAF2). Activated Akt up-regulates kinase activity of the IKK complex triggering NF-κB mediated cell survival.

In another process, HSPs also induce survival signals through the inhibition of the apoptosome which plays an important role in apoptosis pathways induced by a heat treatment (Fig. 5.6). HSP90 enhances cell survival through its involvement in steps which occur in the formation of active NF-κB. HSP90 is an important factor in the generation of a stable state for the receptor-interacting protein (RIP) which is then recruited by the activated TNF receptor 1 (TNFR-1) after that protein binds with its ligand TNF, and this leads to sustained NF-κB activity. HSP90 also directly interacts with and preserves the activity of Akt by preventing its dephosphorylation. The HSP90-Akt complex can also indirectly promo In addition, HSP90 and its co-chaperone cell division cycle protein (cdc) 37 participate in the formation of active IκB kinase (IKK) and Akt complexes, each of which can phosphorylate IκB and trigger the dissociation of NF-κB from its inhibitor. The cell survival through

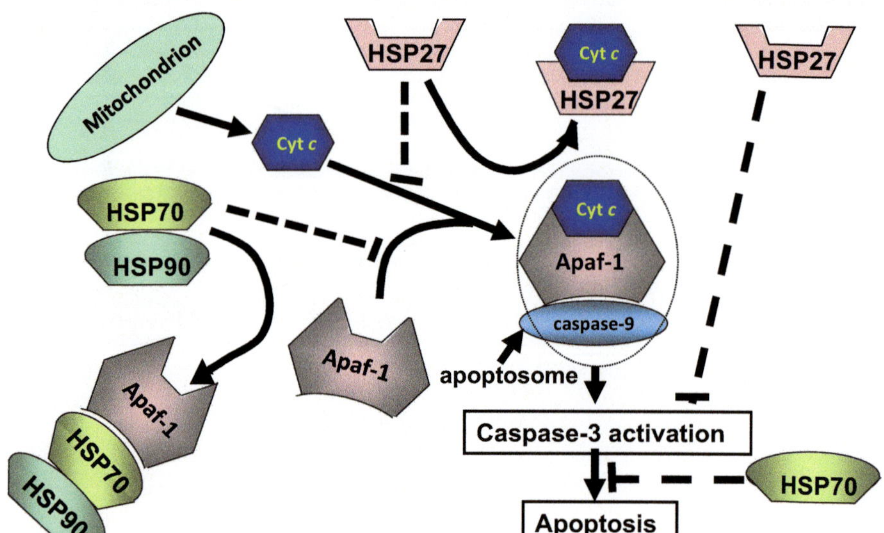

Fig. 5.6 Interference of apoptosis signal pathways by HSPs. *Solid arrows* indicate promotion; *dot lines* and "**T**" symbols and *dot lines* indicate inhibition

the suppression of JNK-mediated apoptosis. It leads to the inhibition of apoptosis through the phosphorylation and resulting inactivation of apoptosis signal-regulating kinase (ASK)-1, which is one of the activators of the c-Jun N(H2)-terminal kinase (JNK). HSP90 also prevents the formation of an active apoptosome complex by suppressing the oligomerization of apoptosis protease-activating factor (Apaf)-1, and plays a role in modulating apoptosis (Fig. 5.6). HSP27 regulates apoptosis in neutrophils through an interaction with Akt. HSP27 is phosphorylated by Akt which results in the dissociation of HSP27 and in the stabilization of Akt. Disruption of interactions between Akt and HSP27 epresses Akt activation, which triggers enhanced constitutive apoptosis in neutrophils.

Protein microarray analysis showed over-expression of PI(3)K (Fig. 5.2), TNF receptor-associated factor (TRAF)-2, NF-κB and HSP90 in heat-treated m*p53* cells in this pathway. This indicates that the Akt-mediated pathway may work effectively after heat shock, if *p53* does not function normally. HSP70 acts primarily as an anti-apoptotic factor. It inhibits apoptosis through chaperone dependent and chaperone independent activities. HSP70 protects cells from cytotoxicity which is induced by TNF, monocytes, oxidative stress, chemotherapeutic drugs, ceramide, and radiation. The apoptosis cascade initiated by heat stress triggers the translocation of Bax from the cytoplasm to the mitochondria, and this cascade is suppressed by an increased expression of HSP70. Moreover, downstream, HSP70 inhibits the formation of a functional apoptosome complex through a direct interaction with Apaf-1. HSP70 inhibits late Caspase-dependent events such as the activation of cytosolic phospholipase A2 (cPLA2) and changes in nuclear morphology; it can rescue cells from apoptosis which is induced by the enforced expression of Caspase-3. HSP70 can suppress JNK mediated apoptosis by inhibiting JNK phosphorylation either directly, and/or through the upstream SEK kinase (also known as MKK4 and JNK kinase) independently of its chaperone activity. In addition, independent of its chaperone function, HSP70 can also regulate the activation of Bid. Furthermore, various death-inducing stimuli, TNF-α, Fas, and many others are known to cause apoptosis *via* ASK-1 activation. HSP70 prevents TNF mediated cell death by suppressing ASK-1. HSP70 binds with the apoptosis inducing factor (AIF) released by mitochondria following death-inducing stimuli, and restricts its translocation into the nucleus, thereby inhibiting apoptosis. Cytosolic HSP60 in rat cardiac myocytes inhibits the triggering of the apoptotic machinery. Increases in HSP60 act primarily in anti-apoptotic roles because it binds with the pro-apoptotic Bax and Bak proteins. Increased expression levels of HSP27 during stress responses correlate with improved survival rates after exposure to cytotoxic stresses. HSP27 negatively regulates the activation of procaspase-9 by blocking cytosolic cytochrome *c* from Apaf-1, after its release from mitochondria, and suppresses the assembly of apoptosomes. HSP27 can prevent the release of cytochrome *c* from mitochondria in cells exposed to staurosporine, etoposide or cytochalasin D. It also mediates the suppression of procaspase-3. In addition, HSP27 maintains the integrity of the actin network and inhibits tBid (*i.e.* translocation of pro-apoptotic factors like activated Bid) from reaching the mitochondrial membrane.

5.6 Classic MAP Kinase Pathways

The cascade from ceramide to Ras/Raf/ MAP kinase-ERK kinase (MEK)/extracellular signal-regulated kinases (ERKs) is considered to exemplify a classic pathway for MAP kinases. This pathway is influenced by heat shock and transcription factors such as Ets-like protein 1 (Elt-1), c-Jun, and activating transcription factor 2 (ATF2) [14]. Heat shock produced a rapid activation of ERK1/2 kinases. As another p38-MAP kinases mediated pathway, it was reported that the pathway from ceramide to MEKKs/ASK-1/MKKs/p38 is also activated by heat shock. This pathway is being studied as a potential molecular target for thermal sensitzation because p38 is related to cell proliferation. Protein microarray data indicated up-regulated p38 activity was present in heat-treated m*p53* cells [6].

References

1. Hollstein M, et al. *p53* mutations in human cancers. Science. 1991;253:49–53.
2. Slee EA, et al. To die or not to die: how does *p53* decide? Oncogene. 2004;23:2809–18.
3. Kirita T, et al. A new strategy for cancer therapy based on a predictive indicator. Hum Cell. 2001;14:1–6.
4. Yasumoto J, et al. Apoptosis related gene expression after hyperthermia in human tongue squamous cell carcinoma cells harboring wild-type or mutated-type p53. Cancer Lett. 2004;204:41–51.
5. Tabuchi Y, et al. Genes and gene networks in the apoptosis induced by heat stress in human leukemia U937 cells. Therm Med. 2011;27:31–40.
6. Kajihara A, et al. Protein microarray analysis of apoptosis-related protein expression following heat shock in human tongue squamous cell carcinomas containing different *p53* phenotypes. Int J Hyperthermia. 2008;24:605–12.
7. Ota I, et al. Transfection with a mutant *p53* gene inhibits heat-induced apoptosis in a head and neck cell line of human squamous cell carcinoma. Int J Radiat Oncol Biol Phys. 2000;47:495–501.
8. Durocher D, et al. DNA-PK, ATM and ATR as sensors of DNA damage: variations on a theme? Curr Opin Cell Biol. 2001;13:225–31.
9. Takahashi A, et al. ATM is the predominant kinase involved in the phosphorylation of histone H2AX after heating. J Radiat Res. 2010;51:417–22.
10. Wang X, et al. *p53*-Dependent signal transduction induced by stress. J Radiat Res. 1997;38:179–94.
11. Ohnishi K, et al. Heat-induced *p53*-dependent signal transduction and its role in hyperthermic cancer therapy. Int J Hyperthermia. 2001;17:415–27.
12. Ohnishi T, et al. Signal transduction of radiation and/or hyperthermic cancer therapies. Curr Signal Transduct Ther. 2010;5:231–6.
13. Nakagawa Y, et al. Depression of *p53*-independent *Akt* survival signals in human oral cancer cells bearing a mutated *p53* gene after exposure to high-LET radiation. Biochem Biophys Res Commun. 2012;423:654–60.
14. Kajihara A, et al. Thermal medicine review heat-induced signal transduction pathways leading to cell death and cell survival in cancer cells. Termal Med. 2009; 255:1–11.

Chapter 6
Cell Cycle Responses to Hyperthermia

Yukihiro Furusawa

Abstract For decades, it has been well known that hyperthermia induced by heat stress (HS) exerts anti-tumor activities through the inhibition of DNA repair, induction of cell death, and arrest or delay of the cell cycle. HS as well as ionizing radiation and anti-cancer agents exert multiple effects on cell proliferation, likely by activating the cell-cycle checkpoint pathway. One of the most important molecular mechanisms by which HS drives the cell-cycle checkpoint machinery is by the activation of the DNA damage response pathway probably by generating reactive oxygen species or through protein denaturation. DNA damage initiator and effector proteins in this pathway are modulated by HS as well as by anti-cancer agents and ionizing radiation, thus resulting in the delay of progression to the S-phase and cell cycle arrest at the G_1 and/or G_2/M phases of the cell cycle. Here, I summarize the cell cycle response provoked by HS with an aim to provide significant information for the potential of hyperthermia for use in cancer treatment.

Keywords Cell-cycle checkpoint • Replication • DNA damage response • Cyclin-dependent kinase inhibitor • Checkpoint kinase

6.1 Introduction

For a better understanding of how heat stress affects the cell cycle, I first describe the cell cycle and cell-cycle regulating proteins. Since there are many excellent textbooks and reviews describing the cell cycle [1], I will only briefly describe it with a focus on mammalian cells (in particular human cells).

A cell has the capability to duplicate itself. When a cell reproduces, it needs to duplicate its contents, including DNA and other organelles, and then divides in two cells. Cells also need time to check the quality of the contents being duplicated for cell division. After successful cell division, the newly reproduced cells perform these steps again when they receive appropriate signals for proliferation. This

Y. Furusawa (✉)
Department of Liberal Arts and Sciences, Faculty of Engineering, Toyama Prefectural University, 5180 Kurokawa, Toyama 939-0398, Japan
e-mail: furusawa@pu-toyama.ac.jp

S. Kokura et al. (eds.), *Hyperthermic Oncology from Bench to Bedside*,
DOI 10.1007/978-981-10-0719-4_6

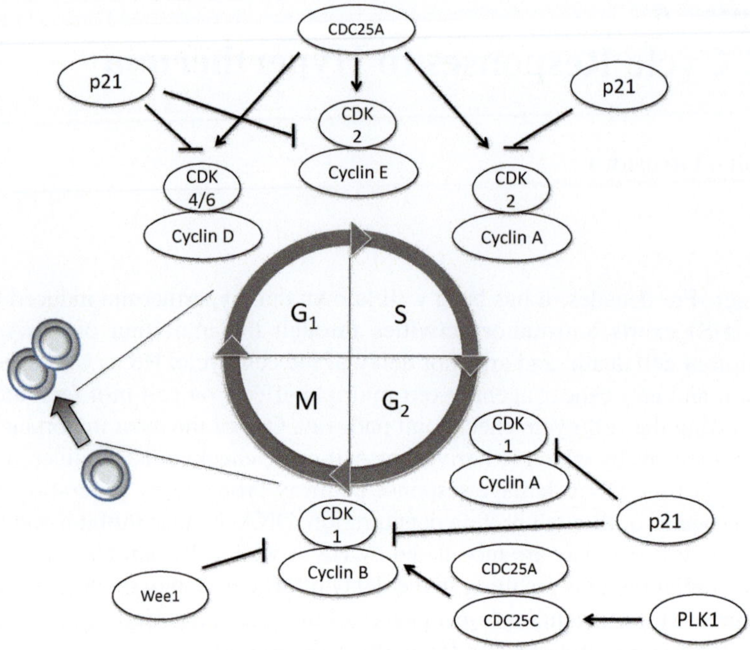

Fig. 6.1 Cell cycle regulation by CDK-cyclin complexes, CDC25 phosphatases, and p21 CKI. CDK4/6-cyclinD, CDK2-cyclin E, CDK2/1-cyclin A, and CDK1-cyclinB complexes are associated with the G_1, G_1/S, S-G_2, and G_2/M phases of the cell cycle, respectively. CDC25 phosphatases promote cell cycle progression by dephosphorylating inhibitory phosphates on CDKs, whereas p21 inhibits cell cycle progression by directly binding to CDK-cyclin complexes

orderly sequence of events is known as the "cell cycle," which usually consists of 4 phases—the G_1, S, G_2, and M phases (Fig. 6.1), and each phase has its own checkpoint machinery [2]. The most dynamic phase is the "M phase" in which cell division occurs ("M" stands for mitosis). The chromosome is not stabilized through the M phase and cells in the M phase thus show high sensitivity to heat stress (HS) and ionizing radiation (IR). Previous studies have shown that HS causes microtubule disorganization and multipolar mitosis, resulting in mitotic catastrophe, which is a cell death mode represented by cell destruction during mitosis (including pre-mature mitosis) [3]. Heat shock protein 70, a 70-kDa molecular chaperone protein, localized to the mitotic centrosome and protected against abnormalities in cell division, thus suggesting that the denaturation of proteins induced by HS may cause damage to the mitotic centrosome [4].

The "S phase" is also a dynamic phase of the cell cycle where in chromosome duplication for cell division occurs ("S" stands for synthesis). DNA replication is initiated by the unwinding of the DNA double helix by DNA helicase, and the unwound DNA is then stabilized by the associated single-strand DNA-binding proteins known as replication protein A (RPA). The RPA complex, composed of the RPA70, RPA32, and RPA14 subunits in humans, is an essential component of the DNA replication process. The positions at which the DNA helix is opened are generally called replication origins, and thus this step is called "origin firing." The replication complexes composed of several proteins such as DNA polymerase for nucleotide triphosphate polymerization and an accessary protein that forms a sliding clamp (known as PCNA in mammalian cells) for keeping the polymerase on the DNA strand are loaded onto the replication origin. Previous studies have shown that HS preferentially led to the killing of cells at the S phase, probably due to the thermal denaturation of the replication complex, which led to interference in the initiation and elongation of the DNA [5, 6].

A cell has two gap phases called the G_1 phase occurring between the M and S phases, and the G_2 phase occurring between the S and M phases. During the two gap phases, the cell prepares components for entry into the next phase and monitors whether the cellular condition is suitable for the transition to the next phase. If not, the premature cell in the gap phase operates machineries to block the progression to the next phase of the cell cycle. This step is called the "cell-cycle checkpoint." [1, 2] It should be noted that cell-cycle checkpoints are not restricted to only the G_1 and G_2 phases but are also present in the S phase; the cell-cycle checkpoint of the S phase is called the "intra-S phase checkpoint," but the mechanism of action of the intra-S phase checkpoint is less clear compared to that of the G_1 and G_2 phase checkpoints [7–9]. Cells exposed to HS are also known to activate the G_1, intra-S, and G_2/M checkpoint pathways, as described in following sections.

6.2 Cell Cycle Regulation by Cyclins, Cyclin-Dependent Kinases (CDKs), Cell Cycle Division (CDC) Phosphatase 25, and Cyclin-Dependent Kinase Inhibitors (CKIs)

A cell commonly harbors a cell-cycle control system composed of members of the protein kinase family, called CDK(s) [1]. Increase in CDK activity promotes cell-cycle progression by regulating the phosphorylation of multiple downstream cell-cycle regulators. CDK activity is controlled by the binding of cyclin, and a specific pair of cyclin-CDK complexes contributes to the promotion of each step of entry and progression into the next phase of the cell cycle (Fig. 6.1). For instance, cyclin B, known as M-cyclin, forms a complex with CDK1 and stimulates mitotic entry at the G_2 phase, whereas cyclin D, known as G_1 cyclin, forms a complex with CDK4 or CDK6 and promotes cell-cycle progression into the S phase [11]. Among these, the CDK1-cyclinB complex phosphorylates key proteins such as Polo-like kinases (PLKs) and Aurora kinases (e.g., Aurora-A and -B), which contribute to the assembly and stabilization of the mitotic spindle, respectively [12]. The CDK4-cyclinD complex in the G_1 phase promotes the phosphorylation and inhibition of retinoblastoma protein (Rb), leading to E2F-dependent transcription of genes required for S-phase entry [13].

CDK activity is controlled by the proteolysis or transcription of cyclins. In addition, other cell-cycle regulating molecules such as Wee1 kinase and CDC25 phosphatases contribute to cell cycle regulation through the phosphorylation or dephosphorylation of CDKs [14]. Phosphorylation of the CDK inactivation site by Wee1 changes the conformation of CDK-cyclin complexes, resulting in the inhibition of CDK activity. On the contrary, dephosphorylation by CDC25 counteracts the inhibition. Among CDC25 phosphatases, CDC25A is required for the progression of both the S- and the M-phases, whereas CDC25C is only required for mitotic entry [15]. Phosphorylation of the CDC25 family of proteins by checkpoint kinases (detailed in Sect. 6.3) as well as p38 mitogen-activated kinases (MAPKs) and extracellular signal-regulated kinase (ERK) block S-phase entry through the phosphorylation and following degradation of CDC25A in response to extra-cellular stress [16, 17].

CDK activity is also regulated by conformational changes in protein structure caused by protein-protein interactions. CKIs, such as p21 and p27, bind to CDK-cyclin complexes and inhibit their activities by alteration of the structure of the complexes [18].

Negative regulation of the cell cycle has been well investigated in cells exposed to genotoxic stress. Genotoxic stress caused by IR, hydroxyurea, and hyperthermia induces cell-cycle arrest by activating cell-cycle checkpoint machinery, as described below.

6.3 Cell-Cycle Checkpoint Activation Through Ataxia Telangiectasia Mutated (ATM)-Checkpoint Kinase 2 (Chk2) and Ataxia Telangiectasia Mutated and Rad3-Related (ATR)-Checkpoint Kinase 1 (Chk1) Axes

As shown in Chap. 3, HS itself is known to induce multiple types of DNA damage such as single-strand breaks (SSBs), replicative stress in cells at the S phase, and perhaps DNA double-strand breaks (DSBs) only in cells at the G_1 and G_2 phases [19]. Whether HS induces DSBs in the same way as IR is a highly debated topic [20, 21], and I hence will not refer to the type of DNA damage induced by HS in the current section; however, there is no doubt that HS as well as genotoxic agents activate the DNA damage response (DDR) pathway, thus promoting cell-cycle arrest [22, 23]. To understand how the DDR pathway activated by HS negatively regulates cell-cycle progression, I will shortly introduce the molecular functions of proteins involved in the DDR pathway.

ATM and ATR are members of the phosphatidylinositol 3-kinase-related kinase family that act as sensor proteins for DNA damage (reviewed in [24, 25]). ATM is recruited to DNA DSB sites in conjunction with the MRE11/RAD50/ NBS1 (MRN) complex, which recognizes the double-stranded end of DNA and contributes to DNA repair (Fig. 6.2) [24]. Therefore, ATM is strongly activated by DNA damage agents that directly induce DNA DSBs, such as IR or bleomycin. Auto-phosphorylation of ATM at Ser1981 is often utilized for the activation of ATM

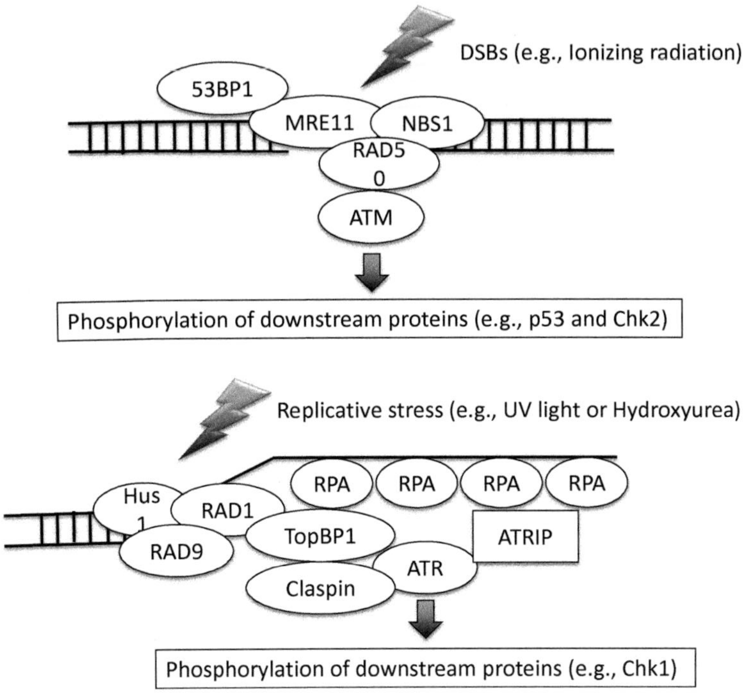

Fig. 6.2 Model of the cellular response to DNA DSBs and replicative stress. In response to DSBs, ATM recognizes DSBs and activates itself, a process associated with the MRN complex and 53BP1. In response to replicative stress, RPA stabilizes the stalled replication fork followed by ATR activation, which is associated with the 9-1-1 complex, ATRIP, TopBP1, and Claspin on the replication fork. See Sect. 6.3 for details

and the full activation of ATM in response to DSBs requires the MRN complex [24].

On the other hand, ATR is strongly activated by replicative stress elicited by ultraviolet light or hydroxyurea [25]. When replication is impeded, DNA polymerase uncouples from DNA helicase, resulting in the formation of ssDNA ends. Under replicative stress, ssDNA is coated with the RPA complex to stabilize the stalled replication fork. ATRIP forms a complex with RPA70, one of the components of the RPA complex, as described above [26]. Association of the ATR-ATRIP complex with the RPA-coated ssRNA at the stalled replication fork is required for the activation of ATR. Activation of ATR also requires the Rad9-Hus1-Rad1 (9-1-1) complex and topoisomerase binding protein (TopBP1) [27]. The 9-1-1 complex, a heterotrimeric ring similar in structure to PCNA, is loaded onto RPA-coated ssDNA by Rad17 [28]. TopBP1 interacts with Rad9, which brings the activation domain of TopBP1 close to the RPA-bound ATR, thus allowing the activation of ATR. Claspin, which associates with the active replication fork, is also required for the activation of ATR [29]. ATR phosphorylation at Ser428 is a marker of ATR activation, but unlike ATM, the auto-phosphorylation of ATR has not yet been demonstrated. ATR is activated by replicative stress, but it should be noted that ATR is also activated in response to DSBs because ssDNA is generated during the repair of DSBs [30]. Conversely, ATM is also activated by replicative stress, probably due to DSB formation at the stalled replication fork [31].

ATM preferentially phosphorylates Chk2 at Thr68 rather than Chk1 at Ser317/345, and both ATM and Chk2 kinases can phosphorylate and stabilize p53 (at Ser15 by ATM and at Ser20 by Chk2) [32]. p53 is well known to up-regulate p21 and thus contribute to G_1/S, intra-S phase, and weak G_2/M checkpoint activation [9, 10]. The primary role of the ATM-Chk2 pathway seems to be p53-dependent arrest at G_1, because mice deficient in Chk2 showed impaired G_1 arrest in response to DNA damage [33].

G_2/M checkpoint activation following IR exposure is also largely dependent on the ATR-Chk1 pathway. ATR phosphorylates Chk1 and phospho-Chk1 facilitates the degradation or inactivation of CDC25 family proteins [34]. Chk1 contributes to intra-S phase checkpoint activation by phosphorylating CDC25A and thus inhibits DNA replication [8]. In addition, Chk1 phosphorylates CDC25C as well as Wee1 kinase and PLK1, leading to the inactivation of CDK1 and consequently to G_2/M arrest. [35, 36]. Chk2 also has the potential to phosphorylate CDC25 family proteins; however, the role of Chk2 in G_2/M arrest is less clear than that of Chk1 because Chk2-deficient MEFs and HCT116 cells do not exhibit defects in G_2/M checkpoint activation [33, 37]. However, inhibition of Chk1 results in G_2/M checkpoint defects, which is not observed in the case of Chk2 inhibition [38]. Therefore, it is considered that Chk2 exists for compensation of G_2/M checkpoint activation by Chk1. Conversely, ATR and Chk1 can phosphorylate p53 [39, 40]; however, the mechanism of p53-mediated G_1 arrest by ATR-Chk1 is less understood than that by ATM-Chk2, because ATR-Chk1 activation seems to be restricted to the S and G_2 phases of the cell cycle [41].

In summary, both the ATM-Chk2 and the ATR-Chk1 pathways redundantly activate downstream molecules associated with cell cycle regulation; however, ATM-Chk2 preferentially regulates the p53-mediated activation of G_1 and intra-S phase checkpoints, whereas ATR-Chk1 regulates intra-S and G_2/M checkpoint activation.

6.4 Heat-Induced G_1/S Checkpoint Activation

It may be difficult to simply compare the evidence showing heat-induced cell cycle responses, because the effect of heat seems to be dependent on the duration of heating at a given temperature and the cell type utilized in the studies. Therefore, the information about both temperature and the duration of heating in addition to the p53 phenotype must be described in order to accurately interpret the results obtained by several groups (also see Table 6.1 and Fig. 6.3).

p53 response is well known to be augmented in cells damaged not only by IR but also by HS [42, 43]. The phosphorylation of p53 at Ser15, leading to the

Table 6.1 Heat-induced cell-cycle arrest associated with temperature, exposure time, and p53 phenotype

Cell	p53 phenotype	Temp. (°C)	Duration	Arrest	References	Note
HEK293	Wild type	42	20 min	G_1	[53]	p38-dependent Cdc25A degradation (even in the absence of pChk2)
Human fibroblast	Wild type	43	45 min	G_1	[46]	p53-dependent p21 induction
HL-60	Null	45	10 min	G_1	[51]	HS for 20 min caused G_2/M arrest
A172	Mutated	45	15 min	G_1	[48]	p53-independent p21 induction
T98G	Mutated	45	15 min	G_1	[48]	p53-independent p21 induction
U373-MG	Mutated	43	120 min	G_2/M	[47]	Transfection of wild-type p53 counteracts HS-induced G_2/M arrest
U251-MG	Mutated	43	120 min	G_2/M	[47]	Transfection of wild-type p53 counteracts HS-induced G_2/M arrest
Jurkat	Mutated	44	30 min	G_2/M	[52]	Activation of ATR-Chk1
HL-60	Null	45	20 min	G_2	[51]	HS for 10 min caused G_1 arrest

stabilization of p53, seems to be dependent on ATM even in cells exposed to HS [44]. HS induces p21 expression markedly in cells harboring wild-type p53, leading to G_1 arrest and a decrease in the population of cells in the S phase [45]. A previous study showed that HS (43 °C, 45 min) induced cell cycle arrest at the G_1 phase in normal human fibroblasts following p21 induction [46].

In general, it is widely believed that p53 deficiency weakens p21 induction and causes preferential G_2 arrest rather than G_1 arrest in cells under genotoxic stress. A previous study showed that U373-MG and U251-MG human glioma cells harboring defective p53 showed increased numbers of cells at the G_2/M phase in response to HS (43 °C, 120 min) [47], but this effect was not observed in cells in which p53 was over-expressed using adenoviral transfection, thus indicating that p53 contributes to G_1 arrest in cells exposed to HS.

However, G_1 arrest but not G_2 arrest was found to be predominant in the human glioblastoma cell lines A172 and T98G harboring defective p53 after HS exposure (44 °C, 15 min) [48]. p53-independent p21 induction, which may be explained by the p53-independent p21 transcription pathway reviewed in [10], might be responsible for the G_1 arrest induced by HS in these p53-deficient cells.

Although the detailed mechanisms underlying the p53-independent induction of p21 expression by HS remain unknown, we have previously observed induction of CDKN1A mRNA (coding p21) expression in p53-deficient U937 cells exposed to

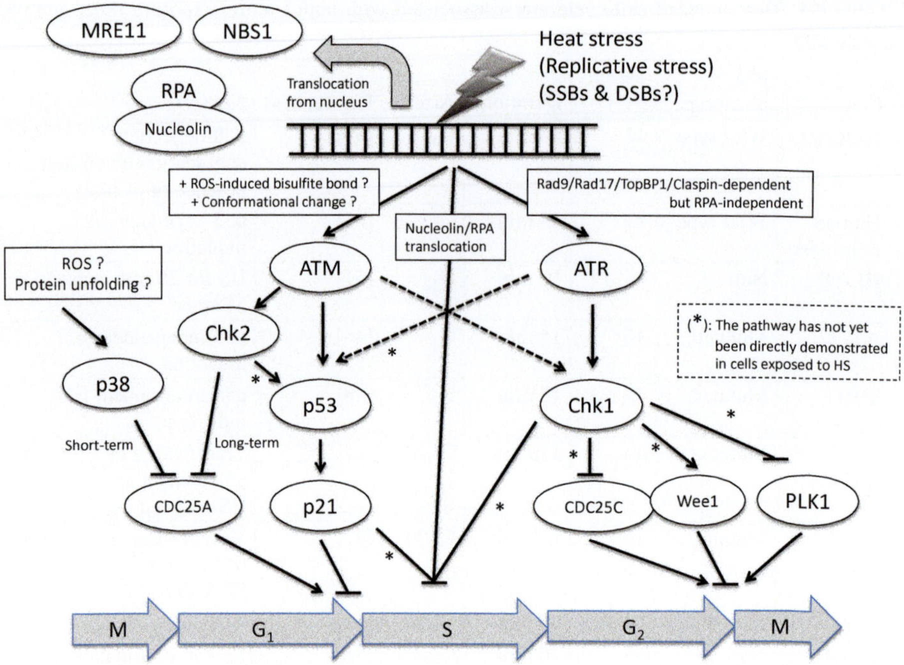

Fig. 6.3 Model of the cellular response to HS. In brief, ATM and ATR are activated even if NBS1, MRE11, and RPA are translocated from the nucleus under HS. ATM phosphorylates and stabilizes p53, perhaps leading to the induction of p21 expression and cell cycle arrest at the G_1 phase. p38 MAPK-dependent G_1 arrest is also induced by CDC25A degradation. Chk2-dependent CDC25A phosphorylation also occurs if the heat treatment is prolonged. HS activates the intra-S phase checkpoint through translocation of the RPA-nucleolin complex, although it has not yet been demonstrated that p21 and Chk1 contributes to the inhibition of S-phase progression in cells exposed to HS. ATR rather than ATM phosphorylates Chk1 and causes G_2/M arrest, probably thorough phosphorylation of downstream cell-cycle regulators such as CDC25C, PLK1, and Wee1, even in cells exposed to HS. See Sects. 6.4, 6.5 and 6.6 for details

HS (44 °C, 15 min) [49]. In addition, Tabuchi et al. showed p53-dependent induction of CDKN1A mRNA expression even in cells exposed to HS at a lower temperature (41 °C, 30 min) [50] (also see GEO Accession: GSE10043). These results indicate that p21 expression can be responsive to HS at a broad temperature range even in the absence of p53.

The duration of heating rather than the p53 phenotype seems to determine the next cell-cycle stage. An impressive study in HL-60 cells showed that heating at 45 °C for 10 min preferentially increased the number of cells at the G_1 phase, whereas heating for 20 min increased the number of cells at the G_2 phase [51], suggesting that prolonged heat exposure switches G_1 arrest to G_2/M arrest. Consistent with this report, another study also showed that the long-term exposure to HS (44 °C, 30 min) causes G_2/M arrest in Jurkat cells harboring mutated p53 [52], whereas hyperthermia for a short duration (42 °C, 20 min or 45 °C, 15 min) induces the accumulation of cells at the G_1 phase but not at the G_2 phase, not only in p53-proficient HEK293 cells but also in p53-mutated A172 and T98G cells [48, 53]. In another study using HEK293 cells, Madlener et al. showed that the p38-dependent degradation of CDC25A is also responsible for the G_1 arrest induced by exposure to HS for 20 min [53]. The authors reported that the degradation of CDC25A was also dependent on Chk2 but not on Chk1 under HS for 60 min. Notably, the phosphorylation of both Chk1 and Chk2 was not observed in cells exposed to HS for 20 min, indicating that short-term exposure to HS induces p38-dependent but Chk2-independent G_1 arrest. Considering that the p38 MAPK is present even in p53-deficient cells, heat-induced G_1 arrest in cells deficient in p53 gene might be explained by p38-dependent CDC25A phosphorylation as well as p53-independent p21 induction.

Taken together, it appears that the heat durations used in Ref. [48] and Ref. [51], showing heat-induced G_1 arrest in p53-mutated or -deficient cells, may be not enough to activate checkpoint kinases and thus failed to activate the following G_2/M checkpoint (See Sect. 6.7). p53 was also found to be important but dispensable for G_1 arrest in cells exposed to HS. The G_1 arrest in p53-mutated or -deficient cells, probably through p53-dependent p21 induction and p38 MAPK-dependent CDC25A degradation, seems to be largely dependent on the duration and temperature of heat treatment.

6.5 Heat-Induced Intra S-Phase Checkpoint Activation

The intra S-phase checkpoint contributes to a transient decrease in the rate of DNA synthesis, probably through slowing down of the elongation of the replication fork or inhibition of origin firing. Among DDR proteins, p21 and Chk1 are implicated in the regulation of the intra-S phase checkpoint. p21 blocks DNA synthesis as well as G_1-S transition by interacting with PCNA, which is necessary for the formation of the replication complex [10]. In addition, Chk1 phosphorylates and thus degrades CDC25A, resulting in the inactivation of the CDK2-Cyclin A complex [8, 54].

To the best of my knowledge, however, direct evidence showing the involvement of p21 and Chk1 in intra-S phase checkpoint activation in cells exposed to HS is not yet available. On the other hand, a previous study revealed that HS inhibited S-phase progression by directly inhibiting the process of replication, including the initiation of new replicons and the elongation of the replication fork [55]. The molecular mechanism underlying heat-induced intra-S phase checkpoint activation is the translocation of RPA from the nucleus. HS causes a translocation of nucleolin, which forms a complex with RPA and leads to RPA dysfunction, resulting in the inhibition of DNA replication [55]. We also confirmed that HS did not cause nuclear RPA32 foci formation, which is a marker of replication fork stalling in cells treated with a replication inhibitor such as hydroxyurea [52], probably due to the release of the RPA-nucleolin complex from the nucleus [55, 56].

These results raised a new question: How does HS activate the ATR pathway even in the absence of RPA? The RPA complex accumulates in the single-stranded end on a stalled replication fork (thus forming foci in the nucleus), not only leading to the stabilization of the replication fork but also providing a platform for ATR phosphorylation [25, 56]. However, HS was found to induce ATR phosphorylation despite RPA foci formation being attenuated [52].

Tuul et al. reported that the heat-induced ATR phosphorylation was also dependent on TopBP1, Rad9, Rad17, and Claspin, suggesting the presence of a mechanism underlying the RPA-independent ATR activation by HS [57]. Supporting this idea, although in a Xenopus egg experiment, it was reported that ATR activation occurred even in the absence of RPA32 but not in the absence of ATRIP or Claspin [58].

Other possible mechanism involved in ATR activation might be the conformational change in proteins or the production of reactive oxygen species (ROS) evoked in cells exposed to HS (see Chap. 2 for ROS production by HS). Hunt et al. proposed that the conformational change in ATM caused by HS led to its phosphorylation because HS induced ATM phosphorylation even in the absence of obvious DNA DSBs [21]. Guo et al. also revealed that an ATM dimer formed disulfide bonds in the presence of massive oxidative stress induced by hydrogen peroxide, thus inducing the conformational change in ATM [59].

Therefore, it might be possible that the activation of ATR as well as ATM occurs as a consequence of conformational changes in proteins caused by HS. The detailed mechanism by which HS activates ATR and the role of p21 and Chk1 in intra-S phase checkpoint activation should be further elucidated.

6.6 Heat-Induced G_2/M Checkpoint Activation

Unlike that of p53-dependent G_1/S checkpoint activation, the molecular mechanism underlying G_2/M checkpoint activation in cells exposed to HS had not been studied until recent years. Recently, based on the DDR-mediated cell-cycle regulation pathway, we aimed to evaluate the role of the ATM-Chk2 and ATR-Chk1 pathways in G_2/M checkpoint activation in cells exposed to HS. We found that HS (44 °C, 30 min) activated the ATR-Chk1 pathway rather than the ATM-Chk2 pathway [52]. ATR phosphorylation at Ser428 rapidly occurred, but ATM phosphorylation at Ser1981 was slightly delayed after heat treatment in human leukemia Jurkat cells harboring mutated p53. In addition, ATR phosphorylation induced by HS was not suppressed in the presence of the ATM inhibitor Ku55933, which was different from that observed in the case of SSBs ends of DNA during DSB repair [41], thus indicating that ATR does not act downstream of ATM in cells exposed to HS.

The molecular mechanisms involved in ATR and ATM phosphorylation induced by HS were similar but slightly different from that induced by genotoxic agents such as IR or hydroxyurea. In the case of ATR activation, as mentioned above, RPA32 was not required in cells exposed to HS. Similarly, heat-induced ATM activation is not accompanied with the formation of the MRN complex and 53BP1 because HS induces the transport of MRE11 and NBS1 from the nucleus into the cytoplasm and suppresses 53BP1 recruitment to the damaged DNA site [21, 60, 61].

Although the mechanism underlying the activation of ATR and ATM needs to be further elucidated, it seems certain that cells can respond to HS through these kinases. In particular, the role of ATR-Chk1 in G_2/M checkpoint activation was obvious, unlike that of ATM-Chk2. An ATR-specific inhibitor, Schisandrin B, preferentially suppressed Chk1 phosphorylation than Ku55933, supporting evidence showing that ATR is the primary kinase of Chk1. Inhibition of Chk1 by a selective Chk1 inhibitor SB218078 abrogated the accumulation of cells at the G_2/M phase, and increased the population of Sub- G_1 cells and the cleavage of caspase-3 (apoptotic markers) not only in Jurkat cells, but also in other p53-deficient human cancer cells such as HeLa (cervical carcinoma), HSC-3 (squamous carcinoma), and PC3 (prostate cancer) cells exposed to HS. This suggested a critical role of Chk1 in G_2/M checkpoint activation and apoptosis evasion. However, whether cells with dysfunctional Chk1 undergo apoptosis because of checkpoint abrogation and following chromosomal aberration needs to be investigated in the future.

The role of the ATM-Chk2 pathway in checkpoint activation and cell survival is less clear than that of the ATR-Chk1 pathway. One study indicated that Chk2 contributes to CDC25A degradation and the subsequent induction of G_1 arrest as described in Sect. 6.4 [53]. However, in our study, inhibition of ATM or Chk2 showed marginal effects on apoptosis in Jurkat cells exposed to HS, indicating that, at least under the conditions of HS employed in our study, G_1 arrest induced by Chk2-dependent CDC25A phosphorylation might not be involved in the promotion of cell survival. One of the possible roles of the ATM-Chk2 pathway in response to HS is in p53-dependent cell-cycle checkpoint activation because both ATM and Chk2 have the potential to phosphorylate and thus stabilize p53.

Currently, to the best of my knowledge, whether the abrogation of ATM and/or Chk2 elicits dysfunction of cell cycle regulation in p53-proficient cells exposed to HS has not been elucidated. It will be interesting to elucidate whether ATM and Chk2 contribute to p53-dependent cell-cycle checkpoint activation under HS. Efforts to elucidate the role of ATM-Chk2 in heat response are currently underway.

6.7 Sensitization of Tumor Cells to Hyperthermia by Targeting Cell-Cycle Checkpoint Machineries: A Future Strategy for Combination Therapy

Hyperthermia has been used to sensitize cells to DNA-damaging agents (combination therapy) because HS has the potential to inhibit DNA repair (see Chaps. 8, 9, and 10 and Chaps. 16, 17, 18, 19, 20, 21, 22, 23, 24, 25, 26, and 27 on combination therapy). Inhibition of the DNA damage response pathway by small molecules, in particular inhibition of DNA repair or cell-cycle checkpoint activation, is currently being investigated as an approach to boost the effect of DNA-damaging agents in cancer therapy [62]. In particular, a combination of DNA-damaging agents and checkpoint kinase inhibitors has been considered as a promising strategy for cancer therapy because many cancer cells exhibit defective p53 thus rely on G_2/M checkpoint for cell survival [54].

Our group and Tuuls et al. showed that abrogation of the G_2/M checkpoint pathway by inhibition of ATR-Chk1 promoted heat-induced apoptosis and loss of clonogenicity [52, 57], indicating that targeting cell-cycle checkpoint machinery may be one of the strategies for hyperthermia. An ATR-specific inhibitor, VE-821 (IC50 is 13 nM in Ref [63]), which inhibits ATR more strongly than does Schisandrin B, was recently developed; however, the clinical use of ATR inhibitors has some problems. First, ATR is essential for accurate DNA replication in normal cells and ATR inhibition can exert unexpected side effects by affecting the ATM-p53 pathway [28]. Although Chk1 is also essential for replication, currently, the small molecule inhibitor of Chk1 rather than that of ATR is being developed for use in combination with chemotherapy [64].

Further, as shown in Table 6.2, several checkpoint kinase 1 inhibitors (some of which are not selective only to Chk1 but also inhibit Chk2) are currently under preclinical or clinical trial phase I/II trials but have not yet been approved for use as drugs. Before considering the application of Chk1 inhibitors in hyperthermia, whether Chk1 inhibition can be applied clinically in the chemotherapy needs to be further investigated.

Table 6.2 Chk1 inhibitors under preclinical (Pre) or clinical trials (-Oct, 2015)

Compound name (Alias)	IC50	Other targets (IC50)	Phase of clinical Trial	Reference
LY2606368 (MK-1775)	1.5 nM	Chk2 (4.7 nM)	II	https://clinicaltrials.gov
SCH900776 (MK-8776)	3 nM	CDK2 (160 nM), Chk2 (1500 nM)	II	https://clinicaltrials.gov
LY2603618	7 nM	PDK1 (893 nM)	II	https://clinicaltrials.gov
GDC-0425	N.D.	N.D.	I	https://clinicaltrials.gov
GDC-0575 (Arry-575)	2 nM	Rks3 (39 nM), MYLK (48 nM)	I	https://clinicaltrials.gov
XL-844	2.2 nM	Chk2 (200 nM)	I	https://clinicaltrials.gov
PF-00477736	0.5 nM	Chk2 (47 nM)	I	https://clinicaltrials.gov
SAR-020106	13 nM	CDK1 (7500 nM)	Pre	[65]
CCT-244747	8 nM	N.D.	Pre	[66]
CEP-3891	4 nM	TrkA (9 nM), MLK1 (42 nM)	Pre	[67]

N.D.: Not disclosed

References

1. Graña X, Reddy EP. Cell cycle control in mammalian cells: role of cyclins, cyclin dependent kinases (CDKs), growth suppressor genes and cyclin-dependent kinase inhibitors (CKIs). Oncogene. 1995;11:211–19.
2. Kastan MB, Bartek J. Cell-cycle checkpoints and cancer. Nature. 2004;18:316–23.
3. Vidair CA, Doxsey SJ, Dewey WC. Heat shock alters centrosome organization leading to mitotic dysfunction and cell death. J Cell Physiol. 1993;154:443–55.
4. Hut HM, Kampinga HH, Sibon OC. Hsp70 protects mitotic cells against heat-induced centrosome damage and division abnormalities. Mol Biol Cell. 2005;16:3776–85.
5. Warters RL, Stone OL. The effects of hyperthermia on DNA replication in HeLa cells. Radiat Res. 1983;93:71–84.
6. Mondovi B, Strom R, Rotilio G, Agro AF, Cavalliere R, Rossi FA. The biochemical mechanism of selective heat sensitivity of cancer cells. Eur J Cancer. 1969;5:129–36.
7. Seiler JA, Conti C, Syed A, Aladjem MI, Pommier Y. The intra-S-phase checkpoint effects both DNA replication initiation and elongation: single-cell and -DNA fiber analyses. Mol Cell Biol. 2007;27:5806–18.
8. Sørensen CS, Syljuåsen RG, Falck J, Schroeder T, Rönnstrand L, Khanna KK, et al. Chk1 regulates the S phase checkpoint by coupling the physiological turnover and ionizing radiation-induced accelerated proteolysis of Cdc25A. Cancer Cell. 2003;3:247–58.
9. Abbas T, Dutta A. p21 in cancer: intricate networks and multiple activities. Nat Rev Cancer. 2009;9:400–14.
10. Jung Y-S, Qian Y, Chen X. Examination of the expanding pathways for the regulation of p21 expression and activity. Cell Signal. 2010;22:1003–12.
11. Johnson DG, Walker CL. Cyclins and cell cycle checkpoints. Annu Rev Pharmacol Toxicol. 1999;39:295–312.
12. Barr FA, Silljé H, Nigg EA. Polo-like kinases and the orchestration of cell division. Nat Rev Mol Cell Biol. 2004;5:429–40.
13. Weinberg RA. The retinoblastoma protein and cell cycle control. Cell. 1995;81:323–30.
14. Coleman TR, Dunphy WG. Cdc2 regulatory factors. Curr Opin Cell Biol. 1994;6:877–82.
15. Donzelli M, Draetta GF. Regulating mammalian checkpoints through Cdc25 inactivation. EMBO Rep. 2003;4:671–7.
16. Isoda M, Kanemori Y, Nakajo N, Uchida S, Yamashita K, Ueno H, et al. The extracellular signal-regulated kinase–mitogen-activated protein kinase pathway phosphorylates and targets Cdc25A for SCFβ-TrCP-dependent degradation for cell cycle arrest. Mol Biol Cell. 2009;20:2186–95.
17. Wagner EF, Nebreda ÁR. Signal integration by JNK and p38 MAPK pathways in cancer development. Nat Rev Cancer. 2009;9:537–49.
18. Xiong Y, Hannon GJ, Zhang H, Casso D, Kobayashi R, Beach D. p21 is a universal inhibitor of cyclin kinases. Nature. 1993;366:701–4.
19. Velichko AK, Petrova NV, Kantidze OL, Razin SV. Dual effect of heat shock on DNA replication and genome integrity. Mol Biol Cell. 2012;23:3450–60.
20. Takahashi A, Matsumoto H, Nagayama K, Kitano M, Hirose S, Tanaka H, et al. Evidence for the involvement of double-strand breaks in heat-induced cell killing. Cancer Res. 2004;64:8839–45.
21. Hunt CR, Pandita RK, Laszlo A, Higashikubo R, Agarwal M, Kitamura T, et al. Hyperthermia activates a subset of ataxia-telangiectasia mutated effectors independent of DNA strand breaks and heat shock protein 70 status. Cancer Res. 2007;67:3010–17.
22. Velichko AK, Markova EN, Petrova NV, Razin SV, Kantidze OL. Mechanisms of heat shock response in mammals. Cell Mol Life Sci. 2013;70:4229–41.
23. Turner T, Caspari T. When heat casts a spell on the DNA damage checkpoints. Open Biol. 2014;4:140008–8.
24. Lavin MF, Kozlov S. ATM activation and DNA damage response. Cell Cycle. 2007;6:931–42.

25. Nam EA, Cortez D. ATR signalling: more than meeting at the fork. Biochem J. 2011;436:527–36.
26. Zou L, Elledge SJ. Sensing DNA damage through ATRIP recognition of RPA-ssDNA complexes. Science. 2003;300:1542–8.
27. Lee J, Kumagai A, Dunphy WG. The Rad9-Hus1-Rad1 checkpoint clamp regulates interaction of TopBP1 with ATR. J Biol Chem. 2007;21:28036–44.
28. Wagner JM, Kaufmann SH. Prospects for the use of ATR inhibitors to treat cancer. Pharmaceuticals. 2010;3:1311–34.
29. Jeong SY, Kumagai A, Lee J, Dunphy WG. Phosphorylated claspin interacts with a phosphate-binding site in the kinase domain of Chk1 during ATR-mediated activation. J Biol Chem. 2003;278:46782–8.
30. Cuadrado M, Martinez-Pastor B, Toledo LI, Gutierrez-Martinez P, Lopez E, Fernandez-Capetillo O. ATM regulates ATR chromatin loading in response to DNA double-strand breaks. J Exp Med. 2006;203:297–303.
31. Unno J, Takagi M, Piao J, Sugimoto M, Honda F, Maeda D, et al. Artemis-dependent DNA double-strand break formation at stalled replication forks. Cancer Sci. 2013;104:703–10.
32. Bartek J, Lukas J. Chk1 and Chk2 kinases in checkpoint control and cancer. Cancer Cell. 2003;3:421–9.
33. Takai H, Naka K, Okada Y, Watanabe M, Harada N, Daito S, et al. Chk2-deficient mice exhibit radioresistance and defective p53-mediated transcription. EMBO J. 2002;21:5195–205.
34. Blasina A, de Weyer IV, Laus MC, Luyten WH, Parker AE, McGowan CH. A human homologue of the checkpoint kinase Cds1 directly inhibits Cdc25 phosphatase. Curr Biol. 1999;9:1–10.
35. O'Connell MJ, Raleigh JM, Verkade HM, Nurse P. Chk 1 is a wee 1 kinase in the G_2 DNA damage checkpoint inhibiting cdc2 by Y15 phosphorylation. EMBO J. 1997;16:545–54.
36. Tang J, Erikson RL, Liu X. Checkpoint kinase 1 (Chk1) is required for mitotic progression through negative regulation of polo-like kinase 1 (Plk1). Proc Natl Acad Sci. 2006;103:11964–9.
37. Jallepalli PV, Lengauer C, Vogelstein B, Bunz F. The Chk2 tumor suppressor is not required for p53 responses in human cancer cells. J Biol Chem. 2003;278:20475–9.
38. Liu Q, Guntuku S, Cui XS, Matsuoka S, Cortez D, Tamai K, et al. Chk1 is an essential kinase that is regulated by Atr and required for the G(2)/M DNA damage checkpoint. Genes Dev. 2000;14:1448–59.
39. Tibbetts RS, Brumbaugh KM, Williams JM, Sarkaria JN, Cliby WA, Shieh SY, et al. A role for ATR in the DNA damage-induced phosphorylation of p53. Genes Dev. 1999;13:152–7.
40. Shieh SY, Ahn J, Tamai K, Taya Y, Prives C. The human homologs of checkpoint kinases Chk1 and Cds1 (Chk2) phosphorylate p53 at multiple DNA damage-inducible sites. Genes Dev. 2000;14:289–300.
41. Jazayeri A, Falck J, Lukas C, Bartek J, Smith GCM, Lukas J, et al. ATM- and cell cycle-dependent regulation of ATR in response to DNA double-strand breaks. Nat Cell Biol. 2006;8:37–45.
42. Ohnishi K, Ohnishi T. Heat-induced p53-dependent signal transduction and its role in hyperthermic cancer therapy. Int J Hyperthermia. 2001;17:415–27.
43. Ohnishi T. The role of the p53 molecule in cancer therapies with radiation and/or hyperthermia. J Cancer Res Ther. 2005;1:147–50.
44. Miyakoda M, Suzuki K, Kodama S, Watanabe M. Activation of ATM and phosphorylation of p53 by heat shock. Oncogene. 2002;21:1090–6.
45. Abe T, Tamiya T, Ono Y, Salker AH, Akiyama K, Ohmonot T. Accumulation of cell cycle regulatory proteins, p21 and p27, induced after hyperthermia in human glioma cells. Int J Hyperthermia. 2001;17:499–507.
46. Niitta M, Okamura H, Aizawa S, Yamaizumi M. Heat shock induces transient p53-dependent cell cycle arrest at G_1/S. Oncogene. 1997;15:561–8.
47. Nashimoto T, Komata T, Kanzawa T, Aoki H, Endo S, Kon T, et al. Mild hyperthermia plus adenoviral p53 over-expression additively inhibits the viability of human malignant glioma cells. Int J Hyperthermia. 2005;21:615–29.

48. Fuse T, Yamada K, Asai K, Kato T, Nakanishi M. Heat shock-mediated cell cycle arrest is accompanied by induction of p21 CKI. Biochem Biophys Res Commun. 1996;225:759–63.
49. Furusawa Y, Tabuchi Y, Takasaki I, Wada S, Ohtsuka K, Kondo T. Gene networks involved in apoptosis induced by hyperthermia in human lymphoma U937 cells. Cell Biol Int. 2009;33:1253–62.
50. Tabuchi Y, Takasaki I, Wada S, Zhao QL, Hori T, Nomura T, et al. Genes and genetic networks responsive to mild hyperthermia in human lymphoma U937 cells. Int J Mol Med. 2011;28:143–51.
51. Lim CU, Zhang Y, Fox MH. Cell cycle dependent apoptosis and cell cycle blocks induced by hyperthermia in HL-60 cells. Int J Hyperthermia. 2006;22:77–91.
52. Furusawa Y, Iizumi T, Fujiwara Y, Zhao Q-L, Tabuchi Y, Nomura T, et al. Inhibition of checkpoint kinase 1 abrogates G_2/M checkpoint activation and promotes apoptosis under heat stress. Apoptosis. 2012;17:102–12.
53. Madlener S, Rosner M, Krieger S, Giessrigl B, Gridling M, Vo TP, et al. Short 42 C heat shock induces phosphorylation and degradation of Cdc25A which depends on p38MAPK, Chk2 and 14.3.3. Hum Mol Genet. 2009;18:1990–2000.
54. Zhang Y, Hunter T. Roles of Chk1 in cell biology and cancer therapy. Int J Cancer. 2013;134:1013–23.
55. Iliakis G, Krieg T, Guan J, Wang Y, Leeper D. Evidence for an S-phase checkpoint regulating DNA replication after heat shock: a review. Int J Hyperthermia. 2004;20:240–9.
56. Wang Y, Guan J, Wang H, Wang Y, Leeper D, Iliakis G. Regulation of DNA replication after heat shock by replication protein A-nucleolin interactions. J Biol Chem. 2001;276:20579–88.
57. Tuul M, Kitao H, Iimori M, Matsuoka K, Kiyonari S, Saeki H, et al. Rad9, Rad17, TopBP1 and claspin play essential roles in heat-induced activation of ATR kinase and heat tolerance. PLoS One. 2013;8:e55361.
58. Kim SM, Kumagai A, Lee J, Dunphy WG. Phosphorylation of Chk1 by ATM- and Rad3-related (ATR) in Xenopus egg extracts requires binding of ATRIP to ATR but not the stable DNA-binding or coiled-coil domains of ATRIP. J Biol Chem. 2005;280:38355–64.
59. Guo Z, Kozlov S, Lavin MF, Person MD, Paull TT. ATM activation by oxidative stress. Science. 2010;330:517–21.
60. Takahashi A, Mori E, Ohnishi T. Phospho-Nbs1 and Mre11 proteins which recognize DSBs co-localize with γH2AX in the nucleus after heat treatment. Ann Cancer Res Ther. 2007;15:50–3.
61. Laszlo A, Fleischer I. Heat-induced perturbations of DNA damage signaling pathways are modulated by molecular chaperones. Cancer Res. 2009;69:2042–9.
62. Jekimovs C, Bolderson E, Suraweera A, Adams M, O'Byrne KJ, Richard DJ. Chemotherapeutic compounds targeting the DNA double-strand break repair pathways: the good, the bad, and the promising. Front Oncol. 2014;4:86.
63. Reaper PM, Griffiths MR, Long JM, Charrier J-D, Maccormick S, Charlton PA, et al. Selective killing of ATM- or p53-deficient cancer cells through inhibition of ATR. Nat Chem Biol. 2011;7:428–30.
64. Toledo LI, Murga M, Fernandez-Capetillo O. Targeting ATR and Chk1 kinases for cancer treatment: a new model for new (and old) drugs. Mol Oncol. 2011;5:368–73.
65. Walton MI, Eve PD, Hayes A, Valenti M, Brandon A, Box G, et al. The preclinical pharmacology and therapeutic activity of the novel CHK1 inhibitor SAR-020106. Mol Cancer Ther. 2010;9:89–100.
66. Walton MI, Eve PD, Hayes A, Valenti MR, De Haven Brandon AK, Box G, et al. CCT244747 is a novel, potent and selective inhibitor of CHK1 with oral efficacy alone and in combination with genotoxic anti-cancer drugs. Clin Cancer Res. 2012;18:5650–61.
67. Syljuåsen RG, Sørensen CS, Nylandsted J, Lukas C, Lukas J, Bartek J. Inhibition of Chk1 by CEP-3891 accelerates mitotic nuclear fragmentation in response to ionizing radiation. Cancer Res. 2004;64:9035–40.

Chapter 7
Thermo-Tolerance

Ken Ohnishi

Abstract It is known that cells acquire a transient resistance to heat when cells are treated with a pre-mild heat. This phenomenon is so-called thermo-tolerance. Thermo-tolerance restricts the effectiveness of hyperthermia and combined therapy of hyperthermia and radio-therapy. Heat shock proteins (HSPs) are reported to play important roles in thermo-tolerance of cells. However, the details of the molecular mechanism in which HSPs restore or repair proteins are still unknown. This chapter proposes possible molecular mechanisms which might play a role in heat sensitivity and thermo-tolerance.

Keywords Thermo-tolerance • HSP • polβ • NO

7.1 Phenomenon of Thermo-Tolerance

When cells are treated with a pre-mild heat, cells acquire a transient resistance to subsequent heat treatment (Fig. 7.1a). Development of heat resistance of cells is also observed as a biphasic cell survival curve during a prolonged mild heat treatment (Fig. 7.1b). Thermo-tolerance is generally observed in both tumors and normal tissues and it clinically develops when fractionated hyperthermia is applied to patients. Thermo-tolerance limits the effectiveness of hyperthermia and combined therapy of hyperthermia and radio-therapy. Thus, appropriate heat treatments are required in hyperthermia. In contrast to tumors, thermo-tolerance would be acceptable in normal tissues due to reduction of lethal damage.

Significant problems concerning thermo-tolerance have been suggested from clinical point of view [1]. The kinetics and magnitude of thermo-tolerance varies among tissues that are consisted of different types of cells and tissue constructs. Thus, it is difficult to predict that thermo-tolerance will develop in the same manner in various tumors or normal tissues. Furthermore, if a targeted tumor is unhomogeneously heated, thermo-tolerance develops with different kinetics,

K. Ohnishi (✉)
Department of Biology, Center for Humanity and Sciences, Ibaraki Prefectural University of Health Sciences, 4669-2 Ami, Inashiki-gun, Ibaraki 300-0394, Japan
e-mail: ohnishi@ipu.ac.jp

S. Kokura et al. (eds.), *Hyperthermic Oncology from Bench to Bedside*,
DOI 10.1007/978-981-10-0719-4_7

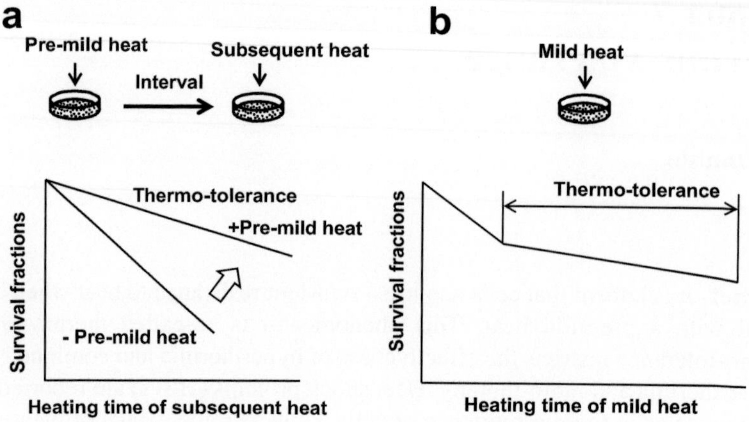

Fig. 7.1 Phenomenon of thermo-tolerance. Thermo-tolerance can be detected as a survival curve that shifts toward radio-resistance (**a**) or as a biphasic survival curve (**b**)

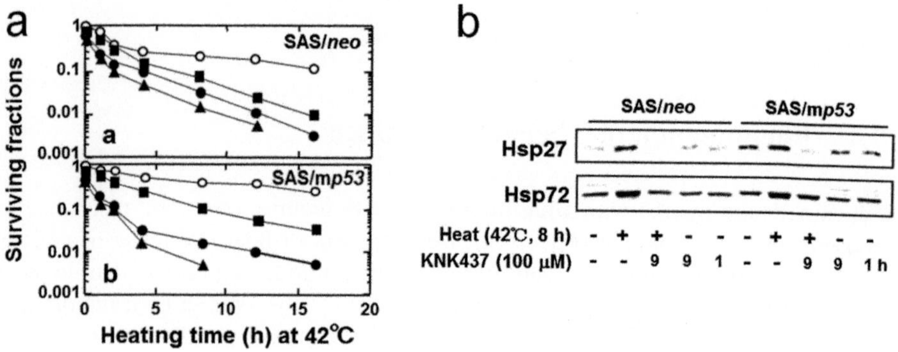

Fig. 7.2 Surviving fractions after heating at 42 °C measured by colony formation assay in SAS/*neo* and SAS/m*p53* cells. (**a**) SAS cells were heated with KNK437 of 0 (*open circles*), 50 (*closed squares*), 100 (*closed circles*) or 300 μM (*closed triangles*). (**b**) accumulation of HSP27 and HSP72 in response to heating at 42 °C [2]

depending on areas of the tumor. These differences in cell reactivity will show different sensitivity of tumor cells to a subsequent hyperthermia.

Advanced therapeutic planning, that considers the interval between fractionated hyperthermia, will prevent tumors from developing thermo-tolerance. However, more fundamental strategy to overcome thermo-tolerance is to elucidate molecular mechanisms contributing to thermo-tolerance and control the mechanisms.

7.2 Molecular Mechanisms of Thermo-Tolerance

Heat-induced HSPs have many biological functions. In particular, many types of HSPs function as molecular chaperons for heat-denatured proteins and enzymes. In addition, HSPs interfere with the apoptosome which consists of the apoptosis protease activating factor-1 (Apaf-1), cytochrome *c*, and caspase-9 during the process leading to heat induced apoptosis. It seems likely that these HSPs may help to lead to thermo-tolerance in cells. However, an important question is what types of proteins are denatured and then restored or repaired by HSPs? It is still unclear which molecules introduce thermo-tolerance. A possibility is that thermo-tolerance is dependent on HSP chaperon activity which can be induced by a mild pre-heat treatment. When cells were treated with mild heat for a long period, thermo-tolerance that is recognized as a biphasic cell survival curve was observed (Fig. 7.2a). Disappearance of the thermo-tolerance was well correlated with depression of mild heat-induced hsp27 and hsp72 accumulation by a HSP inhibitor, KNK437 (Fig. 7.2b) [2]. Interestingly, KNK437 enhanced the heat sensitivity more effectively in squamous cell carcinoma (SAS) bearing mutated *p53* (m*p53*)

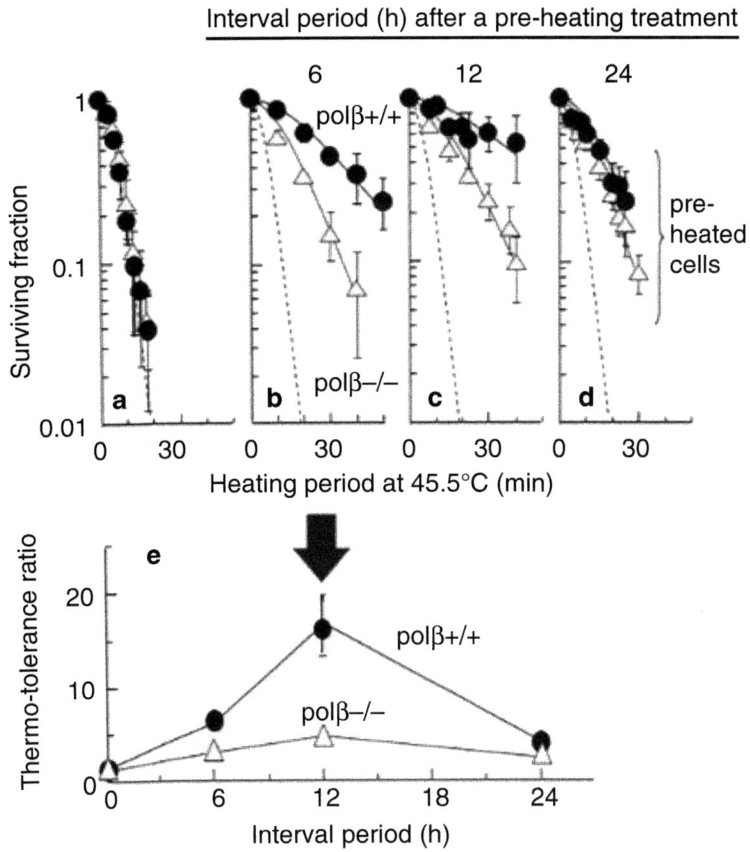

Fig. 7.3 Contribution of polβ to thermo-tolerance. (**a**) the heat sensitivity of polβ-deficient cells (*red triangles*) and the wild-type (*blue circles*) cells were almost the same in human lung cancer cells (H1299 cells). (**b–d**) pre-heat treatments (45.5 °C, 20 min) were delivered to the cells at the indicated intervals and heat sensitivity was then measured. Thermo-tolerance was apparently observed in the wild-type cells at 12 h post-treatment (**e**) but not in polβ-deficient cells [3]

than in SAS bearing wild-type *p53* (wt*p53*). It is unknown how p53 contributes to thermo-tolerance of cells.

Pre-heat-induced HSPs modify some heat-denatured proteins. The details of the mechanism are still unknown. One possible mechanism which has been suggested is that DNA polymeraseβ (polβ) might play a role in heat sensitivity and thermo-tolerance, and that this could be examined with our study [3] using polβ-deficient mutants and the corresponding normal parental cells (Fig. 7.3). This study proposed a hypothesis that polβ may be restored by hsp27 accumulated after pre-heat treatment and repairs the DNA double strand breaks (DSBs) induced indirectly by subsequent heat (Fig. 7.4). It seems that polβ functions as a critical element involved in thermo-tolerance and exerts an important role in heat-induced DSBs. Although many excellent studies have been reported, there is still no clear explanation for molecular mechanisms of thermo-tolerance. These studies seem to be quite relevant to clinical research.

Other reports have examined the basic mechanisms involved in the adaptive response in human cells. Interesting reports have shown that a pre-heat treatment induced heat resistance through nitric oxide (NO) radicals. This phenomenon was first found in the radio-adaptive response. When cells were pre-treated with a low dose of radiation or heat, the cells acquired radiation- or thermo-tolerance, respectively. If the medium from these exposed cells was then used to culture non-exposed cells, it was found that NO radicals were produced in the medium

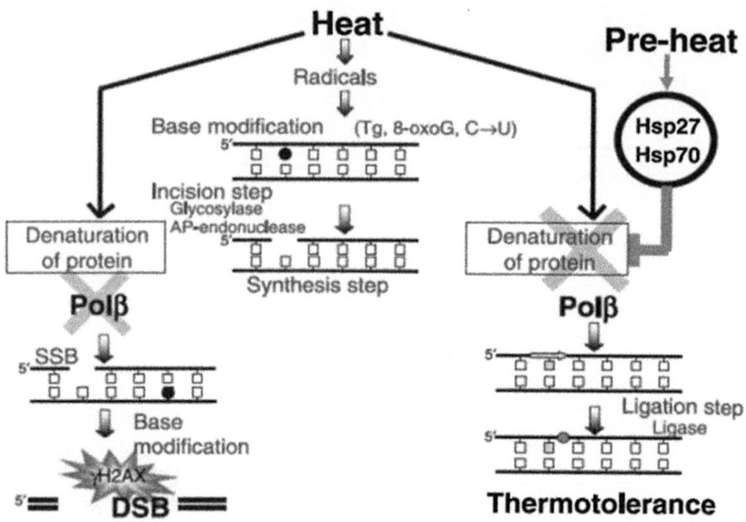

Fig. 7.4 A model for involvement of polymerase-β (polβ) in heat-induced double-strand breaks (DSB). Heat induces base modifications through free radical species. polβ is heat sensitive compared with incision enzymes for excision repair. Heat indirectly induces nicks through inhibition of base excision repair. DSB appears when nicks form in close proximity to each other on both strands through a cell cycle, and a nick is converted into DSB at a DNA replication fork during the S-phase. When cells are preconditioned, polβ is protected or reactivated through the interaction with heat shock protein and fewer DSB are generated [3]

from the pre-treated cells and helped to condition or influence non-exposed cells [4, 5]. Furthermore, it was reported that this phenomenon was found in wt*p53* cells, but not in m*p53* cells. That indicates that these adaptive responses are *p53*-dependent processes.

References

1. Overgaard J, Nielsen OS. The importance of thermotolerance for the clinical treatment with hyperthermia. Radiother Oncol. 1983;1:167–78.
2. Ohnishi K, Takahashi A, Yokota S, Ohnishi T. Effects of a heat shock protein inhibitor KNK437 on heat sensitivity and heat tolerance in human squamous cell carcinoma cell lines differing in p53 status. Int J Radiat Biol. 2004;80:607–14.
3. Takahashi A, Yamakawa N, Mori E, Ohnishi K, Yokota S, Sugo N, Aratani Y, Koyama H, Ohnishi T. Development of thermotolerance requires interaction between polymerase-β and heat shock proteins. Cancer Sci. 2008;99:973–8.
4. Takahashi A, Matsumoto H, Ohnishi T. Hdm2 and NO contribute to the *p53*-dependent radio adaptive response. Int J Radiat Oncol Biol Phys. 2008;71:550–8.
5. Matsumoto H, Takahashi A, Ohnishi T. Nitric oxide radicals choreograph a radioadaptive response. Cancer Res. 2007;67:8574–9.

Chapter 8
Enhancement of Hyperthermia on Radio-Sensitivity

Ken Ohnishi

Abstract Hyperthermia enhances radio-sensitivity of cancer cells when cancer cells are treated with combination of heat and radiation. This chapter describes usefulness of hyperthermia combined with radio-therapy and molecular mechanisms that enable hyperthermia to enhance radio-sensitivity of cancer cells.

Keywords Radio-sensitization • Cell cycle • DNA repair • O_2 concentration

8.1 Cell Cycle and Heat/Radio-Sensitivity

It is well known that radio-sensitivity of cancer cells is associated with cell cycle. In general, cancer cells are relatively sensitive to radiation in M-phase and a transitional phase from G_1- to S-phase and late G_2-phase. In contrast to this, cancer cells are relatively resistant to radiation in G_1-phase and a transitional phase from late S-phase to early G_2-phase. Such cell cycle-dependent radio-sensitivity is closely related to DNA repair systems. Double strand breaks (DSBs) of DNA induced by radiation are repaired by two major repair systems, nonhomologous end-joining (NHEJ) and homologous recombination (HR). DNA damages in G_1 phase are repaired by NHEJ during G_1-phase before the fixation of DNA damages in S-phase. This repair system leads cancer cells to resistant to radiation. The period of G_1 phase varies among cell type. In cells derived from Chinese hamster, the period of G_1 phase is very short. Thus, the radio-resistance in G_1 phase is not observed in those cells [1]. In late G_1 phase and early S-phase, DNA damages of cancer cells are fixed during DNA replication in S-phase and the fixation of DNA damages causes cell death. The cell death results in relatively higher radio-sensitivity of cancer cells in those phases. HR functions in late S-phase to early G_2-phase and DNA repair by HR causes radio-resistance of cancer cells. M-phase shows dynamic morphological changes of cells such as disappearance of nuclear membrane and condensation of chromosome. The morphological changes are

K. Ohnishi (✉)
Department of Biology, Center for Humanity and Sciences, Ibaraki Prefectural University of Health Sciences, 4669-2 Ami, Inashiki-gun, Ibaraki 300-0394, Japan
e-mail: ohnishi@ipu.ac.jp

S. Kokura et al. (eds.), *Hyperthermic Oncology from Bench to Bedside*,
DOI 10.1007/978-981-10-0719-4_8

closely related to radiation-induced lethal DNA damages. Thus, cancer cells are relatively radio-sensitive in M-phase.

Heat sensitivity of cancer cells is also associated with cell cycle. It is reported that nuclear fragmentation and premature chromosome condensation are induced by heat in S-phase cells [2]. The activity of DNA polymerase α functioning in DNA replication in S-phase is depressed by heat [3]. Thus, heat sensitivity is relatively high in cancer cells heated in S-phase. Hyperthermia is effective for growing tumors because they contain many cells of S-phase. The cell cycle dependent heat sensitivity suggests that there are various molecules regulating cell response after hyperthermia. However, the influence of hyperthermia on cell cycle progression is not well understood.

Taking together, when combination treatment of hyperthermia and radio-therapy is applied to cancer cells in S-phase, hyperthermia most effectively enhances radio-sensitivity of cancer cells (Fig. 8.1) [1]. Hyperthermia and radio-therapy complement each other from cell cycle point of view.

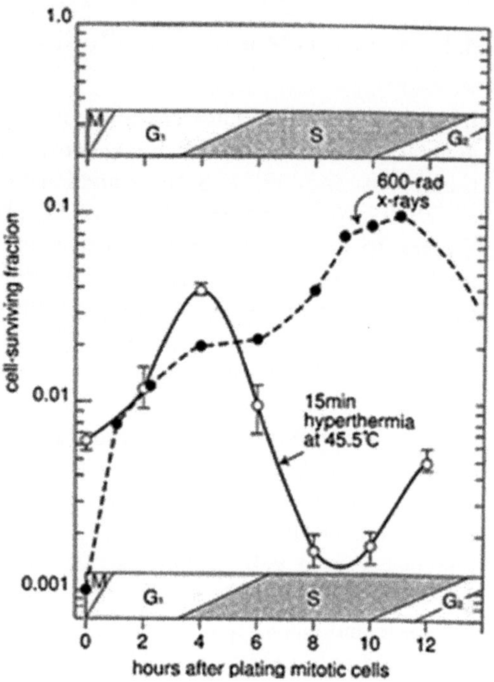

Fig. 8.1 Cell cycle and sensitivity to heat and radiation in Chinese hamster cells. The cells are relatively sensitive to radiation in M-phase and a transitional phase from G_1- to S-phase and late G_2-phase, and relatively resistant to radiation in G_1-phase and a transitional phase from late S-phase to early G_2-phase. Contrarily, heat sensitivity of cells is relatively high in S-phase. Hyperthermia and radio-therapy complement each other especially in S-phase [1]

8.2 Depressive Effects of Hyperthermia on DNA Repair Molecules

When hyperthermia is combined with radio-therapy that induces DNA damages, it is important to understand how heat influences DNA repair systems. It has been reported that heat depresses DNA repair systems. As shown in Table 8.1, the depressive effects of hyperthermia are reported in varied DNA repair factors such as BRCA1/2, RPA and MRN complex in HR, and DNA-PK and Ku 70/80 in NHEJ, and ATM, ATR and H2AX in damage signaling, and MDC1 and 53BP1 in chromatin remodeling [4–8]. These factors involved in DNA repair systems are relatively heat-labile (see Chap. 9 Inhibition of DNA Repair System Activity for further explanation). For example, enhancement of hyperthermia on radio-sensitivity depends on the inhibition of NHEJ repair through depression of DNA-PKcs activities by heat. (Fig. 8.2) [7]. Table 8.1 suggests that enhancement of radio-sensitivity by hyperthermia may be not due to depression of a single DNA repair factor, but rather to multi-processes of DNA repair. The multifaceted effects of heat stress on DNA repair systems may be extremely beneficial for hyperthermia combined with radio-therapy.

Table 8.1 DNA repair factors affected by hyperthermia

HR	NHEJ	Signaling	Chromatin remodeling
BRCA1/2	DNA-PK	ATM	MDC1
RPA	Ku 70/80	ATR	53BP1
MRN complex		H2AX	

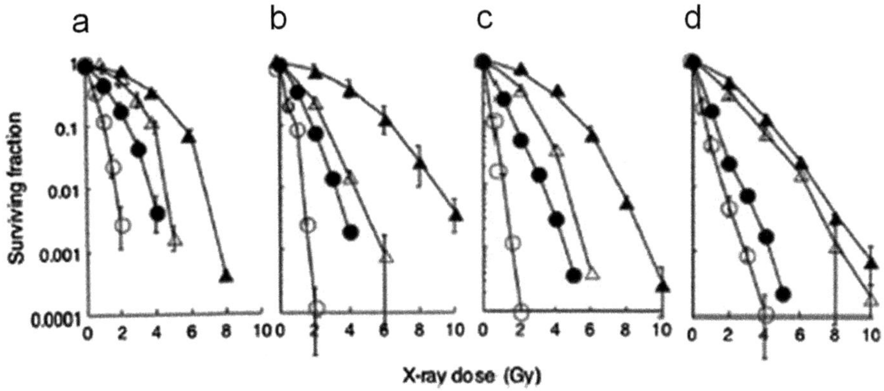

Fig. 8.2 Heat-induced radio-sensitization in DNA-PKcs-deficient cells. Cells were exposed to heat (*open symbols*) or no heat (*closed symbols*) for 15 min at 44 °C and then irradiated by X-ray. (**a**) Mouse scid (DNA-PKcs−/−) and hybrid (wild), (**b**) mouse, PK33N (DNA-PKcs−/−) and CB17 (wild), (**c**) Chinese hamster V3 (DNA-PKcs−/−) and CHO-K1 (wild), (**d**) human M059J (DNA-PKcs−/−) and M059K (wild). *Circles* and *triangles* indicate DNA-PKcs−/− and wild cells, respectively [7]

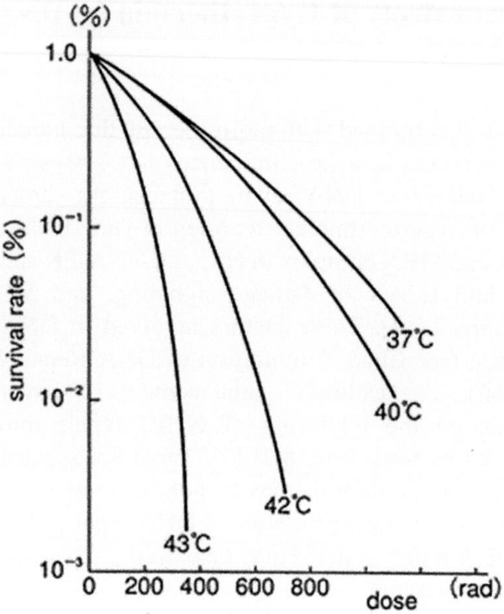

Fig. 8.3 Radio-sensitivity of Chinese hamster cells after combined treatment of heat and radiation. Lethal effects of heat on the cells are higher at temperature of 42–43 °C than that of 37–40 °C. Protein damages are effectively induced over an inflection point (42.5 °C) in Arrhenius plots [9]

Effects of heat on protein damages are relatively higher at temperature over 42.5 °C. An inflection point in Arrhenius plots indicates that a critical action of hyperthermia is to induce protein inactivation. Figure 8.3 shows that lethal effects of heat on cancer cells are higher at temperature of 42–43 °C than that of 37–40 °C [9]. A possible important action of hyperthermia is discussed in Chap. 3 in relation to heat-induced DNA damages.

It is reported that effects of hyperthermia do not depend on quality of radiation, i.e., low LET radiation such as X-rays and γ-rays and high LET radiation such as carbon-beams (Fig. 8.4) [10, 11]. This suggests that hyperthermia depresses DNA repair system regardless of types of DNA damage induced by different quality of radiation.

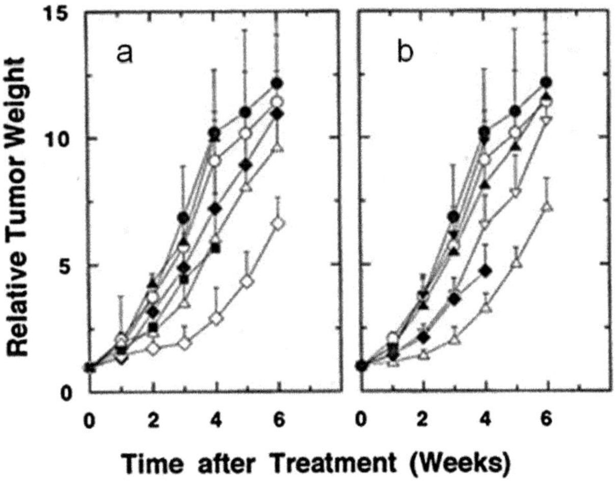

Fig. 8.4 Tumor growth curves of transplantable human esophageal cancer cells in nude mice. (**a**) X-rays and/or hyperthermia. (**b**) C-beams and/or hyperthermia. ●, non-treated group; ○, hyperthermia at 42 °C for 30 min. Other *closed symbols*, radiation alone; other *open symbols*, combination of radiation and hyperthermia at 42 °C for 30 min. ▼▽, 1.25 Gy; ▲△, 2.5 Gy; ◆◇, 5 Gy; ■, 10 Gy. [10]

8.3 Hyperthermia and O_2 Concentration

Angiogenesis does not fully develop in tumors compared with normal tissues. Thus, increased blood flow rate by hyperthermia is relatively low in tumors (Fig. 8.5) [12]. This resultantly produces low increase of O_2 delivery in tumors. As DNA damages induced by radiation are increased with elevation of O_2 concentration (oxygen effect), hyperthermia enhances lethal effects of radiation in hypoxic cells of tumors.

Radio-therapy is relatively low effective to hypoxic cells which exist in the area apart from blood vessel. In contrast to this, hyperthermia is effective to those hypoxic cells, because cooling effect by blood flow is not enough for cells apart from blood vessel. This poor cooling effect produces the elevation of temperature in hypoxic cells and depresses DNA repair systems activated after radio-therapy.

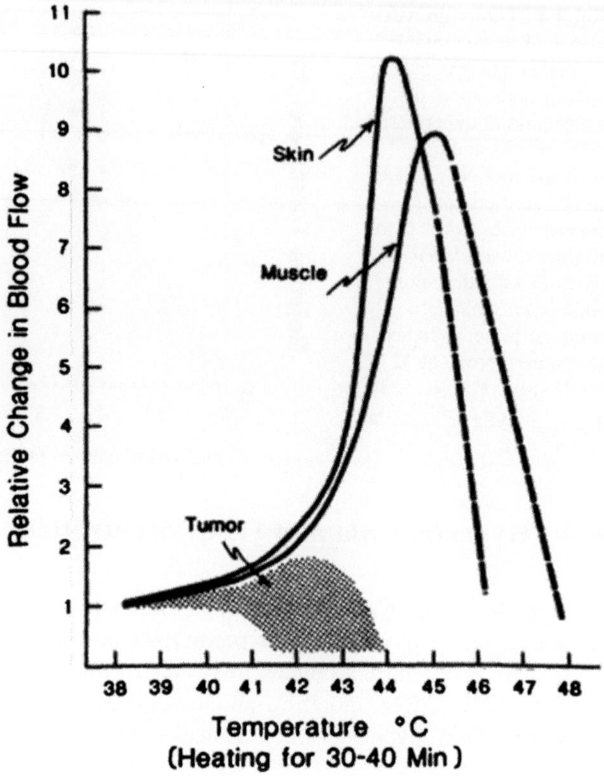

Fig. 8.5 Effects of hyperthermia on blood flow rate in tumors and normal tissues. Angiogenesis does not develop in tumors compared with normal tissues. Hyperthermia somewhat increases blood flow rate in tumors [12]

8.4 Timing of Combined Treatment of Heat and Radiation

Radio-sensitivity of cancer cells is affected by timing of combined treatment of heat and radiation. When cancer cells are simultaneously treated with heat and radiation, the enhancement by heat is most effective, as shown in Fig. 8.6 [13]. The effect becomes lower with the increase of interval between heat and radiation or radiation and heat.

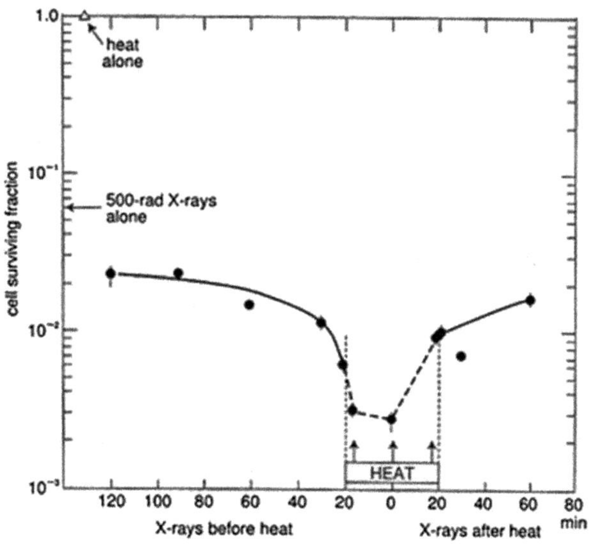

Fig. 8.6 Effects of order of combined treatment of heat and radiation on cell survival. Radio-sensitivity of cells is affected by timing of combined treatment of heat and radiation. When cells are simultaneously treated with heat and radiation, the enhancement by heat is most effective [13]

8.5 Other Systems of Enhancement of Radio-Sensitivity by Hyperthermia

Recent studies have shown that hyperthermia activates immune system [14]. This action of hyperthermia is expected to enhance radio-sensitivity of cancer cells. It is likely that HSPs contributes to immunoreactions leading to radio-sensitization but detailed mechanisms are not well known.

It is clinically difficult to heat up tumors over 42.5 °C. Thus, combined therapy of radio-therapy and mild hyperthermia less than 41.5 °C is recently expected. Mild hyperthermia is reported to inhibit HR through degradation of BRCA2 [15]. It is possible that mild hyperthermia enhances radio-sensitivity of cancer cells through depression of DNA repair systems.

References

1. Westra A, Dewey WC. Variation in sensitivity to heat shock during the cell-cycle of Chinese hamster cells *in vitro*. Int J Radiat Biol Relat Stud Phys Chem Med. 1971;19:467–77.
2. Mackey MA, Morgan WF, Dewey WC. Nuclear fragmentation and premature chromosome condensation induced by heat shock in S-phase Chinese hamster ovary cells. Cancer Res. 1988;48:6478–83.
3. Dube DK, Seal G, Loeb LA. Differential heat sensitivity of mammalian DNA polymerases. Biochem Biophys Res Commun. 1977;76:483–7.
4. Oei AL, Vriend LEM, Crezee J, Franken NAP, Krawczyk PM. Effects of hyperthermia on DNA repair pathways: one treatment to inhibit them all. Radiat Oncol. 2015;10:165.

5. Dewey WC, Esch JL. Transient thermal tolerance: cell killing and polymerase activities. Radiat Res. 1982;92:611–14.
6. Dikomey E, Franzke J. Effect of heat on induction and repair of DNA strand breaks in X-irradiated CHO cells. Int J Radiat Biol. 1992;61:221–33.
7. Ihara M, Takeshita S, Okaichi K, Okumura Y, Ohnishi T. Heat exposure enhances radiosensitivity by depressing DNA-PK kinase activity during double strand break repair. Int J Hyperthermia. 2014;30:102–9.
8. Matsumoto Y, Suzuki N, Sakai K, Morimatsu A, Hirano K, Murofushi H. A possible mechanism for hyperthermic radiosensitization mediated through hyperthermic lability of Ku subunits in DNA-dependent protein kinase. Biochem Biophys Res Commun. 1997;234:568–72.
9. Ben-Hur E, Elkind MM, Bronk B. Thermally enhanced radioresponse of cultured Chinese hamster cells: inhibition of repair of sublethal damage and enhancement of lethal. Radiat Res. 1974;58:38–51.
10. Ohnishi T, Takahashi A, Yano T, Matsumoto H, Wang X, Ohnishi K, Tamamoto T, Tsuji K, Furusawa Y, Yukawo O. Hyperthermic enhancement of tumor growth inhibition by accelerated carbon-ions in transplantable human esophageal cancer. Int J Hyperthermia. 1998;14:195–202.
11. Takahashi A, Ohnishi K, Ota I, Asakawa I, Tamamoto T, Furusawa Y, Matsumoto H, Ohnishi T. *p53*-dependent thermal enhancement of cellular sensitivity in human squamous cell carcinomas in relation to LET. Int J Radiat Biol. 2001;77:1043–51.
12. Song CW. Effect of local hyperthermia on blood flow and microenvironment: a review. Cancer Res. 1984;44:4721s–30.
13. Sapareto SA, Raaphorst GP, Dewey WC. Cell killing and the sequencing of hyperthermia and radiation. Int J Radiat Oncol Biol Phys. 1979;5:343–7.
14. Calderwood SK, Theriault JR, Gong J. How is the immune response affected by hyperthermia and heat shock proteins? Int J Hyperthermia. 2005;21:713–16.
15. Krawczyk PM, Eppink B, Essers J, et al. Mild hyperthermia inhibits homologous recombination, induces BRCA2 degradation, and sensitizes cancer cells to poly (ADP-ribose) polymerase-1 inhibition. Proc Natl Acad Sci U S A. 2011;108:9851–6.

Chapter 9
Inhibition of DNA Repair System Activity

Akihisa Takahashi

Abstract Hyperthermia is a potent sensitizer for cell killing by ionizing radiation and DNA damaging chemotherapeutic agents. Heat mediates these effects because it causes protein denaturation, which strongly depresses DNA repair mechanisms that otherwise contribute to the development of radio- and chemo-resistance. Therefore, hyperthermic therapy has been utilized to augment traditional cancer treatments such as radiotherapy and chemotherapy. Recently research has provided insights into the molecular mechanisms involved in DNA repair pathways, *i.e.* DNA damage responses, base excision repair, DNA single strand break repair, non-homologous end joining repair, and homologous recombination repair, and their effect in responses to hyperthermia. In this chapter, the DNA repair pathways that are relevant in the clinical treatment of cancer are reviewed. The effects of heat on DNA repair and the sensitization of cells to ionizing radiation and DNA damaging agents are also considered. This chapter is intended to provide an overview of the relationship between heat sensitivity and DNA repair.

Keywords Hyperthermia • DNA damage response • DNA repair

9.1 Introduction

Numerous clinical studies of a variety of tumor types have demonstrated synergistic effects in clinical applications of hyperthermia combined with conventional radiotherapy and/or systemic chemotherapy. Observations have also shown that hyperthermia therapy allows clinicians to reduce the doses of radiotherapy and anticancer drugs delivered to patients [1], which can help prevent adverse anti-cancer therapy side effects.

At the molecular level, radiotherapy and chemotherapy cause numerous events resulting in DNA damage. In addition, early research demonstrated that heat itself may directly induce DNA damage [2–4]. However, the toxic consequences of DNA

A. Takahashi (✉)
Gunma University Heavy Ion Medical Center, 3-39-22 Showa-machi, Maebashi, Gunma 371-8511, Japan
e-mail: a-takahashi@gunma-u.ac.jp

S. Kokura et al. (eds.), *Hyperthermic Oncology from Bench to Bedside*,
DOI 10.1007/978-981-10-0719-4_9

damage can be reduced through the activity of several DNA repair pathways that repair such lesions. As a result, efficient DNA repair systems activated by DNA damage can lead to heat-, radio- and chemo-resistance in some cancer cells. Although there is a growing understanding of DNA repair networks and their action in responding to radiation- and anti-cancer-drug-induced DNA damage, the effects of hyperthermia on DNA repair are still not well understood [5]. It is thought that the sensitizing or cytotoxic effects of hyperthermia are probably not caused by the inactivation or inhibition of a single DNA repair pathway, but rather by effects that influence multiple pathways involved in DNA repair.

Understanding how heat interacts with and affects DNA repair networks will lead clinicians to develop improved therapies that combine hyperthermia with radiotherapy and/or chemotherapy. The aim of this chapter is to examine the literature describing the influence of hyperthermia on DNA repair pathways.

9.2 The Nature of DNA Damage that Accumulates after Exposure to Radiation and Chemotherapeutic Agents Combined with Hyperthermia

Hyperthermia sensitizes cells to radiation and this effect is generally thought to result from the inhibition of DNA repair. Early research using classical assays demonstrated that a heat shock at temperatures over 42 °C given before irradiation inhibited the repair of radiation-induced DNA damage (Table 9.1).

Many types of DNA lesions such as base damage, inter- and intra-strand crosslinks, DNA single strand breaks (SSBs) and DNA double strand breaks (DSBs), are induced by many different types of ionizing radiation and chemotherapeutic agents (Table 9.2). DSBs are considered the most important type of damage leading to cell lethality. DSBs can be formed indirectly *via* other lesions through the

Table 9.1 DNA damage accumulated by radiation in combination with hyperthermia

Type of DNA damage	Assay	Ref.
Base damage	High performance liquid chromatography (HPLC)	[6, 7]
DNA-protein crosslinks	Filter binding assay	[8]
DNA strand breaks	Halo assay	[9, 10]
Alkali labile base damage and strand breaks	Alkaline sucrose gradient	[11, 12]
	Alkaline elution	[13, 14]
	Alkaline unwinding	[15–19]
Alkali labile base damage and DSBs	Non-denaturing or neutral elution	[20–23]
DSB	Neutral sucrose gradient	[11]
	Pulsed-field gel electrophoresis (PFGE)	[24–26]
	Constant-field gel electrophoresis	[27]

Table 9.1 is summarized from [28]. *DSB* double stranded breaks

Table 9.2 Types of DNA damage accumulated after exposure to chemotherapeutic agents delivered in combination with hyperthermia

Type of DNA damage	Class	Agent	Ref.
Base damage, Crosslink, SSB, DSB	Radiomimetics	Bleomycin	[29–36]
		Neocarzinostatin	[32]
Crosslink, DSB (indirect)	Antimetabolites	5-fluorouracil	[37]
		Gemcitabine	[38–41]
		2-aminopurine	[42]
		6-thioguanine	[42]
		Methotrexate	[34, 43]
Base damage, Crosslink, DSB (indirect)	Alkylating agents	Mitomycin C	[32, 36, 44–48]
		Cyclophosphamide	[36, 44, 49–54]
		Melphalan	[36, 40, 55–63]
		Temozolomide	[64, 65]
Crosslink, DSB (indirect)	Platinum compounds	Cisplatin	[34–36, 46, 66–76]
		Carboplatin	[77–80]
		Oxaliplatin	[71, 81]
SSB, DSB (indirect)	Top1 poisons	Camptothecin	[82]
		β-lapachone	[83, 84]
		Irinotecan	[40]
DSB	Top2 poisons	Doxorubicin	[30, 44, 52, 85–91]
		Etoposide	[92]

Table 9.2 is summarized from [5]. *DSB* double stranded breaks, *SSB* single stranded breaks, *Top* topoisomerase

inactivation of DNA repair and replication processes. Hyperthermia sensitizes cells to chemotherapeutic agents (Table 9.2), providing support for the hypothesis that the DNA repair pathway is affected by hyperthermia. However, multiple complex additional pathways, resembling networks, participate in the repair of these various lesions [5].

9.3 DNA Repair Networks

Recent progress in understanding the biological basis of the mechanisms used by cells and tissues to respond to ionizing radiation and DNA damaging agents has highlighted some key molecular processes involved in these responses (Fig. 9.1). DNA damage responses (DDR) comprise a complex regulatory and functional network that involves DNA damage sensing, signaling, and repair. The dynamics in DNA repair networks are regulated by post-translational modifications, recruitment of DNA repair factors to the sites of DNA damage, the enzymatic activity of

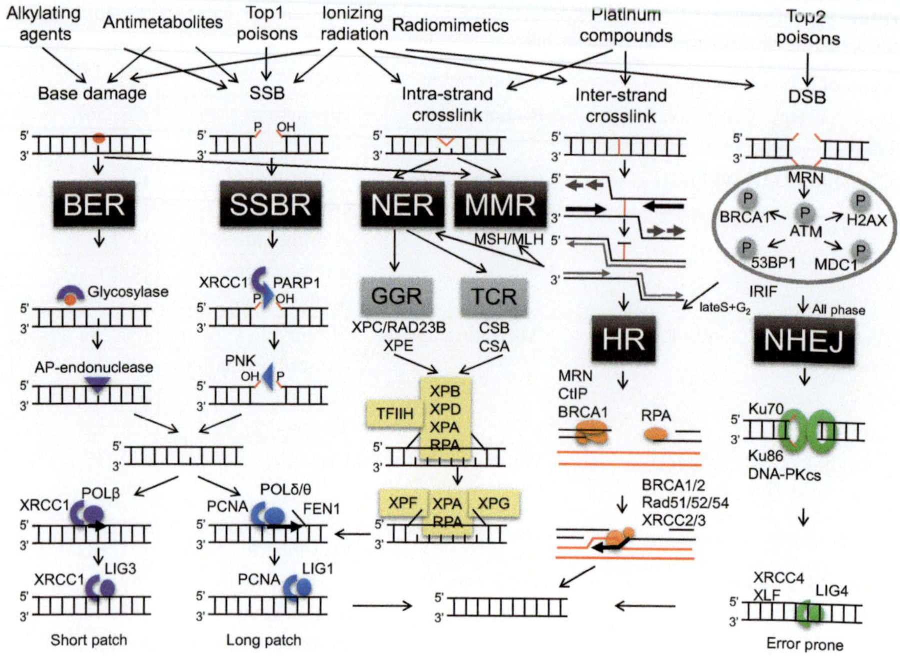

Fig. 9.1 DNA repair networks

these factors and their interactions, and the particular DNA repair pathway active at a particular damage site.

9.3.1 *Excision Repair*

Excision repair occurs in living cells where damaged or mismatched bases in one of the two DNA strands are excised and replaced by DNA synthesis using the undamaged strand as a template. Excision repair can be classified as base excision repair (BER), nucleotide excision repair (NER) or mismatch repair (MMR).

9.3.1.1 BER

BER is the primary repair system for the most frequent types of DNA base damage induced by alkylating agents, antimetabolites, ionizing radiation, radiomimetic agents, and cellular metabolism. BER involves the concerted efforts of several repair proteins, and is initiated by DNA glycosylases that recognize and remove specific damaged or inappropriate bases, thereby forming apyrimidinic/apurinic or abasic (AP) sites in the damaged DNA. To maintain genomic integrity, the AP sites must be repaired, and the main system active in their removal is the BER pathway. These sites are cleaved by an AP endonuclease. The resulting SSB can then be repaired by either short-patch BER (where one nucleotide is replaced) or by long-patch BER (where more than one nucleotide is replaced) [93, 94].

Short-Patch BER In short patch repair, a single nucleotide is replaced by DNA polymerase-β (POLβ) with the help of XRCC1, and the break is rejoined by ligase 3 (LIG3). This is the main pathway used by BER.

Long-Patch BER Proliferating cell nuclear antigen (PCNA), and flap endonuclease 1 (FEN1) are required for the processing of long patches. In long patch repair, up to 13 nucleotides are replaced by DNA polymerase-δ/ε (POLδ/ε) and rejoining is completed by ligase 1 (LIG1).

9.3.1.2 NER

NER also plays a key role in repairing a variety of distorting lesions, notably platinum-induced DNA adducts and intra-strand cross-links. NER is divided into two sub-pathways: global genome NER (GGR) and transcription coupled NER (TCR), depending on the transcription state of DNA. These two sub-pathways differ only by their damaged DNA recognition mechanisms: the XPC-RAD23B complex together with the XPE complex in GGR, or by Cockayne syndrome A (CSA) and Cockayne syndrome B (CSB) in the TCR process. The Xeroderma Pigmentosum (XP) gene family codes for most proteins involved in the successive phases of DNA repair, allowing open access to the double helix (XPD/ERCC2, XPB/ERCC3) and the removal and substitution of the damaged DNA strand (XPA, RPA, XPG/ERCC5, XPF/ERCC4 and ERCC1) [95]. The main role of PCNA during NER is involved in the re-synthesis of the damaged DNA fragment and this is achieved through the interaction of PCNA with DNA POLδ/ε, LIG1 and Fen1, which is similar to its role in BER [96].

9.3.1.3 MMR

The MMR mismatch repair system plays an essential role in the post-replication repair of mis-incorporated bases resulting from base damage, and intra- and inter-strand crosslinks that have escaped the proofreading activity of replication polymerases. In addition to mismatched bases, MMR proteins also correct insertion/deletion loops that result from polymerase slippage during replication of repetitive DNA sequences. The MMR pathway can be divided into three principle steps: a recognition step where mis-paired bases are recognized by a MutS homologue (MSH), an excision step where the error-containing strand is degraded resulting in a gap made by the MutL homolog (MLH)-MSH complex, and a repair synthesis step, where the gap is filled through DNA re-synthesis [97].

9.3.2 SSB Repair (SSBR)

DNA SSBs are induced by alkylating agents, antimetabolites, Topoisomerase (Top) 1 poisons, ionizing radiation, and radiomimetic agents. Poly(ADP-ribose) polymerase (PARP) senses DNA nicks and also binds to SSBs and short gaps in duplex DNA. After the activation of PARP close to SSBs, XRCC1 accumulates at poly (ADP-ribose) (pAR) sites. XRCC1 stimulates human polynucleotide kinase (PNK) activity at damaged DNA termini and accelerates the SSBR processes. SSBR is affected by polymerase β (POLβ)-dependent short-patch repair or by PCNA/ POLδ/ε-dependent long-patch repair [98].

9.3.3 DSB Repair

DNA DSBs are highly toxic because unrepaired or mis-repaired DSBs can lead to genomic instability and cell death [99–101]. There are two major pathways leading to the repair of DSBs in eukaryotes: Non-homologous end-joining (NHEJ) repair [102] and homologous recombination (HR) repair [103].

9.3.3.1 NHEJ Repair

NHEJ is active throughout the cell cycle and rejoins broken DNA ends. NHEJ is a 'quick and dirty' process when compared to HR, and is less accurate, with small deletions or insertions often occurring at the repaired break sites [104, 105]. This can lead to mutations, although they allow the cell to survive. An unrepaired DSB is often lethal through the loss of a piece of chromosome during the next round of mitosis, with a potential loss of tens or hundreds of genes. However, only a few percent of genomic DNA codes for genes or regulatory regions, so the likelihood of a break occurring in such critical regions is low, and these regions may also be silent and/or non-essential. NHEJ is a good repair pathway for the cell to maximize its chance of survival [105]. Although the ligation of DNA DSBs by NHEJ does not require sequence homology or a repair template, the damaged ends of DNA DSBs cannot simply be ligated together.

Classical NHEJ (C-NHEJ) Repair Damaged DNA ends must first be modified before they can be rejoined through a ligation reaction. NHEJ can be divided into five steps: (i) end recognition by a Ku autoantigen protein p70 (Ku70, XRCC6) and p80 (Ku80, XRCC5) heterodimer that binds to and protects DNA ends from degradation by exonucleases; (ii) recruitment of the DNA-dependent protein kinase catalytic subunit (DNA-PKcs), which is a large protein that forms a physical bridge between the two ends; (iii) end processing by Artemis, which possesses endonuclease activity to repair 5′ and 3′ overhangs as well as hairpins, and by PNK which trims 'dirty' ends; (iv) fill-in synthesis or end bridging by DNA polymerase μ or λ;

and (v) ligation by ligase 4 (LIG4), aided by other proteins, such as XRCC4 and the XRCC4-like factor (XLF) [105]. This pathway is also called DNA-PK dependent NHEJ (D-NHEJ) repair.

Backup NHEJ (B-NHEJ) Repair PARP1 plays a role in initiating this process. Without the protection provided by the Ku70/Ku80 proteins, the DNA ends are resected in a reaction promoted by the nuclease activity of Mre11 and C terminus-binding protein (CtBP)-interacting protein (CtIP). It has been proposed that a single-strand DNA resection reveals complementary microhomologies that can be annealed, and gap filling completes the end-joining. Subsequently, XRCC1 and LIG3 (which can be substituted for by LIG1) complete B-NHEJ. Of note, B-NHEJ is always associated with deletions at the repair junctions and frequently involves micro homologies that are distant from the DSB. This pathway is also called alternative backup NHEJ (A-NHEJ or A-EJ) repair [106].

Single-Strand Annealing (SSA) SSA is initiated when a DSB is formed between two repeated sequences oriented in the same direction. SSA involves reannealing of replication protein A (RPA)-covered ssDNA by the Rad52 protein [107].

9.3.3.2 HR Repair

HR repair results in the error-free repair of lesions in the late S-G_2 phases of the cell cycle because HR repairs DSBs using the homologous undamaged DNA strand as a repair template. Briefly, the HR process is described as follows [104, 105]. The initial step in HR is the recognition of the lesion and processing of the double-strand DNA ends into 3′ DNA single strand tails by the Mre11-Rad50-Nbs1 (MRN) complex. These tails are then coated with RPA protein forming a nucleoprotein filament. Specific HR proteins are then recruited to the nucleoprotein filaments. Such proteins include the Rad51 family of proteins including XRCC3, Rad52, and breast cancer susceptibility gene (BRCA) 1 and BRCA2. These single-stranded nucleoprotein filaments then invade the undamaged double-stranded DNA on the neighboring sister chromatid, forming a crossover, or bubble structure. These bubbles are then expanded with specialized enzymes called helicases, which include the Bloom syndrome (BLM) protein and other members of the DNA helicase Q1 (RecQ) family, possibly with the help of Rad54. The object of this process is to provide an undamaged DNA template with the same base sequence around the break site, so that the DNA polymerases can then synthesize new DNA across the missing regions, thereby accurately repairing the break. The crossover structure then is undone or reversed to replace the chromatin with its original configuration. This is achieved with specialized nucleases that cut or resolve the junctions, followed by connecting, or ligating, the adjacent ends with LIG1. The entire process takes several hours to complete. Deletions or mutations in any of the genes involved in this process can severely impair HR.

9.4 Effects of DNA Repair on Heat-Induced Radio-Sensitization

Since the 1990s, there have been many reports describing the process of DNA repair during heat-induced radiosensitization by using DNA repair defective cells (Table 9.3). Abundant data was accumulated using a series of mutant cell lines defective in various steps of the NHEJ pathway, and this showed that heat-induced radiosensitization was the same in NHEJ-deficient cell lines and in DNA repair

Table 9.3 Overview of the effects of DNA repair on heat-induced radio-sensitization

Pathway	Defective gene	Cell line	Relative TER[a]	Ref.
NHEJ	*Ku80*	Xrs-5 CHO	Reduced in plateau	[108]
			Same in exponential	[108]
			Same/Larger	[109]
		Xrs-6 CHO	Same	[109]
		Ku80−/−MEF	Same/Larger	[110, 111]
	Ku70	Ku70−/−DT40	Reduced in sensitive fraction	[111, 112]
			Larger in resistant fraction	[112]
	DNA-PKcs	Irs-20 CHO	Same	[24]
		V3 CHO	Same/Larger	[109, 113, 114]
		Scid MEF	Same	[115]
			Reduced	[116]
		SCGR11 Mouse	Same	[109]
		DNA-PKcs−/−DT40	Same in sensitive fraction	[112]
			Larger in resistant fraction	[112]
	Ligase IV	180 BRM Human	Same	[110]
HR	*Xrcc2*	Irs1 CHO	Same	[117]
	Xrcc3	Irs1SF CHO	Same	[111, 117]
			Reduced	[114]
	Rad51C	Irs3 CHO	Reduced	[114]
	Rad51D	51D1 CHO	Reduced	[114]
	Rad54	Rad54−/−DT40	Same	[111, 112]
		Rad54−/−ES	Same	[118]
NHEJ & HR	*Ku70 & Rad54*	Ku70−/−Rad54−/−DT40	Same	[111, 112]

[a]Relative TER, magnitude of thermal radio-sensitization compared with parental, wild-type cells. Table 9.3 is based on [28]. *TER* thermal enhancement ratio, *CHO* Chinese Hamster Ovary cells, *HR* homologous recombination, *NHEJ* non-homologous end-joining

proficient cell lines. In contrast, it was recently reported that the HR pathway is associated with heat-radio-sensitization [114, 118].

9.5 Effects of Hyperthermia on DNA Repair Factors

The behavior of DNA repair factors is altered after heat treatment and subsequent exposure to radiation (Table 9.4). Many DNA repair factors were degraded, down-regulated, or inactivated after heat treatment. In particular, the MRN complex [129–135], p53 binding protein (53BP1) [128], Rad51 [114, 118] and BRCA2 [118] do not translocate into the nucleus and form foci after heat treatment. However, factors important in the HR process subsequently re-formed foci at damaged sites over time, and the behavior of these factors exhibited a pattern related to the temperature and length of the hyperthermia exposure [114]. It was reported that heat induces the activation of the AP-endonuclease [145] and the

Table 9.4 Overview of the kinetics of DNA repair factors

Pathway	Factor	After heat treatment Early response (0 to several hours)	After heat treatment Late response (at 6–12 h)	Ref.
DDR	ATM	A, P, F	A, P, F	[119–121]
	ATR	A, P, F	A, P, F	[122, 123]
	H2AX	P, F	P, F	[124–129]
	MDC1	F	F	[120, 128]
	Mre11	D, T	R, F	[130–134]
	Rad50	D, T	R, F	[131, 134]
	NBS1	D, T	P, R, F	[129, 131, 133–135]
	53BP1	D	F	[128]
	BRCA1	D	–	[136]
HR	BRCA2	D	–	[118]
	RPA	D, I	F	[137]
	Rad51	D	F	[114, 118]
NHEJ	Ku70/80	D, I	–	[114, 138–142]
	DNA-PKcs	D	P, F	[129]
	DNA-PK	I	–	[140, 142, 143]
SSBR	PARP1	D	–	[143, 144]
BER	AP endonuclease	–	A	[145]
	POLβ	I	–	[146–148]
	XRCC1	D	–	[143]
	OGG1	I	–	[149]

D Degradation/Downregulation, *I* Inactivation, *T* Translocation from the nucleus, *A* Activation, *P* Phosphorylation, *R* Relocation to the nucleus, *F* Foci formation, *DDR* DNA damage response, *HR* homologous recombination, *NHEJ* non-homologous end-joining, *SSBR* single strand break repair, *BER* base excision repair

deactivation of POLβ [146–148]. These experimental results suggest that heat might induce DSBs through the BER process after exposure to radiation [28], but not without exposure to radiation [124].

9.6 Effects of DNA Repair on Heat Sensitivity

Since the 2000s, there have been many reports examining the effects of DNA repair on heat sensitivity by using DNA repair defective cells (Table 9.5). A series of mutant cell lines was used that were defective in various steps of the HR pathway, and these cell lines displayed heat sensitivity when compared with DNA repair proficient normal cell lines. These studies suggest that the HR pathway is associated with heat-induced radio-sensitization [114, 118].

Table 9.5 Overview of the effects of DNA repair on heat sensitivity

Pathway	Defective gene	Cell lines	Treatment	Relative SER[a]	Ref.
DDR	*ATM*	GM 3440		Same	[150]
		ATM−/−DT40		Reduced	[123]
	NBS1	8305c Human cancer cells	si*NBS1*	Reduced	[151]
	53BP1	53BP1−/−MEF		Same	[128]
NHEJ	*Ku80*	Xrs-5 CHO		Reduced	[114]
	Ku70	Ku70−/−DT40		Same	[112]
	DNA-PKcs	DNA-PKcs−/−DT40		Same	[112]
		V3 CHO		Same	[114]
		Scid MEF		Reduced	[152]
HR	*Rad9*	Rad9−/−DT40		Reduced	[123]
	Rad17	Rad17−/−DT40		Reduced	[123]
	Xrcc3	Irs1SF CHO		Reduced	[114]
	Rad51D	51D1 CHO		Reduced	[114]
	Rad54	Rad54−/−DT40		Reduced	[112]
NHEJ & HR	*Ku70& Rad54*	Ku70−/−Rad54−/−DT40		Same	[112]
SSBR	*PARP1*	Human cancer cells	si*PARP*	Reduced	[118]
			NU1025		
			PJ34		
BER	*POLβ*	POLβ−/−MEF		Reduced	[153]

[a]Relative SER, the magnitude of heat sensitivity compared with parental, wild-type cells or untreated cells. *SER* sensitization enhancement ratio, *DDR* DNA damage response, *HR* homologous recombination, *NHEJ* non-homologous end-joining, *SSBR* single strand break repair, *BER* base excision repair

9.7 Conclusion

Currently, hyperthermia is widely used to treat patients with a range of cancer types. Hyperthermia can be used as a powerful adjuvant to enhance radiotherapy and chemotherapy significantly without harmful secondary effects. Positive therapeutic effects observed in combination therapies that include hyperthermia seem to reflect the presence of a multitude of DNA repair pathways as well as other pathways. Another possible mechanism leading to the depression of DNA repair postulates the inhibition of access of repair enzymes to DNA damage sites by the denaturation of nuclear matrix proteins, and/or the translocation and removal of repair related proteins from the nucleus. Future molecular biology studies will continue to contribute to our understanding of the therapeutic advantages of hyperthermia, and a more extensive knowledge of the precise mechanisms leading to the inhibition of DNA repair mechanisms through which hyperthermia kills cells.

References

1. Falk MH, Issels RD. Hyperthermia in oncology. Int J Hyperthermia. 2001;17:1–18.
2. Warters RL, Henle KJ. DNA degradation in Chinese hamster ovary cells after exposure to hyperthermia. Cancer Res. 1982;42:4427–32.
3. Anai H, Maehara Y, Sugimachi K. *In situ* nick translation method reveals DNA strand scission in HeLa cells following heat treatment. Cancer Lett. 1988;40:33–8.
4. Wong RS, Dynlacht JR, Cedervall B, Dewey WC. Analysis by pulsed-field gel electrophoresis of DNA double-strand breaks induced by heat and/or X-irradiation in bulk and replicating DNA of CHO cells. Int J Radiat Biol. 1995;68:141–52.
5. Oei AL, Vriend LE, Crezee J, Franken NA, Krawczyk PM. Effects of hyperthermia on DNA repair pathways: one treatment to inhibit them all. Radiat Oncol. 2015;10:165. doi:10.1186/s13014-015-0462-0.
6. Warters RL, Roti Roti JL. Production and excision of 5', 6'-dihydroxydihydrothymine type products in the DNA of preheated cells. Int J Radiat Biol. 1978;34:381–4.
7. Warters RL, Roti Roti JL. Excision of X-ray-induced thymine damage in chromatin from heated cells. Radiat Res. 1979;79:113–21.
8. Cress AE, Bowden GT. Covalent DNA-protein crosslinking occurs after hyperthermia and radiation. Radiat Res. 1983;95:610–9.
9. Kampinga HH, Wright WD, Konings AWT, Roti Roti JL. The interaction of heat and radiation affecting the ability of nuclear DNA to undergo supercoiling changes. Radiat Res. 1988;116:114–23.
10. Wynstra JH, Wright WD, Roti Roti JL. Repair of radiation-induced DNA damage in thermotolerant and nonthermotolerant HeLa cells. Radiat Res. 1990;124:85–9.
11. Corry PM, Robinson S, Getz S. Hyperthermic effects on DNA repair mechanisms. Radiology. 1977;123:475–82.
12. Clark EP, Dewey WC, Lett JT. Recovery of CHO cells from hyperthermic potentiation to X-rays: repair of DNA and chromatin. Radiat Res. 1981;85:302–13.
13. Bowden GT, Kasunic MD. Hyperthermic potentiation of the effects of a clinically significant X-ray dose on cell survival, DNA damage, and DNA repair. Radiat Res. 1981;87:109–20.
14. Mills MD, Meyn RE. Effects of hyperthermia on repair of radiation-induced DNA strand breaks. Radiat Res. 1981;87:314–28.

15. Lunec J, Hesslewood IP, Parker R, Leaper S. Hyperthermic enhancement of radiation cell killing in HeLa S3 cells and its effect on the production and repair of DNA strand breaks. Radiat Res. 1981;85:116–25.
16. Dikomey E. Effect of hyperthermia at 42 and 45 °C on repair of radiation-induced DNA strand breaks in CHO cells. Int J Radiat Biol. 1982;41:603–14.
17. Jorritsma JB, Konings AW. Inhibition of repair of radiation-induced strand breaks by hyperthermia, and its relationship to cell survival after hyperthermia alone. Int J Radiat Biol. 1983;43:505–16.
18. McGhie JB, Wold E, Pettersen EO, Moan J. Combined electron radiation and hyperthermia. Repair of DNA strand breaks in NHIK 3025 cells irradiated and incubated at 37, 42.5, or 45 °C. Radiat Res. 1983;96:31–40.
19. Dikomey E, Franzke J. Effect of heat on induction and repair of DNA strand breaks in X-irradiated CHO cells. Int J Radiat Biol. 1992;61:221–33.
20. Radford IR. Effects of hyperthermia on the repair of X-ray-induced DNA double-strand breaks in mouse L cells. Int J Radiat Biol. 1983;43:551–7.
21. Warters RL, Lyons BW, Axtell-Bartlett J. Inhibition of repair of radiation-induced DNA damage by thermal shock in Chinese hamster ovary cells. Int J Radiat Biol. 1987;51:505–17.
22. Iliakis G, Seaner R, Okayasu R. Effects of hyperthermia on the repair of radiation-induced DNA single-and double-strand breaks in DNA double-strand break repair-deficient and repair-proficient cell lines. Int J Hyperthermia. 1990;6:813–33.
23. Warters RL, Axtell J. Repair of DNA strand breaks at hyperthermic temperatures in Chinese hamster ovary cells. Int J Radiat Biol. 1992;61:43–8.
24. Kampinga HH, Kanon B, Konings AW, Stackhouse MA, Bedford JS. Thermal radiosensitisation in heat- and radiation-sensitive mutants of CHO cells. Int J Radiat Biol. 1993;64:225–30.
25. Nevaldine B, Longo JA, Hahn PJ. Hyperthermia inhibits the repair of DNA double-strand breaks induced by ionising radiation as determined by pulsed-field gel electrophoresis. Int J Hyperthermia. 1994;10:381–8.
26. Kampinga HH, Hiemstra YS, Konings AW, Dikomey E. Correlation between slowly repairable double-strand breaks and thermal radiosensitisation in the human HeLa S3 cell line. Int J Radiat Biol. 1997;72:293–301.
27. Dahm-Daphi J, Brammer I, Dikomey E. Heat effects on the repair of DNA double-strand breaks in CHO cells. Int J Radiat Biol. 1997;72:171–9.
28. Kampinga HH, Dikomey E. Hyperthermic radiosensitization: mode of action and clinical relevance. Int J Radiat Biol. 2001;77:399–408.
29. Braun J, Hahn GM. Enhanced cell killing by bleomycin and 43 degrees hyperthermia and the inhibition of recovery from potentially lethal damage. Cancer Res. 1975;35:2921–7.
30. Hahn GM, Braun J, Har-Kedar I. Thermochemotherapy: synergism between hyperthermia (42–43 degrees) and adriamycin (of bleomycin) in mammalian cell inactivation. Proc Natl Acad Sci U S A. 1975;72:937–40.
31. Kubota Y, Nishimura R, Takai S, Umeda M. Effect of hyperthermia on DNA single-strand breaks induced by bleomycin in Hela cells. Gann. 1979;70:681–5.
32. Mizuno S, Amagai M, Ishida A. Synergistic cell killing by antitumor agents and hyperthermia in cultured cells. Gann. 1980;71:471–8.
33. Dahl O, Mella O. Enhanced effect of combined hyperthermia and chemotherapy (bleomycin, BCNU) in a neurogenic rat tumour (BT4A) *in vivo*. Anticancer Res. 1982;2:359–64.
34. Herman TS, Sweets CC, White DM, Gerner EW. Effect of heating on lethality due to hyperthermia and selected chemotherapeutic drugs. J Natl Cancer Inst. 1982;68:487–91.
35. Herman TS, Henle KJ, Nagle WA, Moss AJ, Monson TP. Effect of step-down heating on the cytotoxicity of adriamycin, bleomycin, and cisdiamminedichloroplatinum. Cancer Res. 1984;44:1823–6.

36. Takemoto M, Kuroda M, Urano M, Nishimura Y, Kawasaki S, Kato H, et al. The effect of various chemotherapeutic agents given with mild hyperthermia on different types of tumours. Int J Hyperthermia. 2003;19:193–203.
37. Maehara Y, Sakaguchi Y, Takahashi I, Yoshida M, Kusumoto H, Masuda H, et al. 5-Fluorouracil's cytotoxicity is enhanced both *in vitro* and *in vivo* by concomitant treatment with hyperthermia and dipyridamole. Cancer Chemother Pharmacol. 1992;29:257–60.
38. Haveman J, Rietbroek RC, Geerdink A, van Rijn J, Bakker PJ. Effect of hyperthermia on the cytotoxicity of 2',2'-difluorodeoxycytidine (gemcitabine) in cultured SW1573 cells. Int J Cancer. 1995;62:627–30.
39. Shewach DS, Lawrence TS. Radiosensitization of human tumor cells by gemcitabine *in vitro*. Semin Oncol. 1995;22:68–71.
40. Mohamed F, Marchettini P, Stuart OA, Urano M, Sugarbaker PH. Thermal enhancement of new chemotherapeutic agents at moderate hyperthermia. Ann Surg Oncol. 2003;10:463–8.
41. Raoof M, Zhu C, Cisneros BT, Liu H, Corr SJ, Wilson LJ, et al. Hyperthermia inhibits recombination repair of gemcitabine-stalled replication forks. J Natl Cancer Inst. 2014;106. doi:10.1093/jnci/dju183.
42. Azzam EI, Vadasz JA, Raaphorst GP. Thermal sensitivity and radiosensitization in Chinese hamster V79 cells exposed to 2-aminopurine or 6-thioguanine. Radiat Res. 1991;125:223–6.
43. Herman TS, Cress AE, Sweets C, Gerner EW. Reversal of resistance to methotrexate by hyperthermia in Chinese hamster ovary cells. Cancer Res. 1981;41:3840–3.
44. Haas GP, Klugo RC, Hetzel FW, Barton EE, Cerny JC. The synergistic effect of hyperthermia and chemotherapy on murine transitional cell carcinoma. J Urol. 1984;132:828–33.
45. Teicher BA, Kowal CD, Kennedy KA, Sartorelli AC. Enhancement by hyperthermia of the *in vitro* cytotoxicity of mitomycin C toward hypoxic tumor cells. Cancer Res. 1981;41:1096–9.
46. Wallner KE, Li GC. Effect of drug exposure duration and sequencing on hyperthermic potentiation of mitomycin-C and cisplatin. Cancer Res. 1987;47:493–5.
47. Wallner KE, Banda M, Li GC. Hyperthermic enhancement of cell killing by mitomycin C in mitomycin C-resistant Chinese hamster ovary cells. Cancer Res. 1987;47:1308–12.
48. van der Heijden AG, Jansen CF, Verhaegh G, O'donnell MA, Schalken JA, Witjes JA. The effect of hyperthermia on mitomycin-C induced cytotoxicity in four human bladder cancer cell lines. Eur Urol. 2004;46:670–4.
49. Hazan G, Ben-Hur E, Yerushalmi A. Synergism between hyperthermia and cyclophosphamide *in vivo*: the effect of dose fractionation. Eur J Cancer. 1981;17:681–4.
50. Hazan G, Lurie H, Yerushalmi A. Sensitization of combined cis-platinum and cyclophosphamide by local hyperthermia in mice bearing the Lewis lung carcinoma. Oncology. 1984;41:68–9.
51. Hiramoto RN, Ghanta VK, Lilly MB. Reduction of tumor burden in a murine osteosarcoma following hyperthermia combined with cyclophosphamide. Cancer Res. 1984;44:1405–8.
52. Gerad H, van Echo DA, Whitacre M, Ashman M, Helrich M, Foy J, et al. Doxorubicin, cyclophosphamide, and whole body hyperthermia for treatment of advanced soft tissue sarcoma. Cancer. 1984;53:2585–91.
53. Urano M, Kim MS, Kahn J, Kenton LA, Li ML. Effect of thermochemotherapy (combined cyclophosphamide and hyperthermia) given at various temperatures with or without glucose administration on a murine fibrosarcoma. Cancer Res. 1985;45:4162–6.
54. Wiedemann G, Roszinski S, Biersack A, Weiss C, Wagner T. Local hyperthermia enhances cyclophosphamide, ifosfamide andcisdiamminedichloroplatinum cytotoxicity on human-derived breast carcinoma and sarcoma xenografts in nude mice. J Cancer Res Clin Oncol. 1992;118:129–35.
55. Goss P, Parsons PG. The effect of hyperthermia and melphalan on survival of human fibroblast strains and melanoma cell lines. Cancer Res. 1977;37:152–6.
56. Joiner MC, Steel GG, Stephens TC. Response of two mouse tumours to hyperthermia with CCNU or melphalan. Br J Cancer. 1982;45:17–26.

57. Honess DJ, Bleehen NM. Thermochemotherapy with cis-platinum, CCNU, BCNU, chlorambucil and melphalan on murine marrow and two tumours: therapeutic gain for melphalan only. Br J Radiol. 1985;58:63–72.
58. Bates DA, Mackillop WJ. The effect of hyperthermia in combination with melphalan on drug-sensitive and drug-resistant CHO cells *in vitro*. Br J Cancer. 1990;62:183–8.
59. Zaffaroni N, Villa R, Orlandi L, Vaglini M, Silvestrini R. Effect of hyperthermia on the formation and removal of DNA interstrand cross-links induced by melphalan in primary cultures of human malignant melanoma. Int J Hyperthermia. 1992;8:341–9.
60. Petridou M, Mourelatos D, Tsolaki M, Kazis A, Kourakis A, Routsonis K. Cytogenetic damage by melphalan and hyperthermia in patients with an initial epileptic attack. Mutat Res. 1992;280:143–8.
61. Laskowitz DT, Elion GB, Dewhirst MW, Griffith OW, Savina PM, Blum MR, et al. Hyperthermia-induced enhancement of melphalan activity against a melphalan-resistant human rhabdomyosarcoma xenograft. Radiat Res. 1992;129:218–23.
62. Orlandi L, Zaffaroni N, Bearzatto A, Costa A, Supino R, Vaglini M, et al. Effect of melphalan and hyperthermia on cell cycle progression and cyclin B1 expression in human melanoma cells. Cell Prolif. 1995;28:617–30.
63. Urano M, Ling CC. Thermal enhancement of melphalan and oxaliplatin cytotoxicity *in vitro*. Int J Hyperthermia. 2002;18:307–15.
64. Ko SH, Ueno T, Yoshimoto Y, Yoo JS, Abdel-Wahab OI, Abdel-Wahab Z, et al. Optimizing a novel regional chemotherapeutic agent against melanoma: hyperthermia-induced enhancement of temozolomide cytotoxicity. Clin Cancer Res. 2006;12:289–97.
65. Pagani E, Falcinelli S, Pepponi R, Turriziani M, Caporaso P, Caporali S, et al. Combined effect of temozolomide and hyperthermia on human melanoma cell growth and O^6-methylguanine-DNA methyltransferase activity. Int J Oncol. 2007;30:443–51.
66. Meyn RE, Corry PM, Fletcher SE, Demetriades M. Thermal enhancement of DNA damage in mammalian cells treated with cis-diamminedichloroplatinum (II). Cancer Res. 1980;40:1136–9.
67. Cohen JD, Robins HI. Hyperthermic enhancement of cis-diammine-1,1-cyclobutane dicarboxylate platinum (II) cytotoxicity in human leukemia cells *in vitro*. Cancer Res. 1987;47:4335–7.
68. Eichholtz-Wirth H, Hietel B. Heat sensitization to cisplatin in two cell lines with different drug sensitivities. Int J Hyperthermia. 1990;6:47–55.
69. Eichholtz-Wirth H. Restoration of cisplatin sensitivity by mild hyperthermia in radiation-induced cisplatin-resistant mouse fibrosarcoma cells. Int J Oncol. 1995;7:935–9.
70. Raaphorst GP, Doja S, Davis L, Stewart D, Ng CE. A comparison of hyperthermia cisplatin sensitization in human ovarian carcinoma and glioma cell lines sensitive and resistant to cisplatin treatment. Cancer Chemother Pharmacol. 1996;37:574–80.
71. Rietbroek RC, van de Vaart PJ, Haveman J, Blommaert FA, Geerdink A, Bakker PJ, et al. Hyperthermia enhances the cytotoxicity and platinum-DNA adduct formation of lobaplatin and oxaliplatin in cultured SW 1573 cells. J Cancer Res Clin Oncol. 1997;123:6–12.
72. Raaphorst GP, Yang DP. The evaluation of thermal cisplatin sensitization in normal and XP human cells using mild hyperthermia at 40 and 41 degrees C. Anticancer Res. 2005;25:2649–53.
73. Yano T, Nakatani K, Nakano H, Ihara M, Ohnishi T. Hyperthermic enhancement of cis-diamminedichloroplatinum (II)-induced DNA damage in human esophageal cancer transplantable to nude mice. Anticancer Res. 1993;13:1989–97.
74. Raaphorst GP, Li LF, Yang DP, LeBlanc JM. Cisplatin sensitization by concurrent mild hyperthermia in parental and mutant cell lines deficient in homologous recombination and non-homologous end joining repair. Oncol Rep. 2005;14:281–5.

75. Haveman J, Bergs JW, Franken NA, van Bree C, Stalpers LJ. Effect of hyperthermia on uptake and cytotoxicity of cisplatin in cultured murine mammary carcinoma cells. Oncol Rep. 2005;14:561–7.
76. Bergs JW, Haveman J, Ten Cate R, Medema JP, Franken NA, van Bree C. Effect of 41 °C and 43 °C on cisplatin radiosensitization in two human carcinoma cell lines with different sensitivities for cisplatin. Oncol Rep. 2007;18:219–26.
77. Cohen JD, Robins HI, Javid MJ. Sensitization of C6 glioma to carboplatin cytotoxicity by hyperthermia and thymidine. J Neurooncol. 1990;9:1–8.
78. Ohno S, Siddik ZH, Baba H, Stephens LC, Strebel FR, Wondergem J, et al. Effect of carboplatin combined with whole body hyperthermia on normal tissue and tumor in rats. Cancer Res. 1991;51:2994–3000.
79. Murray TG, Cicciarelli N, McCabe CM, Ksander B, Feuer W, Schiffman J, et al. *In vitro* efficacy of carboplatin and hyperthermia in a murine retinoblastoma cell line. Invest Ophthalmol Vis Sci. 1997;38:2516–22.
80. Westermann AM, Grosen EA, Katschinski DM, Jäger D, Rietbroek R, Schink JC, et al. A pilot study of whole body hyperthermia and carboplatin in platinumresistant ovarian cancer. Eur J Cancer. 2001;37:1111–7.
81. Song X, Kim SY, Lee YJ. The role of Bcl-xL in synergistic induction of apoptosis by mapatumumab and oxaliplatin in combination with hyperthermia on human colon cancer. Mol Cancer Res. 2012;10:1567–79.
82. Ng CE, Bussey AM, Raaphorst GP. Sequence of treatment is important in the modification of camptothecin induced cell killing by hyperthermia. Int J Hyperthermia. 1996;12:663–78.
83. Park HJ, Choi EK, Choi J, Ahn KJ, Kim EJ, Ji IM, et al. Heat-induced up-regulation of NAD (P)H:quinone oxidoreductase potentiates anticancer effects of beta-lapachone. Clin Cancer Res. 2005;11:8866–71.
84. Hori T, Kondo T, Lee H, Song CW, Park HJ. Hyperthermia enhances the effect of β-lapachone to cause γH2AX formations and cell death in human osteosarcoma cells. Int J Hyperthermia. 2011;27:53–62.
85. Hahn GM, Strande DP. Cytotoxic effects of hyperthermia and adriamycin on Chinese hamster cells. J Natl Cancer Inst. 1976;57:1063–7.
86. Ohnoshi T, Ohnuma T, Beranek JT, Holland JF. Combined cytotoxicity effect of hyperthermia and anthracycline antibiotics on human tumor cells. J Natl Cancer Inst. 1985;74:275–81.
87. Supino R, Bardella L, Gibelli N, Cairo G, Schiaffonati L. Interaction of heat with chemotherapy *in vitro*: Effect on cell viability and protein synthesis in human and murine cell lines. Tumori. 1987;73:109–16.
88. Averill DA, Su C. Sensitization to the cytotoxicity of adriamycin by verapamil and heat in multidrug-resistant Chinese hamster ovary cells. Radiat Res. 1999;151:694–702.
89. Hermisson M, Weller M. Hyperthermia enhanced chemosensitivity of human malignant glioma cells. Anticancer Res. 2000;20:1819–23.
90. Blasiak J, Widera K, Pertyński T. Hyperthermia can differentially modulate the repair of doxorubicin-damaged DNA in normal and cancer cells. Acta Biochim Pol. 2003;50:191–5.
91. Lee H, Kim S, Choi BH, Park MT, Lee J, Jeong SY, et al. Hyperthermia improves therapeutic efficacy of doxorubicin carried by mesoporous silica nanocontainers in human lung cancer cells. Int J Hyperthermia. 2011;27:698–707.
92. Kampinga HH. Hyperthermia, thermotolerance and topoisomerase II inhibitors. Br J Cancer. 1995;72:333–8.
93. Liu Y, Prasad R, Beard WA, Kedar PS, Hou EW, Shock DD, et al. Coordination of steps in single-nucleotide base excision repair mediated by apurinic/apyrimidinic endonuclease 1 and DNA polymerase β. J Biol Chem. 2007;282:13532–41.
94. Dianov GL, Hübscher U. Mammalian base excision repair: the forgotten archangel. Nucleic Acids Res. 2013;41:3483–90.
95. Marteijn JA, Lans H, Vermeulen W, Hoeijmakers JH. Understanding nucleotide excision repair and its roles in cancer and ageing. Nat Rev Mol Cell Biol. 2014;15:465–81.

96. Sancar A, Lindsey-Boltz LA, Unsal-Kacmaz K, Linn S. Molecular mechanisms of mammalian DNA repair and the DNA damage checkpoints. Ann Rev Biochem. 2004;73:39–85.
97. Martin SA, Lord CJ, Ashworth A. Therapeutic targeting of the DNA mismatch repair pathway. Clin Cancer Res. 2010;16:5107–13.
98. Fortini P, Dogliotti E. Base damage and single-strand break repair: mechanisms and functional significance of short- and long-patch repair subpathways. DNA Repair. 2007;6:398–409.
99. Ohnishi T, Mori E, Takahashi A. DNA double-strand breaks: their production, recognition, and repair in eukaryotes. Mutat Res. 2009;669:8–12.
100. Falk M, Lukasova E, Kozubek S. Higher-order chromatin structure in DSB induction, repair and misrepair. Mutat Res. 2010;704:88–100.
101. Chapman JR, Taylor MR, Boulton SJ. Playing the end game: DNA double-strand break repair pathway choice. Mol Cell. 2012;47:497–510.
102. Wang C, Lees-Miller SP. Detection and repair of ionizing radiation-induced DNA double strand breaks: new developments in nonhomologous end joining. Int J Radiat Oncol Biol Phys. 2013;86:440–9.
103. Jeggo PA, Geuting V, Löbrich M. The role of homologous recombination in radiation-induced double-strand break repair. Radiother Oncol. 2011;101:7–12.
104. Hall EJ, Giaccia AJ. Radiobiology for the radiologist. 7th ed. Philadelphia: Lippincott Williams & Wilkins; 2012.
105. Wouters BG, Begg AC. Irradiation-induced damage and the DNA damage response. In: Joiner M, van der Kogel A, editors. Basic clinical radiobiology. 4th ed. London: Hodder Arnold Publication; 2009. p. 11–26.
106. Bétermier M, Bertrand P, Lopez BS. Is non-homologous end-joining really an inherently error-prone process? PLoS Genet. 2014;10:e1004086. doi:10.1371/journal.pgen.1004086.
107. Ceccaldi R, Rondinelli B, D'Andrea AD. Repair pathway choices and consequences at the double-strand break. Trends Cell Biol. 2015. doi:10.1016/j.tcb.2015.07.009.
108. Iliakis G, Seaner R. A DNA double-strand break repair-deficient mutant of CHO cells shows reduced radiosensitization after exposure to hyperthermic temperatures in the plateau phase of growth. Int J Hyperthermia. 1990;6:801–12.
109. Woudstra EC, Konings AW, Jeggo PA, Kampinga HH. Role of DNA-PK subunits in radiosensitization by hyperthermia. Radiat Res. 1999;152:214–8.
110. Dynlacht JR, Bittner ME, Bethel JA, Beck BD. The non-homologous end-joining pathway is not involved in the radiosensitization of mammalian cells by heat shock. J Cell Physiol. 2003;196:557–64.
111. Raaphorst GP, Maude-Leblanc J, Li L. Evaluation of recombination repair pathways in thermal radiosensitization. Radiat Res. 2004;161:215–8.
112. Yin HL, Suzuki Y, Matsumoto Y, Tomita M, Furusawa Y, Enomoto A, et al. Radiosensitization by hyperthermia in the chicken B-lymphocyte cell line DT40 and its derivatives lacking nonhomologous end joining and/or homologous recombination pathways of DNA double-strand break repair. Radiat Res. 2004;162:433–41.
113. Raaphorst GP, Thakar M, Ng CE. Thermal radiosensitisation in two pairs of CHO wild-type and radiation-sensitive mutant cell lines. Int J Hyperthermia. 1993;9:383–91.
114. Genet SC, Fujii Y, Maeda J, Kaneko M, Genet MD, Miyagawa K, et al. Hyperthermia inhibits homologous recombination repair and sensitizes cells to ionizing radiation in a time- and temperature-dependent manner. J Cell Physiol. 2013;228:1473–81.
115. Komatsu K, Kubota N, Gallo M, Okumura Y, Lieber MR. The scid factor on human chromosome 8 restores V(D)J recombination in addition to double-strand break repair. Cancer Res. 1995;55:1774–9.
116. O'Hara MD, Pollard MD, Wheatley G, Regine WF, Mohiuddin M, Leeper DB. Thermal response and hyperthermic radiosensitization of scid mouse bone marrow CFU-C. Int J Radiat Oncol Biol Phys. 1995;31:905–10.

117. Wachters FM, van Putten JWG, Maring JG, Zdzienicka MZ, Groen HJM, Kampinga HH. Selective targeting of homologous DNA recombination repair by Gemcitabine. Int J Radiat Biol Oncol Phys. 2003;57:553–62.
118. Krawczyk PM, Eppink B, Essers J, Stap J, Rodermond H, Odijk H, et al. Mild hyperthermia inhibits homologous recombination, induces BRCA2 degradation, and sensitizes cancer cells to poly (ADP-ribose) polymerase-1 inhibition. Proc Natl Acad Sci U S A. 2011;108:9851–6.
119. Miyakoda M, Suzuki K, Kodama S, Watanabe M. Activation of ATM and phosphorylation of p53 by heat shock. Oncogene. 2002;21:1090–6.
120. Hunt CR, Pandita RK, Laszlo A, Higashikubo R, Agarwal M, Kitamura T, et al. Hyperthermia activates a subset of ataxia-telangiectasia mutated effectors independent of DNA strand breaks and heat shock protein 70 status. Cancer Res. 2007;67:3010–7.
121. Takahashi A, Mori E, Su X, Nakagawa Y, Okamoto N, Uemura H, et al. ATM is the predominant kinase involved in the phosphorylation of histone H2AX after heating. J Radiat Res. 2010;51:417–22.
122. Furusawa Y, Iizumi T, Fujiwara Y, Zhao QL, Tabuchi Y, Nomura T, et al. Inhibition of checkpoint kinase 1 abrogates G_2/M checkpoint activation and promotes apoptosis under heat stress. Apoptosis. 2012;17:102–12.
123. Tuul M, Kitao H, Iimori M, Matsuoka K, Kiyonari S, Saeki H, et al. Rad9, Rad17, TopBP1 and claspin play essential roles in heat-induced activation of ATR kinase and heat tolerance. PLoS One. 2013;8:e55361. doi:10.1371/journal.pone.0055361.
124. Takahashi A, Matsumoto H, Nagayama K, Kitano M, Hirose S, Tanaka H, et al. Evidence for the involvement of double-strand breaks in heat-induced cell killing. Cancer Res. 2004;64:8839–45.
125. Kaneko H, Igarashi K, Kataoka K, Miura M. Heat shock induces phosphorylation of histone H2AX in mammalian cells. Biochem Biophys Res Commun. 2005;328:1101–6.
126. Takahashi A, Mori E, Somakos GI, Ohnishi K, Ohnishi T. Heat induces γH2AX foci formation in mammalian cells. Mutat Res. 2008;656:88–92.
127. Laszlo A, Fleischer I. The heat-induced gamma-H2AX response does not play a role in hyperthermic cell killing. Int J Hyperthermia. 2009;25:199–209.
128. Laszlo A, Fleischer I. Heat-induced perturbations of DNA damage signaling pathways are modulated by molecular chaperones. Cancer Res. 2009;69:2042–9.
129. Takahashi A, Mori E, Ohnishi T. The foci of DNA double strand break-recognition proteins localize with γH2AX after heat treatment. J Radiat Res. 2010;51:91–5.
130. Zhu WG, Seno JD, Beck BD, Dynlacht JR. Translocation of MRE11 from the nucleus to the cytoplasm as a mechanism of radiosensitization by heat. Radiat Res. 2001;156:95–102.
131. Seno JD, Dynlacht JR. Intracellular redistribution and modification of proteins of the Mre11/Rad50/Nbs1 DNA repair complex following irradiation and heat-shock. J Cell Physiol. 2004;199:157–70.
132. Xu M, Myerson RJ, Xia Y, Whitehead T, Moros EG, Straube WL, et al. The effects of 41 degrees C hyperthermia on the DNA repair protein, MRE11, correlate with radiosensitization in four human tumor cell lines. Int J Hyperthermia. 2007;23:343–51.
133. Takahashi A, Mori E, Ohnishi T. Phospho-Nbs1 and Mre11 proteins which recognize DSBs co-localize with γH2AX in the nucleus after heat treatment. Ann Cancer Res Therapy. 2007;15:50–3.
134. Gerashchenko BI, Gooding G, Dynlacht JR. Hyperthermia alters the interaction of proteins of the Mre11 complex in irradiated cells. Cytometry A. 2010;77:940–52.
135. Ohnishi K, Scuric Z, Yau D, Schiestl RH, Okamoto N, Takahashi A, et al. Heat-induced phosphorylation of NBS1 in human skin fibroblast cells. J Cell Biochem. 2006;99:1642–50.
136. Xian Ma Y, Fan S, Xiong J, Yuan RQ, Meng Q, Gao M, et al. Role of BRCA1 in heat shock response. Oncogene. 2003;22:10–27.
137. Wang Y, Guan J, Wang H, Wang Y, Leeper D, Iliakis G. Regulation of DNA replication after heat shock by replication protein a-nucleolin interactions. J Biol Chem. 2001;276:20579–88.

138. Burgman P, Ouyang H, Peterson S, Chen DJ, Li GC. Heat inactivation of Ku autoantigen: possible role in hyperthermic radiosensitization. Cancer Res. 1997;57:2847–50.
139. Matsumoto Y, Suzuki N, Sakai K, Morimatsu A, Hirano K, Murofushi H. A possible mechanism for hyperthermic radiosensitization mediated through hyperthermic lability of Ku subunits in DNA-dependent protein kinase. Biochem Biophys Res Commun. 1997;234:568–72.
140. Ihara M, Suwa A, Komatsu K, Shimasaki T, Okaichi K, Hendrickson EA, et al. Heat sensitivity of double-stranded DNA-dependent protein kinase (DNA-PK) activity. Int J Radiat Biol. 1999;75:253–8.
141. Beck BD, Dynlacht JR. Heat-induced aggregation of XRCC5 (Ku80) in nontolerant and thermotolerant cells. Radiat Res. 2001;156:767–74.
142. Ihara M, Takeshita S, Okaichi K, Okumura Y, Ohnishi T. Heat exposure enhances radiosensitivity by depressing DNA-PK kinase activity during double strand break repair. Int J Hyperthermia. 2014;30:102–9.
143. Zhu H, Guo FJ, Zhao W, Zhou J, Liu Y, Song F, et al. ATF4 and IRE1α inhibit DNA repair protein DNA-dependent protein kinase 1 induced by heat shock. Mol Cell Biochem. 2012;371:225–32.
144. Vyas S, Chang P. Dual roles for PARP1 during heat shock: transcriptional activator and posttranscriptional inhibitor of gene expression. Mol Cell. 2013;49:1–3.
145. Kenny MK, Mendez F, Sandigursky M, Kureekattil RP, Goldman JD, Franklin WA, et al. Heat shock protein 70 binds to human apurinic/apyrimidinic endonuclease and stimulates endonuclease activity at abasic sites. J Biol Chem. 2001;276:9532–6.
146. Spiro IJ, Denman DL, Dewey WC. Effect of hyperthermia on CHO DNA polymerases alpha and beta. Radiat Res. 1982;89:134–49.
147. Jorritsma JB, Kampinga HH, Scaf AH, Konings AW. Strand break repair, DNA polymerase activity and heat radiosensitization in thermotolerant cells. Int J Hyperthermia. 1985;1:131–45.
148. Mivechi NF, Miyachi H, Scanlon KJ. Heat radiosensitization and the level of DNA polymerases alpha and beta of human colony-forming unit-granulocyte-macrophage and myeloid leukemias sensitive and resistant to chemotherapeutic agents. Cancer Res. 1990;50:2044–8.
149. Fantini D, Moritz E, Auvré F, Amouroux R, Campalans A, Epe B, et al. Rapid inactivation and proteasome-mediated degradation of OGG1 contribute to the synergistic effect of hyperthermia on genotoxic treatments. DNA Repair. 2013;12:227–37.
150. Raaphorst GP, Azzam EI. The thermal sensitivity of normal and ataxia telangiectasia human fibroblasts. Int J Radiat Oncol Biol Phys. 1982;8:1947–50.
151. Okamoto N, Takahashi A, Ota I, Ohnishi K, Mori E, Kondo N, et al. siRNA targeted for NBS1 enhances heat sensitivity in human anaplastic thyroid carcinoma cells. Int J Hyperthermia. 2011;27:297–304.
152. Ohnishi T, Komatsu K, Tauchi H, Wang X, Takahashi A, Ohnishi K, et al. Brief communication: heat-induced accumulation of p53 and hsp72 is suppressed in lung fibroblasts from the SCID mouse. Int J Radiat Biol. 2000;76:711–5.
153. Takahashi A, Yamakawa N, Mori E, Ohnishi K, Yokota S, Sugo N, et al. Development of thermotolerance requires interaction between polymerase-beta and heat shock proteins. Cancer Sci. 2008;99:973–8.

Chapter 10
Enhancement of Hyperthermia on Anti-tumor Drug Sensitivity

Takeo Ohnishi

Abstract Anti-tumor drugs have been utilized in combination with hyperthermia, synergistic effects have been observed, and efforts have been made to explain the basis for these observations. It is thought that the initial extent of DNA damage is enhanced by hyperthermia effects, and this outcome is associated with tissue blood flow. If this is true, then an associated effect would be an additional increase in DNA damage induced by anti-tumor drugs. Another hypothesized mechanism is that DNA repair enzyme activity required for repair of DNA damage is depressed by a heat exposure. A third hypothesis is that more efficient cell killing is induced by the induction of cell death signaling pathways and the depression of cell survival signaling pathways through hyperthermia (Fig. 10.1). In this chapter, the main focus is on the increase in activity of anti-tumor drugs induced by hyperthermia. DNA repair inhibition induced by hyperthermia has been described in other chapters.

Keywords Anti-tumor drug • Nanotechnology • Liposome • Nanomedicine

10.1 Increase of DNA Damage by Hyperthermia

Blood vessels in tumors are not as well developed as those in normal tissue. Since blood flow in tumors is not the same or as effective as in normal tissue, specific local tumor environments can exhibit low pH, hypoxia and low nutrient levels. When heat treatments are applied to tumors, however, blood flow is promoted, even in the tumors. In the presence of injected anti-tumor drugs, such drugs may be more efficiently incorporated into cancer cells. When attempting to achieve highly effective results with anti-tumor drugs, it is important to focus on the heating temperature and heating period. In addition, specific heating techniques are required for specific local applications. Physiological measurements are also needed to estimate local temperatures. It is expected that the number of DNA

T. Ohnishi (✉)
Department of Radiation Oncology, Nara Medical University, 840 Shijo-cho, Kashihara, Nara 634-8521, Japan
e-mail: tohnishi@ares.eonet.ne.jp

S. Kokura et al. (eds.), *Hyperthermic Oncology from Bench to Bedside*,
DOI 10.1007/978-981-10-0719-4_10

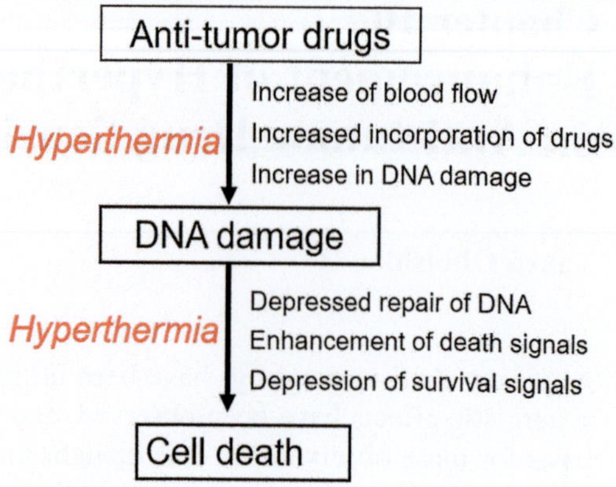

Fig. 10.1 Possible mechanism of hyperthermic enhancement of anti-tumor drugs [5]

lesions will increase after exposure to anti-tumor drugs. In the case of the transplanted tumors *in vivo*, basic experimental work has shown that there is a delay in tumor growth. Anti-tumor drugs which have applied in conjunction with hyperthermia include bleomycin, bis-chloroethylnitrosourea (BCNU), amphotericin, Adriamycin, 1,3-bis(2-chloroethyl)-1-nitrosourea, *cis*-diamminedichloroplatinum (CDDP), mitomycin C (MMC), 5-fluorouracil (5-FU), fluorescein, docetaxel (DTX), near-infrared dye-12 (NIRD-12), S-(2-aminoethyl) isothiouronium bromide, and others [1]. Among these, CDDP [2, 3] and DTX have been frequently studied. In particular, the cytotoxic effects of CDDP were shown to be strongly stimulated by hyperthermia. The molecular mechanisms leading to this hyperthermia effect have been investigated using alkaline elution or labeled platinum in CDDP to measure the degree of DNA-DNA crosslinking. Hyperthermia enhances drug uptake into cancer cells and leads to an increase in the number DNA-DNA cross-links. Tumor growth in the transplant-nude mice system was apparently depressed through synergistic effects [3, 4]. The intra-tumor concentration of CDDP in heated tumors derived from a transplantable human esophageal cancer (ESO-2) was 1.4 times higher than in unheated tumors, but the amount of DNA-bound CDDP in the heated tumor was 2.2 times higher than in the non-heated controls. Therefore, the enhancement of CDDP effects in tumors after a hyperthermia treatment follows the increase in the initial concentration of CDDP present in the tumor [4].

Newly developed applications will be described below which utilize nanotechnology to allow drugs to work more effectively. Progress and new developments in utilizing anti-cancer drugs is still required in order to improve the effectiveness of combined therapies using hyperthermia and chemotherapy in the clinic.

10.2 Progress of Nanomedicine for Anti-tumor Drugs

To improve the quality and effectiveness of anti-tumor drugs, advanced nanotechnology materials are being investigated. Such materials include bioconjugated nanoparticles and liposomes to be used in the delivery of anti-cancer therapeutic drugs (Fig. 10.2). An excellent review by Xu et al describes recent progress in nanomedicine techniques [5]. Since the late 1970s, this technology has been advancing with the goal of delivering drugs more precisely to tumor cells, and in maintaining the drugs at therapeutic concentrations over a long period. Recently, more advanced nanomedicine techniques have been developed using many types of organic and inorganic materials, and these materials are being designed to allow simple modification of targeting molecules, drugs, and other molecules, and to

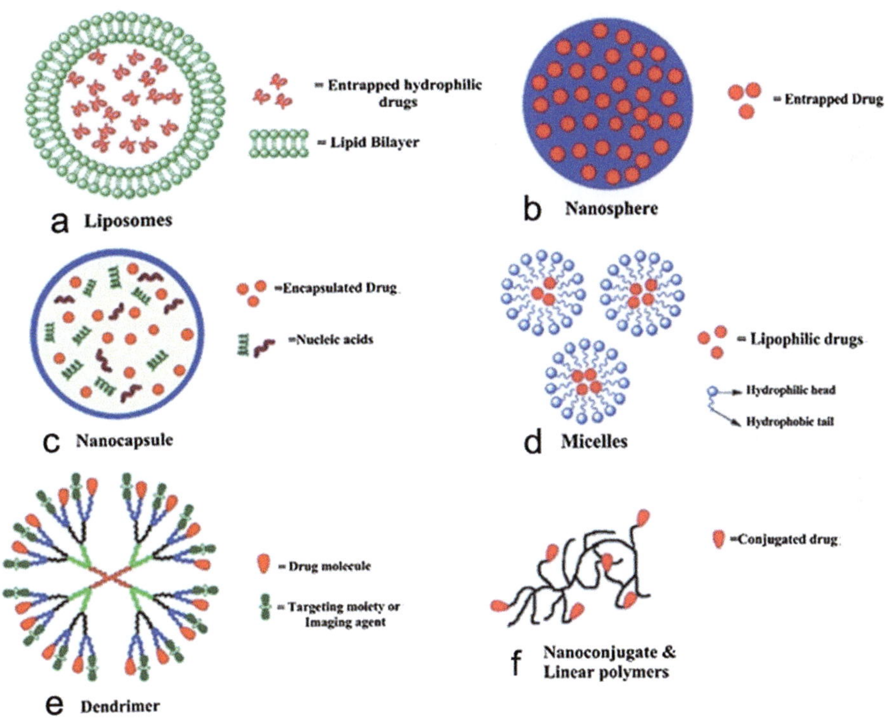

Fig. 10.2 A schematic representation of some nanoparticle platforms used as ocular drug delivery systems. (**a**) Liposomes (~100–400 nm) are small spherical artificial vesicles typically made with lipid bilayers. (**b**) Nanoshere (~20–200 nm) are typically made with biodegradable polymers for sustained drug release. (**c**) Nanocapsules (~10–1000 nm) can encapsulate relatively large amounts of drugs and nucleic acids such as DNA, microRNA, siRNA and shRNA. (**d**) Micelles (~10–100 nm) are self-assembled amphiphilic particles that can encapsulate both lipophilic or lipophobic drugs stabilized by surfactants. (**e**) Dendrimers (~3–20 nm) are monodisperse macromolecules that can be used to encapsulate or covalently conjugate drugs, targeting moieties & imaging drugs. (**f**) Nanoconjugates are polymers to which drug molecules are covalently conjugated

permit an effective delivery to target sites. These developments should result in a high therapeutic efficacy and provide good control of drug release using external and internal stimuli. Some of these nanoparticles and liposomes are shown in Fig. 10.2. Drug delivery at the target site is achieved by using macromolecular soft matrixes or inorganic solids which can incorporate the drugs, and also colloidal particles with a size range of 10–1000 nm, which is similar to that of biological macromolecules such as proteins and DNA. These nanocarriers can be precisely fabricated with desired sizes, shapes, surface charge density, stability and various other characteristics for use with *in vivo* applications. To further enhance their functionality, additional techniques such as surface modifications along with the introduction of a targeting moiety are included in these approaches. For effective anti-tumor treatment, the drugs must penetrate the tumor tissue efficiently and studies have suggested strategies to improve drug penetration into tumor tissues in order to enhance therapeutic effectiveness through the use of several delivery systems with differing abilities to penetrate tissue. In addition, an understanding of the physiological aspects of cancer cells provides aid in designing these new agents and materials. For example, the vascular architecture around cancer cells is poorly organized with a reduced vascular density, an irregular blood flow and a compression of blood and lymphatic vessels by cancer cells. In addition, numerous biomolecular markers have been identified which are specifically expressed in cancer cells, which can help in designing targeted cancer therapies. In order to transport anti-tumor drugs into cancer cells, immunological characteristics of cancer cells have been studied. With these developments, clinical approaches using nanotechnology have shown that drug delivery systems can exhibit an enhanced efficacy while simultaneously reducing side effects. Such improvements are derived from improved methods to target delivery to tumors and to utilize active cellular uptake. New nanotechnology developments will include improvements in the preparation of drug nanocarriers and strategies to increase their accumulation in cancer cells.

10.3 Combination with Nanomedicine and Hyperthermia

Thermally responsive poly(*N*-isopropylacrylamide-co-acrylamide) (P(NIPA-co-AAm)nanohydrogel (NHG) with a particle diameter of about 50 nm and a low critical solution temperature (LCST) of about 40 °C was synthesized using precipitation polymerization methods. Four hydrophobic model drugs (5-FU, fluorescein, DTX and NIRD-12) with different hydrophilicities were embedded in the nanoparticles. The anti-tumor efficacy of DTX-loaded NHG was enhanced with local hyperthermia treatments to tumor tissue in transplanted mice, and DTX-loaded NHG showed low toxicity during the therapeutic procedure [6]. The superparamagnetic iron oxide nanoparticles were electrostatically activated with either nucant multivalent pseudo-peptide or DOX [7]. Cytotoxic potentials were assessed in female athymic nude mice bearing breast adenocarcinomas. Nanoparticles have been used as drug carriers because some formulations can reduce unwanted side effects. Recently, stimuli-responsive, nanomaterials which can be activated are being tested, and these are capable of undergoing chemical or physical changes after exposure to external stimuli [8].

Among these nanoparticles, thermo-responsive bubble-generating liposomal system is largely expected for cancer therapy with hyperthermia (Fig. 10.2a). Liposomes are small bilayer vesicles enclosing aqueous compartment that can be produced using amphiphilic phospholipids and cholesterol. Liposomes are spherical, vary in size most are 400 nm or less. Liposomes are effective platforms for drug/diagnostic delivery owing to their size, hydrophobic and hydrophilic character, biocompatibility, biodegradability, low toxicity and immunogenicity. Commonly, liposomes with smaller diameter with neutral or negative charge are used for theranostic purpose. Liposomes has an affinity to bind to, ocular surfaces, and release contents at optimal rates. Carbopol1342-coated pilocarpine containing liposomes were shown to produce a longer duration of action. Ciprofloxacin (CPFX) was also formulated in liposomal environmental which lowered tear-driven dilution in the conjunctival sac. Multilamellar vesicles from lecithin and alpha-L-dipalmithoyl-phosphatidylcholine were used to prepare liposome containing CPFX. This approach produced sustained release of the drug depending on the nature of the lipid composition selected [9].

To examine thermo-stability or thermo-instability during hyperthermic therapy, the properties of liposomes were investigated at different temperatures. Tumor drug concentrations, drug distribution, and the therapeutic efficacies achieved by three fundamentally different liposomes were studied [10]. These included non-thermo-sensitive liposomes (NTSL), traditional thermosensitive liposomes (TTSL), low temperature sensitive liposomes (LTSL) were examined along with free DOX, and saline in combination with hyperthermia (HT). These agents were directly compared in a human tumor xenograft model. NTSL is a non-thermo-sensitive liposome in the physiological temperature range; TTSL is triggered in the range of ~42–45 °C and releases drugs over approximately 30 min while LTSL triggers in the range of 39–40 °C and releases drugs in a matter of seconds. Hyperthermia triggered liposomes can provide effective cytotoxicity, drug interactions and drug release. In athymic nude mice bearing the FaDu human tumor xenograft, a single *i. v.* injection was given with a dose of 5 mg/kg of DOX (free drug or liposome encapsulated), and the tumors were then heated to either 34 °C or 42 °C for 1 h. All treatment groups were similar, resulting in low concentrations of DOX (0–4.5 ng/mg). The LTSL (25.6 ng/mg) achieved the highest DOX concentration with hyperthermia at 42 °C. LTSL and hyperthermia significantly increased the amount of DNA-bound DOX. Tumor tissue sections were visualized for DOX fluorescence intensity. There was relatively little fluorescence seen in the treatment groups at 34 °C. Maximum DOX fluorescence was seen at 42 °C with LTSL. At 42 °C, all three liposomal formulations were more effective than DOX alone. LTSL plus hyperthermia resulted in the longest growth time and the largest number of local controls. With hyperthermia, the DOX concentrations and fluorescence were tightly correlated with tumor growth delay, indicating that improved drug delivery could lead to improved therapeutic effects. Thus, hyperthermia-triggered liposomal drug release appeared to account for the largest differential therapeutic effect observed, and demonstrated the importance of rapid drug release from the drug carriers in the tumor.

Thermo-responsive bubble-generating liposomes are being investigated for tumor-specific chemotherapy [11, 12]. The key component in this liposomal formulation is its encapsulated ammonium bicarbonate (ABC) which is used to actively load DOX into liposomes and trigger drug release when heated locally. Incubating ABC liposomes with whole blood results in a significantly smaller decrease in retained encapsulated DOX than observed with lysolipid liposomes, indicating a superior plasma stability. Biodistribution analysis indicated that the ABC formulation circulated longer than its lysolipid counterpart. Following the injection of ABC liposome suspensions into mice with locally heated tumors, the decomposition of the ABC encapsulated in the liposomes facilitated the immediate thermal activation and CO_2 bubble generation, subsequently increasing intratumoral DOX accumulations. The antitumor efficacy of ABC liposomes demonstrated that this thermo-responsive bubble-generating liposomal system is a highly promising carrier for tumor-specific chemotherapy, and especially for local drug delivery mediated by hyperthermia temperatures.

References

1. Hahn GM. Potential for therapy of drugs and hyperthermia. Cancer Res. 1979;39:2264–8.
2. Ning S, Hahn GM. Combination therapy: lonidamine, hyperthermia, and chemotherapy against the RIF-1 Tumor *in vivo*. Cancer Res. 1991;51:5910–14.
3. Meyn RE, et al. Thermal enhancement of DNA damage in mammalian cells treated with *cis*-diamminedichloroplatinum(II). Cancer Res. 1980;40:1136–9.
4. Yano T, et al. Hyperthermic enhancement of *cis*-diamminedichloroplatinum (II)-induced DNA damage in human esophageal cancer transplantable to nude mice. Anticancer Res. 1993;13:1989–97.
5. Xu Q, et al. Nanotechnology approaches for ocular drug delivery. Middle East Afr J Ophthalmol. 2013;20:26–37.
6. Zhang J, et al. *In vivo* anti-tumor efficacy of docetaxel-loaded thermally responsive nanohydrogel. Nanotechnology. 2009;20:325102.
7. Kossatz S, et al. Efficient treatment of breast cancer xenografts with multi-functionalized iron oxide nanoparticles combining magnetic hyperthermia and anti-cancer drug delivery. Breast Cancer Res. 2015;17:66.
8. Mura S, et al. Stimuli-responsive nanocarriers for drug delivery. Nat Mater. 2013;12:991–1003.
9. Puri A, et al. Lipid-based nanoparticles as pharmaceutical drug carriers: from concepts to clinic. Crit Rev Ther Drug Carrier Syst. 2009;26:523–80.
10. Kong G, et al. Efficacy of liposomes and hyperthermia in a human tumor xenograft model: importance of triggered drug release. Cancer Res. 2000;60:6950–7.
11. Chen KJ, et al. Hyperthermia-mediated local drug delivery by a bubble-generating liposomal system for tumor-specific chemotherapy. ACS Nano. 2014;8:5105–15.
12. Ponce AM, et al. Hyperthermia mediated liposomal drug delivery. Int J Hyperthermia. 2006;222:205–13.

Chapter 11
Inhibition of Epithelial-to-Mesenchymal Transition (EMT) by Hyperthermia

Satoshi Kokura

Abstract Hyperthermia therapy suppresses tumor growth through deliberate heating. To further enhance the antitumor effect, it is often combined with radiation or chemotherapy. In addition, hyperthermia may reduce the metastatic potential of cancer cells and inhibit tumor metastasis. However, the underlying mechanisms through which hyperthermia inhibits cancer metastasis have yet to be fully elucidated. Epithelial-to-mesenchymal transition (EMT) plays a key role in tumor metastasis; therefore, the aim of this chapter is to summarize the effects and underlying mechanisms of EMT in cancer cells. The results suggest that hyperthermia not only inhibits tumor growth, but also mediates the expression of EMT-related genes—including E-cadherin and vimentin—and a number of transcription factors important for cancer cell motility, invasiveness, and metastasis during tumorigenesis. Furthermore, hyperthermia may also reverse alterations in cell morphology induced by transforming growth factor-β (TGF-β) and/or hypoxia, as well as the increased cell invasiveness characteristic of EMT. Taken together, these data indicate that hyperthermia suppresses cancer migration and invasion through the inhibition of EMT.

Keywords EMT • E-cadherin • Vimentin • TGF-β

11.1 Hyperthermia and Metastases

Hyperthermia is a therapeutic method that suppresses tumor growth through the deliberate heating of a primary tumor. The antitumor effect of hyperthermia is enhanced when used in combination with radiation or chemotherapy. Moreover, a previous study in an animal model demonstrated that local hyperthermia can inhibit tumor metastasis [1], and a clinical trial demonstrated that local hyperthermia could be effective for treating cervical lymph node metastasis in oral cancer [2]. In addition, in our previous study, we reported the potential of hyperthermia for

S. Kokura (✉)
Kokura Lab. Faculty of Health and Medical Sciences, Kyoto Gakuen University, 18, Gotanda-cho, Yamanouchi, Ukyo-ku, Kyoto-shi, Kyoto 615-8577, Japan
e-mail: kokura@kyotogakuen.ac.jp

S. Kokura et al. (eds.), *Hyperthermic Oncology from Bench to Bedside*,
DOI 10.1007/978-981-10-0719-4_11

overcoming gemcitabine resistance in pancreatic cancer [3–5]. The results from these previous reports suggest that hyperthermia may inhibit cell metastasis and thereby improve the prognosis of pancreatic cancer [3–5].

11.2 EMT

EMT plays a key role in a process vital for morphogenesis during embryonic development, and is one of the major molecular mechanisms through which invasion and metastasis are promoted during the oncogenic process [6, 7]. The EMT phenotype is characterized by (1) the loss of cell-to-cell adhesion with the disintegration of tight, adherens, and gap junctions, and (2) a phenotypic change in which cells shift from an epithelial morphology to an elongated fibroblast-like morphology that is reportedly associated with increased motility and tumor invasion [8]. The EMT process involves the upregulation of mesenchymal markers such as vimentin, N-cadherin, and fibronectin, and the downregulation of epithelial adhesion molecules such as E-cadherin and cytokeratins [9, 10]. EMT is triggered by the interplay of extracellular signals-such as collagen-and many secrete soluble factors such as Wnt, transforming growth factor-β (TGF-β), fibroblast growth factor, epidermal growth factor, hepatocyte growth factor, and platelet-derived growth factor. Numerous transcription factors-including zinc finger homologs (*e.g.*, Snail), Hedgehog, Notch, basic helix-loop-helix transcription factors (*e.g.*, Twist), ZEB family members (ZEB1, ZEB2/SIPI), Smad2 and NF-κB signaling pathways—have been found to play important roles in EMT induction [9–12].

Coupled with its involvement in cancer metastasis, emerging lines of evidence also suggest the presence of a molecular link between the EMT phenotype and chemo- or radio-resistance [13–15], as well as a link between EMT and drug resistance in pancreatic cancer cells [16, 17]. These finding suggest that controlling EMT may inhibit cancer metastasis and help cancer patients overcome chemo- and radio-resistance, and is therefore essential for improving clinical outcomes.

11.3 EMT Inducers

In what is considered to be a unique characteristic of cancer, a diverse array of extracellular or intracellular stimuli in cancer tissue may lead to EMT. For example, TGF-β is widely known to induce EMT [17]. TGF-β is mainly produced by cancer cells and fibroblasts within cancer tissue. Moreover, angiogenesis in cancer tissue is necessary for tumor growth. However, the growth rate of the tumor exceeds the vascularization rate, which leads to hypoxia, which in turn induces EMT [18–20]. Reactive oxygen species (ROS) are also thought to participate in the induction of EMT as a result of this hypoxic condition [18]. These EMT inducers activate various transcription factors and cause EMT in cancer cells, resulting in chemoresistance and the induction of cancer metastasis [21, 22]. Therefore, the clinical application of pharmacological or other methods such as hyperthermia is critically important to inhibit and/or to prevent EMT and thereby chemoresistance and the induction of cancer metastasis.

11.4 Hyperthermia Inhibits TGF-β-Induced EMT in Pancreatic Cancer Cell Lines

Because hyperthermia has been shown to prevent cancer metastasis [1, 2], we hypothesized that the principal mechanism through which hyperthermia deters the progression of pancreatic cancer is the inhibition of EMT. In a study by Reiko K. et al. [23], hyperthermia was conducted at 40 °C for 60 min prior to TGF-β-induced EMT. After treatment with TGF-β1, the morphology in pancreatic cancer cells changed from a typical epithelium to a mesenchymal spindle shape. Hyperthermia for 1 h prior to TGF-β1 exposure inhibited morphologic changes consistent with EMT (Fig. 11.1). They also investigated E-cadherin and vimentin expression by immunocytochemistry, detecting E-cadherin expression in the BxPC-3 pancreatic cell line. After exposure to TGF-β1 for 48 h, E-cadherin expression in BxPC-3 cells was decreased, showing a loss of cell-to-cell adhesion with the disintegration of adherens by E-cadherin; however, this phenomenon was reversed when pre-treated with hyperthermia (Fig. 11.2). To further confirm this observation, E-cadherin protein expression levels were assessed by western blotting (Fig. 11.3). E-cadherin expression slightly decreased after exposure to TGF-β1 and markedly increased after hyperthermia.

The upregulation of mesenchymal markers such as vimentin is also associated with the induction of EMT. As shown in Fig. 11.4, vimentin expression was

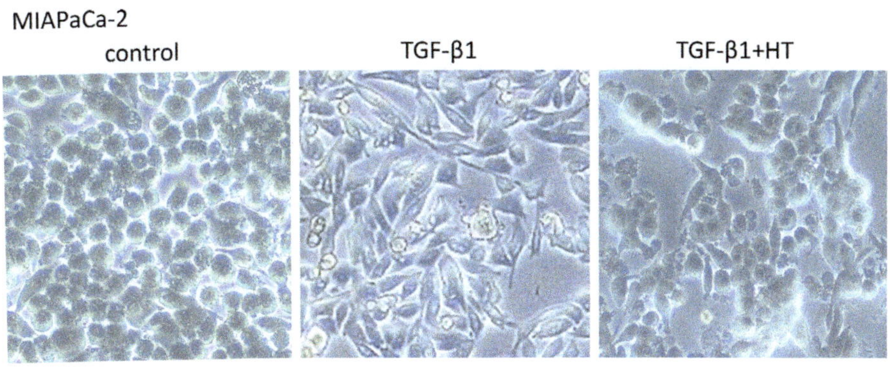

Fig. 11.1 Morphological changes in MIAPaCa-2 cells after treatment with transforming growth factor-β1 (TGF-β1) and hyperthermia. Control, untreated cells; TGF-β1, cells were treated with TGF-β1 (10 ng/mL) for 48 h; TGF-β1+HT (hyperthermia pre-treatment), cells were treated at 43 °C for 1 h and then stimulated with TGF-β1 [23]

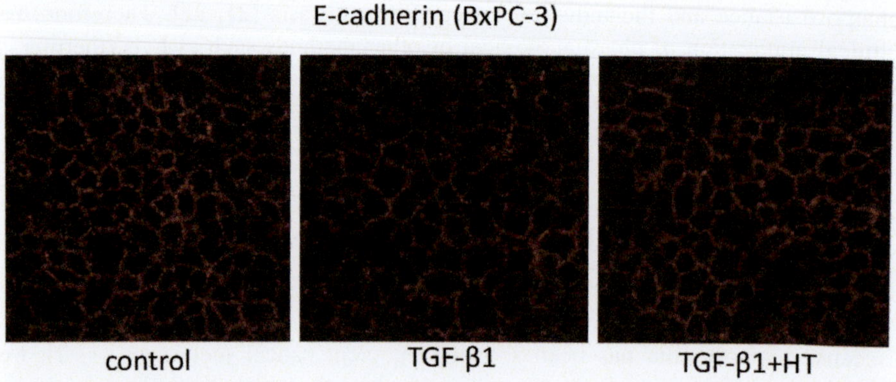

Fig. 11.2 Immunofluorescence staining of E-cadherin in BxPC-3 cells. Although exposure to TGF-β1 (10 ng/mL) for 48 h suppressed E-cadherin expression in BxPC-3 cells, this was reversed for 1 h of hyperthermia pre-treatment. HT, hyperthermia pretreatment [23]

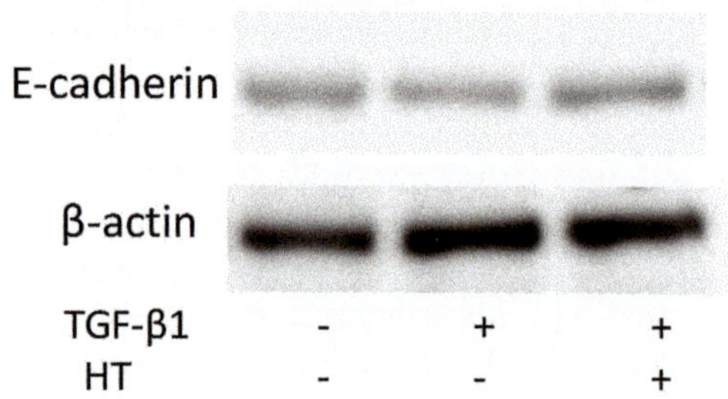

Fig. 11.3 Western blotting analysis of E-cadherin in BxPC-3 cells [23]

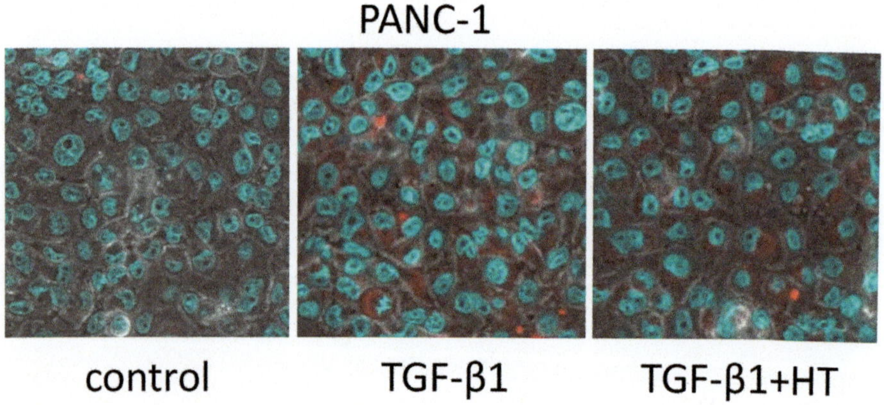

Fig. 11.4 Immunofluorescence staining for Vimentin (*red*) and nucleus (*green*) in PANC-1 cancer cells [23]

observed in BxPC-3 cells that acquired a mesenchymal spindle-shape. Hyperthermia reduced the expression of vimentin induced by TGF-β1 (Fig. 11.4). The mRNA levels of vimentin of three pancreatic cancer cell lines-PANC-1, MIAPaCa-2, and BxPC-3-were assessed by quantitative reverse transcription-polymerase chain reaction. In BxPC-3 and MIAPaCa-2 cells, although vimentin expression was significantly upregulated by TGF-β1 treatment; it was blocked after hyperthermia (Fig. 11.5).

Regarding the transcription level, TGF-β1 upregulates the expression of Snail, which suppresses the expression of E-cadherin [2, 24]. The expression of Snail was upregulated by TGF-β1 treatment in PANC-1 and MIAPaCa-2 cells; this upregulation was then significantly blocked by hyperthermia in MIAPaCa-2 cells. In PANC-1 cells, although hyperthermia tended to attenuate TGF-β1-induced Snail expression, this was not the case in MIAPaCa-2 cells (Fig. 11.6) [25]. In BxPC-3 cells, no significant change in Snail expression was observed after exposure to TGF-β1 or hyperthermia; this caused us to investigate ZEB-1 expression in BxPC-3 cells [26–28]. We observed that ZEB-1 expression was upregulated by TGF-β1 treatment in BxPC-3 cells; this upregulation was then significantly blocked by hyperthermia (Fig. 11.7); however, an inverse correlation was observed with

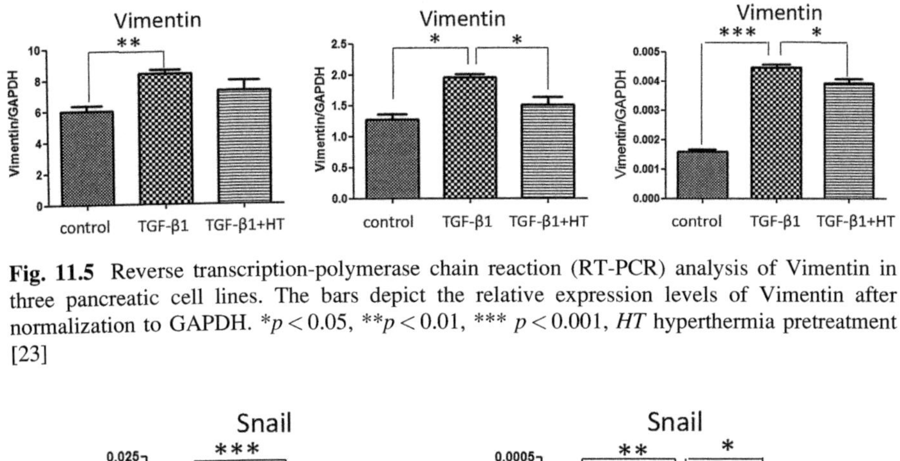

Fig. 11.5 Reverse transcription-polymerase chain reaction (RT-PCR) analysis of Vimentin in three pancreatic cell lines. The bars depict the relative expression levels of Vimentin after normalization to GAPDH. *$p < 0.05$, **$p < 0.01$, *** $p < 0.001$, *HT* hyperthermia pretreatment [23]

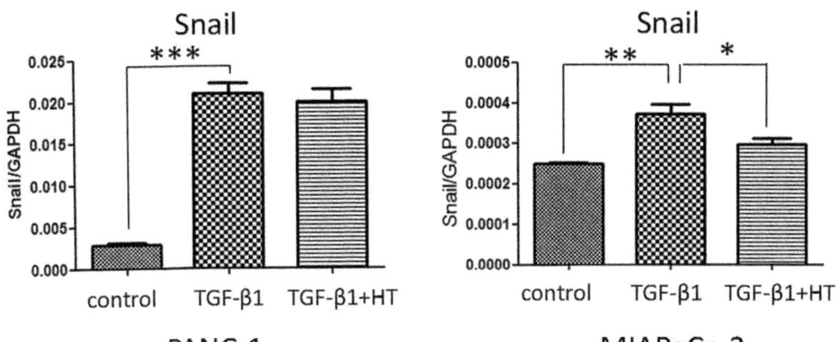

Fig. 11.6 RT-PCR analysis of Snail in two pancreatic cell lines. The *bars* depict the relative expression levels of Snail after normalization to glyceraldehyde-3-phosphate dehydrogenase (GAPDH). *, $p < 0.05$, **, $p < 0.01$, ***, $p < 0.001$. *HT* hyperthermia pretreatment [23]

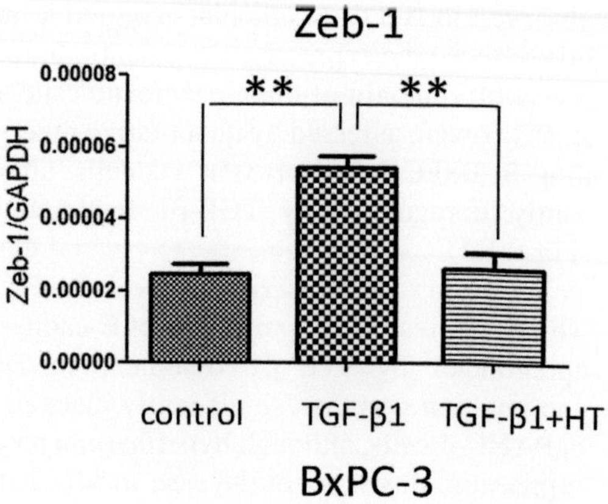

Fig. 11.7 RT-PCR analysis of ZEB-1 in BxPC-3 cells. The *bars* depict the relative expression levels of ZEB-1 after normalization to GAPDH. **, $p < 0.01$. *HT* hyperthermia pretreatment [23]

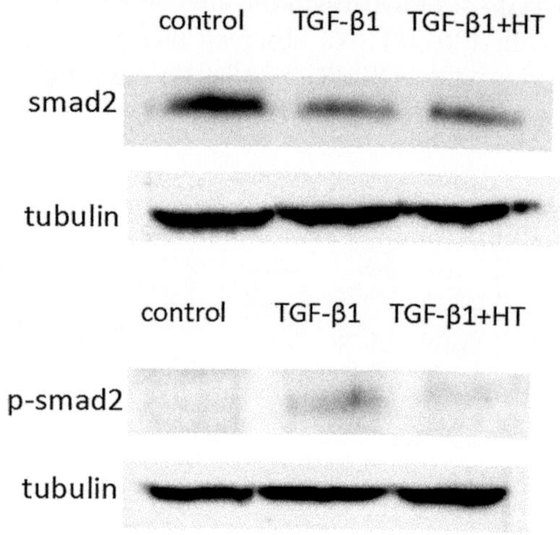

Fig. 11.8 Western blotting analysis of Smad2 and p-Smad2 in PANC-1 cells. *HT* hyperthermia pretreatment [23]

E-cadherin expression. Therefore, although cancer cells each have induction signals for EMT in response to a TGF-β1 stimulus, they are by no means identical.

EMT is characterized by a breakdown of cell junctions and the loss of epithelial characteristics and cell polarity, which promote cancer progression. In addition to the increased acquisition of mesenchymal markers, EMT also provides cancer cells with the ability to migrate and invade into surrounding tissues, thereby promoting the subsequent formation of metastases [29]. Although the role of TGF-β1 in induced EMT in cancer progression has been thoroughly investigated, substantial evidence for the involvement of downstream TGF-β1 signaling pathways in EMT has yet to be presented.

Smads are important intracellular effectors of the TGF-β1 signaling superfamily [30]. Therefore, Reiko K. et al. [23] also investigated Smad2 expression in PANC-1 cells by western blotting (Fig. 11.8), finding that TGF-β1 induced Smad2 phosphorylation; however, the TGF-β1-induced phosphorylation of Smad2 was subsequently blocked by hyperthermia. Therefore, the TGF-β1/Smad signaling pathway is an EMT-inducing pathway and can be inhibited by hyperthermia.

Some studies [31–33] have reported that resveratrol inhibits EMT in pancreatic cancer via suppression of the NF-κB pathway. Indeed, NF-κB is known to be associated with in EMT [34, 35]. It is widely accepted that antioxidants such as resveratrol also control EMT through inhibiting NF-κB *in vitro*. However, it is difficult to inhibit NF-κB in an animal model using these substances. Although we previously reported that hyperthermia can inhibit NF-κB in cancer-bearing mice [36], we did not investigate EMT at that time.

Heat stress elicits a wide spectrum of stress responses in mammalian cells, including an induction of heat shock proteins (HSPs), the production of ROS, and DNA and RNA damage. Most HSPs play important roles as protein chaperones, regulators of protein folding, regulators of protein degradation, and supporters of protein complex formation [37, 38]. Recently, several HSPs have been shown to be involved in EMT in renal tubular epithelial cells [39–41], lung cells [42], and prostate cancer cells [43]. While there is some agreement regarding the characteristics of HSPs, their effect on EMT remains controversial. For example, HSP72 has been shown to inhibit TGF-β1-induced EMT in renal epithelial cells by preventing TGF-β1-induced phosphorylation and nuclear translocation of Smad and p-Smad [39, 40]. In contrast, Noh H. et al. [41] reported that HSP90 inhibitor blocks TGF-β1-induced Smad phosphorylation and induces the degradation of the TGF-β type II receptor, thereby preventing TGF-β-stimulation from inducing EMT. HSP27 is considered to be a component of several pathways known to induce EMT in prostate cancer, including the IL-6/STAT3 pathway [43]. In addition, the inhibition of HSP27 blocks EMT features by promoting Snail degradation in lung cells. However, different effects of HSPs on EMT were reported in a number of studies; this may have been due in part to not only the different kind of HSPs studied, but also the different types of cells that were used. Since heat can strongly induce HSP72, it is possible that heat-induced HSP72 was involved in the blocking of TGF-β1-induced Smad phosphorylation observed in the present study. However, hyperthermia can induce not only a multitude of HSPs, but also various cellular responses. We have demonstrated that hyperthermia inhibits the activation of NF-κB which can regulate EMT-inducing transcription factors, in pancreatic cancer cells [unpublished data]. Therefore, the inhibitory effect of hyperthermia on EMT may involve several underlying molecular mechanisms; however, additional studies are needed to clarify the precise mechanisms underlying these inhibitory effects.

11.5 Hyperthermia Inhibits Hypoxia-Induced EMT in Hepatocellular Carcinoma (HCC)

Hypoxia is a common characteristic of solid cancers such as liver cancer. Therefore, it is important to investigate phenomena in cancer cells under hypoxic conditions. Several studies have shown that hypoxic conditions can induce EMT in cancer cells, the mechanisms of which may involve ROS generated during hypoxia, hypoxia-inducible factor-1α, and phosphoinositide-3-kinase/Akt signaling [18–20]. Yuan, G.-J. et al. [44] reported that hypoxia induced morphological (spindle shape and loss of cell contact), molecular (decreased E-cadherin but increased vimentin expressions), and functional (increased cell migration) changes in HepG2 HCC cells; all of these changes are consistent with EMT. As previously described [1], hyperthermia has been shown to inhibit tumor metastasis in an animal model. Furthermore, hyperthermia inhibits the expression of some genes associated with metastasis, including vascular endothelial growth factor and membrane type 1-matrix metalloproteinase [45–47]. In addition, Yuan G.-J. et al. [44] reported that hyperthermia can inhibit hypoxia-induced EMT. According to their data, Snail was found to play an important role in hypoxia-induced EMT because hypoxia induced Snail mRNA and protein expression in HepG2 HCC cells; however, the hypoxia-induced Snail expression was inhibited by pre-treatment with hyperthermia. Therefore, the mechanism by which hyperthermia inhibits hypoxia-induced EMT may involve a reduction in Snail expression.

11.6 Conclusion

Hyperthermia can help prevent tumor metastasis by inhibiting EMT. Several studies have reported that hyperthermia can inhibit a number of transcription factors that play important roles in EMT. However, the mechanisms by which hyperthermia inhibits EMT may be cancer cell-dependent. Further investigations are needed before hyperthermia can be effectively used with other treatments.

References

1. Nagashima K, Takagi R, Hoshina H. Effect of local hyperthermia on metastases in oral squamous cell carcinoma. Int J Oral Maxillofac Surg. 2002;31:84–9.
2. Hoshina H, Takagi R, Tsurumaki H, et al. Clinical result of thermochemoradiotherapy for advanced head and neck cancer. Gan To Kagaku Ryoho. 2001;28:331–6.
3. Ishikawa T, Kokura S, Sakamoto N, et al. Phase II trial of combined regional hyperthermia and gemcitabine for locally advanced or metastatic pancreatic cancer. Int J Hyperthermia. 2012;28:597–604.
4. Satoko A, Kokura S, Yoshikawa T. Effect of heating prior to gemcitabine exposure on therapeutic outcomes in combination therapies. Therm Med. 2010;26:75–85.
5. Adachi S, Kokura S, Okayama T, et al. Effect of hyperthermia combined with gemcitabine on apoptotic cell death in cultured human pancreatic cancer cell lines. Int J Hyperthermia. 2009;25:210–19.
6. Thiery JP, Acloque H, Huang RY, et al. Epithelial-mesenchymal transitions in development and disease. Cell. 2009;139:871–90.
7. Kalluri R, Weinberg RA. The basics of epithelial-mesenchymal transition. J Clin Invest. 2009;119:1420–8.

8. Hugo H, Ackland ML, Blick T, et al. Epithelial-mesenchymal and mesenchymal-epithelial transitions in carcinoma progression. J Cell Physiol. 2007;213:374–83.
9. Min C, Eddy SF, Sherr DH, et al. NF-kappaB and epithelial to mesenchymal transition of cancer. J Cell Biochem. 2008;104:733–44.
10. Larue L, Bellacosa A. Epithelial-mesenchymal transition in development and cancer: role of phosphatidylinositol 3' kinase/AKT pathways. Oncogene. 2005;24:7443–54.
11. Wu Y, Zhou BP. New insights of epithelial-mesenchymal transition in cancer metastasis. Acta Biochim Biophys Sin. 2008;40:643–50.
12. Sabbah M, Emami S, Edeuilh G, et al. Molecuar signature and therapeutic perspective of the epitelial-to-mesenchymal transition in epithelial cancers. Drug Resist Updat. 2008;11:123–51.
13. Hiscox S, Jiang WG, Obermeier K, et al. Tamoxifen resistance in MCF7 cells promotes EMT-like behaviour and involves modulation of beta-catenin phosphorylation. Int J Cancer. 2006;118:290–301.
14. Wang Z, Li Y, Kong D, et al. Acquisition of epithelial-mesenchymal transition phenotype of gemcitabine-resistant pancreatic cancer cells is linked with activation of the notch signaling pathway. Cancer Res. 2009;69:2400–7.
15. Ponti D, Zaffaroni N, et al. Breast cancer stem cells: an overview. Eur J Cancer. 2006;42:1219–24.
16. Arumugam T, Ramachandran V, Fournier KF, et al. Epithelial to mesenchymal transition contributes to drug resistance in pancreatic cancer. Cancer Res. 2009;69:5820–8.
17. Massague J. How cells read TGF-beta signals. Nat Rev Mol Cell Biol. 2000;1:169–78.
18. Cannito S, Novo E, Compagnone A, et al. Redox mechanisms switch on hypoxia-dependent epithelial-mesenchymal transition in cancer cells. Carcinog. 2008;29:2267–78.
19. Krishnamachary B, Zagzag D, et al. Hypoxia-inducible factor-1-dependent repression of E-cadherin in von Hippel-Lindau tumor suppressor-null renal cell carcinoma mediated by TCF3, ZFHX1A, and ZFHX1B. Cancer Res. 2006;66:2725–31.
20. Yan W, Fu Y, Tian D, et al. PI3 kinase/Akt signaling mediates epithelial-mesenchymal transition in hypoxic hepatocellular carcinoma cells. Biochem Biophys Res Commun. 2009;382:631–6.
21. Huang J, Li H, Ren G. Epithelial-mesenchymal transition and drug resistance in breast cancer (Review). Int J Oncol. 2015;47:840–8.
22. Liu Y, Du F, Zhao Q, et al. Acquisition of 5-fluorouracil resistance induces epithelial-mesenchymal transitions through the Hedgehog signaling pathway in HCT-8 colon cancer cells. Oncol Lett. 2015;9:2675–9.
23. Reiko K, Takeshi I, Kokura S, et al. The inhibitory effect of heat treatment against Epithelial-Mesenchymal Transition (EMT) in human pancreatic adenocarcinoma cell lines. J Clin Biochem Nutr. 2014;55:56–61.
24. Wang YL, Zhao XM, Shuai ZF, et al. Snail promotes epithelial-mesenchymal transition and invasiveness in human ovarian cancer cells. Int J Clin Exp Med. 2015;8:7388–93.
25. Zhang Y, Fan N, Yang J. Expression and clinical significance of hypoxia-inducible factor 1α, Snail and E-cadherin in human ovarian cancer cell lines. Mol Med Rep. 2015;12:3393–9.
26. Qiu X, Cheng JC, Klausen C, et al. Transforming growth factor-α induces human ovarian cancer cell invasion by down-regulating E-cadherin in a snail-independent manner. Biochem Biophys Res Commun. 2015;461:128–35.
27. Galván JA, Zlobec I, Wartenberg M, et al. Expression of E-cadherin repressors SNAIL, ZEB1 and ZEB2 by tumour and stromal cells influences tumour-budding phenotype and suggests heterogeneity of stromal cells in pancreatic cancer. Br J Cancer. 2015;112:1944–50.
28. Pu H, Horbinski C, Hensley PJ, et al. PARP-1 regulates epithelial-mesenchymal transition (EMT) in prostate tumorigenesis. Carcinog. 2014;35:2592–601.
29. Guarino M. Epithelial-mesenchymal transition and tumour invasion. Int J Biochem Cell Biol. 2007;39:2153–60.
30. Zimowska M. Signaling pathways of transforming growth factor beta family members. Postepy Biochem. 2006;52:360–6.

31. Gao Q, Yuan Y, Gan HZ, Peng Q. Resveratrol inhibits the hedgehog signaling pathway and epithelial-mesenchymal transition and suppresses gastric cancer invasion and metastasis. Oncol Lett. 2015;9:2381–7.
32. Ji Q, Liu X, Han Z, Zhou L, et al. Resveratrol suppresses epithelial-to-mesenchymal transition in colorectal cancer through TGF-β1/Smads signaling pathway mediated snail/E-cadherin expression. BMC Cancer. 2015;15:97.
33. Li J, Chong T, Wang Z, et al. A novel anti-cancer effect of resveratrol: reversal of epithelial-mesenchymal transition in prostate cancer cells. Mol Med Rep. 2014;10:1717–24.
34. Li J, Deng Z, Wang Z, et al. Zipper-interacting protein kinase promotes epithelial-mesenchymal transition, invasion and metastasis through AKT and NF-kB signaling and is associated with metastasis and poor prognosis in gastric cancer patients. Oncotarget. 2015;610:8323–38.
35. Karamitopoulou E, Zlobec I, Gloor B, et al. Loss of Raf-1 kinase inhibitor protein (RKIP) is strongly associated with high-grade tumor budding and correlates with an aggressive phenotype in pancreatic ductal adenocarcinoma (PDAC). J Transl Med. 2013;11:311.
36. Nakabe N, Kokura S, Shimozawa M, et al. Hyperthermia attenuates TNF-alpha -induced up regulation of endothelial cell adhesion molecules in human arterial endothelial cells. Int J Hyperthermia. 2007;23:217–24.
37. Lindquist S, Craig EA. The heat-shock proteins. Annu Rev Genet. 1988;22:631–77.
38. Hartl FU. Molecular chaperones in cellular protein folding. Nature. 1996;381:571–9.
39. Mao H, Li Z, Zhou Y, et al. HSP72 attenuates renal tubular cell apoptosis and interstitial fibrosis in obstructive nephropathy. Am J Physiol Renal Physiol. 2008;295:F202–14.
40. Zhou Y, Mao H, Li S, et al. HSP72 inhibits Smad3 activation and nuclear translocation in renal epithelial-to-mesenchymal transition. J Am Soc Nephrol JASN. 2010;21:598–609.
41. Noh H, Kim HJ, Yu MR, et al. Heat shock protein 90 inhibitor attenuates renal fibrosis through degradation of transforming growth factor-beta type II receptor. Laboratory investigation. J Tech Meth Pathol. 2012;92:1583–96.
42. Wettstein G, Bellaye PS, Kolb M, et al. Inhibition of HSP27 blocks fibrosis development and EMT features by promoting snail degradation. FASEB J Off Publ Fed Am Soc Exp Biol. 2013;27:1549–60.
43. Shiota M, Bishop JL, Nip KM, et al. Hsp27 regulates epithelial mesenchymal transition, metastasis, and circulating tumor cells in prostate cancer. Cancer Res. 2013;73:3109–19.
44. Yuan G-J, Li Q-W, et al. Hyperthermia inhibits hypoxia-induced epithelial-mesenchymal transition in HepG2 hepatocellular carcinoma cells. World J Gastroenterol. 2012;18:4781–6.
45. Sawaji Y, Sato T, et al. Anti-angiogen- ic action of hyperthermia by suppressing gene expression and production of tumour-derived vascular endothelial growth factor *in vivo* and *in vitro*. Br J Cancer. 2002;86:1597–603.
46. Sawaji Y, Sato T, et al. Heat shock-mediated tran- sient increase in intracellular 3',5'-cyclic AMP results in tumor specific suppression of membrane type 1-matrix metal- loproteinase production and progelatinase. Act Clin Exp Metastasis. 2000;18:131–8.
47. Liang X, Zhou H, Liu X, et al. Effect of local hyperthermia on lymphangiogenic factors VEGF-C and -D in a nude mouse xenograft model of tongue squamous cell carcinoma. Oral Oncol. 2010;46:111–15.

Part II
Basic Science in Whole Body

Chapter 12
Potentiating Immune System by Hyperthermia

Hiroshi Terunuma

Abstract Hyperthermia enhances the host immune responses against cancer through several mechanisms; activating immune cells (e.g., natural killer cells, dendritic cells, and cytotoxic T lymphocytes), canceling immune suppression, altering cell-surface molecules on cancer cells, and modifying adhesion molecules on immune cells and endothelial cells. This chapter discusses the positive effects of hyperthermia on the host immune system.

Keywords Hyperthermia • Fever-range hyperthermia • Mild hyperthermia • NK cell • Dendritic cell • Cytotoxic T cell • Regulatory T cell

12.1 Introduction

The immune system distinguishes between self and non-self to identify and remove foreign substances, including cancer cells. Innate immunity, unlike acquired immunity, can immediately respond to foreign substances without prior exposure to antigens. Natural killer (NK) cells are at the front line of innate immune response in attacking cancer cells. Dendritic cells (DCs) capture, process, and display cancer antigens in order to transmit activation signals to the antigen-specific acquired immune response, which cytotoxic T lymphocytes (CTLs) are major players in. Regulatory T cells (Treg) counteract the anti-cancer immune responses described above.

Hyperthermia treatment raises the temperature of target tissues to 42.5 °C or above when destroying cancer cells, and to 38–41 °C (fever-range hyperthermia or mild hyperthermia) when increasing blood flow and activating immune reactions. When a cancer mass is heated over 42.5 °C, both cancer and immune cells in the center part of the lesion are destroyed, whereas the surrounding areas form a

H. Terunuma (✉)
Tokyo Clinic, 1F Shin-Otemachi Building, 2-2-1 Otemachi Chiyoda-ku, Tokyo 100-0004, Japan

Biotherapy Institute of Japan, 2-4-8 Edagawa Koutou-ku, Tokyo 135-0051, Japan
e-mail: terunuma@bij-net.com

S. Kokura et al. (eds.), *Hyperthermic Oncology from Bench to Bedside*,
DOI 10.1007/978-981-10-0719-4_12

thermal gradient to generate a field of 38–41 °C, where immune reactions are activated as fever-range hyperthermia.

The effects of fever-range hyperthermia on immune cells and their networks are discussed in this chapter.

12.2 Enhancing Anti-Cancer Activity of NK Cells Using Hyperthermia

NK cells express many activating and inhibitory receptors, which recognize specific ligands expressed on target cells (Fig. 12.1) [1]. Most activating ligands on target cells are induced by genomic damage from viral infections and malignant transformations. The expression and processing of ligands are enhanced when thermal stress is applied to the cells before they are recognized by NK cell-activating receptors. On the other hand, major histocompatibility complex (MHC) class I antigens, which are expressed constitutively on normal cells, are engaged by NK cell-inhibitory receptors. A balance of these activating and inhibitory receptor signals regulates the effector function of NK cells. Interestingly, the expression of MHC class I antigens frequently becomes reduced or lost when cells become malignant or metastatic (Table 12.1) [1]. This decrease or loss of MHC class I antigens hides malignant cells from CTLs, but exposes them to attacks by NK cells due to the attenuated inhibitory signals [2].

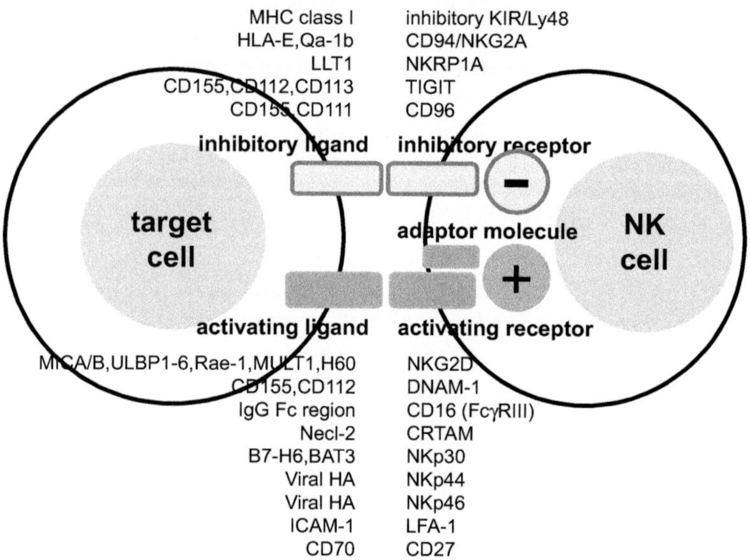

Fig. 12.1 The effector function of NK cells is regulated by a balance of activating and inhibitory signals between NK cells and target cells

Table 12.1 Frequencies of reduced MHC class I expression in surgically removed tumors

Cancer	Low or negative cases/total cases	Percentage (%)	Reference
Melanoma, primary	66/414	16	[23]
Melanoma, metastases	287/495	58	[23]
Grade 2 astrocytoma	3/18	17	[24]
Glioblastoma multiforme	22/47	47	[24]
Head & neck cancer	20/41	49	[25]
Laryngeal cancer	25/70	36	[26]
Breast cancer	356/439	81	[27]
Lung cancer	35/93	38	[28]
Hepatocellular carcinoma	24/57	42	[29]
Pancreatic cancer	32/37	76	[30]
Gastric cancer	45/141	32	[31]
Colorectal cancer	107/452	24	[32]
Renal cell cancer	17/45	38	[33]
Bladder cancer	18/72	25	[34]
Prostatic cancer	311/419	74	[35]
Penile cancer	138/168	82	[36]
Cervical cancer	27/30	90	[37]
Ovary cancer	195/486	41	[38]
Bone and soft tissue sarcoma	46/74	62	[39]
Osteosarcoma, primary	13/25	52	[39]
Osteosarcoma, metastases	7/8	88	[39]

Activated NK cells directly attack cancer cells using cytotoxic molecules (e.g., perforin and granzyme), death receptors (e.g., FasL, TRAIL, TNF-α), and antibody-dependent cell-mediated cytotoxicity (ADCC) [1, 2]. They also secrete cytokines such as IFN-γ, TNF-α, GM-CSF and IL-2 to stimulate other immune cells (i.e., DC, T and B lymphocytes), which reinforce acquired immunity by enhancing antigen-specific CTL induction and immunoglobulin production [1, 2].

Hyperthermia increases the distinct clustering of NK cell-activating receptors such as NKG2D on the surface of NK cells [3]. Hyperthermia also increases the expression of NK cell-activating ligands such as the major histocompatibility complex class I-related chain A (MICA but not MHC class I [3]. We also reported previously that hyperthermia can enhance surface expressions of MICA but not of MHC class I in several cancer cell lines (Fig. 12.2a and b) [4]. Furthermore, the increased expression of MICA/B in heated cancer cells correlates with the increased NK cell cytotoxicity against them (Fig. 12.2c and d) [4].

Hyperthermia also increases the MICA mRNA levels in tumors, which correlate with increased sensitivity to cytolysis [3]. The interaction between NKG2D and MICA/B is just an example, and interactions between other NK cell-activating or inhibitory receptors and ligands are also important in enhancing the cytotoxicity of NK cells against cancer cells [5]. Hyperthermia enhances the anti-cancer activity of NK cells through these molecular events.

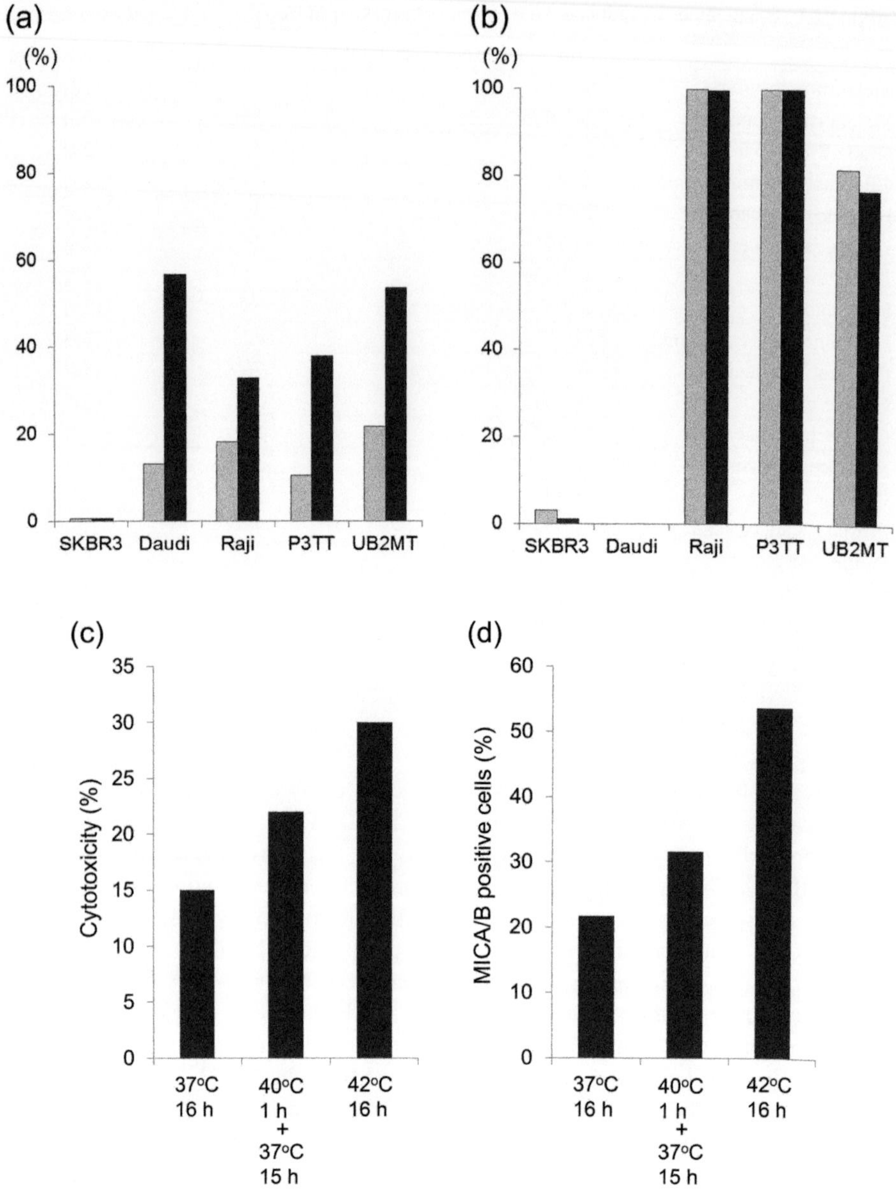

Fig. 12.2 Frequencies of cells expressing MICA/B (**a**) and MHC class I (**b**). Human cancer cell lines were kept at 37 °C (*gray column*) or exposed to 42 °C (*black column*) for 16 h. SKBR3 were derived from breast carcinoma. Daudi and Raji were derived from Burkitt lymphoma. C1AK and UB2MT were established by us from carcinomas of colon and uterine body, respectively. Both the cytotoxicity of NK cells to UB2MT (**c**) and the frequency of MICA/B-positive UB2MT cells (**d**) were increased by thermal treatment. The number and cytotoxicity of NK cells were increased by 14-day cultivation using BINKIT (Biotherapy Institute of Japan, Tokyo, Japan) [22]. The cytotoxicity of the expanded NK cells was measured against UB2MT cells at an effector-to-target ratio of 6:1 using a calcein-AM release assay measured by TERASCAN VP (Minerva Tech., Tokyo,

12.3 Activating DC and CTL by Hyperthermia

DCs are key players in antigen-specific CTL induction for acquired immunity against cancer cells. DCs capture and process cancer antigens originating from damaged cancer cells before presenting them on the cell surface along with MHC antigens. The complex made up of cancer antigens and MHC class II antigens stimulates helper T cells, while the complex made up of cancer antigens and MHC class I antigens induces CTLs via cross-presentation. The resulting CTLs recognize cancer cells through the cancer antigen presented on the MHC class I antigen, and thus attacks them.

Hyperthermia causes the generation of heat shock proteins (HSPs), which serve as molecular chaperones, in cancer cells. HSPs enhance the intracellular transport of cancer antigens generated in proteasomes and are presented on the cell surface along with the MHC class I antigen to induce antigen-specific CTLs [6–8].

The complex of HSPs and cancer antigens can also be released to the extracellular space when cancer cells are destroyed. Such complexes are captured by DCs, which express HSP receptors, and are processed for cross-presentation to efficiently induce antigen-specific CTLs [7].

DCs and CTLs are activated further by cytokines released from T lymphocytes, which are also activated by hyperthermia [9].

In order to make T cells interact with DCs for CTL induction in lymphoid organs, it is crucial to recruit T cells into lymphoid organs. Thermal stress up-regulates the expression of adhesion molecules such as L-selectin and α4β7 integrin on lymphocytes, and CCL21+ and ICAM-1 on high endothelial venules, which serve as gatekeepers for lymphocyte recruitment into lymphoid organs [10]. Interestingly, CD3-negative CD56-bright NK cells also highly express L-selectin, and these immunomodulatory NK cells may activate DCs to induce CTLs.

Collectively, hyperthermia potentiates the anti-cancer acquired immunity [11–13].

12.4 Release from Immune Suppression and Hyperthermia

The balance between immunity and tolerance has to be coordinated in a finely tuned manner within the body. Tolerance is critically important in protecting oneself from autoimmune response. Cancer cells, however, take advantage of the tolerance mechanisms to escape immune surveillance; e.g., by utilizing Tregs, which negatively regulate both innate and acquired immunity at least in vitro by suppressing

---

Fig. 12.2 (continued) Japan) as previously described [22]. Representative results are shown; similar results were obtained in three independent experiments. Based on [4]

the function of immune cells including T lymphocytes, B lymphocytes, DCs, and NK cells [14, 15].

The number of Tregs within the peripheral blood as well as in cancer tissues is higher in cancer patients compared to that of non-cancer individuals. Moreover, cancer patients with increased numbers of Tregs in cancer tissues have poorer prognoses (cervical, renal, hepatocellular, gastric, and breast cancers and malignant melanoma) and higher risks of recurrence (lung, gastric, and hepatocellular carcinomas and malignant melanoma) [16]. These observations imply that cancer patients are prone to cancer growth due to the immune suppression caused by Tregs.

Malignant melanoma patients who were treated with a combination of hyperthermia and intratumoral injection of immature DCs exhibited lower numbers of Tregs and higher numbers of CTLs compared to patients treated without hyperthermia [17]. Furthermore, patients with hyperthermia showed significant delays in cancer growth. These results suggest that hyperthermia exerts therapeutic benefits at least partly by decreasing the number of Tregs in cancer tissues. We also observed a decrease in the number of Tregs within the peripheral blood after applying fever-range hyperthermia to the upper abdomen of healthy individuals [18]. Hyperthermia could be reducing the number of Tregs by inducing the apoptosis of Tregs, enhancing NK cytotoxicity against Tregs, or inhibiting the induction of Tregs. Further studies are needed to clarify the precise mechanisms of the hyperthermia-induced decrease of Tregs.

Additionally, immune checkpoint modulation is a recent breakthrough in cancer therapy. Therapeutic use of the anti-PD-1 antibody is one of them. We have studied the effects of hyperthermia on the PD-1/PD-L1 immune checkpoint, and observed that the expression of PD-L1 in some cancer cell lines is reduced by exposure to temperatures between 40 and 43 °C (Fig. 12.3). This observation implies the potential of hyperthermia in inhibiting immune checkpoints to sensitize cancer cells to anti-cancer immune reactions. Further studies on this matter shall establish yet another strategy of cancer therapy using hyperthermia.

Fig. 12.3 Thermal stress reduces the PD-L1 expression in cancer cell lines

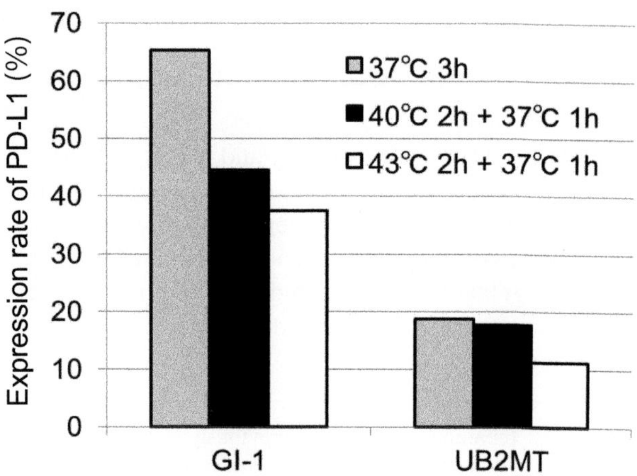

12.5 Summary

Hyperthermia activates the host immune system against cancer whereas the overall mechanisms are yet to be fully elucidated. So far, several mechanisms of anticancer activity by hyperthermia have been postulated (Fig. 12.4): increasing cytotoxicity of effector cells such as NK cells against cancer [18, 19]; enhancing maturation of immature DCs via the production of the HSP-tumor antigen complex in tumors [11–13]; facilitating antigen presentation to CD8+ T cells for CTL induction [6]; decreasing the numbers of Tregs in tumor tissues and blood [18, 20]; increasing the expression levels of adhesion molecules in endothelial venules [10]; enhancing the trafficking of lymphocytes, such as T cells and cytokine-producing NK cells, to secondary lymphoid tissues [10, 21]; and sensitizing cancer cells to immune effector cells such as NK cells via MICA/B molecules [3, 4].

Better uses of hyperthermia will be achieved by further understanding of its effects on the immune system. Additionally, clinical data on cancer treatments using hyperthermia in combination with other therapies need to be accumulated and evaluated for their respective clinical efficacies. These efforts will lead to the development of efficient combinatory therapies using immune therapy, chemotherapy, radiation, and hyperthermia.

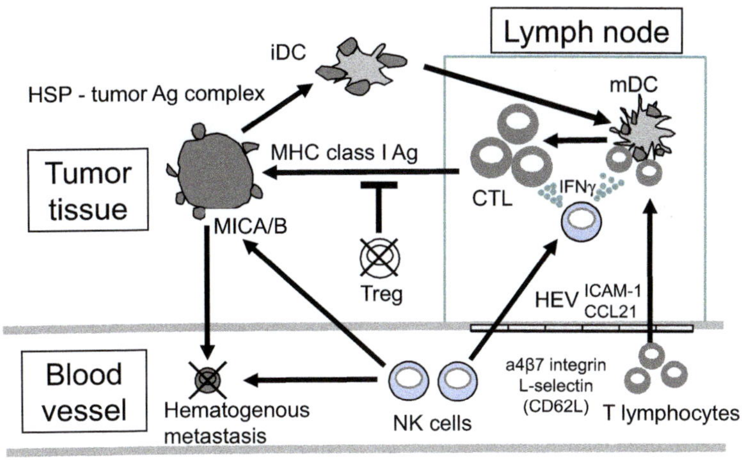

Fig. 12.4 Regional hyperthermia induces the activation of immune system against cancer. *NK cells:* natural killer cells, *iDC:* immature dendritic cells, *mDC:* mature dendritic cells, *CTL:* cytotoxic T lymphocytes, *Treg:* regulatory T cells, *HEV:* high endothelial venules, *HSP:* heat shock protein, *Ag:* antigen, *MHC:* major histocompatibility complex, *MICA/B:* major histocompatibility complex class I-related chain A/B, *IFN:* interferon, *ICAM-1:* intercellular adhesion molecule 1, *CCL:* chemokine (C-C motif) ligand, *CD:* cluster of differentiation. Based on [4]

References

1. Terunuma H, Deng X, Dewan Z, et al. Potential role of NK cells in the induction of immune responses: Implications for NK cell-based immunotherapy for cancers and viral infections. Int Rev Immunol. 2008;27:93–110.
2. Cheng M, Chen Y, Xiao W, et al. NK cell-based immunotherapy for malignant diseases. Cell Mol Immunol. 2013;10:230–52.
3. Ostberg JR, Dayanc BE, Yuan M, et al. Enhancement of natural killer (NK) cell cytotoxicity by fever-range thermal stress is dependent on NKG2D function and is associated with plasma membrane NKG2D clustering and increased expression of MICA on target cells. J Leukoc Biol. 2007;82:1322–31.
4. Terunuma H, Deng X, Toki A, et al. Effects of hyperthermia on the host immune system: from NK cell-based science to clinical application. Therm Med. 2012;28:1–9.
5. Farjadian S, Norouzian M, Younesi V, et al. Hyperthermia increases natural killer cell cytotoxicity against SW-872 liposarcoma cell line. Iran J Immunol. 2013;10:93–102.
6. Srivastava PK, Udono H, Blachere NE, Li Z. Heat shock proteins transfer peptides during antigen processing and CTL priming. Immunogenetics. 1994;39:93–8.
7. Binder RJ, Srivastava PK. Peptides chaperoned by heat-shock proteins are a necessary and sufficient source of antigen in the cross-priming of CD8+ T cells. Nat Immunol. 2005;6:593–9.
8. Takahashi T. Effects of hyperthermia and modification of thermosensitivity. Therm Med. 2007;23:171–9.
9. Kida Y, Tsuji-Kawahara S, Ostapenko V, et al. Increased liver temperature efficiently auguments human cellular immune response: T-cell activation and possible monocyte translocation. Cancer Immunol Immunother. 2006;55:1459–69.
10. Vardam TD, Zhou L, Appenheimer MM, et al. Regulation of a lymphocyte-endothelial-IL-6 trans-signaling axis by fever-range thermal stress: hot spot of immune surveillance. Cytokine. 2007;39:84–96.
11. Chen CH, Wang TL, Hung CF, et al. Enhancement of DNA vaccine potency by linkage of antigen gene to an *HSP70* gene. Cancer Res. 2000;60:1035–42.
12. Castelli C, Ciupitu AM, Rini F, et al. Human heat shock protein 70 peptide complexes specifically activate antimelanoma T cells. Cancer Res. 2001;61:222–7.
13. Tanaka K, Ito A, Kobayashi T, et al. Intratumoral injection of immature dendritic cells enhances antitumor effect of hyperthermia using magnetic nanoparticle. Int J Cancer. 2005;116:624–33.
14. Sakaguchi S, Wing K, Onishi Y, et al. Regulatory T cells: how do they suppress immune responses? Int Immunol. 2009;21:1105–11.
15. Smyth MJ, Teng MWL, Swann J, et al. CD4 + CD25+ T regulatory cells suppress NK cell-mediated immunotherapy of cancer. J Immunol. 2006;176:1582–7.
16. Shang B, Liu Y, Jiang S, Liu Y. Prognostic value of tumor-infiltrating FoxP3+ regulatory T cells in cancers: a systematic review and meta-analysis. Sci Rep. 2015. doi:10.1038/srep15179.
17. Guo J, Zhu J, Sheng X, et al. Intratumoral injection of dendritic cells in combination with local hyperthermia induces systemic antitumor effect in patients with advanced melanoma. Int J Cancer. 2007;120:2418–25.
18. Terunuma H, Wada A, Deng X, et al. Mild hyperthermia modulates the relative frequency of lymphocyte cell subpopulations: an increase in a cytolytic NK cell subset and a decrease in a regulatory T cell subset. Therm Med. 2007;23:41–7.
19. Dayanc BE, Beachy SH, Ostberg JR, Repasky EA. Dissecting the role of hyperthermia in natural killer cell mediated anti-tumor responses. Int J Hyperthermia. 2008;24:41–56.
20. Kokura S, Okayama T, Adachi S, et al. The effects of hyperthermia on immune surveillance in tumor. Thermal Med. 2011;27 Supplement:79.

21. Atanackovic D, Pollok K, Faltz C, et al. Patients with solid tumors treated with high-temperature whole body hyperthermia show a redistribution of naïve/memory T-cell subtypes. Am J Physiol Regul Integr Comp Physiol. 2006;290:R585–94.
22. Deng X, Terunuma H, Terunuma A, et al. *Ex vivo*-expanded natural killer cells kill cancer cells more effectively than *ex vivo* expanded γδT cells or αβT cells. Int Immunopharmacol. 2014;22:486–91.
23. Ferrone S, Marincola FM. Loss of HLA class I antigens by melanoma cells: molecular mechanisms, functional significance and clinical relevance. Immunol Today. 1995;16:487–94.
24. Facoetti A, Nano R, Zelini P, et al. Human leukocyte antigen and antigen processing machinery component defects in astrocytic tumors. Clin Cancer Res. 2005;11:8304–11.
25. Feenstra M, Veltkamp M, van Kuik J, et al. HLA class I expression and chromosomal deletions at 6p and 15q in head and neck sequamous cell carcinomas. Tissue Antigens. 1999;54:235–45.
26. Maleno I, Lopez-Nevot MA, Cabrera T, et al. Multiple mechanisms generate HLA class I altered phenotypes in laryngeal carcinomas: high frequency of HLA haplotype loss associated with loss of heterozygosity in chromosome region 6p21. Cancer Immunol Immunother. 2002;51:389–96.
27. Madjd Z, Spendlovel PSE, et al. Total loss of MHC class I is an independent indicator of good prognosis in breast cancer. Int J Cancer. 2005;117:248–55.
28. Korkolopoulou P, Kaklamanis L, Pezzella F, et al. Loss of antigen-presenting molecules (MHC class I and TAP-1) in lung cancer. Br J Cancer. 1996;73:148–53.
29. Kutomi G, Tamura Y, Torigoe SN. Effective immunotherapy by HSP-cancer peptide complex and immune escape of HLA class I antigen down regulation. Igaku No Ayumi (Japanese). 2007;221:627–30.
30. Pandha H, Rigg A, John J, Lemoine N. Loss of expression of antigen-presenting molecules in human pancreatic cancer and pancreatic cancer cell lines. Clin Exp Immunol. 2007;148:127–35.
31. Ishigami S, Natsugoe S, Nakajo A, et al. HLA-class I expression in gastric cancer. J Surg Oncol. 2008;97:605–8.
32. Watson NFS, Ramage JM, Madjd Z, et al. Immunosurveillance is active in colorectal cancer as downregulation but not complete loss of MHC class I expression correlates with a poor prognosis. Int J Cancer. 2006;118:6–10.
33. Kitamura H, Honma I, Torigoe T, et al. Down-regulation of HLA class I antigens is independent prognostic factor for clear cell renal cell carcinoma. J Urol. 2007;177:1269–72.
34. Romero JM, Jimenez P, Cabrera T, et al. Coordinated downregulation of the antigen presentation machinery and HLA class I/β2-microglobulin complex is responsible for HLA-ABC loss in bladder cancer. Int J Cancer. 2005;113:605–10.
35. Kitamura H, Torigoe T, Asanuma H, et al. Down-regulation of HLA class I antigens in prostate cancer tissues and up-regulation by histone deacetylase inhibition. J Urol. 2007;178:692–6.
36. Djajadiningrat RS, Horenblas S, Heideman DA, et al. Classic and nonclassic HLA class I expression in penile cancer and relation to HPV status and clinical outcome. J Urol. 2015;193:1245–51.
37. Koopman LA, Corver WE, van der Slik AR, et al. Multiple genetic alterations cause frequent and heterogeneous human histocompatibility leukocyte antigen class I loss in cervical cancer. J Exp Med. 2000;191:961–75.
38. De Jong RA, Boerma A, Boezen HM, et al. Loss of HLA class I and mismatch repair protein expression in sporadic endometrioid endometrial carcinomas. Int J Cancer. 2012;131:1828–36.
39. Tsukahara T, Kawaguchi S, Torigoe T, et al. Prognostic significance of HLA class I expression in osteosarcoma defined by anti-pan HLA class I monoclonal antibody, EMR8-5. Cancer Sci. 2006;97:1374–80.

Chapter 13
Magnetic Nanoparticle-Mediated Hyperthermia and Induction of Anti-Tumor Immune Responses

Takeshi Kobayashi, Akira Ito, and Hiroyuki Honda

Abstract Magnetic nanoparticle-mediated hyperthermia (MNHT) can heat tumor tissue to the desired temperature without damaging surrounding normal tissue. The MNHT system consists of targeting tumor with functional magnetic nanoparticles (MNPs) and then applying an external alternating magnetic field (AMF) to generate heat in the MNPs. Temperature in the tumor tissue is increased to above 43 °C, which causes necrosis of cancer cells but does not damage surrounding normal tissue. Among available MNPs, magnetite has been extensively studied. Recent years have seen remarkable advances in MNHT; both functional MNPs and AMF generators have been developed. By applying MNHT, heat shock proteins (HSPs) are highly expressed within and around tumor tissue, which causes intriguing biological responses such as tumor-specific immune response. These results suggest that MNHT is able to kill not only tumors exposed to heat treatment, but also unheated metastatic tumors at distant sites. Currently, some researchers have started clinical trials, suggesting that the time has come for clinical applications.

Keywords Hyperthermia • Magnetic nanoparticles • Magnetic field • Immune responses

---

T. Kobayashi (✉)
School of Bioscience and Biotechnology, Chubu University, 1200 Matsumoto-cho, 487-8501 Kasugai, Aichi, Japan
e-mail: kobatake@isc.chubu.ac.jp

A. Ito
Department of Chemical Engineering, Faculty of Engineering, Kyushu University, Fukuoka, Japan

H. Honda
Department of Biotechnology, School of Engineering, Nagoya University, Aichi, Nagoya, Japan

S. Kokura et al. (eds.), *Hyperthermic Oncology from Bench to Bedside*,
DOI 10.1007/978-981-10-0719-4_13

13.1 Introduction

Hyperthermia is a promising approach to cancer therapy, in part, because hyperthermia is a physical treatment and can result in fewer side effects than chemotherapy or radiotherapy [1, 2]. A major technical problem with the currently available hyperthermia is the difficulty of heating a local tumor region without damaging normal tissue. A consensus landmark is the threshold temperature of 42.5 °C [3], because heat cytotoxicity is markedly higher than 42.5 °C. Although such high temperatures can kill a great number of tumor cells, normal tissues are also damaged, especially under the conventional hyperthermia treatments. Therefore, the development of a novel hyperthermia system which can specifically target tumor tissue and heat the tumor above 42.5 °C has been highly desired.

MNHT is a largely experimental modality for hyperthermia which may overcome these shortcomings [4–7]. The MNHT system consists of targeting functional MNPs to the tumor, and then applying an external AMF to generate heat in the MNPs by mechanisms called as hysteresis loss and Néel relaxational loss. Recently, remarkable advances have been seen in MNHT using both tumor-targeted MNPs and AMF generators.

MNHT increases temperature of tumor tissue above 42.5 °C without damaging normal tissue. A substantial amount of HSPs are expressed in tumor tissues, which in turn causes unexpected biological responses, including an antitumor immune response induced by expression of HSPs [4, 6, 7]. Yanase et al. [8] observed antitumor immune response induced by MNHT in an experimental T-9 rat glioma model in which a tumor was transplanted into each femur of a rat as shown in Fig. 13.1. Functional MNPs were directly injected in the left tumor. When the rat was irradiated with AMF, temperature at the left tumor increased, whereas temperature at the right tumor did not increase. Interestingly, although only one tumor was subjected to MNHT, the other tumor also disappeared completely. Immunohistochemical assay revealed that NK cells, $CD8^+$ T cells and $CD4^+$ T cells migrated into both tumors after the hyperthermia treatment. These results suggest that MNHT is a powerful treatment, because in addition to killing tumors with heat, MNHT induces a host immune response. Similar phenomena have been reported as the abscopal effect in radiotherapy [9], and the study by Yanase et al. [8] was the first report in hyperthermia. This chapter covers recent advances in MNHT and induction of immune responses, and discusses future directions.

13.2 Magnetic Nanoparticles for MNHT

Various heating materials have been used for hyperthermia. However, heating only the tumor region to the required temperature (above 42.5 °C) without damaging surrounding normal tissue is an inevitable technical problem. MNPs less than 100 nm in diameter have been applied for this purpose [4, 5, 7]. The most important

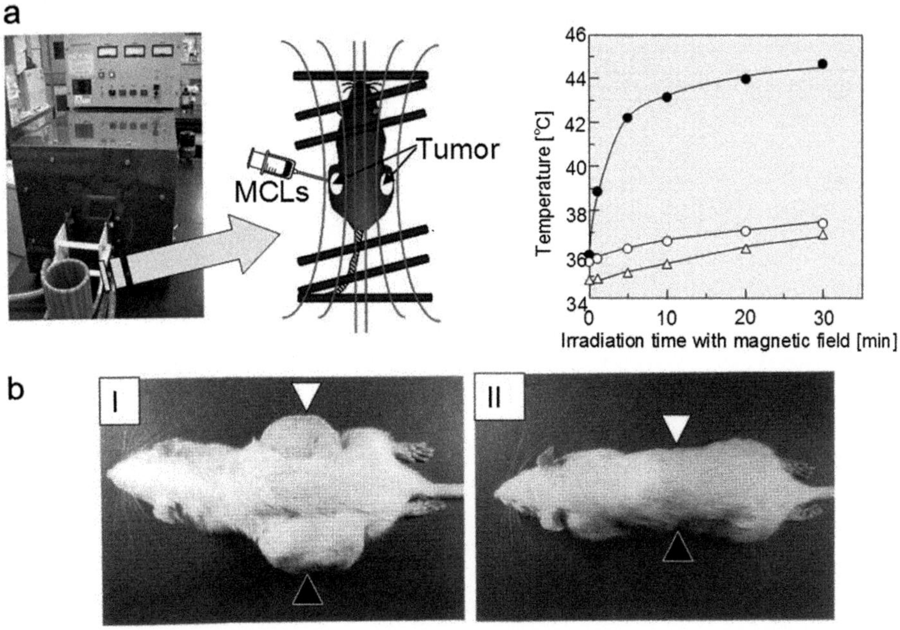

Fig. 13.1 Immune response induced by hyperthermia using MCLs [8]. Rats bearing tumors on each side of the body were prepared. MCLs were injected into the *left* tumor only, and the rats were irradiated with an AMF using the apparatus shown in *a (left)*. Temperature of *left* tumor at which MCLs were injected (*closed circles*), increased specifically above 44 °C, whereas temperature of *right* tumor (*open circles*) and rectum (*open triangles*) remained below 37 °C (*a, right*). The MNHT treatment induced an anti-tumor immune response, and both tumors had disappeared on the 28th day after the MNHT treatment. (I) Control rat without AMF irradiation; (II) rat with AMF irradiation. *Open triangles* in *b*, the side without MCLs; *closed triangle* in *b*, the side with MCLs

criterion is that MNPs be non-toxic. Due to these requirements, magnetite (Fe_3O_4) and maghemite (γ-Fe_2O_3) nanoparticles have been focused. Maghemite nanoparticles are synthesized by the oxidation of magnetite nanoparticles above 300 °C, and the procedure to produce magnetite nanoparticles is simpler than that of maghemite nanoparticles. Since the heating properties of magnetite and maghemite are comparable, magnetite nanoparticles have been used in most studies of MNPs for MNHT.

As previously reviewed [10, 11], heat generation through AMF exposure is caused by Néel relaxation loss and hysteresis loss. Néel relaxation occurs in a superparamagnetic nanoparticle with a single giant magnetic moment and the relaxation loss under AMF causes by flip of magnetic moment. Magnetic hysteresis observes in a magnetic material with residual magnetism and hysteresis loss under AMF occurs when magnetized materials are demagnetized. Methods for MNPs preparation have been reviewed [4, 7, 12–14], and representative MNPs developed for MNHT are shown in Table 13.1.

Table 13.1 Representative magnetic nanoparticles for MNHT

Name	Core size	Characters	References
Magnetite	a few μm	The first report using magnetic particles	[15]
Dextran magnetite	6 nm	The first report using magnetite in nm size	[16]
Aminosilane-coated magnetite	15 nm	Enhances the uptake by cancer cells, prevents intracellular digestion	[5]
Magnetite cationic liposome (MCL)	10 nm	Enhances the uptake by cancer cells, stabilizes the colloidal solution	[6, 17]
NPrCAP-conjugated magnetite (NPrCAP/M)	10 nm	Targets to melanoma cells, exerts chemotherapeutic effects	[22, 23]
Oligosaccharides-conjugated magnetoliposome (OML)	20 nm	Targets to macrophages	[24]
Antibody-conjugated magnetite	10 nm	Targets to human breast cancer, conjugated with radioactive indium	[25]
Antibody-conjugated magnetoliposome	10 nm	Targets to tumor cells, stabilizes the colloidal solution	[26–29]
Magnetoliposomes with encapsulated antitumor drug	20–30 nm	Controlled drug release	[31, 32]
MNPs in needle shape	10 nm	Organs with high blood flow	[33, 34]

The history of MNHT started in 1957: Gilchrist et al used magnetite particles (diameter: a few μm) to induce heat generation in the lymph nodes of dog [15]. In 1979, Gordon et al first developed a dextran-magnetite nanoparticles (diameter: 6 nm) to improve the stability of the colloidal MNPs solution [16]. For drug delivery system, liposomal coatings provide a promising approach, and magnetite cationic liposomes (MCLs) were developed to improve accumulation in tumor tissues [17, 18]. The efficacy of MNHT using MCLs on cancer therapy has been examined and reviewed [6, 19, 20]. MCLs were used in animals with several types of tumors, including mouse B16 melanoma, mouse MM46 mammary carcinoma, human PC3 and LNCaP prostate cancer cells in nude mice, spontaneously occurring primary melanoma in transgenic mice, rat T-9 glioma, rat PLS10 prostate cancer, hamster Os515 osteosarcoma, VX-7 squamous cell carcinomas in rabbit tongue, and human BT474 (HER2-positive) breast cancer cells in nude mice. In these experiments, repeated hyperthermia was conducted and complete tumor regression was achieved totally 96 % (46/48) of the animals. Recently, Cheraghipour et al. [21] also developed cationic albumin-conjugated MNPs to stabilize the colloidal MNPs solution and their MNPs improved adsorption to cells.

Another promising approach for applying MNHT is selective accumulation of MNPs to tumor tissue by utilizing biological materials and/or metabolites uniquely expressed in tumor cells. For example, melanoma possesses tyrosinase as a melanin-forming enzyme which plays a role in the biosynthesis of melanin pigments. A melanoma-targeted MNP was developed by directly conjugating a tyrosine homolog, N-propionyl-cysteaminylphenol (NPrCAP) that is a tyrosinase substrate, or indirectly doing NPrCAP with an end of the polyethylene glycol chain on the surface of dextran-coated MNPs [22, 23]. When the NPrCAP-conjugated MNPs (NPrCAP/M) were added to cultures of various cell lines *in vitro*, no substantial uptake by non-melanoma cells was observed, whereas the MNPs uptake by melanoma cell lines was observed in higher percentage [23]. In an *in vivo* study, mouse B16F1 melanoma and EG7 lymphoma cells were injected into the left and right flank of mice, respectively. The mice received daily intraperitoneal administration of NPrCAP/M. On the 15th day after injection of tumor cells, black brown MNPs were observed in B16F1 tumors whereas no substantial accumulation of the MNPs was observed in EG7 non-melanoma tumors [23].

In 2006, Ikehara et al reported that they developed oligosaccharides-conjugated magnetoliposomes (OMLs) by encapsulating MNPs in oligosaccharide-conjugated liposomes [24]. When mannotriose is conjugated with magnetoliposomes, macrophages specifically recognize OMLs through carbohydrate receptors such as the macrophage-mannose-receptor (CD206), and the OMLs can be used for targeting macrophages.

Using monoclonal antibody (mAb) against tumors is a convincing approach for tumor targeting. MNPs were conjugated with mAb against human breast cancer cells [25], human mAb against renal cell carcinoma [26], humanized mAb against human epidermal growth factor receptor-2 (HER2) [27] or its aptamer [28], and mAb against epidermal growth factor receptor (EGFR) [29]. When such nanoparticles were administrated systemically, they could accumulate specifically within tumors in animal models [25, 26, 29]. Recent advances of such targeted MNPs were reviewed by Yang et al. [30]. They mentioned wide application of MNPs in drug delivery, targeted therapy, molecular imaging, and therapeutic decision-making and monitoring for tumor [30].

Some groups developed the thermo-sensitive liposomes encapsulating both MNPs and anti-cancer drug [31, 32]. When a temperature at the target region was increased to 42–44 °C, the liposome would be caused to "melt" and the entrapped drug would be rapidly released from the liposomes.

When tumors are located in organs where the blood flow is high, it is difficult to increase the tumor temperature against the cooling effect of blood flow. To overcome this problem, a magnetite needle, in which MNPs were molded with carboxymethyl cellulose into a needle shape, was developed [33, 34]. The needle was able to prick tumors easily, and the operator could precisely indwell the needle at the tumor tissue. Then, the temperature of tumors in organs with a high blood flow increased very rapidly in MNHT due to the high MNPs concentration. After the hyperthermia treatment, MNPs were distributed within the tumor. These results indicate that the magnetite needles are very simple and effective particulate heating mediators.

One of the characteristics of MNPs is the spatial distribution of MNPs within tumor tissues after repeated hyperthermia as shown in Fig.13.2 [35]. When MCLs were injected directly into tumors, they remained at the injected site due to the electrostatic interactions. When AMF was applied, the temperature of the MNPs increased and the tumor tissues near the MNPs were destroyed. The MNPs then diffused and spread in this necrotic area. Additional tumor destruction occurred

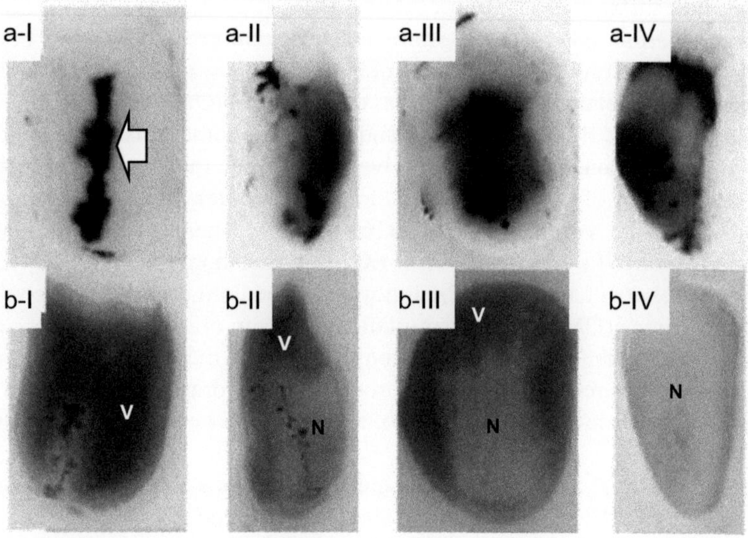

Fig. 13.2 Photographs of tumor specimens [35]. Tumors were resected at 24 h after MNHT treatment using MCLs and were (**a**) paraffinized and (**b**) histologically stained with hematoxylin-eosin. I: without AMF irradiation, II: irradiated once for 30 min, III: irradiated twice for 30 min, IV: irradiated three times for 30 min. An arrow indicates MCLs. N and V in the photographs indicate necrotic tissues and viable tissues, respectively

after repeated hyperthermia, and the MNPs expanded more into the entire tumor tissue. The removal of MNPs is not necessary after repeated hyperthermia, because MNPs are carried away by blood flow after several hyperthermic treatments.

13.3 Magnetic Field Applicators for MNHT

Some companies and institutes have developed AMF applicators for MNHT. MagForce Nanotechnologies in Germany developed MFH 300 F [36]. The MFH 300 F is a 100 kHz "gap type" magnetic field applicator with a C-type toroidal core, and patients are placed inside the gap adjustable from 21 to 45 cm. The MFH 300 F generates a magnetic field intensity of up to 18 kA/m in a cylindrical treatment space 20 cm in diameter and an aperture height of up to 30 cm [36]. In Japan, Mitsumori et al developed a 100-kHz gap-type AMF generator (the gap: 25 cm) with a C-type toroidal core [37]. The magnetic field was uniform in the 10 cm × 10 cm area between the cores.

Some companies and academic researchers developed solenoid type applicators in Japan. Yamamoto Vinyter Co. and Tazawa et al developed the 500 kHz inductive heating system TY-1 with four rolls of water-cooled hollow coils [38]. In 1996, Dai-Ichi High Frequency Co. and Kobayashi et al developed a 118 kHz solenoid type applicator with a horizontal coil (inner diameter, 7 cm; length, 7 cm) for

MNHT [17]. In preclinical studies, animals were placed inside of the coil and an AMF was irradiated [18].

For clinical application, an inevitable technical problem is the difficulty in the scale up of the coil size in the solenoid type applicator, because the large coil is accompanied by a risk associated with high voltage. A new device called a "ferrite core-inserted solenoid type" was thus developed [39]. A MnZn ferrite core was inserted into solenoid to concentrate magnetic fields generated by a coil. An AMF was generated by a vertical coil (inner diameter, 7 cm; length, 9 cm) driven by a transistor inverter at 360 kHz. Using the ferrite core-inserted solenoid type applicator, Kawai et al. reported that a MCL-embedded phantom 10 mm distant from the ferrite core-inserted solenoid showed an increase in temperature of 8 °C within 5 min [39], suggesting that the ferrite core concentrated and emitted magnetic field. These results suggested that the ferrite core-inserted solenoid type applicator was safe and suitable for heating a target positioned outside of the coils.

13.4 Induction of Antitumor Immunity by MNHT

Conventional hyperthermia modulates both innate and adaptive immune responses. Terunuma in Chap. 12 of this book [40] and Frey et al. [41] reviewed hyperthermia-induced modulations of the immune system. The immune responses in conventional hyperthermia may be essentially the same as those in MNHT. However, the temperatures in tumor tissues are normally below the threshold temperature of 42.5 °C in conventional hyperthermia, whereas those are above the threshold temperature in MNHT. Therefore, there are some differences between them. The expression of HSPs is induced depending on the temperature of tumor tissues. HSPs are abundant intracellular proteins that function as molecular chaperons which prevent the aggregation of proteins [42]. HSP expression has been considered to be a complicating factor that causes thermotolerance in conventional hyperthermia, since the expression of HSPs protects cells from heat-induced apoptosis [43]. Therefore, conventional hyperthermia treatments have been performed only once or twice per week at an interval of more than 48 h in order to prevent the thermotolerance [44]. Yanase et al. [18] and Ito et al. [45], however, found that thermotolerance is overcome in MNHT and repeated hyperthermia treatment at an interval of 24 h is very effective for the complete regression of tumor tissues and causes mainly tumor necrosis [45]. They also reported that a large amount of HSP70 was induced in MNHT.

In addition to direct killing effects of heating, Yanase et al [8] observed a very strong induction of anti-tumor immune response by MNHT in an experimental T-9 rat glioma model. When the rats that had been cured by MNHT were re-challenged with T-9 cells 3 months later, tumor take was observed in the rats but the tumor size decreased 12 days later and then disappeared. Tumor-infiltrating T lymphocytes such as $CD8^+$ and $CD4^+$ T cells were observed both right and left tumors and tumor-specific cytotoxic T-lymphocyte (CTL) activity was augmented in the rats treated

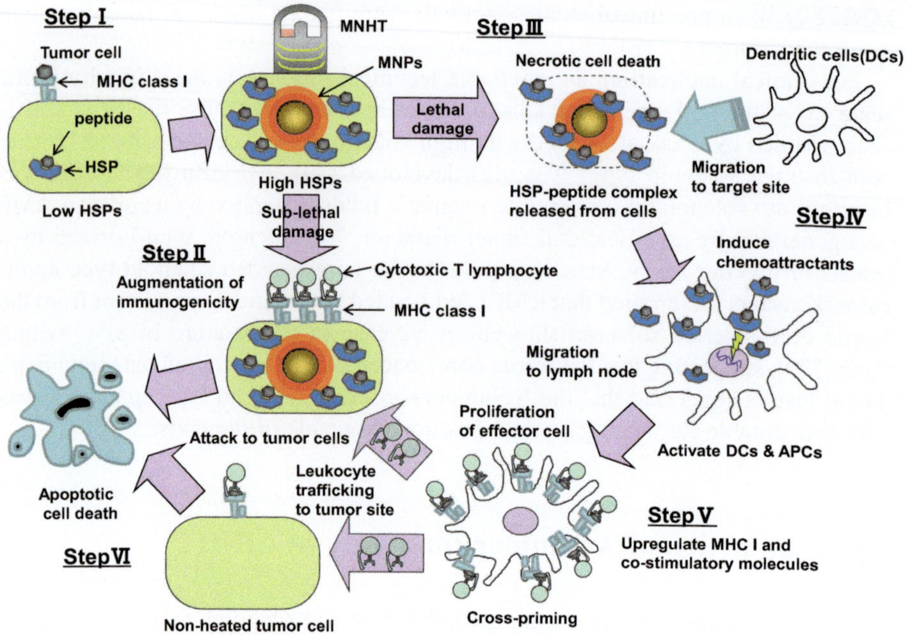

Fig. 13.3 Mechanism of induction of anti-tumor immune response by MNHT. (Step I) Impaired tumor immune responses could be due to poorly immunogenic tumor cells with a low concentration of intracellular HSP-peptide complexes. (Step II) Increased levels of intracellular HSP-peptide complexes, enhance processing of endogenous antigens and increase in the density of MHC class I-peptide complexes. (Step III) Dying tumor cells release their intracellular contents, including HSP-peptide complexes. (Step IV) DCs activate neighboring monocytes in order to produce pro-inflammatory cytokines and recruit APCs. (Step V) The HSP-peptide complexes are taken up by DCs and are in turn presented to T cells via MHC class I (cross-priming) and/or II antigens. (Step VI) Tumor-specific CTLs generated by cross-priming attack to non-heated tumor cells as well as sub-lethal damaged tumor cells, and result in apoptotic cell death. *APC* antigen-presenting cell, *DC* dendritic cell, *HSP* heat-shock protein

with MNHT. Jimbow et al. also reported the anti-tumor immune responses induced by MNHT in mouse B16 melanoma models [22, 23]. Thus, MNHT can kill not only heated tumors, but also non-heated tumors, including metastatic cancer cells. Clinically, it is very important that the abscopal effects reported in radiotherapy [9] occur also in MNHT.

The anti-tumor effects of MNHT may consist of both direct killing by heating and anti-tumor immune response. A proposed mechanism [46–48] of HSPs-mediated anti-tumor responses after MNHT is as follows (Fig. 13.3);

Step I) Poorly immunogenic tumor cells could be due to low concentrations of HSP-peptide complexes, which causes insufficient endogenous antigen processing, resulted in low levels of antigen-presenting by major histocompatibility complexes (MHC) class I molecules at the cell surfaces.

Step II) Sublethal heat stress by MNHT induces HSPs expression at a high level, resulted in increased level of HSP-peptide complexes, which causes the enhancement of endogenous antigen processing and augmentation of antigen-presenting by MHC class I at cell surfaces. These tumor cells are recognized by tumor-infiltrating CTLs, and then resulted in apoptotic cell death.

Step III) Lethal heat stress by MNHT induces necrotic cell death, and intracellular molecules including HSP-peptide complexes are released from dying tumor cells.

Step IV) The HSPs released from the dying tumor cells induce the production of pro-inflammatory cytokines and the recruitment of antigen-presenting cells (APCs). Asea et al. reported that HSP70 stimulated monocytes to secrete cytokines and induced maturation of dendritic cells (DCs) through CD14 and/or CD91 receptors [49], suggesting that HSP70 is a natural adjuvant [50]. This cytokine-like activity stimulating the innate immune response did not depend on whether HSP70 chaperoned antigenic peptides or not [51].

Step V) The HSP-peptide complexes released from dying tumor cells are taken up by $CD91^+$ and $CD40^+$ APCs [52, 53]. HSP-peptide complexes are then endocytosed via CD91 and CD40 receptors, and the antigenic peptides are transported to the endoplasmic reticulum to bind to MHC class I and/or class II. Thus, HSPs can allow peptides to enter both MHC class I and class II pathways. When the peptides are processed by the endogenous MHC class I pathway, it is called as "cross-priming". Antigen capturing DCs migrate into lymph nodes, change to matured APCs, and present antigen peptides to T cells.

Step VI) CTLs generated by cross-priming in lymph nodes are expected to infiltrate into the tumor, exert strong anti-tumor activity against the non-heated tumor cells as well as the sub-lethal damaged tumor cells, and result in apoptotic cell death.

Clinically, tumor-bearing hosts are known to be immunosuppressive. To make a stronger anti-tumor immune response, the enhancement of antitumor immunity caused by MNHT should be required. Based on this strategy, successful approaches have been shown in preclinical trials using animals [54–56]. Granulocyte-macrophage colony-stimulating factor (GM-CSF) is a possible candidate to enhance the therapeutic effects of MNHT, because GM-CSF activates APCs [57]. As another possible candidate, interleukin-2 (IL-2) is a potent stimulator of lymphocyte proliferation and augments the activity of CTL [58]. Injection of GM-CSF and IL-2 after MNHT showed significantly improved results [54]. DCs are known to be the most potent APCs in immune response to tumors [59]. To improve antigen-presentation and cross-presentation after MNHT, injection of immature DCs directly into tumors showed significant therapeutic effects [55, 56].

13.5 Clinical Trials for MNHT

Clinical trials of conventional hyperthermia treatment have been conducted [2]. Many of them were combined with chemotherapy or radiotherapy. In the case of MNHT, several clinical trials have been reported as follows.

In 2005, Johannsen et al [60] described the first clinical application of MNHT. In this study, they evaluated the feasibility of MNHT using aminosilane-coated magnetite nanoparticles and MFH 300 F apparatus in locally recurrent prostate cancer. A 67 year-old patient was injected with the MNPs trans-peritoneally into the prostate under trans-rectal ultrasound and fluoroscopy guidance. Later, they reported more detailed clinical results for prostate cancer in 2007 [61, 62] and glioblastoma in 2009 and 2011 [63, 64].

Matsumine et al. [65] developed an injectable calcium phosphate cement containing MNPs for patients with bone metastases in the extremities. Calcium phosphate cement containing MNPs was injected to fill the bone defect following resection of bone tumors, and AMF was applied totally 10 times from day 8 to day 29 after surgical operation for 23 patients with 25 bone metastases. The radiographic results showed an 'Excellent' (reduced with visible bone formation) in 8 lesions (32 %), 'Good' (not progressive for more than 3 months) in 16 lesions (64 %), and 'Poor' (progressive) in one lesion (4 %).

Recently, Jimbow et al [66] reported a phase I clinical trial of MHNT using NPrCAP/M for stage III and IV melanoma patients. Four patients received NPrCAP/M injections directly into a melanoma and an AMF was irradiated three times every other day. They reported that a stage IV patient showed remarkable regression of distant skin metastases and survival of 30 months, and a stage III patient showed regression of local lymph node metastases and survival of more than 32 months.

In 2011, Imai et al. [67] reported results of a phase I clinical trial of MNHT using MCLs for patients with head and neck, breast, or soft tissue malignant tumors refractory to conventional therapy. The patients received intratumoral injection of MCLs, and AMF was irradiated at 43 °C for 30 min twice at a 24 h-interval by monitoring the temperature of the skin overlying the tumor. A pathological assay revealed necrosis changes with more than a third part of the tumor area at 2 weeks after MCL injection. Temperature was successfully increased up to 43 °C in all six patients and no significant adverse effects were observed.

13.6 Closing Comments

Continuous efforts have been made in order to construct a safe and effective MNHT system. Finally, some researchers are opening the door to clinical trials, suggesting that the time has come for MNHT as a promising cancer therapy. Once clinically available functional MNPs and AMF applicators are developed, clinicians will excitedly start MNHT and many potent combination therapies based on MNHT will also be proposed for clinical trials. The authors hope that MNHT will provide a novel effective therapy for cancer patients.

References

1. Dewhirst MW, Prosnitz L, Thrall D, Prescott D, Clegg S, Charles C, et al. Hyperthermic treatment of malignant diseases: current status and a view toward the future. Semin Oncol. 1997;24(6):616–25.
2. van der Zee J. Heating the patient: a promising approach? Ann Oncol. 2002;13(8):1173–84.
3. Bauer KD, Henle KJ. Arrhenius analysis of heat survival curves from normal and thermotolerant CHO cells. Radiat Res. 1979;78(2):251–63.
4. Ito A, Shinkai M, Honda H, Kobayashi T. Medical application of functionalized magnetic nanoparticles. J Biosci Bioeng. 2005;100(1):1–11. doi:10.1263/jbb.100.1.
5. Jordan A, Wust P, Fahling H, John W, Hinz A, Felix R. Inductive heating of ferrimagnetic particles and magnetic fluids: physical evaluation of their potential for hyperthermia. Int J Hyperthermia. 1993;9(1):51–68.
6. Kobayashi T. Cancer hyperthermia using magnetic nanoparticles. Biotechnol J. 2011;6 (11):1342–7. doi:10.1002/biot.201100045.
7. Moroz P, Jones SK, Gray BN. Magnetically mediated hyperthermia: current status and future directions. Int J Hyperthermia. 2002;18(4):267–84. doi:10.1080/02656730110108785.
8. Yanase M, Shinkai M, Honda H, Wakabayashi T, Yoshida J, Kobayashi T. Antitumor immunity induction by intracellular hyperthermia using magnetite cationic liposomes. Jpn J Cancer Res. 1998;89(7):775–82.
9. Postow MA, Callahan MK, Barker CA, Yamada Y, Yuan J, Kitano S, et al. Immunologic correlates of the abscopal effect in a patient with melanoma. N Engl J Med. 2012;366 (10):925–31. doi:10.1056/NEJMoa1112824.
10. Dennis CL, Ivkov R. Physics of heat generation using magnetic nanoparticles for hyperthermia. Int J Hyperthermia. 2013;29(8):715–29. doi:10.3109/02656736.2013.836758.
11. Hergt R, Dutz S, Roder M. Effects of size distribution on hysteresis losses of magnetic nanoparticles for hyperthermia. J Phys Condens Matter. 2008;20(38):385214. doi:10.1088/0953-8984/20/38/385214.
12. Gazeau F, Levy M, Wilhelm C. Optimizing magnetic nanoparticle design for nanothermotherapy. Nanomedicine (Lond). 2008;3(6):831–44. doi:10.2217/17435889.3.6.831.
13. Thiesen B, Jordan A. Clinical applications of magnetic nanoparticles for hyperthermia. Int J Hyperthermia. 2008;24(6):467–74. doi:10.1080/02656730802104757.
14. Wankhede M, Bouras A, Kaluzova M, Hadjipanayis CG. Magnetic nanoparticles: an emerging technology for malignant brain tumor imaging and therapy. Expert Rev Clin Pharmacol. 2012;5(2):173–86. doi:10.1586/ecp.12.1.
15. Gilchrist RK, Medal R, Shorey WD, Hanselman RC, Parrott JC, Taylor CB. Selective inductive heating of lymph nodes. Ann Surg. 1957;146(4):596–606.
16. Gordon RT, Hines JR, Gordon D. Intracellular hyperthermia. A biophysical approach to cancer treatment via intracellular temperature and biophysical alterations. Med Hypotheses. 1979;5 (1):83–102.
17. Shinkai M, Yanase M, Honda H, Wakabayashi T, Yoshida J, Kobayashi T. Intracellular hyperthermia for cancer using magnetite cationic liposomes: in vitro study. Jpn J Cancer Res. 1996;87(11):1179–83.

18. Yanase M, Shinkai M, Honda H, Wakabayashi T, Yoshida J, Kobayashi T. Intracellular hyperthermia for cancer using magnetite cationic liposomes: an in vivo study. Jpn J Cancer Res. 1998;89(4):463–9.
19. Ito A, Kobayashi T. Intracellular hyperthermia using magnetic nanoparticles: A novel method for hyperthermia clinical applications. Thermal Med. 2008;24:113–29.
20. Kobayashi T. Cancer hyperthermia using nanomagnetic particles and induction of immune responses. In: Torchilin V, editor. Handbook of nanobiomedical research: fundamentals, applications and recent developments, Frontiers in nanobiomedical research, vol. 3. Singapore: World Scientific Publishing Company; 2014. p. 465–99. ISSN of vol 3 is 2251–3965 and ISBN is 978-9814520690
21. Cheraghipour E, Javadpour S. Cationic albumin-conjugated magnetite nanoparticles, novel candidate for hyperthermia cancer therapy. Int J Hyperthermia. 2013;29(6):511–9. doi:10.3109/02656736.2013.803605.
22. Jimbow K, Tamura Y, Yoneta A, Kamiya T, Ono I, Yamashita T, et al. Conjugation of magnetite nanoparticles with melanogenesis substrate, NPrCAP provides melanoma targeted, in situ peptide vaccine immunotherapy through HSP production by chemo-thermotherapy. J Biomat Nanobiot. 2012;3(2):140–53.
23. Sato M, Yamashita T, Ohkura M, Osai Y, Sato A, Takada T, et al. N-propionyl-cysteaminylphenol-magnetite conjugate (NPrCAP/M) is a nanoparticle for the targeted growth suppression of melanoma cells. J Invest Dermatol. 2009;129(9):2233–41. doi:10.1038/jid.2009.39.
24. Ikehara Y, Niwa T, Biao L, Ikehara SK, Ohashi N, Kobayashi T, et al. A carbohydrate recognition-based drug delivery and controlled release system using intraperitoneal macrophages as a cellular vehicle. Cancer Res. 2006;66(17):8740–8. doi:10.1158/0008-5472.CAN-06-0470.
25. DeNardo SJ, DeNardo GL, Natarajan A, Miers LA, Foreman AR, Gruettner C, et al. Thermal dosimetry predictive of efficacy of 111In-ChL6 nanoparticle AMF – induced thermoablative therapy for human breast cancer in mice. J Nucl Med. 2007;48(3):437–44.
26. Shinkai M, Le B, Honda H, Yoshikawa K, Shimizu K, Saga S, et al. Targeting hyperthermia for renal cell carcinoma using human MN antigen-specific magnetoliposomes. Jpn J Cancer Res. 2001;92(10):1138–45.
27. Kikumori T, Kobayashi T, Sawaki M, Imai T. Anti-cancer effect of hyperthermia on breast cancer by magnetite nanoparticle-loaded anti-HER2 immunoliposomes. Breast Cancer Res Treat. 2009;113(3):435–41. doi:10.1007/s10549-008-9948-x.
28. Pala K, Serwotka A, Jelen F, Jakimowicz P, Otlewski J. Tumor-specific hyperthermia with aptamer-tagged superparamagnetic nanoparticles. Int J Nanomedicine. 2014;9:67–76. doi:10.2147/IJN.S52539.
29. Hadjipanayis CG, Machaidze R, Kaluzova M, Wang L, Schuette AJ, Chen H, et al. EGFRvIII antibody-conjugated iron oxide nanoparticles for magnetic resonance imaging-guided convection-enhanced delivery and targeted therapy of glioblastoma. Cancer Res. 2010;70(15):6303–12. doi:10.1158/0008-5472.CAN-10-1022.
30. Yang HW, Hua MY, Liu HL, Huang CY, Wei KC. Potential of magnetic nanoparticles for targeted drug delivery. Nanotechnol Sci Appl. 2012;5:73–86. doi:10.2147/NSA.S35506.
31. Ponce AM, Vujaskovic Z, Yuan F, Needham D, Dewhirst MW. Hyperthermia mediated liposomal drug delivery. Int J Hyperthermia. 2006;22(3):205–13. doi:10.1080/02656730600582956.
32. Yatvin MB, Weinstein JN, Dennis WH, Blumenthal R. Design of liposomes for enhanced local release of drugs by hyperthermia. Science. 1978;202(4374):1290–3.
33. Ohno T, Wakabayashi T, Takemura A, Yoshida J, Ito A, Shinkai M, et al. Effective solitary hyperthermia treatment of malignant glioma using stick type CMC-magnetite. In vivo study. J Neurooncol. 2002;56(3):233–9.
34. Shinkai M, Ueno K, Honda H, Kobayashi T. Magnetite needle as heating mediator for intracellular hyperthermia of tumor. Jpn J Hyperther Oncol. 2002;18:191–8.

35. Ito A, Shinkai M, Honda H, Yoshikawa K, Saga S, Wakabayashi T, et al. Heat shock protein 70 expression induces antitumor immunity during intracellular hyperthermia using magnetite nanoparticles. Cancer Immunol Immunother. 2003;52(2):80–8. doi:10.1007/s00262-002-0335-x.
36. Gneveckow U, Jordan A, Scholz R, Bruss V, Waldofner N, Ricke J, et al. Description and characterization of the novel hyperthermia- and thermoablation-system MFH 300 F for clinical magnetic fluid hyperthermia. Med Phys. 2004;31(6):1444–51.
37. Mitsumori M, Hiraoka M, Shibata T, Okuno Y, Nagata Y, Nishimura Y, et al. Targeted hyperthermia using dextran magnetite complex: a new treatment modality for liver tumors. Hepatogastroenterology. 1996;43(12):1431–7.
38. Wada S, Tazawa K, Furuta I, Nagae H. Antitumor effect of new local hyperthermia using dextran magnetite complex in hamster tongue carcinoma. Oral Dis. 2003;9(4):218–23.
39. Kawai N, Ito A, Nakahara Y, Honda H, Kobayashi T, Futakuchi M, et al. Complete regression of experimental prostate cancer in nude mice by repeated hyperthermia using magnetite cationic liposomes and a newly developed solenoid containing a ferrite core. Prostate. 2006;66(7):718–27. doi:10.1002/pros.20394.
40. Terunuma H. Enhancement of hyperthermia on immune systems. In: Ohnishi T, Yoshikawa T, Kokura S, editors. Hyperthermic Oncology from Bench to Bed. London: Springer; 2016.
41. Frey B, Weiss EM, Rubner Y, Wunderlich R, Ott OJ, Sauer R, et al. Old and new facts about hyperthermia-induced modulations of the immune system. Int J Hyperthermia. 2012;28 (6):528–42. doi:10.3109/02656736.2012.677933.
42. Lindquist S. The heat-shock response. Annu Rev Biochem. 1986;55:1151–91. doi:10.1146/annurev.bi.55.070186.005443.
43. Mosser DD, Caron AW, Bourget L, Meriin AB, Sherman MY, Morimoto RI, et al. The chaperone function of hsp70 is required for protection against stress-induced apoptosis. Mol Cell Biol. 2000;20(19):7146–59.
44. Subjeck JR, Sciandra JJ, Johnson RJ. Heat shock proteins and thermotolerance; a comparison of induction kinetics. Br J Radiol. 1982;55(656):579–84. doi:10.1259/0007-1285-55-656-579.
45. Ito A, Tanaka K, Honda H, Abe S, Yamaguchi H, Kobayashi T. Complete regression of mouse mammary carcinoma with a size greater than 15 mm by frequent repeated hyperthermia using magnetite nanoparticles. J Biosci Bioeng. 2003;96(4):364–9. doi:10.1016/S1389-1723(03)90138-1.
46. Ito A, Honda H, Kobayashi T. Cancer immunotherapy based on intracellular hyperthermia using magnetite nanoparticles: a novel concept of "heat-controlled necrosis" with heat shock protein expression. Cancer Immunol Immunother. 2006;55(3):320–8. doi:10.1007/s00262-005-0049-y.
47. Ito A, Kobayashi T, Honda H. A mechanism of anti-tumor immunity induced by hyperthermia. Jpn J Hyperther Oncol. 2005;21:1–19.
48. Kobayashi T, Kakimi K, Nakayama E, Jimbow K. Antitumor immunity by magnetic nanoparticle-mediated hyperthermia. Nanomedicine (Lond). 2014;9(11):1715–26. doi:10.2217/nnm.14.106.
49. Asea A, Kraeft SK, Kurt-Jones EA, Stevenson MA, Chen LB, Finberg RW, et al. HSP70 stimulates cytokine production through a CD14-dependant pathway, demonstrating its dual role as a chaperone and cytokine. Nat Med. 2000;6(4):435–42. doi:10.1038/74697.
50. Vanaja DK, Grossmann ME, Celis E, Young CY. Tumor prevention and antitumor immunity with heat shock protein 70 induced by 15-deoxy-delta12,14-prostaglandin J2 in transgenic adenocarcinoma of mouse prostate cells. Cancer Res. 2000;60(17):4714–8.
51. Srivastava P. Roles of heat-shock proteins in innate and adaptive immunity. Nat Rev Immunol. 2002;2(3):185–94. doi:10.1038/nri749.
52. Basu S, Binder RJ, Ramalingam T, Srivastava PK. CD91 is a common receptor for heat shock proteins gp96, hsp90, hsp70, and calreticulin. Immunity. 2001;14(3):303–13.
53. Binder RJ. CD40-independent engagement of mammalian hsp70 by antigen-presenting cells. J Immunol. 2009;182(11):6844–50. doi:10.4049/jimmunol.0900026.

54. Ito A, Tanaka K, Kondo K, Shinkai M, Honda H, Matsumoto K, et al. Tumor regression by combined immunotherapy and hyperthermia using magnetic nanoparticles in an experimental subcutaneous murine melanoma. Cancer Sci. 2003;94(3):308–13.
55. Tanaka K, Ito A, Kobayashi T, Kawamura T, Shimada S, Matsumoto K, et al. Heat immunotherapy using magnetic nanoparticles and dendritic cells for T-lymphoma. J Biosci Bioeng. 2005;100(1):112–5. doi:10.1263/jbb.100.112.
56. Tanaka K, Ito A, Kobayashi T, Kawamura T, Shimada S, Matsumoto K, et al. Intratumoral injection of immature dendritic cells enhances antitumor effect of hyperthermia using magnetic nanoparticles. Int J Cancer. 2005;116(4):624–33. doi:10.1002/ijc.21061.
57. Steinman RM. The dendritic cell system and its role in immunogenicity. Annu Rev Immunol. 1991;9:271–96. doi:10.1146/annurev.iy.09.040191.001415.
58. Farrar WL, Johnson HM, Farrar JJ. Regulation of the production of immune interferon and cytotoxic T lymphocytes by interleukin 2. J Immunol. 1981;126(3):1120–5.
59. Winzler C, Rovere P, Rescigno M, Granucci F, Penna G, Adorini L, et al. Maturation stages of mouse dendritic cells in growth factor-dependent long-term cultures. J Exp Med. 1997;185 (2):317–28.
60. Johannsen M, Gneveckow U, Eckelt L, Feussner A, Waldofner N, Scholz R, et al. Clinical hyperthermia of prostate cancer using magnetic nanoparticles: presentation of a new interstitial technique. Int J Hyperthermia. 2005;21(7):637–47.
61. Johannsen M, Gneveckow U, Taymoorian K, Thiesen B, Waldofner N, Scholz R, et al. Morbidity and quality of life during thermotherapy using magnetic nanoparticles in locally recurrent prostate cancer: results of a prospective phase I trial. Int J Hyperthermia. 2007;23(3):315–23. doi:10.1080/02656730601175479.
62. Johannsen M, Gneveckow U, Thiesen B, Taymoorian K, Cho CH, Waldofner N, et al. Thermotherapy of prostate cancer using magnetic nanoparticles: feasibility, imaging, and three-dimensional temperature distribution. Eur Urol. 2007;52(6):1653–61. doi:10.1016/j.eururo.2006.11.023.
63. Maier-Hauff K, Ulrich F, Nestler D, Niehoff H, Wust P, Thiesen B, et al. Efficacy and safety of intratumoral thermotherapy using magnetic iron-oxide nanoparticles combined with external beam radiotherapy on patients with recurrent glioblastoma multiforme. J Neurooncol. 2011;103(2):317–24. doi:10.1007/s11060-010-0389-0.
64. van Landeghem FK, Maier-Hauff K, Jordan A, Hoffmann KT, Gneveckow U, Scholz R, et al. Post-mortem studies in glioblastoma patients treated with thermotherapy using magnetic nanoparticles. Biomaterials. 2009;30(1):52–7. doi:10.1016/j.biomaterials.2008.09.044.
65. Matsumine A, Takegami K, Asanuma K, Matsubara T, Nakamura T, Uchida A, et al. A novel hyperthermia treatment for bone metastases using magnetic materials. Int J Clin Oncol. 2011;16(2):101–8. doi:10.1007/s10147-011-0217-3.
66. Jimbow K, Ishii-Osai Y, Ito S, Tamura Y, Ito A, Yoneta A, et al. Melanoma-Targeted Chemothermotherapy and In Situ Peptide Immunotherapy through HSP Production by Using Melanogenesis Substrate, NPrCAP, and Magnetite Nanoparticles. J Skin Cancer. 2013;2013:742925. doi:10.1155/2013/742925.
67. Imai T, Kikumori T, Akiyama M, Yokoyama K, Nishida Y, Fujimoto Y et al. A phase I study of hyperthermia using magnetite cationic liposome and alternating magnetic field for various refractory malignancies. In: 28th annual meeting of Japan society for thermal medicine, Nagoya, Japan, S3-6. 2011. http://www.congre.co.jo/jstm28/. Accessed 9 Sept 2011.

Chapter 14
Tumor Microenvironment and Hyperthermia

Shin-ichiro Masunaga

Abstract Solid tumors have a more acidic and hypoxic microenvironment than normal tissue. This unfavorable microenvironment results from an imbalance in the oxygen supply and demand of the tumor tissue. To overcome hypoxia, the tumor induces a new vascular supply. This new vasculature, however, is inefficient and chaotic, resulting in preserving the factors that stimulated the neovascularization. This review focuses on these processes and particularly on angiogenesis, tumor vascular morphology, hypoxia, pH, and the metabolic-vascular events induced or following tumor tissue heating. The various mechanisms that either modulate tumor microenvironments or blood perfusion during hyperthermia are described, providing also the many clinical modalities that may enhance or sensitize cancer cells to heat.

Keywords Microenvironment • Hypoxia • Angiogenesis • pH • Blood flow • Mild temperature hyperthermia

14.1 Tumor Microenvironment

The tumor microenvironment is the cellular environment in which the tumor exists, including surrounding blood vessels, immune cells, fibroblasts, other cells, signaling molecules, and the extracellular matrix (ECM). The tumor and the surrounding microenvironment are closely related and interact constantly. Tumors can influence the microenvironment by releasing extracellular signals, promoting tumor angiogenesis and inducing peripheral immune tolerance, while the immune cells in the microenvironment can affect the growth and evolution of cancerous cells, such as in immune-editing [1].

Human solid tumors should be regarded as an intricate, yet poorly organized "organoid", whose function is maintained by a dynamic interplay between neoplastic and host cells. This interplay constitutes the tumor metabolic microenvironment

S.-i. Masunaga (✉)
Particle Radiation Biology, Division of Radiation Life Science, Research Reactor Institute, Kyoto University, 2-1010 Asashiro-nishi, Kumatori-cho, Sennan-gun, Osaka 590-0494, Japan
e-mail: smasuna@rri.kyoto-u.ac.jp

S. Kokura et al. (eds.), *Hyperthermic Oncology from Bench to Bedside*,
DOI 10.1007/978-981-10-0719-4_14

as a complex pathophysiological entity resulting from the interactions of self-influencing factors, which go hand in hand: tumor perfusion, tumor oxygenation status, pH distribution and metabolic-bioenergetic status [1, 2].

14.1.1 Hypoxia, HIF-1 and Angiogenesis

Up to a distance of 100–200 μm from the host vessel the initial foci of neoplastic cells receive their nutrients and oxygen by diffusion. Beyond this distance, hypoxia occurs and the need for an adequate blood supply is crucial. However, the establishment of this neovascular supply in the attempt to overcome hypoxia is inefficient and irregular. It may not occur at the same rate as the proliferation of the tumor. The result is the persistence within the tumor mass including the heterogeneous microregions of nonproliferating hypoxic cells, which are surrounded by vital well-nourished and proliferating cells [3, 4]. Hypoxic microenvironments are characterized by low oxygen tension, low extracellular pH (pHe), high interstitial fluid pressure, glucose deficiency, multidrug resistance, increased extracellular lactate concentration and the tendency to cause metastasis [3, 4].

The forming of new blood vessels depends on a balance between angiogenesis inhibitors and promoters. Tumor regional hypoxia and hypoglycemia are the principal stimulators for the expression of local proangiogenic cytokines, especially vascular endothelial growth factor (VEGF). An early response gene that produces hypoxia inducible factor-1 (HIF-1) and its subunits (HIF-1α and -1β) regulates VEGF expression. HIF-1 is a protein of 120 KDa, a member of the basic helix-loop-helix superfamily of transcription factors, and its expression is very sensitive to oxygen concentration (1 % O_2) [5].

The adaptation to hypoxia by earlier proliferating neoplastic cells results in the induction of genes that regulate anaerobic metabolism, nitric oxide synthase and the angiogenesis process. Recent studies have shown that HIF-1and VEGF transcripts are overexpressed by several human neoplastic cells including breast, prostate, gastric, colon, lung, bladder and endometrium and they are more active in hypoxic and necrotic areas [5, 6]. VEGF is correlated to vascular density, especially in brain tumors and it is associated with poor prognosis. VEGF or vascular permeability factor (VPF), is a 32 to 44-KDa multifunctional potent stimulator of endothelial cells. It becomes active through binding to three high affinity tyrosine kinase receptors [VEGFR-1(flt-1), VEGFR-2(KDR/Flk-1) and VEGFR-3(Flt-4)], which are highly expressed on endothelial tumor vessels but not on mature vessels. They exert different effects on endothelial cells (ECs). VEGFR-1 mediates cell motility of the ECs, whereas VEGFR-2 regulates vascular permeability and VEGFR-3 lymphangiogenesis [5, 7].

Hypoxia appears to be the principal stimulus for the production and the stabilization of VEGF and VEGF mRNA, but recent reports suggest that VEGF expression has increased in many tumors, even in the absence of hypoxia. This increase results from at least three factors: (a) loss of function of some tumor suppressor genes (such as the von Hippel-Lindau (VHL), p53, p16); (b) activation of oncogenes (including raf, ras, HER2/erb2 (neu) and src); (c) excessive quantity of growth factors, produced by tumor cells and their supporting network [5, 7].

14.1.2 Tumor Neovascularization

The acquisition of newly-developed vessels may occur by different mechanisms. ECs are normally quiescent and tightly regulated by a delicate balance between proangiogenic and antiangiogenic molecules. In the presence of an excessive secretion of angiogenic molecules in tumors, ECs are stimulated and they organize themselves in vessel structure through multistep sequential and distinct processes, depending on tumor type and anatomic localization. These processes include vessel-cooption, vasculogenesis and angiogenesis [5].

Vessel-Cooption This is the process in which tumors take up preexisting normal blood vessels and use them for their initial growth. This is a limiting process and is irrelevant in the great majority of solid tumors; in fact, cancer cells grow until oxygen demand exceeds supply and the distance from host vessels is less than 100–200 μm.

Vasculogenesis This is the mechanism in which precursors endothelial ECs from bone marrow are recruited by the tumor and aggregate to form new blood vessels. Recent studies have demonstrated this process in experimental animal tumors, but its relevance in human neoplasia is not fully elucidated.

Angiogenesis Upon adequate stimulus, endothelial cells begin to sprout from preexisting capillaries and after the degradation of the extracellular matrix (ECM) by matrix metalloproteases, and the expression of adhesion molecules such as $\alpha v\beta 3$ integrin, migrate and organize themselves in capillary tube formation and ultimately in a vascular network.

14.1.3 Tumor Vascular Morphology-Perfusion and Hypoxia

Blood vessels associated with the tumor tend to be significantly different in architecture from the surrounding normal tissue (Table 14.1) and they show, in the presence of VEGF and other cytokines, a decreased expression of leukocyte-endothelial cell adhesion molecules (ICAM-1, VCAM-1, E-selectin) and an enhanced expression of CD 44. The decreased expression of adhesion molecules significantly reduces leukocytes and natural killer cells (NKs) recruitment, partially contributing to the phenomenon of avoiding the immune system, whereas the enhanced expression of CD44 may confer a growth advantage on many neoplastic cells [5].

Table 14.1 Structural and functional abnormalities of the tumor vasculature

Decreased vascular density
Heterogeneous distribution of microvessels
Increased presence of sinusoids, dead ends and arterio-venous anastomosis
Lack of arteriolar vessels
Decreased perfusion pressure and increased geometric resistance, absence of lymphatics, increased TIF, hemoconcentration, increased viscosity
Incomplete basement membrane and absence of smooth muscle
Decreased stabilization for absence of pericytes and mural cells
Decreased leukocyte-endothelium adhesion
Absence of innervation and regulation according to metabolic demand

Moreover, VEGF induces tumor neovessels to become leakier and to lose a large quantity of fluids (proteins and other circulating macromolecules) towards the interstitium. Fluid accumulated inside the tumor interstitium, known as tumor interstitial fluid (TIF), occupies from 30 to 60 % of the tumor volume and compared with normal fluids it has a different biochemical nature. TIF retention causes an increase in tumor interstitial pressure (TIFP). The TIFP increase goes from the tumor center towards the periphery, reaching a value of 50 mmHg, as Jain and coworkers measured in human and experimental tumors on colon, breast, head-neck carcinomas and metastases specimens from the lung, liver and lymph nodes [5].

Tumor blood flow (TBF) is determined by arteriovenous pressure difference (Δ Pa-v) divided by a factor η for Z, where η expresses the flow resistance or blood viscosity and Z the geometric resistance.

$$\text{TBF} = \Delta\ \text{Pa-v}/\eta Z$$

Normally in tumors, Δ Pa-v is smaller than in normal microcirculation. This is caused by the increased number of arteriovenous fistulae present at the tumor periphery, which creates a low resistance pathway at the tumor surface and diverts blood from entering the tumor mass. Indicator dilution methods, angiography and experimental studies have confirmed this anomalous behaviour. With regard to low perfusional tumor supply, the viscosity parameters Z and η were abnormally high on tissue isolated mammary Adenocarcinoma (R3230AC) and on 22 carcinosarcoma. The geometric resistance Z is a complex function of vascular morphology (i.e., number of blood vessels, branching pattern, diameter, length and volume), and increases according to tumor size, length and vessel tortuosity. Tumor neovessels are tortuous, heterogeneous, inefficient and devoid of hierarchization [5].

Tumor blood viscosity η has been found to be elevated and correlated with the shear rate, hematocrit and circulating blood cells deformability.

Normally capillary diameter ranges from 3.5 to 20 μm, thus at some point along their travel in the capillaries, red blood cells (RBCs) as well as white cells (WBCs) have to undergo deformation to pass through. In contrast to the tumor microvasculature, various factors can modify RBCs and WBCs deformability. Particularly the low oxygen partial pressure, the acidic pH, and the increased concentration of fibrinogen tends to make red blood cells and leucocytes less deformable, and more sticky, thus more easily trapped intravascularly resulting in the blocking of the TBF.

A further modifying factor of viscosity is the local increase of the hematocrit. This phenomenon is to be ascribed to the fluid loss operating in the tumor capillaries with a consequent local hemo-concentration and further worsening of blood flow.

Blood flow measurements in human cancer show heterogeneous values going from highly vascularized organs, such as the brain, and poorly vascularized ones such as adipose tissue. Perfusion flow at the tumor level is higher or lower than in the tissue of origin, depending on the physiological state of the latter. It is higher at the tumor periphery than in the central zone and generally primary tumors are better supplied than metastatic lesions. In most studies, perfusion on rodent tumors decreases with tumor size compared with normal tissue. However, in some reports, the decreased flow was not confirmed even in a similar tumor type. Several pathophysiological mechanisms have been proposed to explain this difference such as the transplantation site, stage of tumor growth, flow registration and recording methods [2, 8].

TBF is not regulated according to metabolic demand as in the case of normal tissues. This decreased metabolic adaptability to cells related to irregular blood availability (perfusion) produces clusters of cells lacking nutrients and oxygen (hypoxic cells).

Two kinds of hypoxic or clusters of cells in a low energy state have been recognized: (A) Diffusion-limited or chronic hypoxic cells; (B) Transient or acute hypoxic cells. The two types of hypoxia have different origins and coexist together in a well-perfused zone of the tumor mass, causing a functional disturbance in macro- and micro-flow. These two situations change continuously because tumor blood flow fluctuates with time [3].

Chronic hypoxia is the result of the availability of nutrients and oxygen towards the tumor tissue, the diffusion in the extracellular space and the respiration rate of cancer cells. It has been calculated and observed that when cancer cells are >100–200 μm away from a functional blood supply they become hypoxic and suffering.

Cells adjacent to capillaries displayed a mean oxygen concentration of 2 %, those located at 200 μm displayed a mean oxygen concentration of 0.2 %. The distance from the nutritive vessels, the hemoglobin concentration and the blood flow crossing that tumor area are the only parameters responsible for chronic hypoxia. The removal of O_2 by tumor cells has been calculated to be more efficient or better than in normal tissues, showing that neoplastic cells *in vivo* do not have an impaired ability to utilize oxygen [2, 3].

Acute hypoxia is the result of the intermittent opening and reopening of tumor blood vessels. Among the various factors responsible for this temporary blood flow stop, the following seem the most reasonable [2, 3]:

1. TIFP combined with the irregular expansion of the tumor mass, whose three dimensional growth is subjected to a continuous re-modelling in a confined space, causes a temporary compression or occlusion of some tumor capillaries.
2. Transient stopping of tumor blood flow or supply by platelet plugs.

The majority of cancer patients have coagulation abnormalities related to hypoxia. Recently, it has been reported that hypoxia not only induces VEGF, but also stimulates endothelial cells to overexpress tissue factor (TF) and plasminogen activator inhibitor (PAI-I). These factors induce the endothelium to become prothrombotic and cause fibrin formation and platelet activation. Further, VEGF binds to fibrinogen and fibrin by stimulating endothelial cell proliferation. Fibrin has been demonstrated to be essential for supporting the spreading and migration of endothelial cells. The hemostatic system, in a certain sense, becomes a regulator of angiogenesis and this can partially explain its regional appearance and disappearance of acute hypoxia. Hypoxia becomes a self-prolonging mechanism able to trigger angiogenesis, intratumoral fluid accumulation and thrombosis [2, 5, 8].

14.2 Tumor Energetic Status, Hypoxia, and pH

Normally, cancer cells display many altered metabolic abnormalities including an increased capacity to metabolize carbohydrates mainly by anaerobic glycolysis even under aerobic conditions. This metabolic behavior results from the induction of many enzymes involved in the intermediate reactions of glycolysis by HIF-1 genes [9]. The relevance of these alterations is that the oxidation of glucose stops at the stage of pyruvic acid and proceeds anaerobically producing six times more lactic acid and H^+ than normal cells. The excessive lactic acid and H^+ accumulation in the tumor background, matched with the compromised interstitial fluid transport, causes a decrease in the external tumor pH (pHe) [9–11]. Recent studies indicate that both local glucose-glutamine and oxygen availabilities affect tumor acidity independently. These findings can have significant implications for cancer treatment.

Most pH estimations in tumor tissues were obtained by the insertion of microelectrodes [5, 10, 11]. They demonstrated the value of pHe in a range 5.6 through 7.6 (normal tissue pHe is 6.8) and pointed out that tumors grow in an acidic environment. The advent of P magnetic resonance spectroscopy, a noninvasive method of measurement, has permitted the measurement of the internal tumor pH (pHi) and pHe simultaneously and more accurately. Among these measures, other parameters of interest can be obtained, namely tissue perfusion and vessel permeability. These studies have shown that in a majority of animal and human tumors pHe is lower than pHi. The tumor pHi is similar to that of normal tissue or near the neutral status (i.e ± 0.1 to 0.2 pH units), whereas pHe obtained from different human tumors was 0.41 ± 0.27 units lower than normal tissues [5, 10, 11]. The neutral pHi and pH gradient with the acidification of the tumor environment can be explained by the following factors:

1. In normal tissue the lactic acid, accumulated in the interstitial fluid, is rapidly removed through lymphatic drainage, whereas in TIF it is not so easily removed (due to a compromised vasculature and an absent lymphatic drainage).
2. Cancer cells, as do normal cells, express a number of pH regulatory mechanisms to maintain a cytosol pH near neutrality (pHi = 7.4). The mechanisms underlying regulation of intracellular pH have been identified as their inhibitors. A brief description of the most important exchanger mechanisms active in cancer cells and their inhibitors is reported as follows:

   pH regulators and inhibitors.

1. Vacuolar-type H^+ ATP-ase is an ion transporter regulated by an ATP-dependent mechanism.
2. H^+-lactate cotransport Na^+/H^* exchanger [MCT] is an exchanger mechanism reversibly blocked by Amiloride and quercetin.
3. Sodium-proton exchanger [NHE]. The NHE family of ion exchangers includes six isoforms (NHE1—NHE6) that function in an electroneutral exchange of intracellular H^+ for extracellular Na^+. They are blocked by Amiloride and its derivative Cariporide.
4. Na^+dependent Cl^-/HCO_3^- exchanger [BCT] is blocked by disulfonic stilbene derivative 4,4′-diisothiocyanostilbene-2,2′-disulfonic acid [DIDS].

The lactate efflux and formation can be also blocked by metabolic inhibitors such as: lonidamine or by meta-Iodio-benzylguanidine (MIBG) or alpha-cyano-4-hydroxy-cinnamic acid (CNCn). These drugs are active on the mitochondrial generation of lactic acid by blocking the Krebs cycle.

14.2.1 Effect of Hyperthermia on Metabolism

Recent studies have emphasized that metabolic alterations can be caused following hyperthermia. Using bioluminescence and photon counting methods, for imaging *in situ* the metabolites, enhanced glycolytic activity followed by an increase in lactate levels was shown upon heating application in tumors. The higher glucose and lactate concentrations may be the result of a temporary increase in blood flow and an expansion of the interstitial tumor compartment. Furthermore, the intensified acidosis with the enlargement of hypoxic tumor areas that follows the transient decrease in tumor perfusion sensitize and facilitate tumor cell destruction by hyperthermia [2, 12].

14.2.2 Effect of pH on Hyperthermia

Earlier studies, *in vitro* conditions, ascribed this effect to the low pH of the tumor environment (pHe) and showed that thermos-sensitization was more evident at mild temperatures (e.g., <42 °C) rather than at higher temperatures. Cells at low pH and nutritionally deprived (more likely to be in tumors) are more sensitive to heat, although cells can adapt to pH changes and lose their sensitivity to heat. Cells exposed to a low pH for prolonged periods were less sensitive to heat than cells exposed to an acidic medium shortly before heating. For any heating temperature tested, the sensitizing effects of pH were much smaller on *in vivo*-derived cells than on *in vitro*-derived cells.

Although recent experiments proved that thermos-sensitivity is more dependent upon pHi rather than pHe, the pHe value alone has been demonstrated to be a useful prognostic indicator and a reduction in pHe induces a decrease in the intracellular pHi [2, 12].

Since extracellular pHe is the result of lactic acid accumulation, earlier animal and human experiments indeed demonstrated that intraperitoneal or endovenous injection of glucose reduced intratumor pHe and increased the response to thermotherapy. However, recent studies have evidenced that the change of pH is induced by tumor blood flow reduction or nutrition deprivation rather than by increased glycolysis [9].

Although the experimental evidence has demonstrated that hyper-glycaemia is a useful method for enhancing the response to thermos-radiotherapy, its clinical application is not yet routine. This is partly due to a fear of inducing uncontrollable physiologic alterations and to an absence of a standardized protocol.

14.3 Effect of Hyperthermia on Tumor Blood Flow and Endothelium

Hyperthermia has as a principal treatment goal destroying tumor tissue (vasculature included) by achieving a temperature that exceeds the cytotoxic threshold (42.5 °C) and induces cell death in tumor tissue with a selective sparing of the normal surrounding tissue. Although cancer cells can be destroyed by hyperthermia alone, many factors including the cell type and blood perfusion, influence its success. In fact, it has been shown experimentally and theoretically that heat transferred away from a tissue is the result of the rate and volume of blood flowing through that tissue (perfusion) [13, 14].

14.3.1 *Effect on Tumor Blood Flow (TBF) and Normal Circulation*

The relationship between temperature rise and perfusion has been demonstrated by Jain et al. to be inversely correlated: as the perfusion rate decreases the tumor temperature increases [15]. The gap in temperature obtainable between normal and tumor tissue is due to differences in the conduction and convection characteristics between the two tissues.

Tumor blood flow and distribution are different from normal tissue and show regional variations in the same tumor itself. Blood flow in human cancer showed a greater variability as compared to animals; however, a different flow between normal and tumor tissue exists. Tumor blood flow appears to be inferior to that of normal tissue and to have a decreased adaptability to metabolic demands and physical stress (heat). The reasons can be ascribed to the absence of innervation. The vasodilatation that happens after heat application to normal tissues is not present to the same extent in the tumor vasculature. This determines decreased convection and permits the entrapment of heat in the tumor area, raising the temperature in that target area [16–18].

After heat exposition, macroscopic blood flow measurements in normal tissue such as skin or muscle showed a rapid and dynamic vasodilatation with increased permeability of the vascular wall (Fig. 14.1). Song and Vaupel demonstrated that the degree of alteration was temperature- and treatment duration-dependent [13, 14, 17]. The heating changes in the tumor were slightly absent or increased at the beginning of the treatment if the treatment time was prolonged, and decreased if a stasis of blood flow occurred. Jain et al. found and confirmed the above-mentioned observations, but noted that stasis occurred in both normal and tumor tissue [19]. The difference in stasis was dependent on temperature. The stasis in normal tissue occurred later and at a higher temperature (47 °C), whereas stasis in tumor tissue was reached earlier and at a lower temperature (41.0 °C) [18]. Li found that the recovery kinetics of tumor blood flow after heating was temperature-dependent, i.e., blood stasis occurred in the range of 3–5 h after heating, remaining low for 24 h and partially recovering after 48 h. Other than vasodilatation, other complex events follow heat application [20]. They are different biochemical and microcirculatory changes, such as acidosis, RBC stiffening and aggregation, degenerative changes in the endothelium, increased vascular permeability, platelet aggregation, leukocyte sticking and intravascular clotting. These events worsen the tumor microenvironment further and explain why neoplastic cells are damaged more easily by temperature (42–45 °C) than normal cells.

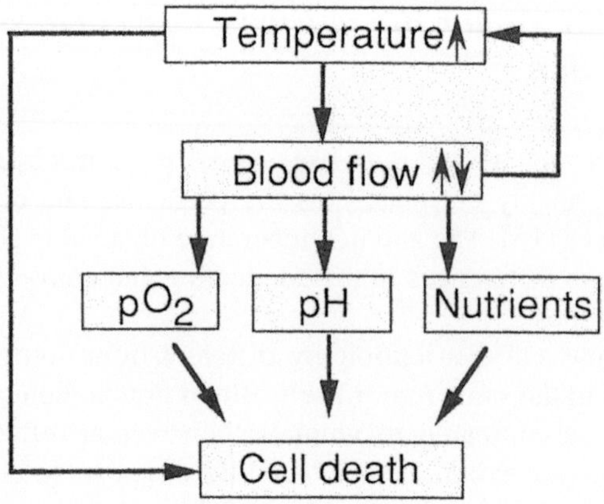

Fig. 14.1 The interplay between hyperthermia and blood flow. An elevated temperature leads to direct cell killing, but it also causes a reduction in blood flow in tumors; this, in turn, causes changes in pH, pO_2, and nutrients, leading to enhanced cell killing. The reduction in blood flow also further elevates the temperature in the tumor, because heat is not being carried away. By contrast, mild hyperthermia causes an increase in blood flow with a concomitant increase in tumor pO_2

14.3.2 Effects on Endothelium

Different *in vivo* and *in vitro* studies have shown that endothelial cells [ECs] and in particular those proliferating ones can be damaged by heat. The damages can be regarded as the integrity of the endothelium or its susceptibility to heat. Histological methods have revealed that after hyperthermia, a rapid reduction and rearrangement of F-actin stress fibers follows. These fibers maintain the junctional integrity among the cells. Their lack determines an increase in vascular permeability, a phenomenon usually observed after hyperthermia [21]. However, the effects of hyperthermia are not similar in all tumor types. The difference has been demonstrated by Nishimura, and it is dependent on the quantity of connective tissues present in the vascular architecture [22]. Further, hyperthermia positively regulates different adhesion molecules on endothelium surfaces providing an increased recruitment of T cells to tumors. This effect, related to the increased vascular permeability and extravasation, partially explains the escape of leukocytes from the vasculature and the peritumoral inflammatory reaction [23].

As described, tumor blood vessels are more vulnerable to heat than normal surrounding blood vessels, probably due to their structurally immaturity [13, 14]. Further, tumor vessels, as neoplastic cells have shown to acquire thermal adaptation (thermotolerance) to reheating [14]. This phenomenon, referred by Song as vascular thermotolerance (VT), differs as regards neoplastic cell thermotolerance (NT). NT refers generally to a resistance to heat-induced cell killing by producing heat shock proteins (HSP), while VT is a thermal adaptation developed by tumor vessels that initially respond to the stress of reheating by increasing blood flow instead of reducing it.

14.3.3 *Antiangiogenic Effect of Hyperthermia*

Hyperthermia has been shown to kill tumor cells by a direct and an indirect mechanism [17]. The indirect killing mechanism has been recognized as an inhibitory effect on angiogenesis. In the 60s, the inhibition of angiogenesis was ascribed to ischemia with a consequent obstruction and destruction of tumor blood vessels followed by an inability to form new vessels [17, 24]. Recently it has been clearly demonstrated that heat application can inhibit angiogenesis by activation of a plasminogen activator inhibitor I-dependent mechanism (PAI-I). The effect has been investigated and found to be operating *in vivo* and *in vitro* with a reduction in the number of microvessels. Endothelial cells viability was not affected.

14.4 Clinical Methods for Improving Response to Hyperthermia

Tumor vascular supply is extremely important for maintaining tumor growth, progression and metastases. Tumor perfusion can affect related microenvironmental parameters, such as oxygenation status, pH distribution, bioenergetic status, nutrient supply and the sensitivity of cancer cells to anticancer treatments, including hyperthermia [13]. Blood flow is the major determinant of heat dissipation and it is responsible for the selective and uniform heating at the tumor target area, paying attention to the normal counterpart [15]. The relationship between blood flow and effective tumor conductivity has been studied by Jain et al., who found that temperature rise in tumor mass is inverse and linked to effective thermal conductivity [15]. Since hypoxic cells and preferential cells exposed to an acute acidification have shown an increased thermosensitivity, many investigators have developed different methods with the aim of increasing thermoresponse as heat-induced cell damage.

Clinically, a modification of tumor thermoresponse could be obtained to modulate tumor blood flow, the tumor microenvironment or in an attempt to improve the heat deposition into the tumor tissue (Table 14.2) [13, 15, 21]. Otherwise expressed, the clinical attempt to increase the heat response can be reached by inducing hypoxia through blood flow manipulation or trying to render tumor cells more susceptible to heat.

Table 14.2 Clinical methods for improving thermoresponse

Tumor blood flow modification	Tumor microenvironment modification
(A) TBF reduction or deprivation: clamping of nutritive vessels and chemoembolization	(A) Hyperglycemia
(B) Drugs able to modify TBF: Vasodilators: [hydralazine, calcium blocking agents, verapamil, flunarizine, serotonin and its analogues]	(B) Drugs active on tumor metabolism: [amiloride, lonidamine, stilbene derivatives (DIDS), MIBG, quercetine]
	(C) Modifiers of thermal sensitivity: [Anesthetics, Lonidamine, Calcium antagonists, EFAs, COX2, Betulinic Acid, aldehydes, Vitamins, Quercetin]
(C) Vascular targeting agents: (VTA) [FAA, DMXXA, CA4DP]	(D) Heat delivery methods – Rapid heating – Mild hyperthermia

14.4.1 *Tumor Blood Flow (TBF) Modulation*

14.4.1.1 Blood Flow Reduction or Deprivation

Cutting off the tumor's blood supply through physical clamping for producing 100 % radiobiological hypoxia leads to cancer cell death through apoptosis [25]. Drastic changes were shown at the isoeffect relationship when deprivation of blood supply through clamping was applied with hyperthermia. Firstly, the tumor tissue becomes more sensitive, equivalent to a factor near two in heating time and secondly the transition at 42.5 °C is eliminated, showing a reduction or abolished thermotolerance [26].

14.4.1.2 Clamping of Nutritive Vessels

Blood flow deprivation by clamping the blood supply has been demonstrated *in vivo* to make tumor cells totally hypoxic and to significantly enhance thermal sensitivity in animals. Thermal enhancement had a ratio of 1.8–2.6 and was dependent on time. The proportion of tumor controlled by hyperthermia increased from 33 to 83 %, depending on whether the clamp was applied immediately through to 60 min before heating. No cures were observed for heat applied immediately before clamping, or immediately after the release of the clamp [27]. Other authors have demonstrated that vascular occlusion by clamping followed by glucose load can reduce tumor pH further and consequently enhance the thermoresponse.

14.4.1.3 Chemoembolization

Since blood flow reduction by clamping is not clinically achievable, embolization has been used as a method for stopping blood flow. Chemoembolization is a combination of two effective therapies, that is, one is high concentration drug delivery to the tumor mass, and the other is the production of hypoxia to inhibit the active efflux of the administered drug. Hypoxia improves the response to heat. Thus, chemoembolization is an efficient method of treatment in association with thermotherapy for treating liver tumors. The liver has two main blood supplies which keep it alive and functioning. The portal vein supplies 75 % of the blood entering the liver and the hepatic arteries supply the remaining 25 %, although they are the ones that provide nearly 100 % of the blood that feeds primitive and secondary liver tumors. The stopping of hepatic arteries concentrates the level of chemotherapy at 10–25 times higher than that of standard chemotherapy. To starve human liver tumors, hepatic arteries are blocked or embolized by appropriate methods [28]. The most common is an oil-based mixture associated with chemotherapy. This embolization consists of lipiodol (ethiodizedoil) or starch microspheres (DSM) alone or associated with chemotherapy. DSM (Spherex, Pharmacia, Sweden) are particles of cross-linked starch, measuring 20–70 μm, degraded by amylases, and able to block the hepatic artery transiently and reversibly. Tanaka et al. blocked the hepatic artery, injecting a mixture of Lipiodol/or DSM plus anticancer drugs [e.g., Mitomycin C (10–20 mg) or 5-FU (500–750 mg)]. Forty eight hours after this block, hyperthermia was performed twice a week for a total of 4–6 treatments. The outcome of this HT schedule was evaluated by CT images and angiograms. The mean maximal temperature (T_{MAX}) was of the order of 41.5 °C and the response rate was 40 % [29].

14.4.1.4 Drugs Modifying Tumor Blood Flow

The goal of TBF modulation by drugs is to make the tumor sufficiently hypoxic/or underfed and consequently more thermosensitive, similar to clamping or chemoembolization. Drugs, which are the less burdensome of the preceding methods of treatment, are of two types: vasodilatators and agents attacking the vascular endothelium (VTA) [30]. Vasodilatators starve the tumor through a steal phenomenon, while VTA gets the tumor hypoxic through disrupting and decreasing the nutritive vessels.

1. Vasodilatators: Hydralazine, Calcium blocking agents (verapamil, flunarizine), Serotonin and its Analogues.
2. Vascular Targeting Agents (VTA): FAA, DMXXA, CA4DP, TNF; IL-1.

14.4.2 Microenvironment Modification

14.4.2.1 Hyperglycemia and Drugs Influencing Tumor Metabolism

Hyperglycemia and drugs act on tumor metabolism. Several animal studies have demonstrated that hyperglycemia can sensitize mammalian cells to heat. Initially the effect was thought to be metabolic induced (pH drop, due to excessive lactic acid production), but subsequent studies showed that the hypoxia and pH reductions are secondary to the tumor blood flow reduction [21]. It was reported that intraperitoneal injection of glucose reduced tumor blood flow by more than 90 % for a few hours [31]. Further, it was also shown that blood flow inhibition and pH reduction are related to the serum concentration of glucose level [31]. The mechanisms responsible for blood flow reduction in tumors during hyperglycemia are multiple. The reduction is a consequence of both systemic and local effects. The systemic effects were ascribed to a significant cardiac output redistribution that consequently reduced tumor blood flow through a steal phenomenon. The local mechanism that contributed to the reduction of tumor blood flow was linked to red blood cell (RBC) deformability [15]. Initially this increase in rigidity was postulated to be pH dependent, but later, it was shown that a long exposure time to a low pH was necessary for obtaining RBC rigidity. Further, it was also shown that RBC deformability followed almost immediately the infusion of glucose and galactose. This rapid local phenomenon is related to the steal effect of cardiac output and is responsible for the initial blood flow reduction. The metabolic interaction with pH decrease can bring about the reduction of blood flow for many hours after glucose administration. Later, another study showed that a difference in pH drop exists between tumor and normal tissue. 30–60 min after glucose administration, the pH rapidly dropped at 0.3–0.6 units in tumor, compared with a decrease of only 0.1 unit in normal tissue. The complete recovery to baseline was also different in the two tissues [32]. Animal studies have demonstrated that the administration of glucose prior to hyperthermia can modify tumor regrowth and thermotolerance. Tumor regrowth delay was greater in the group treated with glucose compared with the group not treated. Thermotolerance disappeared 12 h after heat treatment in the group with glucose administration, whereas in the untreated group thermotolerance disappeared only after 72 h [33]. The majority of experimental studies have been performed *in vitro* and in animals.

After the increased use of hyperthermia combined with radiotherapy, many human studies have been performed. Serum glucose Levels of 400 mg/dl was shown to be the maximum achievable for long periods (24 h) with no modifications of blood count or acid base equilibrium [34]. It was demonstrated that a glucose administration of 500 ml of 10 % glucose by an intravenous route reduced the tumor blood flow, measured by laser Doppler flowmetry, to 66 % of the baseline level at 30 min after the beginning of the infusion. A complete tumor response (CR) of 30 % was obtained in glucose-treated patients compared to a group treated only by radiotherapy and hyperthermia [35]. The importance of extracellular pH as a prognostic indicator of tumor response to thermoradiotherapy was also evaluated [36]. The tumor pHe of 26 human tumors with a needle microelectrode of 2.5 cm in length was measured. It was reported that the difference in pHe exhibited by complete responding (CR) patients and noncompletely responding (NCR), 6.88 ± 0.09 versus 7.24 ± 0.09, was statistically significant ($p \geq 0.08$). On these grounds, it was suggested that the extracellular pHe measurement might be a useful prognostic indicator of tumor response to thermotherapy [36]. Although, a pHe reduction of ≈ 0.2 units is easily obtainable by glucose load [36], greater reduction

≥0.5 units are necessary for inducing acute intracellular acidosis [37]. The degree of reduction in pHi that accompanies acute extracellular acidification is the critical factor for sensitizing cells to hyperthermia and for abrogating heat shock proteins induction. Subsequently it has been demonstrated that for obtaining such a pHe drop in melanoma, metabolic inhibitors such as meta-Iodio-benzylguanidine (MIBG) or alpha-cyano-4-hydroxy-cinnamic acid (CNCn) must be added [37]. In conclusion, the concomitant administration of glucose together with MIBG increases the tumor magnitude and duration of acidification and the oxygen tension [38]. This association has the potential to improve the response to radiation therapy and hyperthermia itself.

14.5 Heat Delivery Methods

Tumor cell killing curves by heat are both time and temperature dependent shapes and similar to those obtained for X-rays. The data *in vitro* are consistent with results *in vivo* and show that relatively small changes in temperature can have a large effect on cell killing. The critical temperature has been demonstrated to be between 42.5 and 43 °C. Unfortunately, heating devices now available are not able to keep this temperature range uniformly for long enough.

14.5.1 Rapid Heating

Rapid heating was a method developed as an attempt to shorten heating treatment while still reaching a temperature sufficient to kill tumor cells and to change tumor blood perfusion [39]. The experiments have been made on C3H mice inoculated with SCC-VII tumor in the thigh, heated with a warm water bath and RF heating devices. The C3H mice were divided into two treatment groups and compared, the former in which the heating temperature was increased to the target temperature in 1 min, and the latter group in which the heating temperature was gradually increased. The following parameters were studied: changes in blood flow in tumor and normal tissue, tumor growth rate, cancer cells apoptosis. Changes in blood flow were not observed in the slow heating group before or after the heat treatment, whereas in the rapid heating group a significant increase in blood flow was observed in the normal tissue followed by a significant decrease after heat treatment in the tumor tissue. Tumor growth delay was more evident in the RF rapid heating group compared with the warm water heating group. Apoptosis and cytokinetic activity modifications were favorable to the rapid heating group, showing that a vascular injury was effectively obtained with a shortage in treatment time in this group [39].

14.5.2 Mild Hyperthermia and Oxygenation

Solid tumors contain regions of low extracellular pH and oxygen that may affect the treatment outcome. Laboratory and clinical data confirm that hyperthermia may enhance the therapeutic index of ionizing radiation. Among these mechanisms, tumor oxygenation improvement after mild hyperthermia (HT with a temperature between 39 and 42.5 °C) is now considered to be of the highest importance. As reported [13], normal and tumor tissue show a different behavior following heat deposition. Blood tumor vessels respond markedly differently to a second heat application showing a greater vulnerability and vasodilatation to heat than normal surrounding blood vessels. These phenomena, as vascular thermotolerance (VT), appears to account for the improvement in the tumor blood flow observed after the reheating at 42.5 °C. As the blood flow increases, an improvement in tumor oxygenation follows which may last for as long as 24–48 h [13, 14, 40, 41]. Tumor oxygenation by mild hyperthermia has been found to be more effective than carbogen breathing at increasing the radiation response of experimental tumors [40, 41]. Clinical studies on 18 patients with locally advanced breast cancer treated with thermo-chemo-radiotherapy have confirmed these experimental results. Tumor oxygenation improvement appeared to be temperature dependent and associated with the lower thermal doses.

It is important to note that in animal studies in which more moderate temperatures are used (41 and 41. 5 °C), hyperthermia has been shown to promote tumor reoxygenation, with the degree of reoxygenation correlating with the level of the radiosensitivity of the tumor. These data have been confirmed in a clinical study of patients with soft tissue sarcomas and breast cancer. It was showed that one HT led to reoxygenation of human soft tissue sarcomas within 24–48 h, whereas there was no measurable reoxygenation during a prior week of standard radiation therapy (RT) [42]. It was also reported that mild hyperthermia (41 and 41.5 °C at 90 % of measured points for 1 h) significantly increased the p02 in hypoxic, but not normoxic, human breast cancers [43]. Such increases in tumor oxygenation could significantly improve tumor response to radiotherapy and is likely to be the primary important effect of local/regional or whole body forms of clinical hyperthermia [20, 40, 41, 44]. If this assumption is correct, then there should be much more attention devoted to optimal scheduling of hyperthermia in future trials to maximize oxygen-dependent radiosensitization. Indeed, instead of aiming to kill cells, mild hyperthermia may have the ability to rapidly stimulate global changes in the tumor microenvironment that collectively could enhance RT; this includes effects on pH, oxygen concentration, metabolism, protein/gene expression, and vascular perfusion [45].

Despite an incomplete understanding of the actual basis of radiation sensitization by hypertherrmia, an overestimation of the role of thermotolerance and suboptimal sequencing of hyperthermia and radiation fractions, multiple trials have been completed revealing a critical role for hyperthermia in enhancing RT. A very active, preclinical, and clinical research effort is under way at the present time to develop the optimal dose and scheduling protocols and to design new clinical trials based on a more complete understanding of the physiologic effects of hyperthermia.

14.6 Conclusion

Thermal ablation technology using temperatures between 50 and 60 °C has significantly advanced the concept of using thermal energy to kill the largest number of tumor cells in the shortest period, whereas lower tumor temperatures may be most effective for simultaneously changing several critical parameters within the tumor microenvironment, including vascular perfusion, oxygenation status, pH, and metabolism. Each of these represents important targets that can greatly affect the efficacy of radiation and chemotherapy. Immune sensitization by heat may also contribute significantly to the adjuvant potential of various forms of hyperthermia and may lead to novel combinations of heat and immunotherapy.

Tumor hypoxia is a problem that makes tumors more resistant to ionizing radiation and chemotherapeutic drugs. Hyperthermia represents a possibility in overcoming this problem in association with other therapies such as radiotherapy and chemoimmunotherapy. Moreover the effect of mild HT on oxygenation is of great relevance. In fact, a temperature of 39–39.5 °C is more easily obtainable in the clinic than the killing temperature of 42.5 °C.

References

1. Vaupel P. Tumor microenvironmental physiology and its implications for radiation oncology. Semin Radiat Oncol. 2004;14(3):198–206.
2. Vaupel PW, Kelleher DK. Pathophysiological and vascular characteristics of tumours and their importance for hyperthermia: heterogeneity is the key issue. Int J Hyperthermia. 2010;26 (3):211–23.
3. Hall EJ, Giaccia AJ. Cell, tissue, and tumor kinetics. In: Hall EJ, Giaccia AJ, editors. Radiobiology for the radiologist. 7th ed. Philadeiphia: Lippencott, Williams & Wilkins; 2012. p. 372–90.
4. Wouters BG, Koritzinsky. The tumour microenvironment and cellular hypoxia responses. In: Joiner M, van der Kogel AJ, editors. Basic clinical radiobiology. 4th ed. London: Hodder Arnold; 2009. p. 217–32.
5. Baronzio G, Barongio A, Crespi E, Freitas I. Effects of tumor microenvironment on hyperthermia, photodynamic and nanotherapy. In: Baronzio G, Fiorentini G, Cogle CR, editors. Cancer microenvironment and therapeutic implications. New York: Springer Science +Business Media B.V; 2009. p. 181–96. doi:10.1007/978-1-4020-9576-4.
6. Hall EJ, Giaccia AJ. The biology and exploitation of tumor hypoxia. In: Hall EJ, Giaccia AJ, editors. Radiobiology for the radiologist. 7th ed. Philadeiphia: Lippencott, Williams & Wilkins; 2012. p. 432–47.

7. Horsman MR, van der Kogel AJ. Therapeutic approachs to tumour hypoxia. In: Joiner M, van der Kogel AJ, editors. Basic clinical radiobiology. 4th ed. London: Hodder Arnold; 2009. p. 233–45.
8. Vaupel P, Mayer A. Hypoxia in tumors: pathogenesis-related classification, characterization of hypoxia subtypes, and associated biological and clinical implications. Adv Exp Med Biol. 2014;812:19–24. doi:10.1007/978-1-4939-0620-8_3.
9. Vaupel P. Metabolic microenvironment of tumor cells: a key factor in malignant progression. Exp Oncol. 2010;32(3):125–7.
10. Mayer A, Vaupel P. Hypoxia, lactate accumulation, and acidosis: siblings or accomplices driving tumor progression and resistance to therapy? Adv Exp Med Biol. 2013;789:203–9. doi:10.1007/978-1-4614-7411-1_28.
11. Kato Y, Ozawa S, Miyamoto C, Maehata Y, Suzuki A, Maeda T, Baba Y. Acidic extracellular microenvironment and cancer. Cancer Cell Int. 2013;13:89. doi:10.1186/1475-2867-13-89.
12. Vaupel PW, Kelleher DK. Blood flow and associated pathophysiology of uterine cervix cancers: characterisation and relevance for localized hyperthermia. Int J Hyperthermia. 2012;28(6):518–27. doi:10.3109/02656736.2012.699134. Epub 2012 Jul 27.
13. Song CW. Effect of local hyperthermia on blood flow and microenvironment: a review. Cancer Res. 1984;44(Suppl):4721–30.
14. Song CW, Chelstrom LM, Sung JH. Effects of a second heating on tumor blood flow. Radiat Res. 1990;122:66–71.
15. Ward KA, Jain RK. Response of tumours to hyperglycaemia: characterization, significance and role in hyperthermia. Int J Hyperthermia. 1988;4:223–50.
16. Vaupel P. Tumor blood flow. In: Molls M, Vaupel P, editors. Blood perfusion and microenvironment of human tumors. Berlin/Heidelberg/New York: Springer; 2000. p. 41–6.
17. Vaupel P, Kallinowski F. Physiological effects of hyperthermia. Recent Results Cancer. 1987;104:71–109.
18. Reinhold HS, Endrich B. Tumor microcirculation as a target for hyperthermia. Int J Hyperthermia. 1986;2:11–137.
19. Dudar TE, Jain RK. Differential response of normal and tumor microenvironment to hyperthermia. Cancer Res. 1984;44:605–12.
20. Sun X, Xing L, Ling CC, Li GC. The effect of mild temperature hyperthermia on tumour hypoxia and blood perfusion: relevance for radiotherapy, vascular targeting and imaging. Int J Hyperthermia. 2010;26(3):224–31.
21. Jain RK, Ward-Hartley K. Tumor blood flow-characterization, modifications and role in hyperthermia. IEEE Trans Sonics Ultrason. 1984;31:504–26.
22. Nishimura Y, Hiraoka M, Jo S, Akuta K, Yukawa Y, Shibamoto Y, et al. Microangiographic and histologic analysis of the effects of hyperthermia on murine tumor vasculature. Int J Radiat Oncol Biol Phys. 1988;15:411–20.
23. Evans SS, Frey M, Scheider DM, Bruce RA, Wang WC, Repasky EA, et al. Regulation of leukocyte-endothelial cell interaction in tumor immunity. In: Mihich E, Croce C, editors. The biology of tumors. New York: Plenum Press; 1998. p. 273–86.
24. Fajardo LF, Prionas SD. Endothelial cells and hyperthermia. Int J Hyperthermia. 1994;3:347–53.
25. Suit H, Shalek RJ. Response of spontaneous mammary carcinoma of the C3H mouse to X-irradiation given under conditions of local tissue anoxia. J Nat Cancer Inst. 1963;31:497–509.
26. Field SB. In vivo aspects of hyperthermic oncology. In: Field SB, Hand JW, editors. An Introduction to the practical aspects of clinical hyperthermia. London/New York/Philadelphia: Taylor & Francis; 1990. p. 55–68.
27. Hill SA, Denekamp J. The effect of vascular occlusion on the thermal sensitisation of a mouse tumour. Br J Radiol. 1978;51:997–1002.
28. Stuart K. Chemoembolization in the management of liver tumors. Oncologist. 2003;8:425–37.

29. Tanaka Y, Yamamoto K, Nagata K. Effects of multimodal treatment and hyperthermia on hepatic tumors. Cancer Chemother Pharmacol (Suppl). 1998;1:111–14.
30. Horsman MR, Siemann DW. Significance of the tumour microenvironment in radiotherapy. In: Baronzio G, Fiorentini G, Cogle CR, editors. Cancer microenvironment and therapeutic implications. New York: Springer Science+Business Media B.V; 2009. p. 137–56. doi:10.1007/978-1-4020-9576-4.
31. Dickson JA, Calderwood SK. Thermosensitivity of neoplastic tissues in vivo. In: Storm FK, editor. Hyperthermia in cancer therapy. Boston: GK Hall Medical; 1983. p. 63–140.
32. Hasegawa T, Gu YH, Takahashi T, Hasegawa T, Tanaka Y. Effects of hyperthermia-induced changes in pH value on tumor response and thermotolerance. In: Kosaka M, Sugahara T, Schmidt KL, Simon E, editors. Thermotherapy for neoplasia, inflammation, and pain. Tokyo: Springer Verlag; 2001. p. 433–8.
33. Shem BC, Dahl O. Thermal enhancement of ACNU and potentiation of thermochemotherapy with ACNU by hypertonic glucose in the BT A rat glioma. J Neuroncol. 1991;10:247–52.
34. Lippmann HG, Graichen D. Glucose and K balance during high dosage intravenous glucose infusion. Infusionsther Klin Ernahr. 1977;4:166–78.
35. Nagata K, Murata T, Shiga T, Isoda H, Aoki Y, Yamamoto K, et al. Enhancement of thermoradiotherapy by glucose administration for superficial malignant tumours. Int J Hyperthermia. 1998;14:157–67.
36. Engin K, Leeper DB, Cater JR, Thistlethwaite AJ, Tupchong L, McFarlane JD. Extracellular pH distribution in human tumors. Int J Hyperthermia. 1995;11:211–16.
37. Han JS, StorcK CW, Wachsberger PR, Leeper DB, Berd D, Wahl ML, et al. Acute extracellular acidification increases nuclear associated protein levels in human melanoma cells during 42 °C hyperthermia and enhances cell killing. Int J Hyperthermia. 2002;18:404–15.
38. Burd R, Lavorgna SN, Daskalakis C, Wachsberger PR, Wahl ML, Biaglow JE, et al. Absence of Crabtree effect in human melanoma cells adapted to growth at low pH: reversal by respiratory inhibitors. Cancer Res. 2001;61:5630–5.
39. Hasegawa T, Gu YH, Takahashi T, Hasegawa T, Yamamoto I. Enhancement of hyperthermic effects using rapid heating. In: Kosaka M, Sugahara T, Schmidt KL, Simon E, editors. Thermotherapy for neoplasia, inflammation, and pain. Tokyo: Springer Verlag; 2001. p. 439–44.
40. Song CW, Park H, Griffin RJ. Improvement of tumor oxygenation by mild hyperthermia. Radiat Res. 2001;155:512–28.
41. Song CW, Park HJ, Lee CK, Griffin R. Implications of increased tumour blood flow and oxygenation caused by mild temperature hyperthermia in tumor treatment. Int J Hyperthermia. 2005;21:761–7.
42. Brizel DM, Scully SP, Harrelson JM, Layfield LJ, Dodge RK, Charles HC. Radiation therapy and hyperthermia improve the oxygenation of human soft tissue sarcomas. Cancer Res. 1996;56:5347–50.
43. Jones EL, Prosnitz LR, Dewhirst MW, Marcom PK, Hardenbergh H, Marks LB, et al. Thermochemoradiotherapy improves oxygenation in locally advanced breast cancer. Clin Cancer Res. 2004;10:4287–93.
44. Masunaga S, Nishimura Y, Hiraoka M, Abe M, Takahashi M, Ono K. Efficacy of mild temperature hyperthermia in combined treatments for cancer therapy. Thermal Med. 2007;23(3):103–12.
45. Hall EJ, Giaccia AJ. Hyperthermia. In: Hall EJ, Giaccia AJ, editors. Radiobiology for the radiologist. 7th ed. Philadeiphia: Lippencott, Williams & Wilkins; 2012. p. 490–511.

Part III
Clinical Science in Cancer Patients

Chapter 15
Hyperthermia Alone

Takayuki Ohguri

Abstract The use of hyperthermia in cancer treatment has shown excellent effectiveness in experimental studies. Electromagnetic wave-induced heating devices have been developed considering these evidences. In early clinical reports, hyperthermia alone for cancer has also been investigated. Moderate objective tumor response rates have been observed, particularly for superficial malignant tumors, which can be treated with a higher intra-tumor temperature. In this chapter, the clinical results of hyperthermia alone in the treatment of malignant neoplasms were reviewed.

Keywords Hyperthermia • Monotherapy • Superficial tumors

15.1 Background

In the 1960s, the cytotoxic effect of heat was confirmed using cultured cells, and subsequent experimental studies were conducted to evaluate hyperthermia for cancer treatment. Many basic studies have demonstrated the following biological merits of hyperthermia in the treatment of cancer [1, 2]: (i) in most cells, applying heat of 42–43 °C lowers the survival rate of cells depending on the duration of heat application, and heat sensitivity does not vary with tissue type; (ii) tumor tissue is more easily heated than normal tissue; (iii) hypoxic cells are more susceptible to heat than aerobic cells; (iv) the recovery of cells from radiation or chemotherapeutic agents is prevented by heat; and (v) sensitivity to heat and radiation differs with the cell cycle.

An electromagnetic wave-induced heating device was developed based on the biological evidence showing its excellent effectiveness in cancer treatment. In the 1980s, many clinical trials were initiated to evaluate the efficacy and toxicity of hyperthermia.

T. Ohguri (✉)
Department of Radiology, University of Occupational and Environmental Health,
1-1 Iseigaoka, Yahatanishi-ku, Kitakyushu 807-8555, Japan
e-mail: ogurieye@med.uoeh-u.ac.jp

S. Kokura et al. (eds.), *Hyperthermic Oncology from Bench to Bedside*,
DOI 10.1007/978-981-10-0719-4_15

15.2 Clinical Results for Hyperthermia Alone

In the 1980s and 1990s, early clinical trials investigated the efficacy of hyperthermia alone, although the clinical effects of hyperthermia as a radiation-sensitizer and chemo-sensitizer have been evaluated in other studies. Table 15.1 summarizes the efficacy and toxicity of hyperthermia alone in patients with malignant tumors. In these studies, superficial tumors were treated with local hyperthermia, and a measurement of the direct intra-tumor temperature was performed to evaluate the thermal parameters.

In 1978, Marmor et al. reported the treatment results of ultrasound hyperthermia alone for superficial malignant tumors that recurred in irradiated fields in 16 patients who previously had received radiotherapy [3]. The tumors reached temperatures of 43.5–45.0 °C. No skin adverse effects were noted in the majority of patients, and no recall radiation damage by heat was observed. The objective tumor response was observed in 69 % of the tumors. Luk et al. showed the treatment results for 37 patients with superficial malignant tumors treated with microwave local hyperthermia with or without radiotherapy in 1981 [4]. The complete response (CR) rates to hyperthermia alone and hyperthermia combined with radiotherapy were 18 % and 41 %, respectively, and the partial response (PR) rates were 18 % and 37 %, respectively. A relationship between the maximum treatment temperature and tumor responses was observed. They also demonstrated that the maximum tumor temperature correlated with complications. When the maximum tumor temperatures were less than 42.5 °C, blisters rarely occurred. When the maximum tumor temperature ranged from 42.6 °C to 43.9 °C, blisters were observed in 54 % of the tumors, and these blisters usually healed spontaneously within 10 days. When the

Table 15.1 Clinical results of hyperthermia alone for superficial malignant tumors

Study	Year	n	Tumors	Intra-tumor temperature (°C)	Range of the No. of hyperthermia sessions	Objective tumor response (CR + PR) (%)
Marmor	1978	16	Reccurence of head and neck cancer in previously irradiated sites.	43.5–45.0	3–25	69 (13 + 56)
Luk	1981	11	Reccurence of breast and head and neck cancer cancer	40.5–44.1	5–12	36 (18 + 18)
Manning	1982	11	Reccurence of head and neck cancer and breast cancer	43–44	2–22	45 (18 + 27)
Gabriele	1990	57	Reccurence of head and neck cancer and chest wall cancer	≥42 °C (58 % of the tumors)	6–10	40 (17 + 23)

maximum tumor temperatures were more than 44 °C, burns and blisters occurred 56 % and 33 % of the tumors, respectively. The burns required careful daily nursing care, and up to 1 month was required for healing in many cases.

In 1982, Manning et al. showed the results of a phase I trial for hyperthermia alone or hyperthermia combined with radiotherapy in patients with superficial tumors [5]. The objective tumor response rate to hyperthermia alone in 11 patients was 45 % (18 % with CR and 27 % with PR), whereas that to hyperthermia combined with external beam radiotherapy in 16 patients was 80 % (27 % with CR and 53 % with PR) and that to interstitial thermoradiotherapy in 17 patients was 100 % (71 % with CR and 29 % with PR).

In 1990, Gabriele et al. reported the treatment results of hyperthermia alone in 57 patients with recurrent or metastatic tumors, most of whom had been already treated with radiotherapy [6]. Measurements of the intra-tumor temperature were performed for all 60 lesions, and 35 (58 %) lesions had reached the desired goal of 42 °C. CR was obtained in 10 (17 %) cases, and PR was obtained in 14 (23 %) cases. Higher rates of CR were observed in the chest wall (39 %) compared with the head and neck and other areas. Interestingly, adenocarcinoma was the most responsive histologic type, and the CR rate was 40 %, whereas the CR rate for squamous cell carcinoma was only 8 %. Side effects and complications of the treatment were minimal. They concluded that hyperthermia alone was an effective and feasible option for the palliative treatment of patients with locoregional recurrences that were not treatable otherwise.

To the best of my knowledge, the clinical results of hyperthermia alone using deep regional heating techniques have not been evaluated. We supposed that the efficacy of deep regional hyperthermia alone is limited because in most cases, the tumor temperature in deep regional hyperthermia cannot be elevated above 42.5 °C; however, this temperature can be elevated during local superficial hyperthermia.

15.3 Closing Comments for Hyperthermia Alone

The early clinical effects for hyperthermia alone have been reported for superficial malignant tumors. Although the objective response rates of the tumors treated with hyperthermia alone were poorer than those treated with hyperthermia combined with radiotherapy, hyperthermia alone provided an objective tumor response rate of approximately 40–60 % in superficial tumors when the tumors were heated to 42–44 °C. Therefore, superficial tumors, such as recurrent chest wall tumors and recurrent head and neck tumors, which are treatable at higher temperatures and which cannot be treated with further radiotherapy, can be selected for hyperthermia alone in the current strategy of cancer treatments.

References

1. Dewhirst MW, Viglianti BL, Lora-Michiels M, Hanson M, Hoopes PJ. Basic principles of thermal dosimetry and thermal thresholds for tissue damage from hyperthermia. Int J Hyperthermia. 2003;19(3):267–94. doi:10.1080/0265673031000119006 UDW7W1EPFQD1L140 [pii].
2. Horsman MR, Overgaard J. Hyperthermia: a potent enhancer of radiotherapy. Clin Oncol (R Coll Radiol). 2007;19(6):418–26. doi:S0936-6555(07)00587-0 [pii] 10.1016/j.clon.2007.03.015.
3. Marmor JB, Hahn GM. Ultrasound heating in previously irradiated sites. Int J Radiat Oncol Biol Phys. 1978;4(11–12):1029–32. doi:0360-3016(78)90016-0 [pii].
4. Luk KH, Purser PR, Castro JR, Meyler TS, Phillips TL. Clinical experiences with local microwave hyperthermia. Int J Radiat Oncol Biol Phys. 1981;7(5):615–19. doi:0360-3016(81)90376-X [pii].
5. Manning MR, Cetas TC, Miller RC, Oleson JR, Connor WG, Gerner EW. Clinical hyperthermia: results of a phase I trial employing hyperthermia alone or in combination with external beam or interstitial radiotherapy. Cancer. 1982;49(2):205–16.
6. Gabriele P, Orecchia R, Ragona R, Tseroni V, Sannazzari GL. Hyperthermia alone in the treatment of recurrences of malignant tumors. Experience with 60 lesions. Cancer. 1990;66(10):2191–5.

Chapter 16
Combination Therapy with Hyperthermia and Chemoradiotherapy: Oral Cancer

Kenji Mitsudo and Iwai Tohnai

Abstract The major prognostic factors for survival in patients with oral cavity cancer are local control, neck control, and distant metastasis. Oral cancer patients with advanced cervical lymph node metastases have poor prognosis. Hyperthermia has generally been used for management of cervical lymph node metastases that are accessible with a radiofrequency system using external application, and in combination with synergistic chemoradiotherapy. Thermochemoradiotherapy using retrograde superselective intra-arterial infusion is used in patients with advanced cervical lymph node metastases, and can be used to treat the primary lesion as well as cervical lymph node metastases.

In this chapter, the therapeutic results of retrograde superselective intra-arterial chemoradiotherapy and thermochemoradiotherapy for patients with advanced oral cancer and advanced cervical lymph node metastases are described.

Keywords Oral cancer • Chemoradiotherapy • Retrograde superselective intra-arterial infusion • Organ preservation • Cervical lymph node metastases

16.1 Introduction: Epidemiology, Frequency of Oral Cancer in Japan

Oral cancer is a malignant neoplasm that occurs in the oral cavity. Squamous cell carcinoma comprises for over 90 % of oral cancers in Japan. The number of cases of oral cancer patients in Japan was 2100 in 1975 and 6900 in 2005 and further is estimated to be 10,000 patients in 2015, which is 1.6 times higher than current number [1–3]. The frequency of oral cancer by site differs according to patient ethnicity, country, region, lifestyle, and practice. According to a 2002 report by the Japanese Society of Oral and Maxillofacial Surgeons, tongue carcinoma is the most common oral cancer in Japan, comprising 40 % of cases, followed by cancers of the

K. Mitsudo (✉) • I. Tohnai
Department of Oral and Maxillofacial Surgery, Yokohama City University Graduate School of Medicine, 3-9 Fukuura, Kanazawa-ku, Yokohama, Kanagawa 236-0004, Japan
e-mail: mitsudo@yokohama-cu.ac.jp

S. Kokura et al. (eds.), *Hyperthermic Oncology from Bench to Bedside*,
DOI 10.1007/978-981-10-0719-4_16

mandibular gingiva (20.3 %), maxillary gingiva (12.0 %), buccal mucosa (10.3 %), and oral floor (9.2 %) [4].

16.2 Treatment of Oral Cancer

For patients with locally advanced head and neck cancer, including oral cavity cancer, surgery with or without radiotherapy is widely accepted as the standard treatment and is thought to be the most effective curative therapy. Radical surgery for patients with locally advanced oral cancer causes various dysfunctions such as dysmasesis, speech disorders and dysphagia. Furthermore, the cosmetic result after surgery is of greater concern than with other cancers, such as gastric cancer and lung cancer. Indeed, extensive surgery for locally advanced oral cancer affects a patient's social life and reduces their quality of life (QOL). To preserve function while maintaining or improving locoregional control and survival rates, concurrent chemoradiotherapy (CRT) represents a standard treatment modality for definitive treatment of locoregionally advanced squamous cell carcinoma of the head and neck, particularly in resectable advanced cases [5].

16.3 Organ Preservation with Superselective Intra-Arterial Chemotherapy

Organ preservation without surgery is desired for patients with advanced oral cancer. One strategy of organ-preservation therapy is intra-arterial chemotherapy combined with radiotherapy. Intra-arterial CRT for head and neck cancer is historically classified into three groups, as detailed below.

Intra-arterial chemotherapy for head and neck cancer was first reported by Klopp et al. [6] and by Sullivan et al. [7] in the 1950s. Those investigators described insertion of a straight catheter into the external carotid artery (ECA) via the superficial temporal artery (STA) or the superior thyroid artery (Fig. 16.1a, b) [8–11].

The second method is catheterization into the tumor-feeding artery via the femoral artery using the Seldinger method (Figs. 16.1c, 16.2a). Lee et al. [12, 13] first described superselective intra-arterial chemotherapy via the femoral artery using the Seldinger method, and Robbins et al. [14–16] developed a cisplatin (CDDP) delivery system in which extremely large amounts of an anticancer agent with radiotherapy can be administered to patients with advanced head and neck cancer. Given the acronym "RADPLAT [14]" this strategy consists of rapid superselective intra-arterial infusions combined with intravenous sodium thiosulfate for systemic CDDP neutralization. There have been many reports of RADPLAT for head and neck cancer [17–25], and RADPLAT is widely accepted

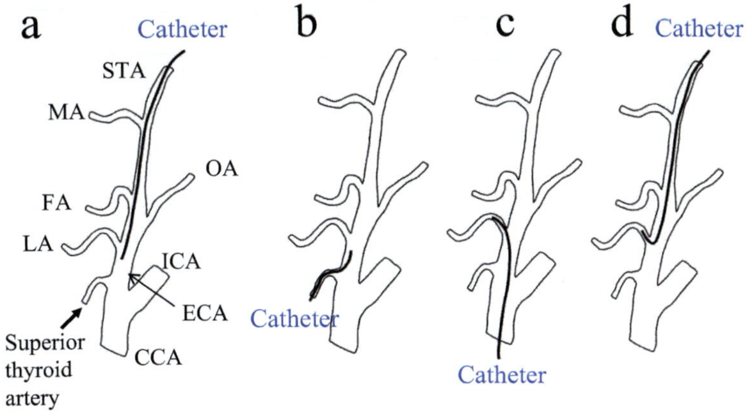

Fig. 16.1 The catheterization method for intra-arterial infusion. (**a**, **b**) Catheterization into the ECA near a tumor-feeding artery via the STA (**a**) or the superior thyroid artery (**b**) with a straight catheter. (**c**) Catheterization into a tumor-feeding artery via the femoral artery using the Seldinger method. (**d**) Catheterization into a tumor-feeding artery via the STA with a hook-shaped catheter. *Abbreviations*: *CCA* common carotid artery, *ICA* internal carotid artery, *ECA* external carotid artery, *LA* lingual artery, *FA* facial artery, *OA* occipital artery, *MA* maxillary artery, *STA* superficial temporal artery

worldwide. Superselective intra-arterial CRT using this method was found to be beneficial for advanced head and neck cancer.

The third method is catheterization into a tumor-feeding artery via the STA using a hook-shaped catheter (Figs. 16.1d, 16.2b, 16.3a). A catheter is inserted from the STA in a retrograde fashion and into a tumor-feeding artery. The utility of this method in combination with radiotherapy has been reported [26–30]. Retrograde superselective intra-arterial infusion using this method can be used to provide daily concurrent radiotherapy and chemotherapy for patients with advanced oral cancer, and it can be given safely without major complications, such as cerebral infarction.

16.4 Retrograde Superselective Intra-Arterial Infusion

The superficial temporal approach is technically simple and is probably the easiest method of inserting a catheter into the lingual, facial, or maxillary artery [29]. Before treatment, three-dimensional computed tomography angiography (3D-CTA) of the carotid artery is necessary to identify the main tumor-feeding arteries and to determine the morphology of the tumor-feeding artery originating from the ECA (Fig. 16.4). Catheterization from the STA is done according to the method of Hattori, Fuwa and Tohnai [26–28] (HFT method). When the tumor has two or more feeding arteries, catheters are inserted into the two arteries via the STA and the occipital artery (OA) or bilaterally. Catheterization from the OA is performed using the method of Iwai et al. [31]. When catheterization with a

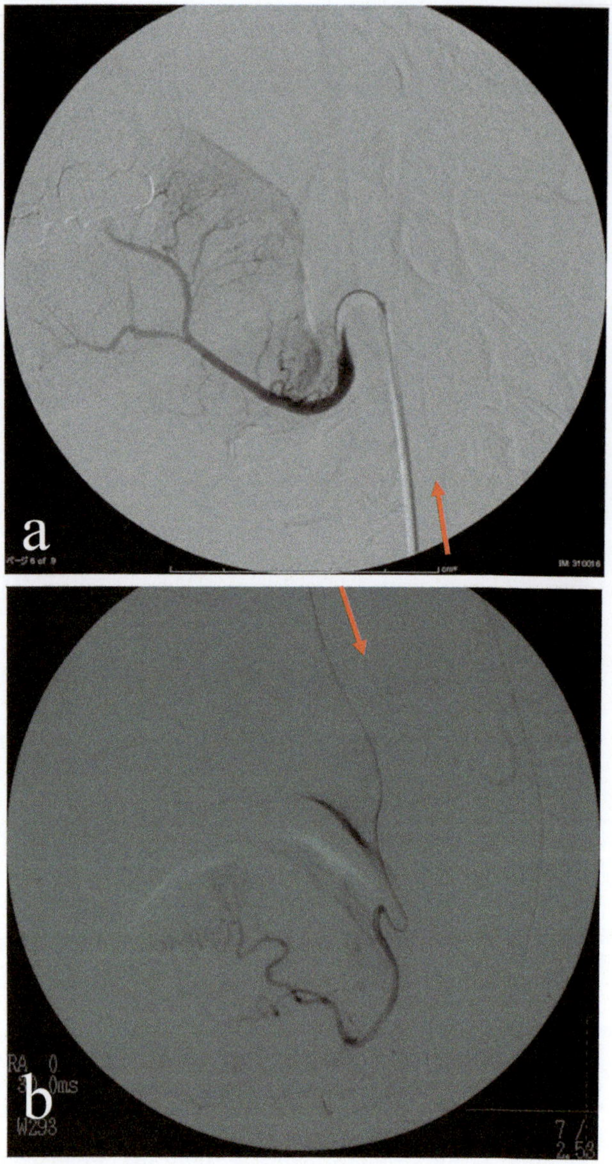

Fig. 16.2 Superselective intra-arterial infusion via the femoral artery by the Seldinger method (**a**) and via the STA (HFT method) (**b**). Tumor staining of the tongue from the LA can be seen with use of contrast medium on digital subtraction angiography (DSA) (Fig. 16.2a was provided by Dr. Takamatsu S. at Fukui Prefectural Hospital)

hook-shaped catheter is not stable, the guidewire exchange method is used to replace it with a polyurethane straight catheter (Fig. 16.3b) [32]. After catheterization, flow check digital subtraction angiography (DSA) and angio-CT are performed in all cases (Fig. 16.5a–d). Angio-CT is helpful for detecting tumors by confirming the enhancement of the feeding area and to allow the catheter to be placed at the appropriate position. Furthermore, weekly confirmation of the feeding

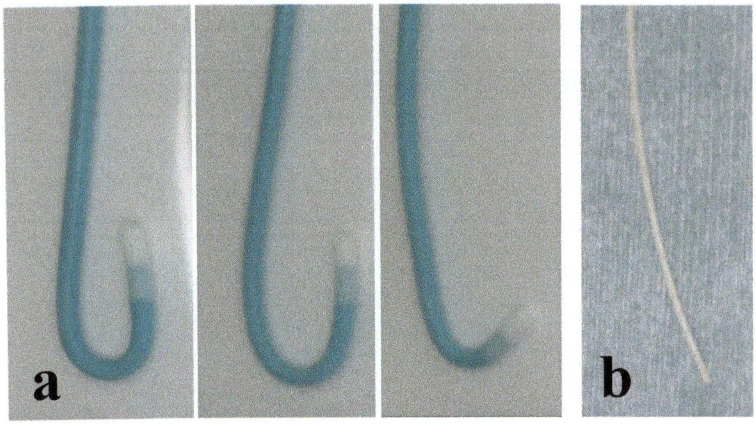

Fig. 16.3 (**a**) A hook-shaped catheter (NECK, Medikit Corp., Tokyo, Japan) (light: Neck 1G, middle: Neck 2G, right: Neck M), (**b**) A polyurethane straight catheter (ANTHRON P-U catheter, Toray Medical Co., Ltd., Tokyo, Japan)

artery by injection of a small amount of indigotindisulfonate sodium is important (Fig. 16.5e, f).

16.5 Retrograde Superselective Intra-Arterial CRT

Radiotherapy is planned for all patients after appropriate immobilization using a thermoplastic mask and three-dimensional CT-based techniques. Conventional radiotherapy is performed with 4 or 6 MV at 2 Gy/fr/day. The irradiation field is changed according to the lymph node status. In cases of N0 disease, the field contains the primary site and levels I to III of the neck on the ipsilateral side. The dose is delivered to 40 Gy/20fr. The portal is then reduced to only the primary site to spare the spinal cord. The total dose delivered to the primary tumor is 60 Gy/30fr. In cases of N1, N2a and N2b disease, the field contains the primary site and levels I to V of the neck on the ipsilateral side. The dose is delivered at 40 Gy/20fr. The portal is then reduced to the primary site and lymph node metastases. The dose for the spinal cord ranges from 40 to 45 Gy. The total dose delivered to the primary tumor is 60 Gy/30fr, and that to the metastatic lymph node sites is to 50 Gy/25fr. In cases of N2c disease, the field contains the primary site and levels I to V of the neck on bilateral sides. The dose is delivered at 40 Gy/20fr. The portal is then reduced to the primary site and lymph node metastases. The dose to the spinal cord ranges from 40 to 45 Gy. The total dose delivered to the primary tumor is 60 Gy/30fr, and, if at all possible, the total dose delivered to the metastatic lymph node sites is 50 Gy/25fr.

The anticancer agent is injected in a bolus through the intra-arterial catheter when radiotherapy is performed. The total dose of docetaxel (DOC) is 60 mg/m^2

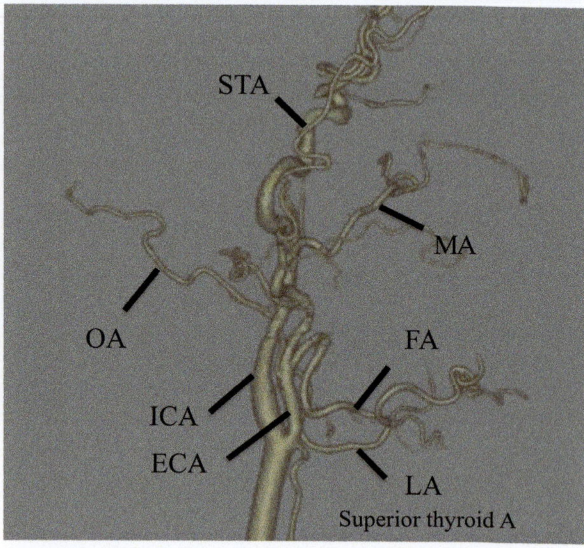

Fig. 16.4 Three-dimensional computed tomography angiography (3D-CTA) of the carotid artery. *Abbreviations*: *ICA* internal carotid artery, *ECA* external carotid artery, *LA* lingual artery, *FA* facial artery, *OA* occipital artery, *MA* maxillary artery, *STA* superficial temporal artery

(10 mg/m^2/week), and that of CDDP is 150 mg/m^2 (5 mg/m^2/day) (Fig. 16.6). Sodium thiosulfate (1 g/m^2) is administered intravenously to provide effective CDDP neutralization after the anticancer agent is given. All patients are given a 5HT-3-receptor antagonist before administration of the anticancer agent.

We evaluated the therapeutic results and rate of organ preservation in patients with advanced oral cancer treated with retrograde superselective intra-arterial chemotherapy and daily concurrent radiotherapy. One hundred and twelve patients with stage III or IV oral cancer were treated with retrograde superselective intra-arterial CRT, and complete response (CR) of the primary site was achieved in 98 of 112 cases (87.5 %). Five-year survival and local control rates were 71.3 % and 79.3 %, respectively. In particular, 5-year local control rates for stage III and stage IV oral cancer patients were 85.1 % and 76.1 %, respectively, indicating that this method provided good local control even for locally advanced oral cancer. This result suggests that our method can preserve organs and minimize functional disturbance [33]. Our treatment strategy for locally advanced oral cancer is to preserve organs and minimize functional disturbance [34, 35]. This treatment has also been reported to be effective for sarcoma and adenoid cystic carcinoma of the head and neck [36, 37].

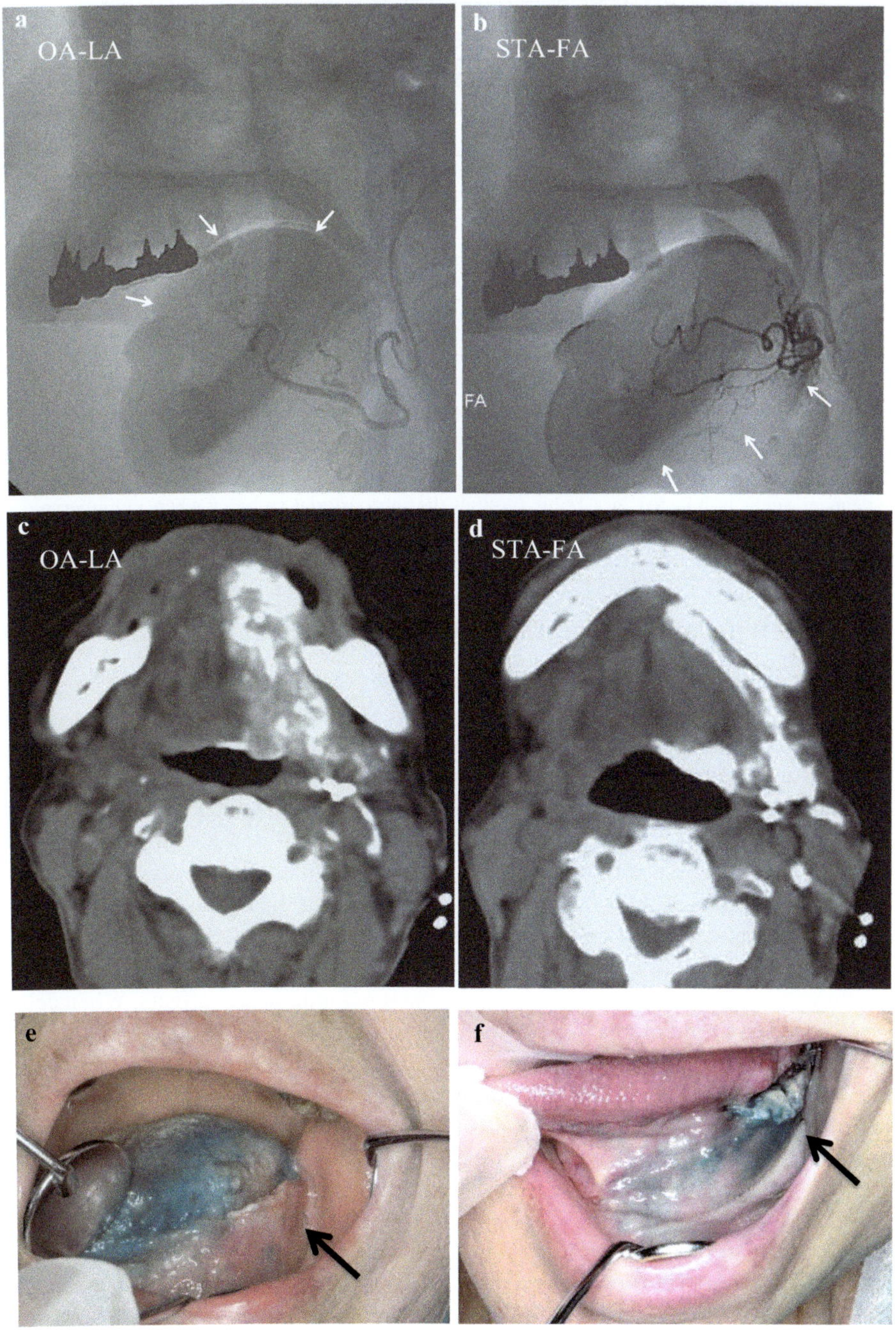

Fig. 16.5 Squamous cell carcinoma of the tongue (T3N1M0). (**a**, **b**) Digital subtraction angiography of retrograde superselective intra-arterial infusion. Two catheters are superselectively inserted into the left LA via the OA (OA-LA) (**a**) and into the left FA via the STA (STA-FA) (**b**). Tumor stain is seen with the use of contrast medium on flow check DSA (**a**, **b**: arrowhead). (**c**, **d**) Axial views of angio-CT. Angio-CT images show tumor staining of the left tongue from the left

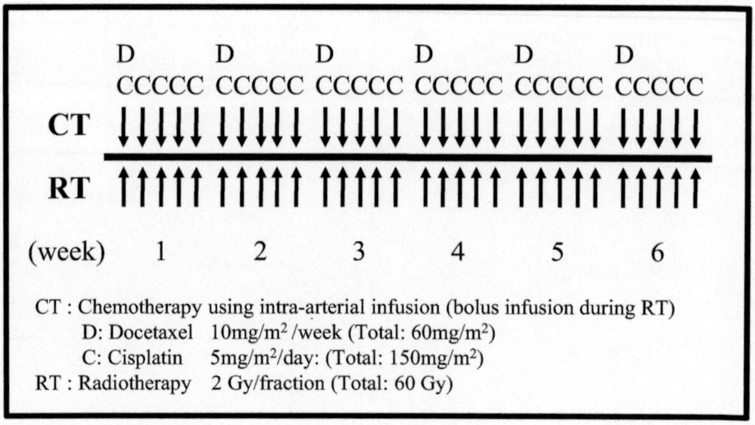

Fig. 16.6 Treatment schedule for chemoradiotherapy using intra-arterial infusion. External irradiation is performed five times a week at 2 Gy per fraction, to a total of 60 Gy, for 6 weeks. The anticancer agent is injected in a bolus through the intra-arterial catheter when radiotherapy is performed. The total dose of DOC is 60 mg/m^2 (15 mg/m^2/week), and that of CDDP is 150 mg/m^2 (5 mg/m^2/day). *Abbreviations*: *DOC* docetaxel, *CDDP* cisplatin

16.6 Retrograde Superselective Intra-Arterial CRT Combined with Hyperthermia

Patients with head and neck squamous cell carcinoma, including oral cavity cancer, with advanced cervical metastases represent a treatment dilemma because their prognosis is generally poor. When there are multiple lymph node metastases (N2b, 2c) or cervical lymph node metastases that exceed 6 cm in their largest diameter (N3), treatment of these metastases is extremely difficult and is often associated with poor prognoses [38]. The use of hyperthermia (HT) has generally been confined to patients with cervical lymph node metastases that accessible with a radiofrequency system employing external application, and HT is used in combination with synergistic CRT. HT treatment itself has anti-cancer effects. When cells are exposed to temperatures over 40 °C, lethal damage is inflicted, predominantly to proteins [39]. In addition, HT is the most potent cellular radiosensitizer. Heat interacts with radiation and potentiates the cellular action of radiation by interfering with the cells' capability to deal with radiation-induced DNA damage [40]. It is well known that hypoxic tumor cells are less responsive to both chemotherapeutic

---

Fig. 16.5 (continued) LA (**c**) and of the left mouth floor from the left FA (**d**) with the use of contrast medium. (**e, f**) The left side of the tongue tumor extends to the floor of the mouth. The left side of the tongue is dyed by injection of a small amount of indigotindisulfonate sodium from the left LA (**e**, arrowhead). On the other hand, the inferior of the tongue and the floor of the mouth are dyed from the left FA (**f**, arrowhead). *Abbreviations*: *LA* lingual artery, *FA* facial artery, *OA* occipital artery, *STA* superficial temporal artery

agents and radiation. Radioresistant cells such as hypoxic cells, low pH cells, and cells in late S phase are thermosensitive [41]. Moreover, radiotherapy and numerous chemotherapeutic agents display complementary effects with HT that lead to synergism. Plataniotis et al. reported tumor control probabilities (TCPs) calculated for hypothetical combined trimodality (radiotherapy + chemotherapy + HT) treatment, and trimodality treatment could give a considerably higher TCP for low-intermediate radio-sensitive tumors [42]. These effects of HT are temperature-dependent. Large volume lymph node metastases usually display poor perfusion that results in central hypoxic conditions, anaerobic glycolysis and low pH. Therefore, combining CRT with HT should improve clinical outcomes for patients with large lymph node metastases, and superficial metastatic neck nodes seem to be suitable for HT.

Our strategy for patients with advanced cervical metastases (N2, N3) is to use thermochemoradiotherapy with retrograde superselective intra-arterial infusion. Treatment consists of superselective intra-arterial CRT (DOC total 60 mg/m^2; CDDP total 150 mg/m^2, total 60 Gy), and HT for 6 weeks. Radiofrequency (RF) capacitive heating equipment (8 MHz, maximum RF output of 60–1500 W, Thermotron RF-8; Yamamoto VINITA Co., Ltd., Osaka, Japan) was used for HT. Two opposing 10-cm electrodes were generally used to heat the cervical lymph node metastases. The electrodes were covered with a water pad, and one was placed along the neck metastatic node, while the other was used for the contralateral site (Fig. 16.7f). HT was applied once or twice a week, and was administered for 50 min within 30 min after each session of radiotherapy. Thermometry of the central skin surface of the neck tumor was performed using thermocouples.

16.7 Thermochemoradiotherapy for Oral Cancer with N3 Cervical Lymph Node Metastases

Head and neck squamous cell carcinoma patients with metastatic lymph nodes measuring <6 cm (N1–N2) have a high rate of clinical and pathologic CR to concurrent CRT, and exhibit a low rate of neck relapse [43]. Outcomes for patients with nodal metastases >6 cm (N3) treated with definitive CRT are poorly defined. This is because patients with N3 disease are generally considered unresectable due to adhesions between metastatic nodes and surrounding tissues, and because patients with N3 treated with CRT experience a very high rate of distant failure [44–46].

In our previous report, nine patients with N3 cervical lymph node metastases of oral squamous cell carcinoma underwent thermochemoradiotherapy using superselective intra-arterial infusion. The primary lesion and metastatic cervical lymph nodes were assessed at 4 weeks after completion of all treatments. Patients were scheduled to undergo neck dissection at 5–8 weeks after the end of

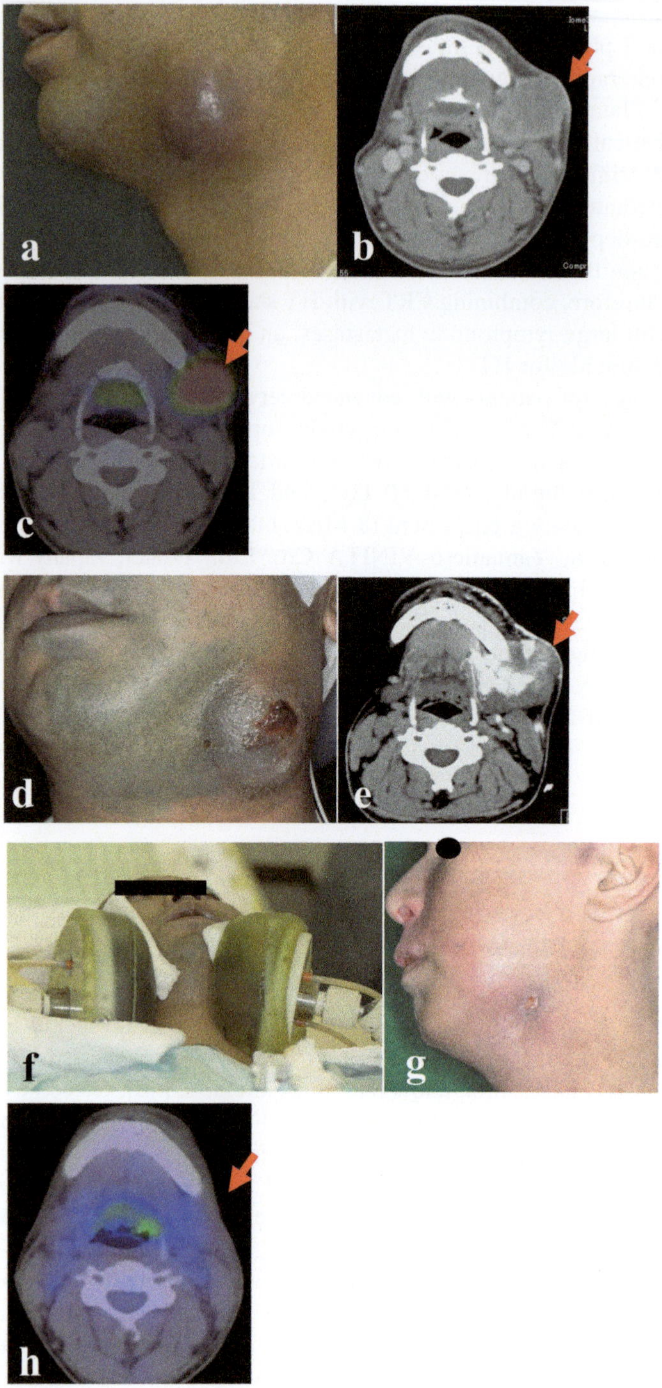

Fig. 16.7 Squamous cell carcinoma of the upper gingiva (T2N3M0). The patient has a large neck mass (85 × 32 mm) on the left side of the neck (**a**), and enhanced-CT reveals a ring-enhancing mass in the left level II of the neck (**b**, arrowhead). PET-CT demonstrates high FDG uptake at the

thermochcmoradiotherapy, unless distant metastases were found. Six of ninc patients underwent neck dissection at 5–8 weeks after treatment. In four of these six patients, all metastatic lymph nodes, including N3, showed pathological CR. The other three patients did not undergo neck dissection due to distant metastasis. Five-year survival and locoregional control rates were 51 % and 88 %, respectively [47, 48].

16.8 Cases of Thermochemoradiotherapy for Oral Cancer with N3 Cervical Lymph Node Metastases

Case 1: Squamous cell carcinoma of the upper gingiva (T2N3M0) (Fig. 16.7) [49]

Superselective intra-arterial CRT combined with HT (DOC total 60 mg/m^2; CDDP total 150 mg/m^2, total 60 Gy, HT: five sessions) for 6 weeks. After the intra-arterial CRT was combined with HT, pathological CR was achieved at the primary site. Radical neck dissection was carried out, and pathological CR was recognized at the N3 cervical lymph node metastasis.

Case 2: Squamous cell carcinoma of the tongue (T4aN3M0) (Fig. 16.8) [50]

The patient was treated with a combination of radiotherapy (2 Gy/day, total 60 Gy), superselective intra-arterial chemotherapy via the STA and the femoral artery (docetaxel: total 124 mg, cisplatin: total 135 mg), and four sessions of HT for cervical lymph node metastases. The tumor responded well to the therapy. Pathological efficacy after thermochemoradiotherapy was CR.

Fig. 16.7 (continued) left cervical lymph nodes (SUV max: 13.2) (**c**, arrowhead). Two catheters are superselectively inserted into the left FA and MA via the left STA and OA. Dying of the skin surface of the N3 disease is confirmed by the injection of indigotindisulfonate sodium (**d**), and tumor stain is seen with the use of contrast medium via the FA for the N3 cervical lymph node metastases on the flow check angio-CT (**e**, arrowhead). Treatment consists of superselective intra-arterial CRT (DOC total 60 mg/m^2; CDDP total 150 mg/m^2, total 60 Gy), and HT (**f**) for 6 weeks. Five sessions of HT are given for cervical lymph node metastases. HT is administered for 50 min within 30 min after each session of CRT The temperature of the central and peripheral skin surface of the neck tumor is over 42 °C. After the completion of treatment, N3 disease is much smaller than before treatment (**g**), and pathological complete response has been achieved at the primary site. FDG uptake on PET-CT of the N3 lymph node metastases has disappeared (**h**, arrowhead). Radical neck dissection has been carried out, and pathological complete response is recognized in the N3 cervical lymph node metastasis. *Abbreviations: PET-CT* positron emission tomography-computed tomography, *FDG* fluoro-2-deoxy-D-glucose, *SUV* standardized uptake value, *FA* facial artery, *OA* occipital artery, *MA* maxillary artery, *STA* superficial temporal artery, *DOC* docetaxel, *CDDP* cisplatin, *HT* hyperthermia

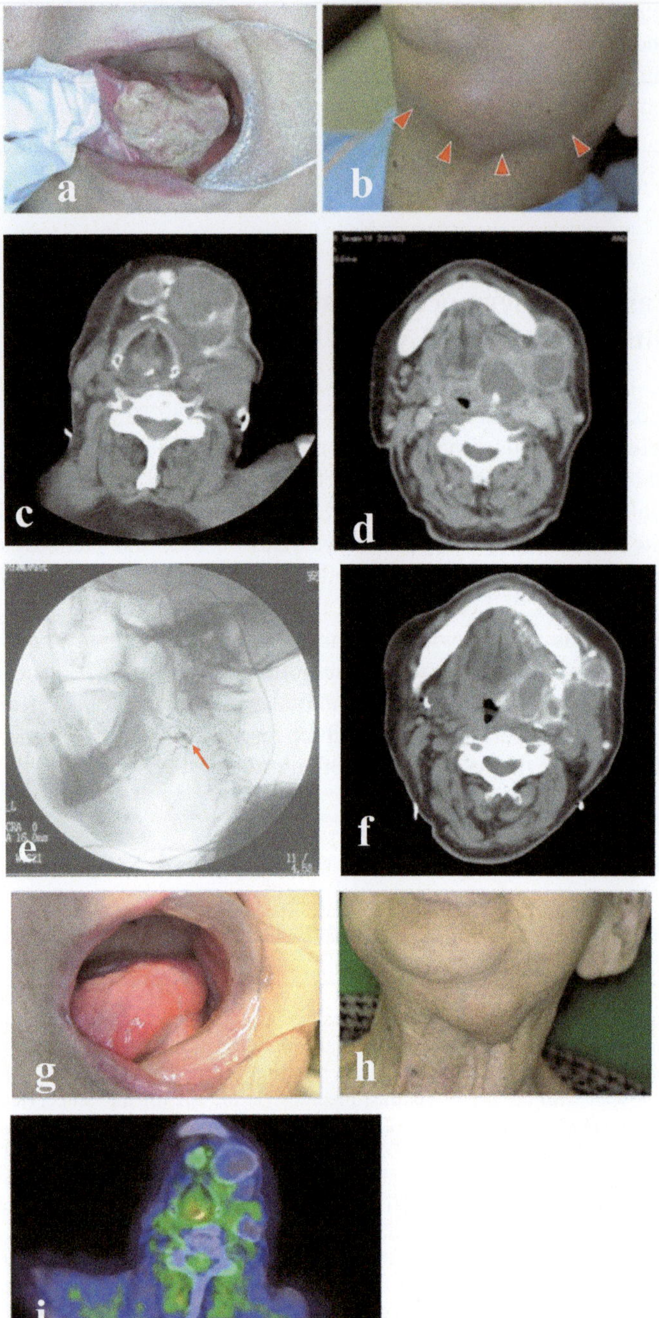

Fig. 16.8 Squamous cell carcinoma of the tongue (T4aN3M0). An ulcerative mass with induration, measuring 55 × 42 mm, at the left lateral border of the tongue (**a**). The patient also had a large neck mass measuring 80 × 45 mm on the left side (**b**). Enhanced CT reveals multiple ring-enhancing masses in the levels I, II, III, and IV of the left cervical area (**c**, **d**). Two catheters are

16.9 Conclusion

Retrograde superselective intra-arterial chemotherapy and daily concurrent radiotherapy for advanced oral cancer provided good overall survival and local control, and thermochemoradiotherapy using intra-arterial infusion provided good histopathologic effects and locoregional control rates for patients with N2, 3 cervical lymph node metastases.

References

1. Tomonaga S, Oshima A, Ishiguro T, et al., editors. White paper affection, death, prognosis. Tokyo: Shinoharashinsha Publisher Inc; 1999. p. 159–70.
2. Foundation for Promotion of Cancer Research. Cancer statistics 2008. Tokyo: Fundation for Promotion of Cancer Research; 2009. p. 26–9.
3. Sobue T. Update on cancer prevention and screening. Curr Treat Cancer. 2005;12:5–23.
4. Ariyoshi Y, Shimobara M, Omura K, et al. Epidemiological research 2002 on malignant tumor of stomatognathic area visited Japanese society of oral and maxillofacial surgery designated training facilities designated. Jpn J Oral Maxillofac Surg. 2006;52:401–10.
5. Salama JK, Seiwert TY, Vokes EE. Chemoradiotherapy for locally advanced head and neck cancer. J Clin Oncol. 2007;25(26):4118–26.
6. Klopp CT, Alford TC, Bateman J, et al. Fractionated intra-arterial cancer; chemotherapy with methyl bis amine hydrochloride; a preliminary report. Ann Surg. 1950;132:811–32.
7. Sullivan RD, Miller E, Sikes MP. Antimetabolite-metabolite combination cancer chemotherapy. Effects of intraarterial methotrexate-intramuscular Citrovorum factor therapy in human cancer. Cancer. 1959;12:1248–62.
8. Ramsden CH, Duff JK. Continuous arterial infusion of head and neck tumors; improvements in technique by retrograde temporal artery catheterization. Cancer. 1963;16:133–5.
9. Burn JI, Johnston ID, Davies AJ, et al. Cancer chemotherapy by continuous intra-arterial infusion of methotrexate. Br J Surg. 1966;53:329–36.
10. Bertino JR, Mosher MB, DeConti RC. Chemotherapy of cancer of the head and neck. Cancer. 1973;31:1141–9.
11. Auersperg M, Furlan L, Marolt F, et al. Intra-arterial chemotherapy and radiotherapy in locally advanced cancer of the oral cavity and oropharynx. Int J Radiat Oncol Biol Phys. 1978;4:273–7.
12. Lee YY, Wallace S, Dimery I, et al. Intraarterial chemotherapy of head and neck tumors. Am J Neuroradiol. 1986;7:343–8.
13. Lee YY, Dimery IW, Van Tassel P, et al. Superselective intra-arterial chemotherapy of advanced paranasal sinus tumors. Arch Otolaryngol Head Neck Surg. 1989;115:503–11.

Fig. 16.8 (continued) inserted superselectively into the left thyrolinguofacial trunk and right FA via the bilateral STAs. The tip of the catheter (*arrow*) is inserted superselectively into the left thyrolinguofacial trunk artery via the left STA (**e**). Angio-CT showed that the tumor corresponds to the enhancement of the feeding area (**f**). The tumor responds well (**g**, **h**), and FDG uptake of PET-CT disappears after treatment (**i**). The patient has no clinical or radiological evidence of local recurrence or distant metastasis for over 7 years after the treatment. *Abbreviations*: *FA* facial artery, *STA* superficial temporal artery, *PET-CT* positron emission tomography-computed tomography, *FDG* fluoro-2-deoxy-D-glucose

14. Robbins KT, Storniolo AM, Kerber C, et al. Rapid superselective high-dose cisplatin infusion for advanced head and neck malignancies. Head Neck. 1992;14:364–71.
15. Robbins KT, Storniolo AM, Kerber C, et al. Phase I study of highly selective supradose cisplatin infusions for advanced head and neck cancer. J Clin Oncol. 1994;12:2113–20.
16. Robbins KT, Kumar PV, Regine WF, et al. Efficacy of target supradose cisplatin and concurrent radiation therapy for advanced head and neck cancer; The Memphis Experience. Int J Radiat Oncol Biol Phys. 1997;38:263–71.
17. Ackerstaff AH, Tan IB, Rasch CR, et al. Quality-of-life assessment after supradose selective intra-arterial cisplatin and concomitant radiation (RADPLAT) for inoperable stage IV head and neck squamous cell carcinoma. Arch Otolaryngol Head Neck Surg. 2002;128:1185–90.
18. Valentino J, Spring PM, Shane M, et al. Interval pathologic assessments in patients treated with concurrent hyperfractionated radiation and intraarterial cisplatin (HYPERRADPLAT). Head Neck. 2002;24:539–44.
19. Balm AJ, Rasch CR, Schornagel JH, et al. High-dose superselective intra-arterial cisplatin and concomitant radiation (RADPLAT) for advanced head and neck cancer. Head Neck. 2004;26:485–93.
20. Robbins KT, Doweck I, Samant S, et al. Factors predictive of local disease control after intra-arterial concomitant chemoradiation (RADPLAT). Laryngoscope. 2004;114(3):411–17.
21. Robbins KT, Kumar P, Harris J, et al. Supradose intra-arterial cisplatin and concurrent radiation therapy for the treatment of stage IV head and neck squamous cell carcinoma is feasible and efficacious in a multi-institutional setting: results of Radiation Therapy Oncology Group Trial 9615. J Clin Oncol. 2005;23:1447–54.
22. Rabbani A, Hinerman RW, Schmalfuss IM, et al. Radiotherapy and Concomitant Intraarterial Cisplatin (RADPLAT) for advanced Squamous cell carcinomas of the head and neck. Am J Clin Oncol. 2007;30:283–6.
23. Doweck I, Robbins KT, Samant S, et al. Intra-arterial chemoradiation for T3-4 oral cavity cancer: treatment outcomes in comparison to oropharyngeal and hypopharyngeal carcinoma. World J Surg Oncol. 2008;14(6):2.
24. Homma A, Oridate N, Suzuki F, et al. Superselective high-dose cisplatin infusion with concomitant radiotherapy in patients with advanced cancer of the nasal cavity and paranasal sinuses: a single institution experience. Cancer. 2009;115:4705–14.
25. Kano S, Homma A, Oridate N, et al. Superselective arterial cisplatin infusion with concomitant radiation therapy for base of tongue cancer. Oral Oncol. 2011;47:665–70.
26. Hattori T, Hirano T, Toyoda S, et al. Superselective intra-arterial chemotherapy via the superficial temporal artery for head and neck tumor. Jpn J Radiat. 1985;45:1056–8.
27. Fuwa N, Ito Y, Kato E, et al. Superselective intra-arterial continuous chemotherapy with CBDCA for advanced head and neck cancer. Jpn J Head Neck Cancer. 1996;22:139–43.
28. Tohnai I, Fuwa N, Hayashi Y, et al. New superselective intra-arterial infusion via superficial temporal artery for cancer of the tongue and tumour tissue platinum concentration after carboplatin (CBDCA) infusion. Oral Oncol. 1998;34:387–90.
29. Fuwa N, Ito Y, Matsumoto A, et al. A combination therapy of continuous superselective intraarterial carboplatin infusion and radiation therapy for locally advanced head and neck carcinoma. Cancer. 2000;89:2099–105.
30. Tohnai I. Chemotherapy using intra-arterial infusion for oral cancer. Nagoya J Med Sci. 2006;68:101–8.
31. Iwai T, Fuwa N, Hirota M, et al. Secure surgical method for catheter placement via the occipital artery to achieve retrograde superselective intra-arterial chemotherapy for advanced oral cancer: alternative to approach via the superficial temporal artery. Indian J Otolaryngol Head Neck Surg. 2014;66(2):205–20.
32. Fuwa N, Kodaira T, Furutani K, et al. A new method of selective intra-arterial infusion therapy via the superficial temporal artery for head and neck cancer. Oral Surg Oral Med Oral Pathol Oral Radiol Endod. 2008;105:783–9.

33. Mitsudo K, Koizumi T, Iida M, et al. Retrograde superselective intra-arterial chemotherapy and daily concurrent radiotherapy for stage III and IV oral cancer: analysis of therapeutic results in 112 cases. Radiother Oncol. 2014;111:306–10.
34. Mitsudo K, Shigetomi T, Fujimoto Y, et al. Organ preservation with daily concurrent chemoradiotherapy using superselective intra-arterial infusion via a superficial temporal artery for T3 and T4 head and neck cancer. Int J Radiat Oncol Biol Phys. 2011;79:1428–35.
35. Mukai Y, Hata M, Mitsudo K, et al. Radiation therapy with concurrent retrograde superselective intra-arterial chemotherapy for gingival carcinoma. Strahlenther Onkol. 2014;190(2):181–5.
36. Mitsudo K, Tohnai I, Fujimoto Y, et al. Leiomyosarcoma of the maxilla: effective chemotherapy with docetaxcel (DOC) and cisplatin (CDDP) using superselective intra-arterial infusion via superficial temporary artery. Oral Oncol Extra. 2006;42:258–62.
37. Adachi M, Mitsudo K, Yamamoto N, et al. Chemoradiotherapy for maxillary sinus adenoid cystic carcinoma using superselective intra-arterial infusion via a superficial temporal artery. Head Neck. 2013;35:e89–93.
38. Jacobs C, Makuch R. Efficacy of adjuvant chemotherapy for patients with resectable head and neck cancer: a subset analysis of the head and neck contracts program. J Clin Oncol. 1990;8:838–47.
39. Lepock JR, Cheng KH, AL-Qysi H, et al. Thermotropic lipid and protein transitions in Chinese hamster lung cell membranes: relationship to hyperthermic cell killing. Can J Biochem Cell Biol. 1983;61:421–7.
40. Kampinga HH, Dikomey E. Hyperthermic radiosensitization: mode of action and clinical relevance. Int J Radiat Biol. 2001;77:399–408.
41. Hahn GM, Shiu EC. Adaptation to low pH modifies thermal and thermo-chemical responses of mammalian cells. Int J Hyperthermia. 1986;2:379–87.
42. Plataniotis GA, Dale RG. Use of the concept of equivalent biologically effective dose (BED) to quantify the contribution of hyperthermia to local tumor control in radiohyperthermia cervical cancer trials, and comparison with radiochemotherapy results. Int J Radiat Oncol Biol Phys. 2009;73:1538–44.
43. Forest VI, Nguyen-Tan PF, Tabet J, et al. Role of neck dissection following concurrent chemoradiation for advanced head and neck carcinoma. Head Neck. 2006;28:1099–105.
44. Ahmed KA, Robbins KT, Wong F, et al. Efficacy of concomitant chemoradiation and surgical salvage for N3 nodal disease associated with upper aerodigestive tract carcinoma. Laryngoscope. 2000;110:1789–93.
45. Carvalho AL, Kowalski LP, Agra IM, et al. Treatment results on advanced neck metastasis (N3) from head and neck squamous carcinoma. Otolaryngol Head Neck Surg. 2005;132:862–8.
46. Ballonoff A, Raben D, Rusthoven KE, et al. Outcomes of patients with N3 neck nodes treated with chemoradiation. Laryngoscope. 2008;118:995–8.
47. Mitsudo K, Koizumi T, Iida M, et al. Thermochemoradiation therapy using superselective intra-arterial infusion via superficial temporal and occipital arteries for oral cancer with N3 cervical lymph node metastases. Int J Radiat Oncol Biol Phys. 2012;83:e639–45.
48. Mitsudo K, Koizumi T, Iida M, et al. Thermochemoradiotherapy for oral cancer with N2, 3 cervical lymph node metastases using retrograde superselective intra-arterial infusion. Themal Med. 2012;28(2):23–8.
49. Mitsunaga S, Mitsudo K, Iwai T, et al. Hyperthermia combined with chemoradiotherapy using superselective intra-arterial infusion via superficial temporal artery and occipital artery for treatment of advanced upper gingival cancer with N3 lymph node metastasis. J Jpn Soc Oral Oncol. 2010;22:157–63.
50. Nishiguchi H, Mitsudo K, Yamoamoto N, et al. Thermochemoradiotherapy using superselective intra-arterial infusion for N3 cervical lymph node metastases of tongue cancer. J Cancer Res Ther. 2013;9(4):718–20.

Chapter 17
Combination by Hyperthermia and Radiation (and Chemotherapy): Lung Cancer

Takayuki Ohguri

Abstract In this chapter, the clinical results of radiotherapy or chemoradiotherapy with hyperthermia for the treatment of lung cancer were reviewed. Especially for the patients with superior sulcus tumors or lung cancer with chest wall invasion, the intra-tumor temperature could be elevated to therapeutic levels by the 8-MHz RF capacitive heating device. The correlations between the clinical outcomes and thermal parameters were observed. A randomized trial demonstrated that an adding of hyperthermia to radiotherapy showed a significant improvement in the local tumor control rate in patients with non-small cell lung cancer.

Keywords Hyperthermia • Radiotherapy • Lung cancer • Chemoradiotherapy

17.1 Heating Technique for the Lung Cancer

The use of regional deep heating has been previously investigated, particularly for the treatment of pelvic tumors and soft tissue sarcomas [1, 2]. Limited information on the use of regional hyperthermia for the treatment of patients with lung cancer exists in the literature, probably because most available devices for regional deep heating are structurally difficult to apply to the thoracic region; however, some reports have described the feasibility and efficacy of regional hyperthermia using an 8-MHz radiofrequency (RF) capacitive heating device for patients with lung cancer [3–6].

A capacitive regional heating device operating at 8 MHz of the Thermotron-RF8 (Yamamoto VINITA Co., Ltd., Osaka, Japan) has been previously utilized for its physical features and thermal distribution characteristics [7, 8]. When using this device, the patient is placed between two electrodes connected to a high-power RF generator. In most patients with lung cancer, both the upper and lower electrodes

T. Ohguri (✉)
Department of Radiology, University of Occupational and Environmental Health,
1-1 Iseigaoka, Yahatanishi-ku, Kitakyushu 807-8555, Japan
e-mail: ogurieye@med.uoeh-u.ac.jp

S. Kokura et al. (eds.), *Hyperthermic Oncology from Bench to Bedside*,
DOI 10.1007/978-981-10-0719-4_17

Fig. 17.1 Overlay boluses with cooled circulating liquid for deep regional hyperthermia using the 8-MHz RF capacitive heating device should be applied to minimize contact between the electrode edge and body surface to decrease pain in patients with lung cancer. The prone position is also recommended to diminish subcutaneous pain. Cited from [10]

were 30 cm in diameter and placed on the opposite sides of the thoracic region. In patients with the invasion of the superficial tissues, such as the chest wall and ribs, the smaller electrode was selected to obtain favorable temperature distribution. Treatment should be administered with the patient in the prone position to decrease the pain caused by heating because the degree of pain in the prone position was significantly smaller than that in the supine position (Fig. 17.1) [9, 10].

Although Asian patients are considered to be relatively suitable for capacitive heating due to their slender constitution, the overheating of subcutaneous fat tissue and pain near the electrode edge are major problems for deep regional capacitive heating devices. Overlay boluses have been applied to minimize contact between the electrode edge and body surface and could decrease the pain associated with this technique (Fig. 17.1) [11]. The circulating liquid of the overlay boluses should be cooled at close to the freezing point of water to achieve a good temperature rise of the deep-seated tumor [12–14]. The treatment results of the 8-MHz RF capacitive heating device for hyperthermia of the entire thoracic region indicated that higher intraesophageal temperature was significantly achieved in patients with better performance status, younger age, and less subcutaneous fat (Fig. 17.2) [15].

Prior studies have reported deep regional hyperthermia-related toxicity in the form of subcutaneous fat burns in 3–12 % of patients [16, 17]. Generally, these burns spontaneously healed and did not result in the discontinuation of treatment. Furthermore, a large randomized phase III study of deep regional hyperthermia did not show any increase in acute or late toxicity from radiotherapy in patients with pelvic tumors [18].

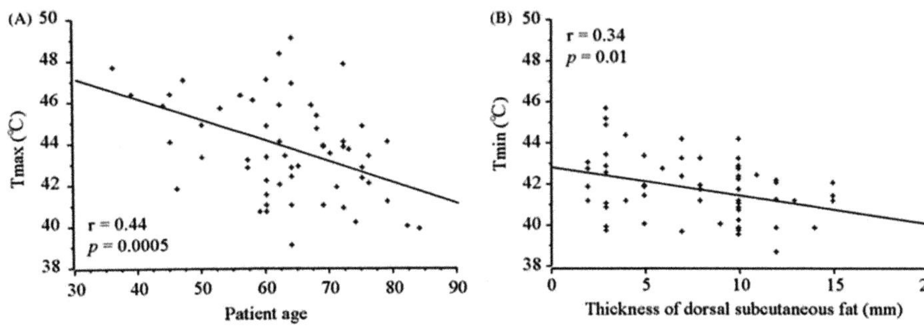

Fig. 17.2 Correlations between the intraesophageal temperatures and clinical characteristics. (**a**) The patient age was inversely correlated with T_{max} as thermal parameters of intraesophageal temperature. (**b**) The thickness of dorsal subcutaneous fat at the levels of the tracheal bifurcation was inversely correlated with T_{min}. Cited from [15]

17.2 Superior Sulcus Tumors and Lung Cancer with Chest Wall Invasion

Lung cancers that occur in the apex of the chest and invade the apical chest wall structures are called superior sulcus tumors (SSTs) or Pancoast tumors. The addition of hyperthermia to radiotherapy for SSTs or lung cancer with chest wall invasion to improve local tumor control has been investigated because the delivery of heat and the measurement of the temperature in these tumors were relatively feasible. In 1992, Hiraoka et al. demonstrated an initial experience with radiotherapy and hyperthermia to treat lung cancer with chest wall invasion [3]. The thermal parameters for intra-tumor temperature were analyzed, and the means for T_{max}, T_{av}, and T_{min} were 42.9 °C, 41.6 °C, and 39.7 °C, respectively. Higher thermal parameters were closely related to the appearance of low-density areas in the tumors.

Karasawa et al. also reported the results for direct tumor temperature monitoring in 11 patients with locally advanced NSCLC and chest wall invasion; the T_{max}, T_{av}, and T_{min} were 42.6 °C, 41.3 °C, and 40.3 °C, respectively, and relationships between tumor response and intra-tumor temperature were observed [4]. Sakurai et al. also reported the results of a retrospective study for NSCLC with direct bony invasion. The 2-year overall survival rate in the radiotherapy plus hyperthermia group and the rate in the radiotherapy alone group were 44 % and 15 %, respectively.

Moon et al. reported on 24 patients with SSTs treated with definitive radiotherapy plus regional hyperthermia. Twelve of the 24 (50 %) patients also underwent chemotherapy. In this study, 88 % of the patients with Stage IIB SSTs achieved local control [19]. Table 17.1 summarizes the clinical treatment results for radiotherapy with or without regional hyperthermia. Figure 17.3 shows one case of NSCLC with direct bony invasion treated with definitive chemoradiotherapy combined with regional hyperthermia at our institution, and complete response was achieved after the combined therapy.

Table 17.1 Previous studies of multimodalty therapy including radiotherapy for superior sulcus tumors

Study	n	Stage	Modality	RT (Gy)	LC rate (%)	5-y OS rate (%)
Van Houtte	31	NA	RT	20–70	55 %	23 %
Rusch	95	III	CRT → S	45	pCR: 36 %	54 %
Kunitoh	75	IIB-IIIB	CRT → S	45	pCR: 16 %	56 %
Komaki	32	IIB-IIIB	CRT	64 (19–70)	44 %	36 % (with ≧66 Gy)
Kappers	28	IIB-IIIB (inoperable)	CRT	66–88	68 %	20 %
Hiraoka	20*	IIIA-IV	RHT	50.4–70	OTR: 75 %	–
Sakurai	13*	III	RHT	60–70	76 %	44 % (2-y)
Ebara	5	IIIB	RHT or CRHT	70 (68–70)	60 %	–
Moon	24	IIB-IIIB	RHT or CRHT	70 (60–84)	55 %	47 %

*Patients with NSCLC with direct bony invasion. *Abbreviations: NA* not available, *RT* radiotherapy, *CRT* chemoradiotherapy, *S* surgery, MST median survival time, *RHT* radiotherapy plus hyperthermia, *CRHT* chemoradiotherapy plus hyperthermia, *OTR* objective tumor response

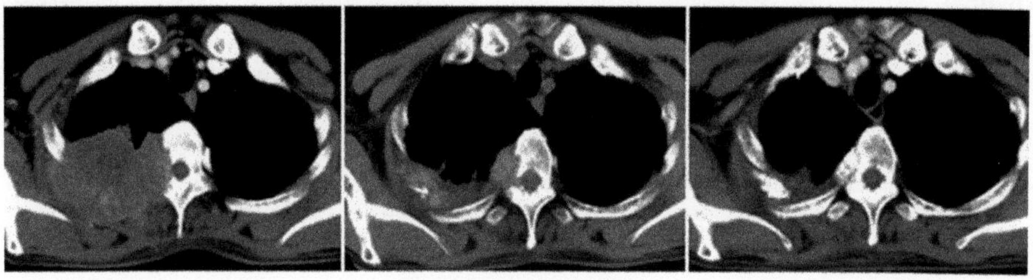

Fig. 17.3 An unresectable case of lung cancer with direct bony invasion (T4N0M0, squamous cell carcinoma) was treated with definitive radiotherapy (76 Gy given in 2.0 Gy per fraction), chemotherapy (carboplatin and paclitaxel), and regional hyperthermia (a total of seven sessions). Regional hyperthermia of the entire thoracic region was immediately administered after radiotherapy. Complete response was achieved 10 months after the combined treatments

17.3 NSCLC Including the Tumor that Do Not Invade the Chest Wall

There have been limited clinical results regarding regional hyperthermia for lung cancers that do not invade the chest wall [6, 20]. A small randomized trial regarding radiotherapy in combination with hyperthermia investigated patients with locally advanced NSCLC [20]. The authors measured the intra-esophageal temperature to estimate the tumor temperatures. The results for the thermometry of the intra-esophageal temperature showed an average T_{max}, T_{min} and T_{av} of 41.3 °C, 39.5 °C, and 40.3 °C, respectively. The thermoradiotherapy group showed a significant improvement in the local tumor control rate, although no improvement in overall survival was observed (Fig. 17.4).

Ohguri et al. investigated the efficacy of radiotherapy combined with regional hyperthermia guided by radiofrequency-output power and intra-esophageal temperature in patients with Stage III NSCLC [6]. The intra-esophageal temperature measurements significantly correlated with the radiofrequency-output power, and higher radiofrequency-output power was a statistically significant prognostic factor for overall and local recurrence-free survival. The authors concluded that the radiofrequency-output power may be used as a promising parameter to assess the treatment of deep regional hyperthermia for lung cancer. This strategy is less invasive and well tolerated, and may be suitably incorporated into the combined clinical therapeutic modality.

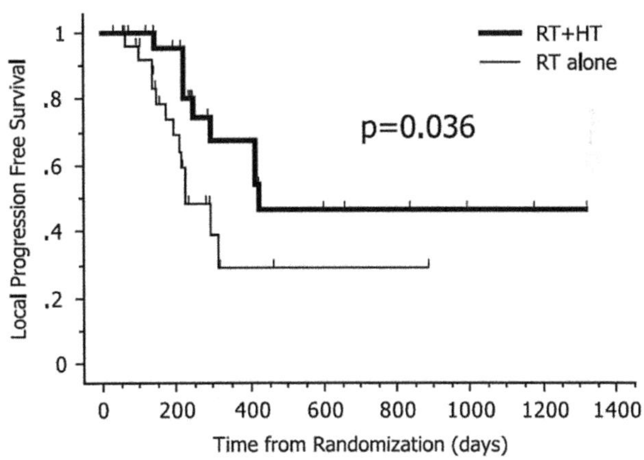

Fig. 17.4 In a multi-institutional, prospective, randomized trial for locally advanced NSCLC, local progression-free survival in the patients treated with radiotherapy and hyperthermia was significantly better than that in patients with radiotherapy alone ($p = 0.036$). Cited from [20]

17.4 Re-irradiation Plus Regional Hyperthermia for Recurrent NSCLC

Although concurrent chemoradiotherapy improved survival in patients with locally advanced NSCLC, the locoregional recurrence rate remained high at approximately 30–40 % [21]. There are few curative or palliative treatment options for recurrent NSCLC after definitive radiotherapy, although local recurrence at the primary site and the presence of a residual tumor continue to be major problems. No general therapeutic guidelines have yet been established for these conditions. The combined therapy of re-irradiation and hyperthermia showed higher local control rates and a good palliative effect in patients with recurrent breast and rectal cancer; this therapy may be ideal to obtain a sufficient antitumor effect for in-field recurrence after radiotherapy because the tolerated dose of the organs at risk limits the prescribed dose for re-irradiation [22–24]. A preliminary result for re-irradiation plus regional hyperthermia in 33 patients with recurrent NSCLC has been reported [25]. Eight patients achieved a long-term survival (more than 3 years after re-irradiation). The use of hyperthermia with a higher radiofrequency-output power tended to result in a better local control rate, and this combination therapy may be promising with acceptable toxicity.

References

1. van der Zee J. Heating the patient: a promising approach? Ann Oncol. 2002;13(8):1173–84.
2. Jones EL, Samulski TV, Leonard RP, Dewhirst MW. Hyperthermia. In: Principles and practice of radiation oncology. 4th ed. Philadelphia: Lippincott Williams & Wilkins; 2003.
3. Hiraoka M, Masunaga S, Nishimura Y, Nagata Y, Jo S, Akuta K, et al. Regional hyperthermia combined with radiotherapy in the treatment of lung cancers. Int J Radiat Oncol Biol Phys. 1992;22(5):1009–14.
4. Karasawa K, Muta N, Nakagawa K, Hasezawa K, Terahara A, Onogi Y, et al. Thermoradiotherapy in the treatment of locally advanced nonsmall cell lung cancer. Int J Radiat Oncol Biol Phys. 1994;30(5):1171–7.
5. Sakurai H, Hayakawa K, Mitsuhashi N, Tamaki Y, Nakayama Y, Kurosaki H, et al. Effect of hyperthermia combined with external radiation therapy in primary non-small cell lung cancer with direct bony invasion. Int J Hyperthermia. 2002;18(5):472–83. doi:10.1080/02656730210146917.
6. Ohguri T, Imada H, Yahara K, Morioka T, Nakano K, Terashima H, et al. Radiotherapy with 8-MHz radiofrequency-capacitive regional hyperthermia for stage III non-small-cell lung cancer: the radiofrequency-output power correlates with the intraesophageal temperature and clinical outcomes. Int J Radiat Oncol Biol Phys. 2009;73(1):128–35. doi:S0360-3016(08)00648-2 [pii] 10.1016/j.ijrobp.2008.03.059.
7. Song CW, Rhee JG, Lee CK, Levitt SH. Capacitive heating of phantom and human tumors with an 8 MHz radiofrequency applicator (Thermotron RF-8). Int J Radiat Oncol Biol Phys. 1986;12(3):365–72. doi:0360-3016(86)90352-4 [pii].
8. Abe M, Hiraoka M, Takahashi M, Egawa S, Matsuda C, Onoyama Y, et al. Multi-institutional studies on hyperthermia using an 8-MHz radiofrequency capacitive heating device

(Thermotron RF-8) in combination with radiation for cancer therapy. Cancer. 1986;58 (8):1589–95.
9. Imada T, Nomoto S, Tomimatsu A, Kosaka K, Kusano S, Ostapenk VV, et al. Importance of patient positioning in hyperthermia for deep-seated intrathoracic tumors using an 8 MHz RF capacitive heating device. Jpn J Hyperthermic Oncol. 1999;15(1):15–9.
10. Ohguri T, Yahara K, Murakami M, Imada H, Terashima H, Korogi Y. Current status and future directions of hyperthermia at the University of Occupational and Environmental Health. Thermal Med. 2010;26(3):87–96.
11. Tomimatsu A, Imada H, Kosaka K, Nomoto S, Kusano S, Ostapenko VV. Refinement of circulating liquid of overlay bolus in hyperthermia using an 8 MHz RF capacitive heating device. Jpn J Hyperthermic Oncol. 1999;15(2):71–7.
12. Ohguri T, Imada H, Yahara K, Kakeda S, Tomimatsu A, Kato F, et al. Effect of 8-MHz radiofrequency-capacitive regional hyperthermia with strong superficial cooling for unresectable or recurrent colorectal cancer. Int J Hyperthermia. 2004;20(5):465–75. doi:10.1080/02656730310001657729. H74GQRD0GEFG59ME [pii].
13. Tomimatsu A, Imada H, Kosaka K, Nomoto S, Kusano S, Ostapenko VV, et al. Advantage of an external cooling unit in deep hyperthermia using an 8 MHz RF capacitive heating device. Jpn J Hyperthermic Oncol. 1999;15(2):65–70.
14. Kroeze H, van de Kamer JB, de Leeuw AA, Kikuchi M, Lagendijk JJ. Treatment planning for capacitive regional hyperthermia. Int J Hyperthermia. 2003;19(1):58–73. doi:905C1MH2DYR4XEF4 [pii].
15. Ohguri T, Yahara K, Moon SD, Yamaguchi S, Imada H, Terashima H, et al. Deep regional hyperthermia for the whole thoracic region using 8 MHz radiofrequency-capacitive heating device: relationship between the radiofrequency-output power and the intra-oesophageal temperature and predictive factors for a good heating in 59 patients. Int J Hyperthermia. 2011;27(1):20–6. doi:10.3109/02656736.2010.500644.
16. Feldmann HJ, Seegenschmiedt MH, Molls M. Hyperthermia--its actual role in radiation oncology. Part III: clinical rationale and results in deep seated tumors. Strahlenther Onkol. 1995;171(5):251–64.
17. van der Zee J, Vujaskovic Z, Kondo M, Sugahara T. The Kadota Fund International Forum 2004--clinical group consensus. Int J Hyperthermia. 2008;24(2):111–22. doi:790655589 [pii] 10.1080/02656730801895058.
18. van der Zee J, Gonzalez Gonzalez D, van Rhoon GC, van Dijk JD, van Putten WL, Hart AA. Comparison of radiotherapy alone with radiotherapy plus hyperthermia in locally advanced pelvic tumours: a prospective, randomised, multicentre trial. Dutch Deep Hyperthermia Group. Lancet. 2000;355(9210):1119–25. doi:S0140673600020596 [pii].
19. Moon SD, Ohguri T, Imada H, Yahara K, Yamaguchi S, Hanagiri T, et al. Definitive radiotherapy plus regional hyperthermia with or without chemotherapy for superior sulcus tumors: a 20-year, single center experience. Lung Cancer. 2011;71(3):338–43. doi:S0169-5002(10) 00276-X [pii] 10.1016/j.lungcan.2010.06.007.
20. Mitsumori M, Zeng ZF, Oliynychenko P, Park JH, Choi IB, Tatsuzaki H, et al. Regional hyperthermia combined with radiotherapy for locally advanced non-small cell lung cancers: a multi-institutional prospective randomized trial of the International Atomic Energy Agency. Int J Clin Oncol. 2007;12(3):192–8. doi:10.1007/s10147-006-0647-5.
21. Fournel P, Robinet G, Thomas P, Souquet PJ, Lena H, Vergnenegre A, et al. Randomized phase III trial of sequential chemoradiotherapy compared with concurrent chemoradiotherapy in locally advanced non-small-cell lung cancer: Groupe Lyon-Saint-Etienne d'Oncologie Thoracique-Groupe Francais de Pneumo-Cancerologie NPC 95-01 Study. J Clin Oncol. 2005;23(25):5910–17. doi:JCO.2005.03.070 [pii] 10.1200/JCO.2005.03.070.
22. Vernon CC, Hand JW, Field SB, Machin D, Whaley JB, van der Zee J, et al. Radiotherapy with or without hyperthermia in the treatment of superficial localized breast cancer: results from five randomized controlled trials. International Collaborative Hyperthermia Group. Int J Radiat Oncol Biol Phys. 1996;35(4):731–44. doi:036030169600154X [pii].

23. Juffermans JH, Hanssens PE, van Putten WL, van Rhoon GC, van Der Zee J. Reirradiation and hyperthermia in rectal carcinoma: a retrospective study on palliative effect. Cancer. 2003;98 (8):1759–66. doi:10.1002/cncr.11719.
24. Milani V, Pazos M, Issels RD, Buecklein V, Rahman S, Tschoep K, et al. Radiochemotherapy in combination with regional hyperthermia in preirradiated patients with recurrent rectal cancer. Strahlenther Onkol. 2008;184(3):163–8. doi:10.1007/s00066-008-1731-8.
25. Ohguri T, Imada H, Yahara K, Moon SD, Yamaguchi S, Yatera K, et al. Re-irradiation plus regional hyperthermia for recurrent non-small cell lung cancer: a potential modality for inducing long-term survival in selected patients. Lung Cancer. 2012;77(1):140–5. doi: S0169-5002(12)00107-9 [pii] 10.1016/j.lungcan.2012.02.018.

Chapter 18
Efficacy of Hyperthermia in Combination with Radiation Therapy for Breast Cancer

Daigo Yamamoto, Chizuko Yamamoto, Satoru Iwase, and Hiroki Odagiri

Abstract Hyperthermia continues to show clinical benefits in randomized trials across a spectrum of malignancies, with generally well-tolerated side effects when administered as multimodality therapy. Hyperthermia contributes to direct tumor cell killing, and may also enhance the antitumor effects of chemotherapy and radiation therapy. Particularly, hyperthermia has been shown to be beneficial for treating superficial malignancies, such as chest wall recurrence of breast cancer. Although much evidence has been gained from clinical trials and studies of hyperthermia, regular use has not yet been established for breast cancer treatment. In this review article, we will highlight the benefits of therapeutic hyperthermia and the combined use of hyperthermia and radiation, and overview the results reported from clinical trials.

Keywords Breast • Hyperthermia • Radiation therapy

18.1 Introduction

Patients with chest wall/superficial breast cancer recurrence are a heterogeneous group unified by the trait of having failed standard therapy. Local recurrence rates after mastectomy range from 5 to 45 %, thus prompting the consideration of adjuvant radiation therapy [1–7]. When patients fail locally, morbidities include

D. Yamamoto (✉)
Department of Surgery, Kansai Medical University, Takii, Osaka, Japan

Department of Internal Medicine, Seiko Hospital, Osaka, Japan
e-mail: yamamotd@takii.kmu.ac.jp

C. Yamamoto
Department of Internal Medicine, Seiko Hospital, Osaka, Japan

S. Iwase
Department of Palliative Medicine, Institute of Medical Science, University of Tokyo Hospital, Tokyo, Japan

H. Odagiri
Department of Surgery, Hirosaki National Hospital, Hirosaki, Japan

S. Kokura et al. (eds.), *Hyperthermic Oncology from Bench to Bedside*,
DOI 10.1007/978-981-10-0719-4_18

pain, ulceration, bleeding, lymphoedema, brachial plexopathy, as well as the psychological distress of having visible local disease [8, 9]. A significant proportion of patients who experience a chest wall failure will also develop distant metastatic disease, prompting some to treat patients with local recurrence palliatively [5, 6, 10–14]. However, with aggressive local therapy, some patients are able to maintain long disease-free intervals [15–21].

Hyperthermia, one of the oldest forms of cancer treatment, involves the selective heating of tumor tissues to temperatures ranging between 39 and 45 °C. The earliest report of its use in the treatment of breast cancer was more than 5000 years ago, described on an Egyptian papyrus [22].

Westermark and Coley first demonstrated the clinical applications of hyperthermia [23, 24]. Afterwards, recent remarkable developments based on the thermoradiobiological rationale of hyperthermia indicated that it is also a potent radio- and chemosensitizer. This has been further corroborated through positive clinical outcomes in various tumor sites using thermoradiotherapy or thermoradiochemotherapy approaches [25]. According to the 2015 National Comprehensive Cancer Network (NCCN) guidelines, hyperthermia is included in clinical practice for the treatment of breast recurrence [26]. The guidelines include consideration of the addition of hyperthermia with irradiation for treatment of localized recurrence or metastasis (category 3) [26]. There have been several prospective randomized trials comparing radiation to radiation plus hyperthermia in the treatment of locally advanced/recurrent cancers, primarily breast cancer chest wall recurrence. While there is heterogeneity among the study results, a patient series with strict quality assurance demonstrated a statistically significant increase in local tumor response and greater duration of local control with the addition of hyperthermia to radiation, compared with radiation alone. The predominant molecular target of hyperthermia appears to be protein [27]. The rationale for combining hyperthermia and radiation is multifold; hyperthermia's mechanism of action is complimentary to the effects of radiation, with regard to DNA damage repair, hypoxia, and cell cycle sensitivity [28–30]. Hyperthermia induces direct cytotoxicity, particularly in cells that are acidotic and nutrient-deprived [29–31]. In addition, hyperthermia has influence on the tumor blood flow and oxygenation that may enhance tumor radiation response [32].

18.2 Efficacy

Lindholm et al. first reported a clinical study of hyperthermia in 1987 [33]. In their study, patients with superficial recurrent malignant tumors received low dose radiation therapy with or without hyperthermia. The majority of the tumors (53 %) were mammary adenocarcinomas. In the combined treatment group, the complete response rate was 65 % and partial response rate was 29 %, while in the radiation alone group, the complete response rate was 35 % and partial response rate was 18 %. Furthermore, in order to explore the prognostic factors for tumor response and skin damage, Lindholm et al. [34] performed radiation therapy and hyperthermia for a total of 3, 4, or 6 heat treatments in 59 patients, reporting a response rate (CR + PR) of 100 % independent of the number of heating sessions performed, with acceptable toxicities.

There are numerous studies detailing the results of combined hyperthermia and radiation therapy for chest wall or superficial recurrence of breast cancer [35, 36]; however, most are single institution and retrospective studies. The first randomised trial was performed by the Radiation Therapy Oncology Group (RTOG), protocol 8104, and included 307 patients with superficially measurable tumors, 245 of whom had single lesions available for analysis. Their treatment consisted of a radiation dose of 32 Gy, given in 4 Gy, twice weekly fractions, with two hyperthermia sessions (goal 42.5 °C, 45–60 min). Of the patients, 30 % (68 patients) had superficial disease in the breast or chest wall. Although there was no significant increase in overall survival, the patients who received radiation therapy and hyperthermia showed a CR rate of 62 %, while those who had radiation alone had a CR rate of 40 %. Furthermore, patients with a lesion diameter <3 cm had a CR rate of 52 % in those receiving combination treatment vs 39 % in those receiving radiation alone ($p < 0.05$). In addition, a recent Phase III trial enrolling 122 patients was performed [37], 109 (89 %) of whom were deemed able to undergo hyperthermia treatment and randomly assigned to the treatment groups. The CR rate was 66.1 % in the hyperthermia and radiation arm and 42.3 % in the radiation arm. The odds ratio for CR was 2.7 (95 % CI, 1.2 to 5.8; $P = .02$). Previously irradiated patients had the greatest incremental gain in CR: 23.5 % in the radiation arm versus 68.2 % in the hyperthermia and radiation arm. No overall survival benefit was seen. A more recent study [38] was performed with 127 patients who received hyperthermia and radiation treatment to 167 sites, including the intact breast (24.4 %), chest wall/skin (67.7 %), and breast/chest wall and nodes (7.9 %). The authors reported that at the median follow-up of 13 months (mean 30 ± 38), improved overall survival was significantly associated with increasing radiation dose ($p < 0.0001$), median thermal equivalent dose (TED) 42.5 °C ≥ 200 min ($p = 0.003$), and local control ($p = 0.0002$). Local control at last follow-up was observed in 55.1 % of patients, and CR was significantly associated with median TED 42.5 °C ≥ 200 min ($p = 0.002$) and median TED 43 °C ≥ 100 min ($p = 0.03$). Therefore, while this study was retrospective, hyperthermia and radiation therapy were shown to be effective for locally advanced or recurrent breast cancer in patients that had been historically difficult to treat by radiation therapy alone; over 50 % of patients achieved control of locoregional disease, and overall survival was improved with local control.

Furthermore, the efficacy of re-irradiation combined with hyperthermia has been reported [39] in a study of 248 patients with macroscopic breast cancer recurrence. Radiotherapy was applied at a dose of 32 Gy in 4 Gy fractions, twice weekly, and hyperthermia was performed once weekly after the radiation. Primary endpoints for this analysis were CR and local control (LC). Secondary endpoints were overall survival (OS), and toxicity. The authors reported a CR rate of 70 %, and at 1, 3, and 5 years, LC was 53 %, 40 % and 39 %, and OS was 66 %, 32 %, and 18 %, respectively. OS after 10 years was 10 %. Thermal burns developed in 23 % patients, which were treated with conservative measures. The incidence of 5-year late grade 3 toxicity was 1 %. A few patients survived more than 10 years without evidence of disease. The combination of re- radiation and hyperthermia may be useful for breast cancer recurrence after postmastectomy radiation therapy. We experienced a case of a 66-year-old woman with recurrent breast cancer who underwent radiation and chemotherapy, but failed to respond. However, hyperthermia and re-irradiation induced reduction of the tumor size and disappearance of the previously present foul odor. Eight weeks later, the tumor became nearly flat, and the patient was deemed to have complete response [40].

Although much evidence has been gained from clinical trials and studies of hyperthermia, regular use has not yet been established for breast cancer treatment [41–43]. Reasons that have been suggested to explain this lack of wide use include the argument that application of hyperthermia is a waste of time, it is technologically complex, and it may be demanding for the patient. Others have challenged the evidence of the positive hyperthermia trials from the last century by criticizing these as being small or technically outdated, and claiming that very few new clinical studies are being performed [42–45]. On the other hand, Phase III studies have been extensively discussed in multiple reviews [44–50], and authorities in many countries have accepted the use of adjuvant hyperthermia for radiotherapy and chemotherapy, as summarized by Sauer et al. [46].

18.3 Conclusion

The reported response rates are high, and the comparison of radiation therapy with or without the addition of hyperthermia has shown that hyperthermia may improve response rates and local control rates. Therefore, the combination of hyperthermia and radiation therapy may be effective for the treatment of recurrent breast cancer.

Conflict of Interest None.

References

1. Fisher B, Slack NH, Cavanaugh PJ, Gardner B, Ravdin RG. Postoperative radiotherapy in the treatment of breast cancer: results of the NSABP clinical trial. Ann Surg. 1970;172:711–32.
2. Lee MC, Jagsi R. Postmastectomy radiation therapy: indications and controversies. Surg Clin North Am. 2007;87:511–26.
3. Taghian A, Jeong JH, Mamounas E, Anderson S, Bryant J, Deutsch M, Wolmark N. Patterns of locoregional failure in patients with operable breast cancer treated by mastectomy and adjuvant chemotherapy with or without tamoxifen and without radiotherapy: results from five national surgical adjuvant breast and bowel project randomized clinical trials. J Clin Oncol. 2004;22:4247–54.

4. Katz A, Strom EA, Buchholz TA, Thames HD, Smith CD, Jhingran A, Hortobagyi G, Buzdar AU, Theriault R, Singletary SE, et al. Locoregional recurrence patterns after mastectomy and doxorubicin-based chemotherapy: implications for postoperative irradiation. J Clin Oncol. 2000;18:2817–27.
5. Nielsen HM, Overgaard M, Grau C, Jensen AR, Overgaard J. Study of failure pattern among high-risk breast cancer patients with or without postmastectomy radiotherapy in addition to adjuvant systemic therapy: long-term results from the Danish Breast Cancer Cooperative Group DBCG 82 b and c randomized studies. J Clin Oncol. 2006;24:2268–75.
6. Ragaz J, Olivotto IA, Spinelli JJ, Phillips N, Jackson SM, Wilson KS, Knowling MA, Coppin CM, Weir L, Gelmon K, et al. Locoregional radiation therapy in patients with high-risk breast cancer receiving adjuvant chemotherapy: 20-year results of the British Columbia randomized trial. J Natl Cancer Inst. 2005;97:116–26.
7. MacDonald SM, Abi-Raad RF, Alm El-Din MA, Niemierko A, Kobayashi W, McGrath JJ, Goldberg SI, Powell S, Smith B, Taghian AG. Chest wall radiotherapy: middle ground for treatment of patients with one to three positive lymph nodes after mastectomy. Int J Radiat Oncol Biol Phys. 2009;75:1297–303.
8. Perez LA, Brady LW. Breast cancer: locally advanced, inflammatory and recurrent tumors. Philadelphia: Lippincott; 2003.
9. De Vita V, Hellman S, Rosenberg SA. Cancer of the breast. In: De Vita V, Hellman S, Rosenberg SA, editors. Cancer. Philadelphia: Lippincott; 2001.
10. Dunst J, Steil B, Furch S, Fach A, Lautenschlager C, Diestelhorst A, Lampe D, Kolbl H, Richter C. Prognostic significance of local recurrence in breast cancer after postmastectomy radiotherapy. Strahlenther Onkol. 2001;177:504–10.
11. Bedwinek JM, Lee J, Fineberg B, Ocwieza M. Prognostic indicators in patients with isolated localregional recurrence of breast cancer. Cancer. 1981;47:2232–5.
12. Chu FC, Lin FJ, Kim JH, Huh SH, Garmatis CJ. Locally recurrent carcinoma of the breast. Results of radiation therapy. Cancer. 1976;37:2677–81.
13. Bedwinek J. Natural history and management of isolated local-regional recurrence following mastectomy. Semin Radiat Oncol. 1994;4:260–9.
14. Beck TM, Hart NE, Woodard DA, Smith CE. Local or regionally recurrent carcinoma of the breast: results of therapy in 121 patients. J Clin Oncol. 1983;1:400–5.
15. Willner J, Kiricuta IC, Kolbl O. Locoregional recurrence of breast cancer following mastectomy: always a fatal event? Results of univariate and multivariate analysis. Int J Radiat Oncol Biol Phys. 1997;37:853–63.
16. Chagpar A, Meric-Bernstam F, Hunt KK, Ross MI, Cristofanilli M, Singletary SE, Buchholz TA, Ames FC, Marcy S, Babiera GV, et al. Chest wall recurrence after mastectomy does not always portend a dismal outcome. Ann Surg Oncol. 2003;10:628–34.
17. Kamby C, Sengelov L. Pattern of dissemination and survival following isolated locoregional recurrence of breast cancer. A prospective study with more than 10 years of follow up. Breast Cancer Res Treat. 1997;45:181–92.
18. Halverson KJ, Perez CA, Kuske RR, Garcia DM, Simpson JR, Fineberg B. Survival following locoregional recurrence of breast cancer: univariate and multivariate analysis. Int J Radiat Oncol Biol Phys. 1992;23:285–91.
19. Hsi RA, Antell A, Schultz DJ, Solin LJ. Radiation therapy for chest wall recurrence of breast cancer after mastectomy in a favorable subgroup of patients. Int J Radiat Oncol Biol Phys. 1998;42:495–9.
20. Schmoor C, Sauerbrei W, Bastert G, Schumacher M. Role of isolated locoregional recurrence of breast cancer: results of four prospective studies. J Clin Oncol. 2000;18:1696–708.
21. Mora EM, Singletary SE, Buzdar AU, Johnston DA. Aggressive therapy for locoregional recurrence after mastectomy in stage II and III breast cancer patients. Ann Surg Oncol. 1996;3:162–8.
22. Breasted JH. The Edwin Smith surgical papyrus. Chicago: University of Chicago Oriental Institute Publications; 1930.

23. Westermark F. Uber die Behandlung des Ulceriereden Cervixcarcinoma mittels konstanter warme. Zbl Gynakol. 1898;22:1335–9.
24. Coley WB. The treatment of malignant tumors by repeated inoculations of erysipelas. With a report of ten original cases. 1893. Clin Orthop Relat Res. 1991;262:3–11.
25. Kouloulias V, Triantopoulou S, Uzunoglou N, Pistevou-Gompaki K, Barich A, Zygogianni A, Kyrgias G, Kardamakis D, Pectasidis D, Kouvaris J. Hyperthermia is now included in the NCCN clinical practice guidelines for breast cancer recurrences: an analysis of existing data; Greek society of hyperthermic oncology. Breast Care (Basel). 2015;10:109–16.
26. Gradishar WJ, Anderson BO, Balassanian R, Blair SL, Burstein HJ, Cyr A, Elias AD, Farrar WB, Forero A, Giordano SH, Goetz M, Goldstein LJ, Hudis CA, Isakoff SJ, Marcom PK, Mayer IA, McCormick B, Moran M, Patel SA, Pierce LJ, Reed EC, Salerno KE, Schwartzberg LS, Smith KL, Smith ML, Soliman H, Somlo G, Telli M, Ward JH, Shead DA, Kumar R. Breast cancer version 1.2016. J Natl Compr Canc Netw. 2015;13(12):1475–85.
27. Lepock JR. Cellular effects of hyperthermia: relevance to the minimum dose for thermal damage. Int J Hyperthermia. 2003;19:252–66.
28. MivechiNF DWC. DNApolymerase alpha and beta activities during the cell cycle and their role in heat radiosensitization in Chinese hamster ovary cells. Radiat Res. 1985;103:337–50.
29. Armour EP, McEachern D, Wang Z, et al. Sensitivity of human cells to mild hyperthermia. Cancer Res. 1993;53:2740–4.
30. Koutcher JA, Barnett D, Kornblith AB, et al. Relationship of changes in pH and energy status to hypoxic cell fraction and hyperthermia sensitivity. Int J Radiat Oncol Biol Phys. 1990;18:14291435.
31. Engin K, Leeper D, Thistlethwaite A, et al. Tumor extracellular pH as a prognostic factor in thermoradiotherapy. Int J Radiat Oncol Biol Phys. 1994;29:125–32.
32. Song CW. Effect of local hyperthermia on blood flow and microenvironment: a review. Cancer Res. 1984;44:4721s–30.
33. Lindholm CE, Kjellen E, Nilsson P, Hertzman S. Microwave-induced hyperthermia and radiotherapy in human superficial tumours: clinical results with a comparative study of combined treatment versus radiotherapy alone. Int J Hyperthermia. 1987;3(5):393–411.
34. Lindholm CE, Kjellén E, Nilsson P, Weber L, Hill S. Prognostic factors for tumour response and skin damage to combined radiotherapy and hyperthermia in superficial recurrent breast carcinomas. Int J Hyperthermia. 1995;11(3):337–55.
35. Moros EG, Peñagaricano J, Novàk P, Straube WL, Myerson RJ. Present and future technology for simultaneous superficial thermoradiotherapy of breast cancer. Int J Hyperthermia. 2010;26 (7):699–709.
36. Zagar TM, Oleson JR, Vujaskovic Z, Dewhirst MW, Craciunescu OI, Blackwell KL, Prosnitz LR, Jones EL. Hyperthermia combined with radiation therapy for superficial breast cancer and chest wall recurrence: a review of the randomised data. Int J Hyperthermia. 2010;26 (7):612–17.
37. Jones EL, Oleson JR, Prosnitz LR, Samulski TV, Vujaskovic Z, Yu D, Sanders LL, Dewhirst MW. Randomized trial of hyperthermia and radiation for superficial tumors. J Clin Oncol. 2005;23(13):3079–85.
38. Refaat T, Sachdev S, Sathiaseelan V, Helenowski I, Abdelmoneim S, Pierce MC, Woloschak G, Small Jr W, Mittal B, Kiel KD. Hyperthermia and radiation therapy for locally advanced or recurrent breast cancer. Breast. 2015;24(4):418–25.
39. Linthorst M, Baaijens M, Wiggenraad R, Creutzberg C, Ghidey W, van Rhoon GC, van der Zee J. Local control rate after the combination of re-irradiation and hyperthermia for irresectable recurrent breast cancer: results in 248 patients. Radiother Oncol. 2015;117 (2):217–22.
40. Yamamoto D, Inui T, Tsubota Y, Sueoka N, Yamamoto C, Kuwana K, Yamamoto M. The utility of hyperthermia for local recurrence of breast cancer. World J Surg Oncol. 2012;10:201.
41. Overgaard J. The heat is (still) on – the past and future of hyperthermic radiation oncology. Radiother Oncol. 2013;109(2):185–7.

42. Januszewski A, Stebbing J. Hyperthermia in cancer: is it coming of age? Lancet Oncol. 2014;15(6):565–6.
43. Mathis S, Johansson T. Hyperthermia. Decision support document 36. Vienna: Ludwig Boltzmann Institute for heath technology assessment 2010, 1992–0496. Available at http://esprints.hta.lbg.ac.at
44. Wild C. Should hyperthermia be included in the benefit catalogue for oncologic indications? Commercial interests are presumed behind the editorial of R Sauer et al. Strahlenther Onkol. 2013;189(1):81–6.
45. van der Zee J, Vujaskovic Z, Kondo M, Sugahara T. The kadota fund international forum 2004 – clinical group consensus. Int J Hyperthermia. 2008;24(2):111–22.
46. Sauer R, Crezee H, Hulshof M, et al. The letter of the Ludwig Boltzamann Institute is unnecessarily polarizing to the discussion on whether or not hyperthermia is evidence based. Strahlenther Onkol. 2013;189:84–6.
47. Issels RD. Hyperthermia adds to chemotherapy. Eur J Cancer. 2008;44(17):2546–54.
48. Horsman MR, Overgaard J. Hyperthermia: a potent enhancer of radiotherapy. Clin Oncol (R Coll Radiol). 2007;19(6):418–26.
49. Lammers RJ, Witjes JA, Inman BA, Leibovitch I, Laufer M, Nativ O, Colombo R. The role of a combined regimen with intravesical chemotherapy and hyperthermia in the management of non-muscle-invasive bladder cancer: a systematic review. Eur Urol. 2011;60(1):81–93.
50. Mi DH, Li Z, Yang KH, Cao N, Lethaby A, Tian JH, Santesso N, Ma B, Chen YL, Liu YL. Surgery combined with intraoperative hyperthermic intraperitoneal chemotherapy (IHIC) for gastric cancer: a systematic review and meta-analysis of randomised controlled trials. Int J Hyperthermia. 2013;29(2):156–67.

Chapter 19
Combination by Hyperthermia and Radiation: Esophageal Cancer

Takayuki Ohguri

Abstract Radiotherapy has an important role in the treatment of esophageal cancer. For locally advanced esophageal cancer in particular, definitive or preoperative chemoradiotherapy is often selected; however, there remains room for improvement in terms of treatment results. Initially, intra-cavitary local hyperthermia was combined with chemoradiotherapy, and a phase III trial showed that the addition of intra-cavitary hyperthermia to preoperative chemoradiotherapy provided a significant improvement in the histopathologic effects. In recent years, clinical treatment results of phase II clinical trials using regional hyperthermia combined with preoperative chemoradiotherapy have been reported, and the results have been promising. In this chapter, the clinical results of radiotherapy or chemoradiotherapy combined with hyperthermia for the treatment of esophageal cancer were reviewed.

Keywords Hyperthermia • Chemotherapy • Esophageal cancer • Recurrence

19.1 Background

The efficacy of radiotherapy with hyperthermia for several types of cancer, including head and neck cancer, breast cancer, melanoma, rectal cancer and cervical cancer of the uterus, has been demonstrated and confirmed by phase III clinical trials [1, 2]. Definitive chemoradiotherapy has been a standard of care in the non-surgical treatment of potentially curable locally advanced esophageal cancer; however, the rates of locoregional failures are still high. Recently, preoperative chemoradiotherapy is often selected in patients with locally advanced esophageal cancer, and there are some reports in which esophageal cancer was treated with preoperative chemoradiotherapy combined with hyperthermia [3, 4]

The use of heating for esophageal cancer is divided in two methods: intra-cavitary local hyperthermia and regional hyperthermia. Generally, regional

T. Ohguri (✉)
Department of Radiology, University of Occupational and Environmental Health, 1-1 Iseigaoka, Yahatanishi-ku, Kitakyushu 807-8555, Japan
e-mail: ogurieye@med.uoeh-u.ac.jp

S. Kokura et al. (eds.), *Hyperthermic Oncology from Bench to Bedside*,
DOI 10.1007/978-981-10-0719-4_19

hyperthermia is less invasive than intra-cavitary local hyperthermia, although the application of regional hyperthermia requires more sophisticated planning, thermometry, and quality assurance than local heating. The heating technique of regional hyperthermia using an 8-MHz radiofrequency capacitive heating device, which is commonly used in Japan, for tumors of the thoracic region is described in the chapter "Combination by Hyperthermia and Radiation (and Chemotherapy): Lung Cancer" of this book. In this chapter, the clinical results of radiotherapy (or chemoradiotherapy) with hyperthermia for the treatment of esophageal cancer are reviewed.

19.2 Clinical Results of Chemoradiotherapy and Hyperthermia for Locally Advanced Esophageal Cancer

In the late 1980s, the results of treatment with hyperthermia for esophageal cancer were initially reported [5, 6]. Sugimachi et al. demonstrated that chemoradiotherapy combined with intra-cavitary local hyperthermia was prescribed for patients with resectable or unresectable squamous cell carcinoma of the esophagus [6]. In 1995, Kitamura et al. demonstrated a small randomized phase III trial for preoperative chemoradiotherapy, examining whether the addition of intra-cavitary hyperthermia could improve the clinical outcomes in patients with advanced esophageal cancer [3]. A total of 66 patients were enrolled, the total dose of preoperative radiotherapy was 30 Gy, and systemic chemotherapy included cisplatin or bleomycin. Intra-cavitary hyperthermia was delivered at 42.5–44 °C at the tumor surface for 30 min per treatment, which was repeated six times, using intraluminal heating. The postoperative histopathologic effects in patients with hyperthermia were significantly improved compared with those in patients without hyperthermia (Fig. 19.1). The 3-year overall survival rate was 50 % in patients with intra-cavitary hyperthermia, which was also higher than that of 24 % in patients without intra-cavitary hyperthermia. The side effects were comparable between the patients with and without intra-cavitary hyperthermia.

Recently, the results of treatment with regional hyperthermia as an external heating method for esophageal cancer have been reported [4, 7, 8]. Regional hyperthermia tends to be less invasive compared with intra-cavitary local hyperthermia and can heat not only the primary tumor tissue but also regional lymph node areas. In 2009, a Phase II study of preoperative chemoradiotherapy with regional hyperthermia was performed using the external heating technique of a home-made academical medical center phased array of four 70 MHz antennas [4]. The dose of radiotherapy was 41.4 Gy in 1.8 Gy daily fractions, and chemotherapy consisted of carboplatin and paclitaxel. Carboplatin was infused during the hyperthermia session. Although only half of the patients reached the intended temperature of at least 41 °C in one probe at the tumor level, pathological complete

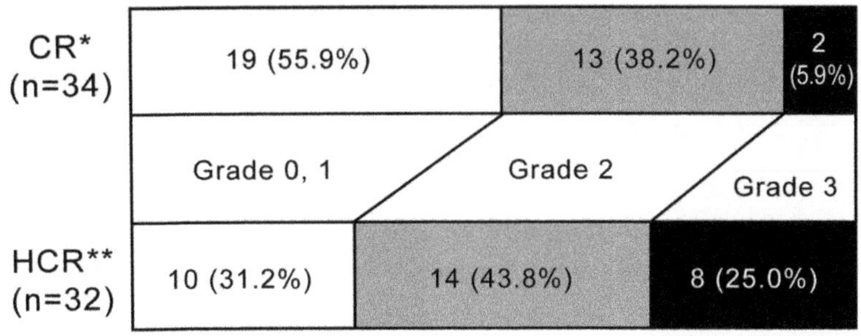

Fig. 19.1 A histopathologic examination revealed the treatment effect of each type of preoperative adjuvant therapy in a randomized phase III trial. The effective rate was 68.8 % in patients treated with the chemoradiotherapy with intra-cavitary hyperthermia and 44.1 % in patients with chemoradiotherapy alone ($p < 0.05$). *CR = chemoradiotherapy, **HCR = chemoradiotherapy with intra-cavitary hyperthermia. Cited from [3]

response (CR) or partial response (PR) was seen in 77 % of patients, and all patients had an R0 resection. The in-field locoregional control during follow up of the operated patients was 100 %. Furthermore, on the basis of the promising results of a phase II study of preoperative thermochemoradiotherapy in patients with resectable esophageal cancer, the European Society for Hyperthermic Oncology (ESHO) is conducting a phase III study. Reports describing the results of phase III studies of preoperative chemoradiotherapy, with or without regional hyperthermia, in esophageal cancer are garnering attention.

In our institution, definitive chemoradiotherapy combined with regional hyperthermia using an 8-MHz radiofrequency capacitive heating device has been performed in some patients with locally advanced esophageal cancer. Figure 19.2 shows one case of T4 esophageal cancer treated with definitive chemoradiotherapy combined with regional hyperthermia, in which a complete response after the combined therapy was achieved.

19.3 Clinical Results of Chemoradiotherapy and Hyperthermia for Recurrent Esophageal Cancer

Loco-regional recurrence is a common problem in patients with esophageal cancer. Radiotherapy and/or chemotherapy are typically selected for postoperative locoregional recurrence after surgery. Yahara et al. reported a retrospective case series involving treatment with radiotherapy with regional hyperthermia in 16 patients with recurrent esophageal cancer [9]. The objective tumor response rate was promising at 88 %. The median survival durations for CR and PR cases were 73 months and 7 months, respectively. Locoregional tumor control may be important for clinical outcomes in these tumors; therefore, further clinical investigation of the combined treatments of radiotherapy with regional hyperthermia is justified.

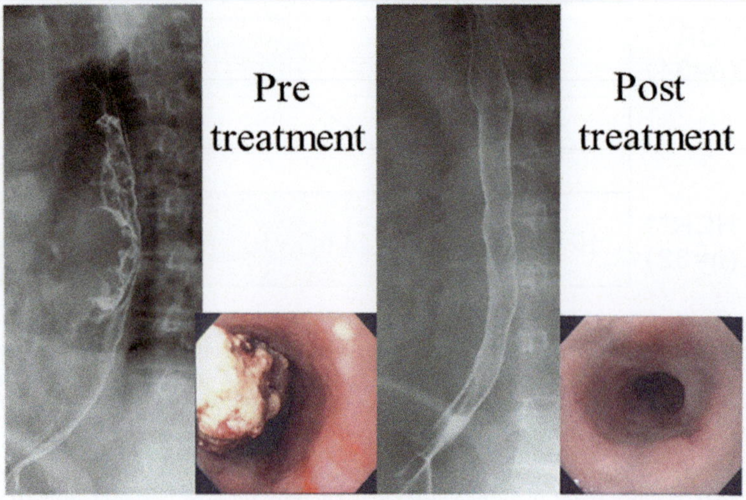

Fig. 19.2 An unresectable case of thoracic esophageal cancer (T4N0M0, squamous cell carcinoma) was treated with definitive radiotherapy (72 Gy, 1.2 Gy per fraction, twice per day), chemotherapy [cisplatin and fluorouracil (5FU)], and regional hyperthermia for a total of six sessions. Regional hyperthermia of the whole thoracic region was administered immediately after radiotherapy. Complete response was achieved, and no evidence of disease after combined treatment was recognized for 4 years

There are few treatment options for recurrent or persistent esophageal cancer after chemoradiotherapy, and no general therapeutic rules are available. Salvage esophagectomy following definitive chemoradiotherapy is one treatment option; however, this has shown high morbidity and mortality rates [10, 11]. Recently, there have been encouraging results for re-irradiation in many cases of recurrent tumors at various sites, such as head and neck cancer, lung cancer, and rectal cancer [12–14]. Because re-irradiation at relatively small doses may not provide local tumor control and adequate palliative effects, re-irradiation with hyperthermia may improve the clinical outcomes in patients with recurrent or residual tumor previously irradiated. Previous clinical results for re-irradiation with hyperthermia have been reported, in which higher local control rates and favorable palliative effects were demonstrated in patients with recurrent breast cancer and rectal cancer [15, 16]. Yamaguchi et al. reported on multimodal approaches, including re-irradiation, chemotherapy, and regional hyperthermia using an 8-MHz radiofrequency capacitive heating device for patients with recurrent or persistent esophageal cancer after radiotherapy [17]. The intra-esophageal temperatures were also measured in the study, and the maximum, minimum, and average temperatures were 42.2, 39.3, and 41.5 °C, respectively. For patients with early-stage recurrent or persistent esophageal cancer, these approaches appeared feasible, promising, and showed a high objective tumor response rate of 91 %, although advanced T stage at the time of re-irradiation was found to be significantly correlated with Grade 3 or higher toxicity in the esophagus.

References

1. van der Zee J. Heating the patient: a promising approach? Ann Oncol. 2002;13(8):1173–84.
2. Jones EL, Samulski TV, Vujaskovic Z, Leonard RP, Dewhirst MW. Hyperthermia. Principles and practice of radiation oncology. 4th ed. Philadelphia: Lippincott Williams & Wilkins; 2003.
3. Kitamura K, Kuwano H, Watanabe M, Nozoe T, Yasuda M, Sumiyoshi K, et al. Prospective randomized study of hyperthermia combined with chemoradiotherapy for esophageal carcinoma. J Surg Oncol. 1995;60(1):55–8.
4. Hulshof MC, Van Haaren PM, Van Lanschot JJ, Richel DJ, Fockens P, Oldenborg S et al. Preoperative chemoradiation combined with regional hyperthermia for patients with resectable esophageal cancer. Int J Hyperthermia. 2009;25(1):79–85. doi:908713273 [pii] 10.1080/02656730802464078.
5. Matsufuji H, Kuwano H, Kai H, Matsuda H, Sugimachi K. Preoperative hyperthermia combined with radiotherapy and chemotherapy for patients with incompletely resected carcinoma of the esophagus. Cancer. 1988;62(5):889–94.
6. Sugimachi K, Matsuda H, Ohno S, Fukuda A, Matsuoka H, Mori M, et al. Long-term effects of hyperthermia combined with chemotherapy and irradiation for the treatment of patients with carcinoma of the esophagus. Surg Gynecol Obstet. 1988;167(4):319–23.
7. Albregts M, Hulshof MC, Zum Vorde Sive Vording PJ, van Lanschot JJ, Richel DJ, Crezee H et al. A feasibility study in oesophageal carcinoma using deep loco-regional hyperthermia combined with concurrent chemotherapy followed by surgery. Int J Hyperthermia. 2004;20(6):647–59. doi:10.1080/02656730410001714977.
8. van Haaren PM, Kok HP, Zum Vorde Sive Vording PJ, van Dijk JD, Hulshof MC, Fockens P et al. Reliability of temperature and SAR measurements at oesophageal tumour locations. Int J Hyperthermia. 2006;22(7):545–61. doi:10.1080/02656730600931765.
9. Yahara K, Imada H, Nomoto S, Ohguri T, Kato F, Morioka T, et al. Thermoradiotherapy for recurrent esophageal carcinoma. Jpn J Hyperthermic Oncol. 2004;20(1):1–8.
10. Swisher SG, Wynn P, Putnam JB, Mosheim MB, Correa AM, Komaki RR, et al. Salvage esophagectomy for recurrent tumors after definitive chemotherapy and radiotherapy. J Thorac Cardiovasc Surg. 2002;123(1):175–83. doi:S0022522302495116 [pii].
11. Nishimura M, Daiko H, Yoshida J, Nagai K. Salvage esophagectomy following definitive chemoradiotherapy. Gen Thorac Cardiovasc Surg. 2007;55(11):461–4. doi:10.1007/s11748-007-0157-z. discussion 4–5.
12. Langer CJ, Harris J, Horwitz EM, Nicolaou N, Kies M, Curran W, et al. Phase II study of low-dose paclitaxel and cisplatin in combination with split-course concomitant twice-daily reirradiation in recurrent squamous cell carcinoma of the head and neck: results of Radiation Therapy Oncology Group Protocol 9911. J Clin Oncol. 2007;25(30):4800–5. doi:10.1200/JCO.2006.07.9194.
13. Okamoto Y, Murakami M, Yoden E, Sasaki R, Okuno Y, Nakajima T, et al. Reirradiation for locally recurrent lung cancer previously treated with radiation therapy. Int J Radiat Oncol Biol Phys. 2002;52(2):390–6.
14. Mohiuddin M, Marks G, Marks J. Long-term results of reirradiation for patients with recurrent rectal carcinoma. Cancer. 2002;95(5):1144–50. doi:10.1002/cncr.10799.
15. Vernon CC, Hand JW, Field SB, Machin D, Whaley JB, van der Zee J, et al. Radiotherapy with or without hyperthermia in the treatment of superficial localized breast cancer: results from five randomized controlled trials. International Collaborative Hyperthermia Group. Int J Radiat Oncol Biol Phys. 1996;35(4):731–44. doi:036030169600154X [pii].
16. Juffermans JH, Hanssens PE, van Putten WL, van Rhoon GC, van Der Zee J. Reirradiation and hyperthermia in rectal carcinoma: a retrospective study on palliative effect. Cancer. 2003;98(8):1759–66. doi:10.1002/cncr.11719.
17. Yamaguchi S, Ohguri T, Imada H, Yahara K, Moon SD, Higure A et al. Multimodal approaches including three-dimensional conformal re-irradiation for recurrent or persistent esophageal cancer: preliminary results. J Radiat Res. 2011;52(6):812–20. doi:JST.JSTAGE/jrr/11066 [pii].

Chapter 20
Combination by Hyperthermia and Radiotherapy and/or Chemo-radiotherapy: Gyneology Cancer

Yoko Harima

Abstract Six randomized clinical trials comparing therapeutic outcomes of hyperthermia combined with radiotherapy with radiotherapy alone in patients with cervical cancer have been published since 1987, of which four showed significant better complete response, locoregional tumor control, and/or disease-free survival rates in the hyperthermia-combined radiotherapy arms than in the sole radiotherapy arms. One trial showed a trend of better locoregional tumor control in the hyperthermia-combined arm and one did not show any benefit of the hyperthermia-combination therapy. First, I summarize the results of these randomized clinical trials and discuss the effectiveness and safety of hyperthermia-combined treatment for cervical cancer. On the other hand, prognostic factors associated with outcomes of thermo-radiotherapy in advanced cervical carcinoma have also been studied and some promising biomarkers have been identified hitherto. Second, I present our comprehensive molecular genetic approach to identify prognostic biomarkers of thermoradiotherapy for cervical cancer patients and discuss them from an aspect of molecular biology.

Keywords cDNA microarray • Cell biological markers cervical cancer • Chemoradiotherapy • Chemotherapy • Hyperthermia • Prediction • Prognosis • Radiotherapy • Randomization

20.1 Introduction

Cervical cancer (CC) is one of the most common tumors affecting women worldwide [1] accounting for 7.9 % in incidence and 7.5 % in mortality of all female cancers [2]. Definitive radiation therapy (RT) is widely accepted as the treatment of choice for patients with locally advanced carcinoma of the cervix. Pelvic tumor

---

Y. Harima (✉)
Department of Surgery, Kansai Medical University, 10-5 Fumizono-cho, Moriguchi-City, Osaka 570-8507, Japan
e-mail: harima@takii.kmu.ac.jp

S. Kokura et al. (eds.), *Hyperthermic Oncology from Bench to Bedside*,
DOI 10.1007/978-981-10-0719-4_20

control is a pre-requisite for cure in this population [3], however, according to a study Perez et al reported, overall pelvic recurrences in 14, 41, 41 and 72 %, whereas distant recurrences only were found in 12, 18, 16 and 21 % of 1499 patients with FIGO stage I, II, III and IVA disease treated with RT alone [4]. Pilot studies to assess feasibility and efficacy of hyperthermia (HT) combined with RT in this population were conducted and resulted in suggesting potential usefulness of this combination therapy. Based on such findings, since 1987, the results of six randomized clinical trials (RCTs) have been published concerning the combined treatment of RT with HT in patients with CC [5–10]. Local tumor control and overall survival appeared to be improved by the addition of HT, although clinical observations were not unanimous. Possible explanations for the discrepancy may be found in the patient selection (e.g. with respect to stage or tumor volume) and other confounding factors such as treatment parameters (e.g. overall treatment time and HT technique). Therefore, the magnitude of any beneficial effect as well as the proper selection of patients for any combined treatment remains to be demonstrated.

Meanwhile, several molecular markers of thermo-radiosensitivity, for example, *p53* [11], *Bax* [12], and *HSP 70/27* [13] have also been reported. However, although the molecular mechanisms for thermoradiosensitivity of such molecules have been partially clarified, the whole picture remains unknown. To address this research question and identify a set of genes related to thermoradiosensitivity of CC, we comprehensively compared the expression profiles of thermoradiosensitive and thermoradioresistant tumors obtained by punch biopsy before treatment using a cDNA microarray consisting of 23,040 human genes.

This manuscript summarizes those six randomized trials [5–10] comparing the results of RT alone with RT + HT. In addition, this paper demonstrates the molecular research in patients with advanced CC treated with RT + HT including our previous study [14].

20.2 Radiotherapy, Chemotherapy, and Hyperthermia for Cervical Cancer: Randomized Clinical Trials

20.2.1 Randomized Clinical Trials of Radiotherapy vs. Radiotherapy Combined with Chemotherapy for Cervical Cancer

In patients with early stage CC, radical surgery is reported to be successful, but patients who have more extensive lesions are treated with chemoradiation (CRT) of cisplatin-containing regimens. To date, the efficacy of cisplatin-based CRT has been demonstrated in five large, randomized studies which showed a reduction in the risk of recurrence of 40–60 %. [15–19]. In a sixth study, no statistically significant benefit was found for the addition of cisplatin to RT [20]. This discrepancy caused considerable discussion in the literature and speculation for the relevant reason. It could be explained that, in the positive studies, quite a lot of patients in the sole RT arms could not be treated with optimal RT, resulted in considerably poor outcomes, and in such cases cisplatin could at most counteract the negative effects of insufficient RT. It might also be explained that, in the negative study, patients in the cisplatin-combined CRT arm were more anemic than those in the RT arm. Anemia is known poor prognostic factor in RT for CC patients [21], in other words, the arms were not balanced regardless of randomization. From the statistical point of view, it might be no wonder if a negative trial might show up after a series of five positive ones [22]. As a whole, the evidence definitely favors the combination of chemotherapy and RT. Clinical guidelines, as issued by the National Cancer Institute [23], also recommend the combined therapy.

Most reviews concluded that the beneficial effect of CRT is clearly present, at least for patients with FIGO stage I and II malignancies. On the other hand, for the more advanced stage patients, it is rather difficult to achieve local control even with CRT. A recent meta-analysis based on individual data confirms earlier suspicions that the gain in overall survival may only be 3 % at 5 years for patients with FIGO stage III–IVa CC.

20.2.2 Randomized Clinical Trials of Radiotherapy vs. Radiotherapy Combined with Hyperthermia for Cervical Cancer

Six randomized trials [5–10] investigating RT vs. RT combined with HT in patients with advanced CC have been published (Table 20.1). The majority of the patients in these studies had Stage IIIB tumors. Four studies [5–7, 9] report significantly better results following RT + HT compared to RT alone. In one study [8], the 1.5-year pelvic control tended to be higher in the RT + HT group. In the study by Vasanthan et al. [10], no beneficial effect from RT + HT was found.

In 2010, Lutgens L et al. reported systematic review entitled "Combined use of hyperthermia and radiation therapy for treating locally advanced cervix carcinoma" in Cochrane Database Syst Rev [24]. They concluded the available data do suggest that superior local tumor control rates (RR, 0.56; 95 % CI, 0.39–0.79; $p < 0.001$) (Fig. 20.1) and overall survival (HR, 0.67; 95 % CI, 0.45–0.99; $p = 0.05$) (Fig. 20.2) yielded a significantly better outcome for the RT combined with HT than the RT only treatment group. Using acute toxicity as endpoint in the pooled data analysis including 310 study patients (Sharma 1991; van der Zee 2000; Harima 2001; Vasanthan 2005) yielded no difference in acute treatment related toxicity between both treatment groups (RR, 0.99; 95 % CI, 0.30–3.31; $p = 0.99$). Similarly, the pooled data analysis using late toxicity as endpoint including 264 study patients (van der Zee 2000; Harima 2001; Vasanthan 2005) yielded no difference in late

Table 20.1 Results of randomized trials comparing RT with TRT in advanced stages of cervical cancer (Based on [24])

Author's name (number of references)	Numbers of patients (FIGO stage)	Complete response		Pelvic control		Disease-free survival		Overall survival	
		RT	TRT	RT	TRT	RT	TRT	RT	TRT
Harima [5]	40 (IIIB 22, III 81, IV 11)	50	80*	49	80* (3 years)	45	64	48	58 (3 years)
Van der Zee [6]	114 (IIB 22, III 81, IV 11)	57	83*	41	61* (3 years)	–	–	27	51* (3 years)
Datta [7]	64 (IIIB 64)	58	74	46	67 (2 years)	27	59* (2 years)	73	81 (2 years)
Sharma [8]	50 (II 7, III43)	–	–	50	70 (1.5 years)	–	–	–	–
Hong-Wei [9]	120 (IIB 28, IIIB 92)	48	72*	–	–	–	–	–	–
Vasanthan [10]	110 (IIB 56, III 51, IV 3)	–	–	80	70 (3 years)	–	–	73	73 (3 years)

RT radiotherapy, *TRT* thermoradiotherapy
*Significantly better than RT alone with $p < 0.05$

toxicity between both treatment groups (RR, 1.01; 95 % CI, 0.44–2.30; p = 0.98) (Fig. 20.3). They concluded that the available data indicated that superior local tumor control rates and overall survival could be achieved in patients with local advanced CC by adding hyperthermia to standard RT.

20.2.3 Chemoradiotherapy Combined with Hyperthermia for Cervical Cancer

Westermann et al reported the retrospective survival data of the combination of CRT and HT for the 68 patients of stage IIB–III–IVA CC in 2005 [25] and they reported also these same patients' long-term survival results in 2012 [26]. They reported that 68 patients with advanced CC were prospectively registered in the USA, Norway and the Netherlands, and treated with a combination of RT, weekly cisplatin, and loco-regional HT. As a results, 5-year relapse-free survival (RFS) from the day of registration in the study was 57.5 % (95%CI: 46.6–71.0) and 5-year OS was 66.1 % (95%CI: 55.1–79.3). Differences between countries can be explained by patient characteristics. They demonstrated that the long-term results of the addition of whole-pelvic HT to CRT for advanced CC were compatible with similar or improved survival compared to standard CRT.

Fig. 20.1 Complete tumor response (Reproduced from Lutgens L et al. [24])

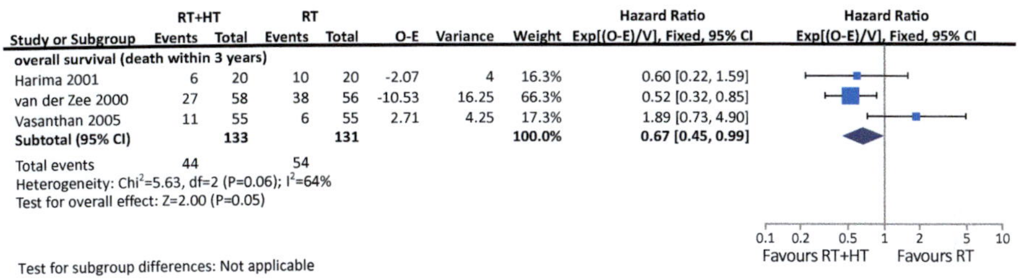

Fig. 20.2 Overall survival (Reproduced from Lutgens L et al. [24])

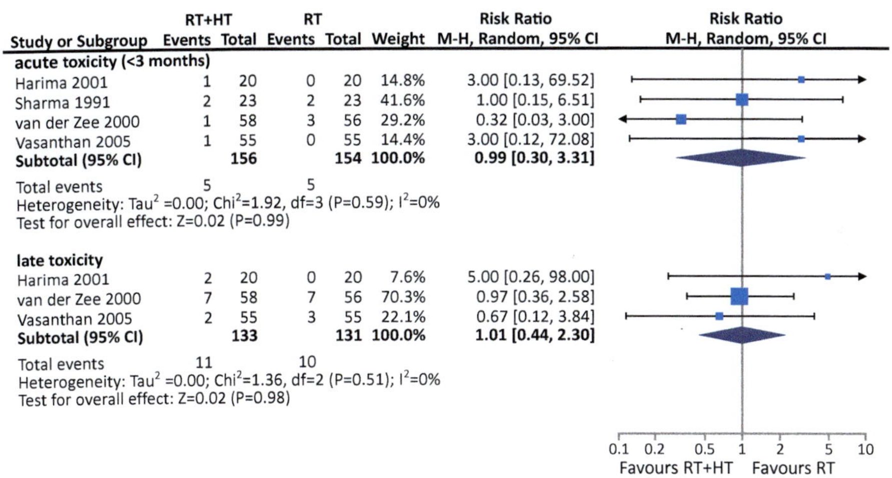

Fig. 20.3 Acute and late phase toxicity (Reproduced from Lutgens L et al. [24])

20.3 Biological Research for the Patients Treated with Hyperthermia and Radiotherapy

20.3.1 Prediction of Advanced Cervical Carcinoma After Thermoradiotherapy According to Expression Profiles Selected by Microarray Analysis in Our Previous Study

20.3.1.1 Materials and Methods

From the subjects of our randomized clinical study mentioned above [5], we selected 19 stage III-IV CC patients (1 stage IIIA, 11 stage IIIB, 5 stage IVA, and 2 stage IVB) to identify a set of genes related to thermo-radiosensitivity or thermo-radioresistancy between May 1995 and August 2001. CC tissues were obtained from all 19 patients who underwent biopsy prior to thermo-radiothrapy (radiotherapy and hyperthermia, TRT), and were snap-frozen at −80 °C.

All patients were treated with TRT according to same protocol after sampling. Details have been described in our previous report [14].

Eight patients are alive and well (no evidence of disease, NED) from 54.9 months to 136.9 months after completion of treatment, and 11 patients have died from recurrent disease (cancer-caused death, CD) between 3.2 and 22.9 months after completion of treatment. When we divided the patients into an NED group and a CD group, the former were classified as the thermoradiosensitive group and the latter as the thermoradioresistant group.

We have already reported the further information of methods of cDNA microarray analysis including RNA extraction, microarray design, hybridization, data analysis, permutation test, and hierarchical clustering in our previous report [14]. Here, we summarized the methods and results of cDNA microarray analysis comparing a NED group with a CD group.

20.3.1.2 Results

Identification of Genes Responding to Thermoradiotherapy

Thermoradiosensitive group survived significantly longer than thermoradioresistant one ($P < 0.001$, Student's t-test). By means of the Mann-Whitney test ($P < 0.05$) and subsequent procedures, we identified a total of 156 genes that were differently expressed between NED and CD groups. Of those 156 genes, 76 revealed increased expression, and 80 showed decreased expression, in carcinomas belonging to the NED group as compared to the CD group.

Permutation Test, Hierarchical Clustering and Cross-Validation Test

To evaluate the validity of the 156 genes selected as thermoradiosensitivity-related genes, permutation testing was performed. Expression levels of each of the 19 samples in both groups for each gene were permuted (randomly scrambled) 10,000 times. After the 10,000 times permutation, the probabilities of the genes being correlated to both NED and CD group distinction were estimated. As a result, all of the selected 35 genes showed P value < 0.05 without exception. To confirm whether the expression patterns of these 156 genes could allow discrimination between the two groups, we performed a hierarchical cluster analysis. This procedure clearly separated the two groups from each other. Genes that were already known to be associated with apoptosis were significantly up-regulated in the NED group compared to the CD group; the former included *Bcl-2*-interacting killer (apoptosis-inducing) (*BIK*; Table 20.2). The latter included testis enhanced gene transcript (*BAX* inhibitor 1) (*TEGT*; Table 20.3), and Hypoxia-inducible gene, hypoxia-inducible factor 1 (*HIF1A*) were up-regulated in the CD group relative to the NED group (Table 20.3). In addition, genes that are considered to be associated with tumor cell invasion and metastasis, including plasminogen activator, urokinase (*PLAU, uPA*) was up-regulated in the CD group relative to the NED group (Table 20.3).

Cross-validation testing was performed to examine whether the 35 genes were crucial in classifying the NED and CD groups.

20.3.1.3 Discussion

In our current research, we profiled the gene expression patterns of cervical cancer cells using a cDNA microarray system containing 23,040 genes. We identified 35 genes that were differentially expressed among patients who showed good response as opposed to poor response to TRT.

Table 20.2 Genes showing relatively higher expression carcinoma cells in the NED group than those in the CD group (Based on [14])

Gene symbol	Gene name
BIK	BCL2-interacting killer (apoptosis-inducing)
SSI-3	STAT induced STAT inhibitor 3
TMSB4X	Thymosin, beta 4, X chromosome
SC5DL	Sterol-C5-desaturase (fungal ERG3, delta-5-desaturase)-like
ALY	Transcriptional coactivator
CKB	Creatine kinase, brain
LIV	LIV protein, estrogen regulated
SNW1	SKI-INTERACTING PROTEIN
UQCRC1	Ubiquinol-cytochrome c reductase core protein I
MLLT4	Myeloid/lymphoid or mixed-lineage leukemia (trithorax (Drosophila) homolog)

Table 20.3 Genes showing relatively higher expression carcinoma cells in the CD group than those in the NED group (Based on [14])

Gene symbol	Gene name
TEGT	Testis enhanced gene transcript (BAX inhibitor 1)
HIF1A	Hypoxia-inducible factor 1, alpha subunit (basic helix-loop-helix transcription factor)
PLAU	Plasminogen activator, urokinase
CA12	Carbonic anhydrase XII
CD44	CD44 antigen (homing function and Indian blood group system)
P63	Transmembrane protein (63kD), endoplasmic reticulum/Golgi intermediate compartment
USP6	Ubiquitin specific protease 6 (Tre-2 oncogene)
CTSB	Cathepsin B
CTSL	Cathepsin L
RPL37A	Ribosomal protein L37a
FLJ22833	Hypothetical protein FLJ22833
RRBP1	Ribosome binding protein 1 (dog 180kD homolog)
D1S155E	NRAS-related gene
SLC31A2	Solute carrier family 31 (copper transporters), member 2
FLJ10422	Hypothetical protein FLJ10422
NRBF-2	Nuclear receptor binding factor-2
ORC3L	Origin recognition complex, subunit 3 (yeast homolog)-like
PBEF	Pre-B-cell colony-enhancing factor
VPS33B	Vacuolar protein sorting 33B (yeast homolog)
FLJ12287	Hypothetical protein FLJ12287 similar to semaphorins
	ESTs
	ESTs

A gene, involved in apoptosis, *BIK*, one of the pro-apoptotic members of the *Bcl-2* family [27], was significantly up-regulated in the NED group compared to the CD group. In contrast, a gene involved in suppression of apoptosis, *TEGT* (testis enhanced gene transcript, *BAX* inhibitor 1), in the tumor cells was significantly up-regulated in the CD group compared to the NED group. The mechanism of tumor response to TRT seems to involve the development of apoptosis through the activation of a pathway of one of the *Bcl-2* family genes, Bax, that we reported previously [12]. In contrast, as it was shown by others, *Bax* inhibitor-1 (*TEGT*) showed significantly higher expression in the CD group that might compensate for *Bcl-2* downregulation and thus promote cell survival by counteracting *Bax* functions [28]. Therefore, it is probable that an additive or synergistic anti-tumor effect of TRT, which was shown to occur through induction of apoptosis, was likely to be involved in one of the pathways with the pro-apoptotic members of the *Bcl-2* family and to inhibit by *TEGT*.

Hypoxia-inducible factors (HIFs) (*HIF1A)* are key proteins regulating the response of a variety of genes to hypoxic stimuli. Because hypoxia has been recognized as a major cause of failure of radiotherapy [29], *HIF* expression may potentially be a potent predictive marker of response to radiation. Several studies also suggest that increased *HIF1A* is related to poor outcome in the patients with cervical cancer after radiotherapy [30, 31].

Among the key players in the proteolytic cascade leading to tumor invasion and metastasis are factors of the plasminogen activation system, such as the serine protease urokinase-type plasminogen activator (*PLAU, uPA*) [32] and its type-1 inhibitor PAI-1. *PLAU* also plays an important role in tumor invasion and metastases by initially catalyzing the conversion of plasminogen to plasmin [33, 34]. *PLAU* is a key step in various cancer metastases including cervical cancer [35], urinary bladder [36], and breast cancer [37].

20.3.1.4 Conclusion

The main conclusion of the HT trials is that in the patients with advanced CC, both pelvic tumor control and survival can be improved by the addition of HT to standard radiotherapy. Our data provided the first evidence that gene-expression profiles can predict sensitivity to TRT for advanced CC, and these findings may eventually lead to the achievement of "personalized therapy" for this disease.

20.3.2 Prognostic Cell Biological Markers in Cervical Cancer Patients Treated with Hyperthermia and Radiotherapy

The biological basis for combining HT and RT and/or chemotherapy includes two types of interaction, namely, direct hyperthermic cytotoxicity (≥42.5 °C) and radio- and/or chemo-sensitization (<43 °C). Most of the tumor volume could be heated to radio- and/or chemo-sensitizing temperatures in the range of 40–42 °C [38]. In the previous study [24], the average temperature was in the range of 40–42 °C in the radio- and/or chemo-sensitizing range. The mechanism of tumor response to RT combined with HT at this range of temperature seems to involve the development of apoptosis through the activation of one of the *bax* pathways that we reported previously [12]. Several molecular markers of thermoradiosensitivity, for example, *p53* [11], *Bax* [12], and *HSP 70/27* [13], have been reported. In recently, Eppink B et al reported that the hyperthermia-induced DNA repair deficiency is enhanced by inhibitors of the cellular heat-shock response [39]. Interestingly, it has become clear that HT not only affects cells in the treatment area, but systemic immune effects may play an important role [40].

Noordhuis MG et al reported reviewed prognostic biological markers in CC patients treated with (chemo) radiation, and they concluded that the most interesting markers that were identified were markers involved in EGFR signaling, angiogenesis, hypoxia, and COX-2. Besides their prognostic significance, EGFR, C-erbB-2, and COX-2 were also associated with poor response to (chemo) radiation [41].

Further research of these markers and their targeting by targeted drugs in combination with CRT and HT in clinical practice will hopefully improve survival rates in advanced-stage CC patients.

References

1. Steven E, Waggoner SE. Cervical cancer. Lancet. 2003;361:2217–25.
2. Ferlay J, Soerjomataram I, Dikshit R, Eser S, Mathers C, Rebelo M, et al. Cancer incidence and mortality worldwide: sources, methods and major patterns in GLOBOCAN 2012. Int J Cancer. 2015;136:E359–86.
3. Rotman M, Pajak TF, Choi K, Clery M, Marcial V, Grigsby PW, et al. Prophylactic extended-field irradiation of para-aortic lymph nodes in stages IIB and bulky IB and IIA cervical carcinomas. Ten-year treatment results of RTOG 79–20. JAMA. 1995;274(5):387–93.
4. Perez CA, Grigsby PW, Chao KS, Mutch DG, Lockett MA. Tumor size, irradiation dose, and long-term outcome of carcinoma of uterine cervix. Int J Radiat Oncol Biol Phys. 1998;41 (2):307–17.
5. Harima Y, Nagata K, Harima K, Ostapenko VV, Tanaka Y, Sawada S. A randomized clinical trial of radiation therapy versus thermoradiotherapy in stage IIIB cervical carcinoma. Int J Hyperthermia. 2001;17:97–105.
6. Van Der Zee J, González González D, Van Rhoon GC, Van Dijk JDP, Van Putten WLJ, Hart AAM. Comparison of radiotherapy alone with radiotherapy plus hyperthermia in locally advanced pelvic tumours: a prospective, randomized, multicentre trial. Lancet. 2000;355:1119–25.
7. Datta NR, Bose AK, Kapoor HK. Thermoradiotherapy in the management of carcinoma of the cervix (IIIB): a controlled clinical study. Ind Med Gaz. 1987;121:68–71.
8. Sharma S, Singhal S, Sandhu AP, Ghoshal S, Gupta BD, Yadav NS. Local thermo-radiotherapy in carcinoma cervix: improved local control versus increased incidence of distant metastasis. Asia Oceania J Obstet Gynaecol. 1991;17:5–12.
9. Chen HW, Jun-Jie F, Wei L. A randomized trial of hyperthermia-radiochemotherapy for uterine cervix cancer. Chin J Clin Oncol. 1997;24:249–51.
10. Vasanthan A, Mitsumori M, Park JH, Zhi-Fan Z, Yu-Bin Z, Oliynychenko P, et al. Regional hyperthermia combined with radiotherapy for uterine cervical cancers: a multi-institutional prospective randomized trial of the international atomic energy agency. Int J Radiat Oncol Biol Phys. 2005;61:145–53.
11. Ohnishi T, Komatsu K, Tauchi H, Wang X, Takahashi A, Ohnishi K, et al. Heat-induced accumulation of p53 and hsp72 is suppressed in lung fibroblasts from SCID mouse. Int J Radiat Biol. 2000;76:711–15.
12. Harima Y, Nagata K, Harima K, Oka A, Ostapenko VV, Shikata N, et al. Bax and Bcl-2 protein expression following radiation therapy versus radiation plus thermoradiotherapy in stage IIIB cervical carcinoma. Cancer. 2000;88:132–8.
13. Samali A, Holmberg CI, Sistonen L, Orrenius S. Thermotolerance and cell death are distinct cellular responses to stress: dependence on heat shock proteins. FEBS Lett. 1999;461:306–10.
14. Harima Y, Sawada S, Togashi A, Horikoshi K, Katagiri T, Tsunoda T, et al. Prediction of outcome of advanced cervical cancer to thermoradiotherapy according to expression profiles of 35 genes selected by cDNA microarray analysis. Int J Radiat Oncol Biol Phys. 2004;60:237–48.

15. Rose PG, Bundy BN, Watkins EB, Thigpen JT, Deppe G, Maiman MA, et al. Concurrent cisplatin-based chemotherapy and radiotherapy for locally advanced cervical cancer. N Engl J Med. 1999;340:1144–53.
16. Peters III WA, Liu PY, Barrett II RJ, Stock RJ, Monk BJ, Berek JS, et al. Concurrent chemotherapy and pelvic radiation therapy compared with pelvic radiation therapy alone as adjuvant therapy after radical surgery in high-risk early-stage cancer of the cervix. J Clin Oncol. 2000;18:1606–13.
17. Morris M, Eifel PJ, Lu J, Grigsby PW, Levenback C, Stevens RE, et al. Pelvic radiation with concurrent chemotherapy compared with pelvic and para-aortic radiation for high-risk cervical cancer. N Engl J Med. 1999;340:1137–43.
18. Keys HM, Bundy BN, Stehman FB, Muderspach LI, Chafe WE, Suggs III CL, et al. Cisplatin, radiation, and adjuvant hysterectomy for bulky Stage IB cervical carcinoma. N Engl J Med. 1999;340:1154–61.
19. Whitney CW, Sause W, Bundy BN, Malfetano JH, Hannigan EV, Fowler Jr WC, et al. Randomized comparison of fluorouracil plus cisplatin versus hydroxyurea as an adjunct to radiation therapy in Stage IIB–IVA carcinoma of the cervix with negative para-aortic lymph nodes: a Gynecologic Oncology Group and Southwest Oncology Group study. J Clin Oncol. 1999;17:1339–48.
20. Pearcey R, Brundage M, Drouin P, Jeffrey J, Johnston D, Lukka H, et al. Phase III trial comparing radical radiotherapy with and without cisplatin chemotherapy in patients with advanced squamous cell cancer of the cervix. J Clin Oncol. 2002;20:966–72.
21. Grogan M, Thomas GM, Melamed I, Wong FL, Pearcey RG, Joseph PK, et al. The importance of haemoglobin levels during radiotherapy for carcinoma of the cervix. Cancer. 1999;86:1528–36.
22. Rose PG, Bundy BN. Chemoradiation for locally advanced cervical cancer: does it help? J Clin Oncol. 2002;20:891–3.
23. National Cancer Institute. Concurrent chemoradiation for cervical cancer [clinical announcement]. 1999.
24. Lutgens L, Van Der Zee J, Pijls-Johannesma M, De Haas-Kock DF, Buijsen J, Mastrigt GA, et al. Combined use of hyperthermia and radiation therapy for treating locally advanced cervix carcinoma. Cochrane Database Syst Rev. 2010;17(3):CD006377.
25. Westermann AM, Jones EL, Schem BC, Van Der Steen-Banasik EM, Koper P, Mella O, et al. First results of triple-modality treatment combining radiotherapy, chemotherapy, and hyperthermia for the treatment of patients with stage IIB, III, and IVA cervical carcinoma. Cancer. 2005;104:763–70.
26. Westermann A, Mella O, Van Der Zee J, Jones EL, Van Der Steen-Banasik EM, Koper P, et al. Long-term survival data of triple modality treatment of stage IIB–III–IVA cervical cancer with the combination of radiotherapy, chemotherapy and hyperthermia – an update. Int J Hyperthermia. 2012;28(6):549–53.
27. Ottilie S, Diaz JL, Horne W, Chang J, Wang Y, Wilson G, et al. Dimerization properties of human BAD. Identification of a BH-3 domain and analysis of its binding to mutant BCL-2 and BCL-XL proteins. J Biol Chem. 1997;272:30866–72.
28. Villalva C, Trempat P, Greenland C, Thomas C, Girard JP, Moebius F, et al. Isolation of differentially expressed genes in NPM-ALK-positive anaplastic large cell lymphoma. Br J Haematol. 2002;118:791–8.
29. Fyles AW, Milocevic M, Wong R, Kavanagh MC, Pintilie M, Sun A, et al. Oxygenation predicts radiation response and survival in patients with cervix cancer. Radiother Oncol. 1998;48:149–56.
30. Hutchison GJ, Valentine HR, Loncaster JA, Davidson SE, Hunter RD, Roberts SA, et al. Hypoxia-inducible factor 1α expression as an intrinsic marker of hypoxia: correlation with tumor oxygen, pimonidazole measurements, and outcome in locally advanced carcinoma of the cervix. Clin Cancer Res. 2004;10:8405–12.

31. Ishikawa H, Sakurai H, Hasegawa M, Mitsuhashi N, Takahashi M, Masuda N, et al. Expression of hypoxic-inducible factor 1α predicts metastasis-free survival after radiation therapy alone in stage IIIB cervical squamous cell carcinoma. Int J Radiat Oncol Biol Phys. 2004;60:513–21.
32. Schmitt M, Wilhelm OG, Reuning U, Krüger A, Harbeck N, Lengyel E, et al. The urokinase plasminogen activation system as a novel target for toumour therapy. Fibrinolysis. 2000;14:114–32.
33. Wang Y. The role and regulation of urokinase-type plasminogen activator receptor gene expression in cancer invasion and metastasis. Med Res Rev. 2001;21:146–70.
34. Andreasen PA, Egelund R, Petersen HH. The plasminogen activation system in tumor growth, invasion, and metastasis. Cell Mol Life Sci. 2000;57:25–40.
35. Horn LC, Pippig S, Raptis G, Fisher U, Kohler U, Hentschel B, et al. Clinical relevance of urokinase-type plasminogen activator and its inhibitor type 1 (PAI-1) in squamous cell carcinoma of the uterine cervix. Aust N Z J Obstet Gynaecol. 2002;42:383–6.
36. Shahrokh F, Shariat M, Monoski A, Andrews B, Wheeler TM, Lerner S. Association of plasma urokinase-type plasminogen activator and its receptor with clinical outcome in patients undergoing radical cystectomy for transitional cell carcinoma of the bladder. Urology. 2003;61:1053–8.
37. Zemzoum I, Kates RE, Ross JS, Dettmar P, Dutta M, Henrichs C, et al. Invasion factors uPA/PAI-1 and HER2 status provide independent and complementary information on patient outcome in node-negative breast cancer. J Clin Oncol. 2003;21:1022–8.
38. Feldmann HJ, Seegenschmiedt MH, Molls M. Hyperthermia—its actual role in radiation oncology Part III: clinical rationale and results in deep seated tumors. Strahlenther Onkol. 1995;171:251–64.
39. Eppink B, Krawczyk PM, Stap J, Kanaar R. Hyperthermia-induced DNA repair deficiency suggests novel therapeutic anti-cancer strategies. Int J Hyperthermia. 2012;28(6):509–17.
40. Frey B, Weiss EM, Rubner Y, Wunderlich R, Ott OJ, et al. Old and new facts about hyperthermia-induced modulations of the immune system. Int J Hyperthermia. 2012;28(6):528–42.
41. Noordhuis MG, Eijsink JJ, Roossink F, De Graeff P, Pras E, Schuuring E, et al. Prognostic cell biological markers in cervical cancer patients primarily treated with (chemo) radiation: a systematic review. Int J Radiat Oncol Biol Phys. 2011;79(2):325–34.

Chapter 21
Combination by Hyperthermia and Radiation and Chemotherapy: Soft Tissue Sarcoma

Hisaki Aiba, Hiroaki Kimura, and Takanobu Otsuka

Abstract Soft tissue sarcoma is an extremely rare disease with an annual incidence of 3 per 100,000 people, accounting for 1 % or less of all malignant tumors. High-grade soft tissue sarcoma requires multidisciplinary therapy including chemotherapy, surgery, and radiotherapy, though surgery remains the primary therapy. Nowadays more than 90 % of patients undergo surgery preserving the affected limb. However, postoperative local recurrence and loss of function after wide excision remains a concern. We have performed radio-hyperthermo-chemotherapy (RHC) for patients with high-grade soft tissue sarcoma with the goal of improving local control, reducing the resection margin, improving postoperative function, and controlling distant metastasis by improving chemotherapy sensitivity. In this chapter, we first reviewed the previous reports of hyperthermia for soft tissue sarcomas, and discuss our treatment protocol and treatment outcomes.

Keywords Radio-hyperthermo-chemotherapy • Soft tissue sarcoma • Regional hyperthermia • Intra-arterial chemotherapy

21.1 Overview of Soft Tissue Sarcoma

Soft tissue sarcoma is a collective term for malignant tumors arising from soft tissues, including muscle and adipose tissue. Soft tissue, in this case, is defined as non-epithelial tissues, which include muscle, adipose, vessels, peripheral nervous, tendon, and ligament, but does not include bone, the reticuloendothelial system, glia, and supporting tissues of parenchymal organs. Most of those are mesoderm-derived tissues, except for peripheral nervous tumors, which are derived from the ectoderm.

H. Aiba • H. Kimura • T. Otsuka (✉)
Department of Orthopedic Surgery, Nagoya City University, 1 Kawasumi, Mizuho-cho, Mizuho-ku, Nagoya 467-8601, Japan
e-mail: t.otsuka@med.nagoya-cu.ac.jp

S. Kokura et al. (eds.), *Hyperthermic Oncology from Bench to Bedside*,
DOI 10.1007/978-981-10-0719-4_21

Soft tissue sarcoma is an extremely rare disease with an annual incidence of 3 per 100,000 people, accounting for 1 % or less of all malignant tumors. There are dozens of types, including extremely rare diseases. Therefore, the treatment plan should be created after adequate evaluation of histological type and malignancy (grade), as making the diagnosis is often challenging. In addition, soft tissue sarcoma does not have a specific imaging finding in many cases, making pathologic evaluation essential for diagnosis. However, soft tissue sarcoma has a wide-ranging classification despite the low number of cases, making pathologic diagnosis difficult. As accurate diagnosis leads to favorable outcome, it is important for the orthopedic surgeons and pathologist to communicate to make an accurate pathologic diagnosis together.

21.2 Treatment of Soft Tissue Sarcoma

After making a pathologic diagnosis, treatment is given according to histologic diagnosis. Soft tissue sarcoma is treated with a combination of surgery, chemotherapy, and radiotherapy, similar to other solid tumors. Surgery is the mainstay, as low-grade soft tissue sarcoma is unlikely to metastasize to distant sites. High-grade soft tissue sarcoma requires multidisciplinary therapy including chemotherapy, surgery, and radiotherapy, though surgery remains the primary therapy. In surgery for soft tissue sarcoma, a wide resection margin is selected, including peripheral normal tissues, to minimize the possibility of local recurrence. Nowadays more than 90 % of patients undergo surgery preserving the affected limb, while amputation may be selected when resection fails to obtain local control. However, postoperative loss of function due to concomitant resection of peripheral normal tissues remains a concern. Therefore, reduction of the resection margin may be attempted by combining radiotherapy or chemotherapy.

Chemotherapy is administered only in the adjuvant setting for non-round cell sarcomas. Anthracycline-based or ifosfamide-based chemotherapy is recommended for high-grade spindle cell sarcoma with the aim of prolonging survival and reducing the rate of local recurrence.

Radiotherapy is not the first choice for therapy for spindle cell sarcoma because of its low sensitivity, but is often used in the adjuvant setting for patients with inappropriate resection margins to prevent recurrence, or to palliate unresectable patients. Preoperative radiation may also be performed with the goal of reducing the resection margin in some cases.

We have performed radio-hyperthermo-chemotherapy (RHC) for patients with high-grade soft tissue sarcoma with the goal of improving local control, reducing the resection margin, improving postoperative function, and controlling distant metastasis by improving chemotherapy sensitivity. The treatment protocol and treatment outcomes are discussed in the following sections.

21.3 Basic Research of Hyperthermia for Soft Tissue Sarcoma

The effects of hyperthermia for bone and soft tissue sarcoma have been demonstrated in various basic investigations. The first experiment of hyperthermia for sarcoma was performed in Germany in the 1960s, showing the efficacy of hyperthermia in rats with sarcoma. Thereafter, several *in vivo* experiments using the Yoshida sarcoma strain have been reported, with the effects of hyperthermia for sarcoma being gradually elucidated. We also first conducted an *in vitro* study of hyperthermia alone using osteosarcoma cells, and demonstrated that DNA synthesis was suppressed in proportion to the processing time when cells were heated to 42–43 °C, and osteosarcoma cell growth was suppressed at 43 °C or higher [1]. Next, we performed a study of the combined effects of hyperthermia and anticancer drugs, and confirmed the increased drug concentration in cells after the permeability of the sarcoma cell membrane was enhanced due to heating. We also showed that hyperthermia suppressed sarcoma cell proliferation, inducing apoptosis as well as causing necrosis of sarcoma cells [2]. Hyperthermia suppressed proliferation of sarcoma cells *in vivo* as well, and the treatment effect was enhanced by administering anticancer drugs before the initiation of thermotherapy and allowing the tumor to take up the drugs as much as possible during hyperthermia, to increase the concentration of anticancer drugs in the tumor [3].

Synergetic combination effects of radiotherapy and hyperthermia were experimentally demonstrated. Those effects are thought to be due to enhancing the effect on tumor cells in late S phase of the cell cycle, which is resistant to radiation, and to increasing radiation sensitivity by elevating the partial pressure of oxygen in hypoxic tumor tissues by increasing blood flow. Radiotherapy can induce a second primary cancer, so the radiation dose should be minimized as far as possible. However, a high dose of radiation is required to treat sarcoma effectively, because it is generally less sensitive to radiation than are other tumors. Therefore, a combination of hyperthermia greatly contributes to reducing the radiation dose [4].

A basic experiment revealed that the triple combination of thermo-radio-chemotherapy was more effective than thermotherapy combined with chemotherapy or radiotherapy. This is because triple combination therapy overcame treatment-resistance at a cellular level while, individually, radiotherapy, chemotherapy, or hyperthermia failed to exert sufficient effect due to treatment-resistance caused by each respective therapy's disadvantages [3].

21.4 Clinical Application of Hyperthermia for Soft Tissue Sarcoma

Hyperthermia for soft tissue sarcoma is divided into whole body hyperthermia (whole body heating) and regional hyperthermia (local heating). Wiedemann et al. reported relatively favorable results of whole body hyperthermia (41.8 °C × 60 min) combined with ICE therapy (ifosfamide, carboplatin, and etoposide) in sarcoma patients with distant metastasis, with a response rate [partial response (PR) + complete response (CR)] of 28 %, and a 63 % response rate when stable disease (SD) was included [5]. However, no further study of whole body hyperthermia in patients with sarcoma has been reported after Wiedemann et al., and regional hyperthermia has become a mainstream therapy.

Several types of regional hyperthermia have been reported, including perfusion-hyperthermia (isolated limb perfusion; ILP), in which the anticancer drug is intra-arterially injected at elevated temperature, and hyperthermia using electromagnetic waves or ultrasound to heat the tumor. Lehti et al. conducted a study in 1986 of perfusion-thermo-chemotherapy using l-phenylalanine mustard and actinomycin D combined with massive resection and radiotherapy in 64 patients with soft tissue sarcoma, reporting a 5-year local recurrence rate and 5-year survival of 11 % and 67 %, respectively, [6]. This result was considered a favorable outcome for the treatment of soft tissue sarcoma at the time. Since Hill et al. first reported in 1993 a study of perfusion-thermo-chemotherapy using a combination of recombinant tumor necrosis factor (TNF) alpha and melphalan [7], several studies followed regarding TNFα-based preoperative perfusion-thermo-chemotherapy in patients with soft tissue sarcoma in whom limb saving surgery was difficult to achieve. Those results were obtained using TNFα to selectively enhance tumor uptake of anticancer drugs to destroy the tumor vessel structure selectively. Grunhagen et al. conducted a study of perfusion-thermo-chemotherapy using a combination of TNFα and Melphalan in 197 patients with locally advanced soft tissue sarcoma of the extremities, and reported CR, PR, and the overall response rates of 26 %, 49 %, and 75 %, respectively, and the affected limbs were preserved in 87 % of patients [8]. Other studies reported response rates of 60–90 % and affected limb preservation rates of 70–90 %. Therefore, this therapy may be a highly effective preoperative therapy for preserving affected limbs. However, no Phase 3 trial has been conducted to date, so no conclusion has been reached as to whether the effects of this therapy are superior to neo-adjuvant chemotherapy or preoperative radiotherapy.

Several studies have reported regional hyperthermia using electromagnetic waves to heat the tumor. Among them, the Klinikum Grosshadern Medical Center sarcoma group had been continuously conducting a Phase 2 study since 1986, and in 2001 reported the result of a combination of EIA therapy (etoposide, ifosfamide, and doxorubicin) and regional hyperthermia in 59 patients with primary advanced or recurrent high-risk soft-tissue sarcoma. The overall response (CR + PR) rate, minor regression rate, SD, and progressive disease (PD) were 17 %, 25 %, 33 %, and 25 %, respectively, in preoperative imaging. However, complete necrosis and favorable histological response (>75 % necrosis) were observed in 14 % and 28 %, respectively, in pathological evaluation using resected specimens, indicating that treatment effects were obtained in some patients despite the lack of reduced tumor size observable by imaging studies. They also reported that regional hyperthermia provided superior local control compared with chemotherapy alone [9]. Based on this result, the European Organization for Research and Treatment of Cancer Soft Tissue and Bone Sarcoma Group (EORTC-STBSG) and the European Society for Hyperthermic Oncology (ESHO) jointly conducted a Phase 3 study from 1997 to 2006. In a multicenter collaborative study conducted in nine facilities in Europe and North America, patients with localized high-risk soft tissue sarcoma (≥5 cm, Fédération Nationale des Centres de Lutte Contre le Cancer [FNCLCC] grade 2 or 3, deep to the fascia) were randomized into the neoadjuvant EIA monotherapy group or the EIA chemotherapy + regional hyperthermia group, to compare treatment outcome. EIA chemotherapy + hyperthermia was superior to EIA monotherapy in disease free survival as well as local control rate; thus, the effect of adding regional hyperthermia to systemic chemotherapy was demonstrated for the first time [10].

21.5 Radio-Hyperthermo-Chemotherapy for Soft Tissue Sarcoma in Our Department

21.5.1 Patients and Methods

Based on our previous basic research and clinical study reports from other departments, we have conducted a clinical study of triple combination therapy using RHC in patients with soft tissue sarcoma of the extremities since the early 1990s. The goals of the study were to improve the local control rate, reduce resection margin to preserve postoperative function of affected limbs, and control distant metastasis. We employed regional hyperthermia as thermotherapy to heat the local tumor using electromagnetic waves.

Patients with highly malignant (Grade2–3) soft tissue sarcoma of the extremities were included, and their treatment was initiated with their consent after making a definitive diagnosis by biopsy. Before initiating treatment, blood was collected to measure blood counts, bone-marrow function, hepatic function, and renal function to evaluate the patients' ability to tolerate chemotherapy. Next, prior to RHC, angiography was performed to assess tumor blood flow, and a catheter was simultaneously inserted with a reservoir placed into the artery feeding the tumor. Anticancer drugs were intra-arterially injected through this reservoir. Placement of a reservoir enabled multiple intra-arterial injections of anticancer drugs, and it had the great advantage of not interfering with daily life, other than the intra-arterial injection therapy itself. Cisplatin, 100 mg/m^2, was intra-arterially injected on Days

1, 15, and 29 and pirarubicin, 30 mg/m^2, was intra-arterially injected on Day 8 and 22 during thermotherapy. Platinum anticancer drugs, including cisplatin, and anthracycline anticancer drugs, including pirarubicin, were reported to have synergetic effects in combination with thermotherapy in a basic study, and the effect of these drugs on bone and soft tissue sarcomas was previously demonstrated. Radiotherapy was initiated on Day 1, and 2 Gy per fraction was delivered 20 times (on Days 1–5, 8–12, 15–19, and 22–26) for a total of 40 Gy. Two weeks after the RHC was completed, ifosfamide, etoposide, and pirarubicin were injected intravenously as systemic chemotherapy. The primary tumor subsequently was resected radically (Fig. 21.1). For thermotherapy, an 8-MHz radiofrequency capacitive heating system (Thermotoron RF-8: Yamamoto VINITA Co., Ltd., Osaka, Japan) (Fig. 21.2) was used. During thermotherapy, the patients' temperature was monitored by a temperature sensor inserted in the tumor to keep temperature at 42.5 °C for 60 min.

21.5.2 Clinical Results

We retrospectively examined patients with soft tissue sarcoma who were treated with RHC therapy prior to surgery, excluding cases with recurrence or after unplanned surgery. From 1994 to 2014, a total of 68 patients were included in this study. The patients' characteristics are summarized in Table 21.1. The mean follow-up was 95 months. The most common histologic subtypes were undifferentiated pleomorphic sarcoma (UPS; $n = 27$), liposarcoma ($n = 22$), synovial sarcoma ($n = 12$), leiomyosarcoma ($n = 3$), alveolar soft part sarcoma ($n = 3$), and others ($n = 1$).

21.5.2.1 Survival

The overall survival in our series was 74.0 % and 67.5 % (5 years and 10 years, respectively). There were 37 patients who were CDF, 6 who were NED, and 3 who were AWD. However, 19 patients died from disease and 3 patients were DOA. The 5-year overall survivals by stage (AJCC 7th edition) were 89.3 % (Stage 2a), 86.5 % (Stage 2b), 50 % (Stage 3), and 0 % (Stage 4) (Fig. 21.3). The mean survival for patients with Stage 4 tumors was 0.7 months. The 5-year overall survivals by histologic subtype were 82.1 % (UPS), 81.1 % (liposarcoma), and 33 % (synovial sarcoma). In multivariate analysis using the Cox-regression model, prognostic factors for overall survival were presence of a metastatic lesion at diagnosis (HR 14.9, 95 % CI 4.6–47.6, $P < 0.001$), histological grade of sarcoma (HR 10.8 for grade 3 over 2, 95 % CI 3.5–33.3, $P < 0.001$) and receipt of adjuvant systemic chemotherapy (HR 3.5, 95 % CI 1.2–10, $P = 0.02$). However, the histological response rate after RHC was not an independent prognostic factor.

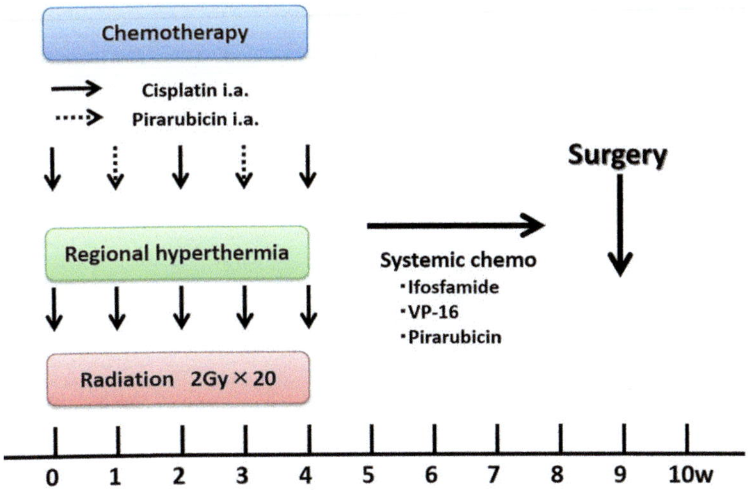

Fig. 21.1 Protocol of neoadjuvant Radio-Hyperthermo-Chemotherapy (RHC)

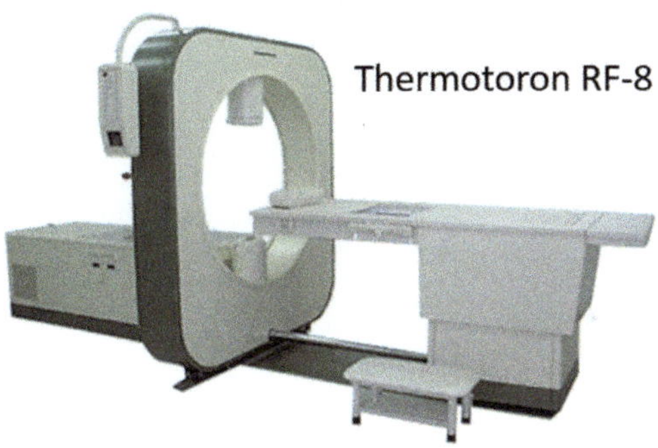

Fig. 21.2 8-MHz radiofrequency capacitive heating system (Thermotoron RF-8: Yamamoto VINITA Co., Ltd., Osaka, Japan)

21.5.2.2 Relapse Free Survival (RFS)

The 5-year RFS was 75.5 % for all patients except for those with Stage 4 disease, and no recurrences were observed after 4.4 years following surgery. The 5-year RFS rates were 83.3 % (Stage 2a), 80.3 % (Stage 2b), and 48.0 % (Stage 3) (Fig. 21.4). By histologic subtypes, the 5-year RFS rates were 82.6 % (UPS), 86.3 % (liposarcoma), and 37.5 % (synovial sarcoma). In multivariate analysis using the Cox-regression model, prognostic factors for RFS rate were histological grade of sarcoma (HR 16.4, for grade 3 over 2, 95 % CI, 3.3–83.3, P = 0.01) and receipt of adjuvant systemic chemotherapy (HR 11.9, 95 % CI 2.3–60.1, P = 0.03). However, the histological response rate following RHC was not an independent prognostic factor.

Table 21.1 Patients characteristics

Characteristics (N = 68)		
Mean age (year)	48.7 ± 16.8 (12.0–81.0)	
Sex		
Male	45 cases	66.2 %
Female	23 cases	33.8 %
Anatomic sites		
Trunk	2 cases	2.9 %
Pelvic area	8 cases	11.8 %
Upper arm	4 cases	5.9 %
Forearm	6 cases	8.8 %
Femur	39 cases	57.4 %
Lower Leg	9 cases	13.2 %
Tumor size		
<5 cm	26 cases	38.2 %
>5 cm	42 cases	61.8 %
Stage (AJCC system, 7th ed)		
Stage 2a	21 cases	30.9 %
Stage 2b	27 cases	39.7 %
Stage 3	10 cases	14.7 %
Stage 4	10 cases	14.7 %

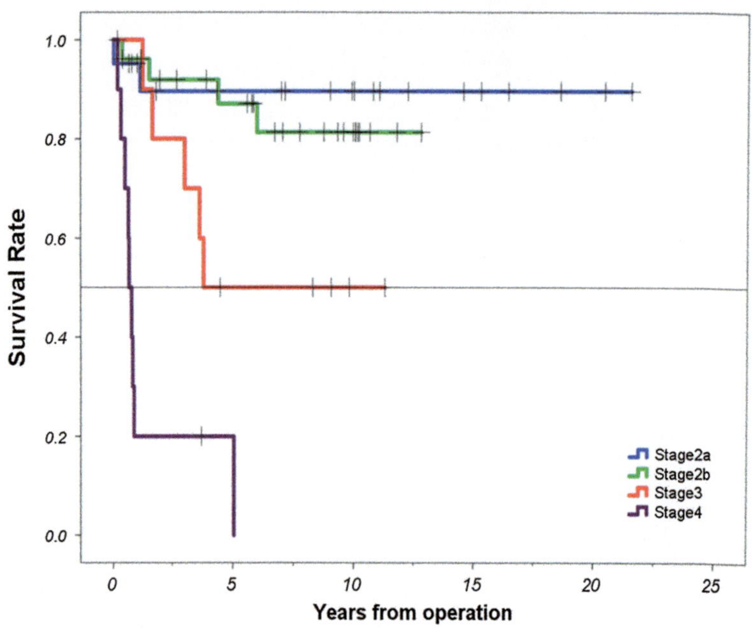

Fig. 21.3 Overall survival rate of each grade of patients

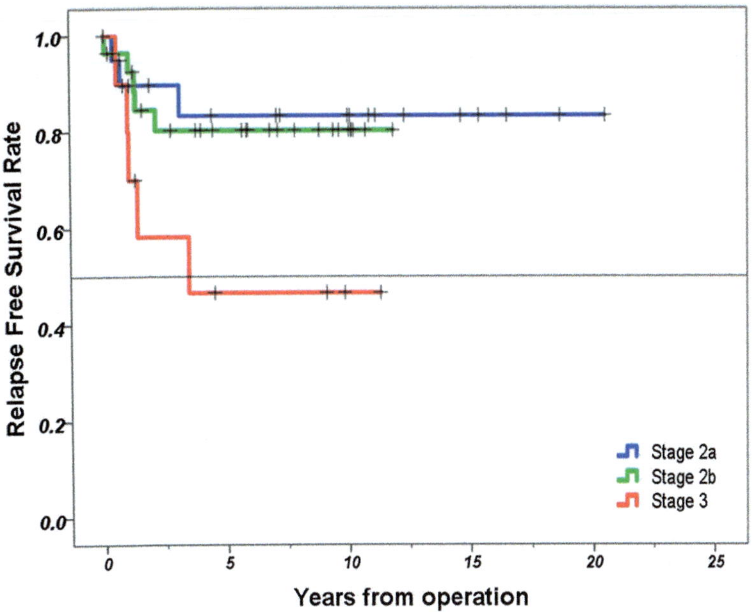

Fig. 21.4 Relapse free survival rate of every grade of patients

21.5.2.3 Local Control Rate

There were no local recurrences in the 68 patients treated, though remote metastasis during follow-up was identified in 12 of them. However, we did have patients that experienced local recurrences after RHC when we performed RHC for residual tumor due to unplanned excision of sarcoma or discrepancy of diagnosis between the diagnostic needle biopsy and the specimen obtained at surgery. All of these patients underwent limb-sparing surgery without amputation. Three patients had involvement of major vessels and nerves, though gentle separation of the tumor capsule made it possible to preserve function and avoid subsequent recurrence.

21.5.2.4 Histologic Tumor Responses (Table 21.2)

Surgical specimens were microscopically reviewed, and overall histologic response to RHC was measured on a 4-point scale using the ratio of necrosis or severe cellular alteration. Ratios of less than 50 % (NC), 50–80 % (PRb), >80 % but <100 % (PRa), and 100 % (CR) were classified as +1, +2, +3, and +4, respectively [11].

Table 21.2 Histopathology and response of RHC

Histopathology				
	Grade 1	Grade2	Grade 3	Grade 4
UPS	13.3 %	33.3 %	20.0 %	33.3 %
Liposarcoma	5.8 %	41.1 %	17.6 %	29.4 %
Synovial sarcoma	9.1 %	27.2 %	27.2 %	36.4 %

21.5.2.5 Complications of RHC

Due to the multidisciplinary approach with intra-arterial chemotherapy, neoadjuvant radiotherapy and hyperthermia for soft tissue sarcoma before surgery, wound healing could be interrupted and several problems could occur. Among study patients, delayed skin healing (over 2 weeks) occurred in 28.9 % patients (20/69 cases), and 8 patients needed other procedures (9 cases of debridement and surgical skin closure and 1 case requiring a skin flap). Additional side effects of RHC included 1 case of secondary leukemia likely caused by chemotherapy, 1 case of femoral head necrosis likely caused by radiotherapy, and 1 case of prolonged fever of unknown origin.

21.5.2.6 Case Presentation

A 61-year-old woman was known to have a soft tissue mass (2–3 cm) on the frontal region of the right knee for 7 years prior to presentation. Although she noticed that the tumor was gradually enlarging, she hesitated to consult a doctor until the tumor reached 7 cm in width (Fig. 21.5). She was referred to our department, and her diagnostic evaluations included needle biopsy (Fig. 21.6), CT-PET (Fig. 21.7), and magnetic resonance imaging (MRI) (Fig. 21.8a–b). Based on the histologic analysis and imaging findings, a diagnosis of undifferentiated pleomorphic sarcoma (UPS) without remote metastasis (Stage 2b) was made.

After hospitalization, an arterial catheter was inserted into the femoral artery (Fig. 21.9) and a reservoir was inserted. RHC therapy was performed as described above. Although during the second session of RHC a mild burn occurred, she completed five sessions of RHC. Based on the preoperative assessment of the tumor extent, a complete response to RHC was suspected (Fig. 21.10, Fig. 21.11a–b).

We performed tumor resection with 1 cm margins (Fig. 21.12). The periosteum of the tibia and fascia of surrounding muscles remained attached to the tumor as a barrier. The skin defect was then reconstructed using a pedicled cutaneous flap from the anterolateral thigh (Fig. 21.13).

Based on the histologic examination, the response to RHC considered to be complete, and the margin of resected tumor was adequate (Fig. 21.14). No adjuvant chemotherapy was administered. After surgery, her function was evaluated using

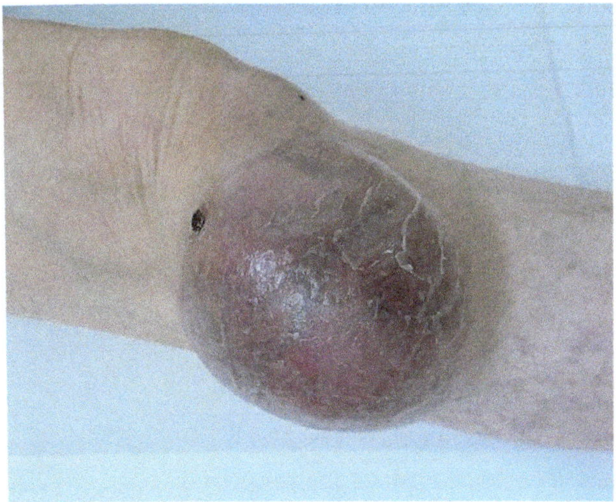

Fig. 21.5 Photograph of the mass at first evaluation. A *dark red* protruding lesion was seen at the lateral aspect of the *right knee*, with distended subcutaneous veins. The width of the tumor was approximately 70 mm

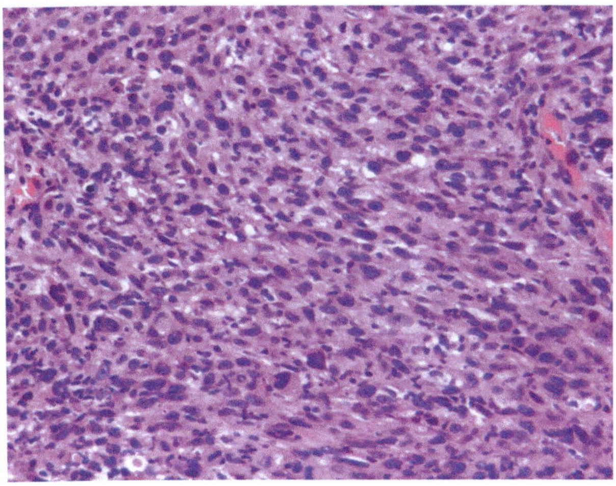

Fig. 21.6 Pathology of the needle biopsy specimen. In the lesion site, polymorphic dysplastic cells with increased chromatin in the nucleus are visible with some inflammatory cells. Immunostaining revealed CD34 (–), S-100 (–), SMA (–), EMA (–), Keratin (–), HMB-45 (–), CD3 (–), CD20 (–), CD 68 (–), CD1a (–), and CD 56 (–), and the MIB1 index was 60 %. The pathologic image was consistent with undifferentiated pleomorphic sarcoma

ISOLS/MSTS scoring system, indicating 93 % function. No recurrence has been observed.

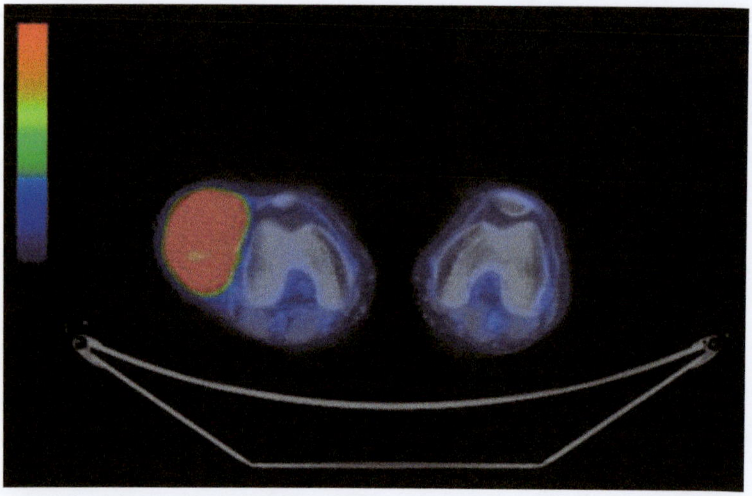

Fig. 21.7 FDG-PET/CT scan before RHC. There was abnormal uptake on FDG-PET-CT at the lateral aspect of the right knee (SUV max = 25.6)

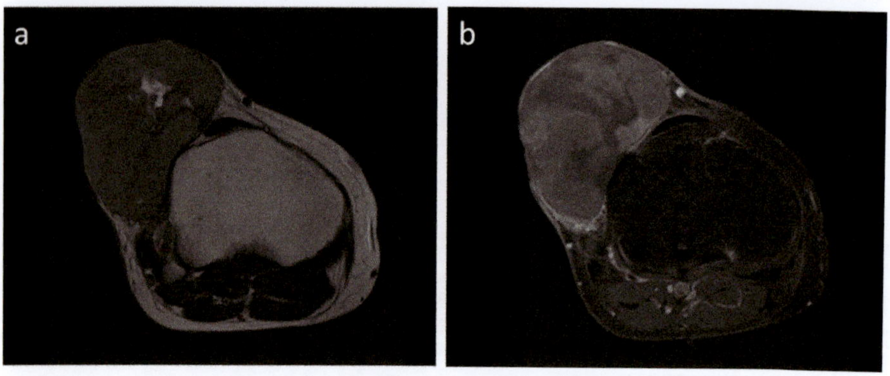

Fig. 21.8 (**a, b**) T2-weighted & Gadolinium-enhanced T1-weighted magnetic resonance image before RHC

21.5.3 Second Look Surgery

The concept of a wide surgical margin has become more sophisticated, including barriers and margins. In some cases, however, soft tissue sarcoma is treated without pretreatment examination or planning. When a cancer diagnosis is revealed after unplanned excision, surgical fields can be contaminated, and an additional wide excision is necessary in most cases. During such procedures, neurovascular structures may be sacrificed and excessive skin and muscle defects may be created.

We established a "second look operation" protocol consisting of whole biopsy of surgical scar tissue following RHC performed after unplanned excision of soft

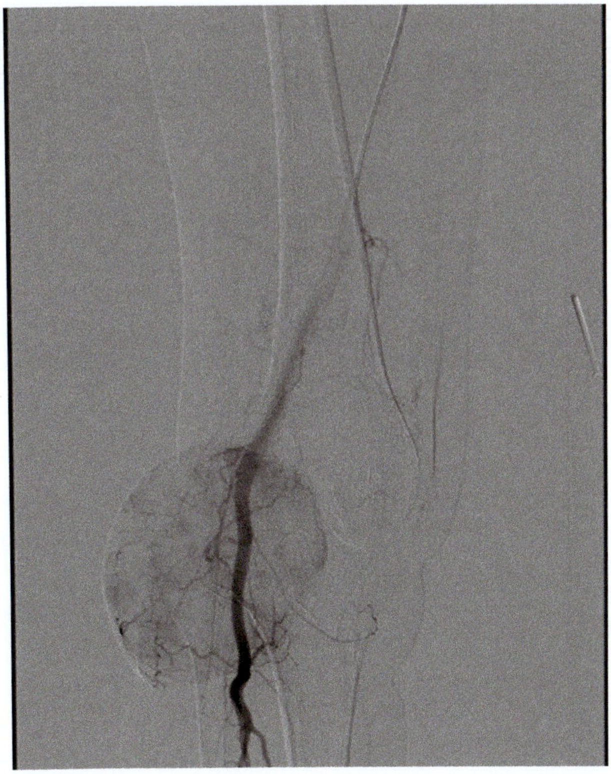

Fig. 21.9 Angiogram & inserted arterial catheter. Strong tumor staining was observed

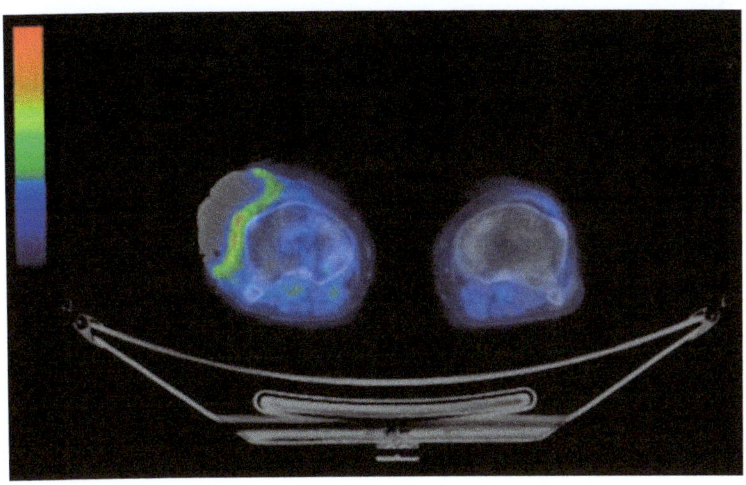

Fig. 21.10 There was no uptake on FDG-PET-CT in the treated tumor. Mild uptake, considered to be a reaction to RHC, was seen around the tumor (SUV max = 3.8)

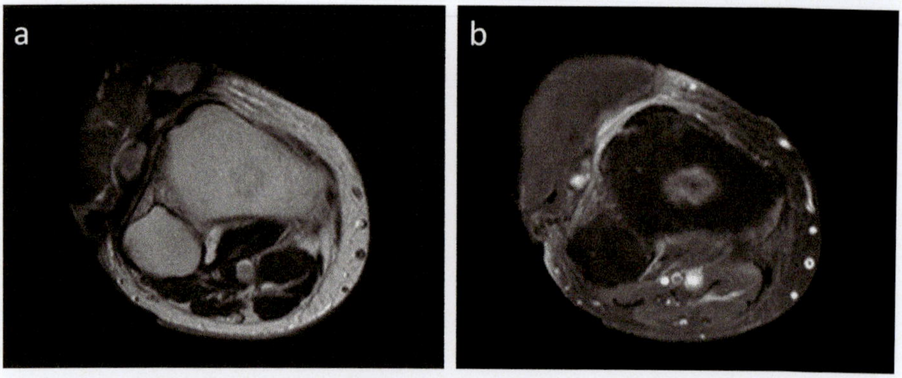

Fig. 21.11 (**a, b**) T2-weighted & Gadolinium-enhanced T1-weighted magnetic resonance image before RHC

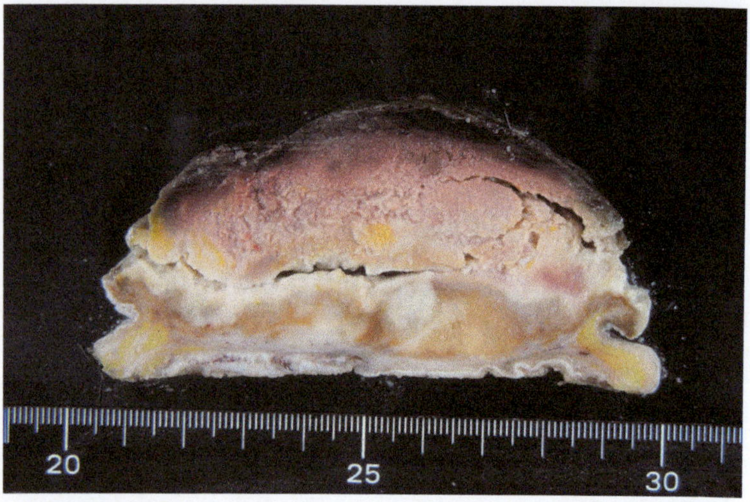

Fig. 21.12 Resected specimen

tissue sarcoma. This protocol is derived from surgical methods for ovarian or colorectal cancer. Following resection of the primary lesion and administration of chemotherapy, a second-look operation is performed to determine whether any cancer cells remain. We modified these procedures for treatment of soft tissue sarcoma after unplanned excision. A second-look operation for soft tissue sarcoma is defined as a procedure used to determine whether a patient is free of disease after initial treatment. If disease is found, an additional surgery is performed at the time of or after the second-look operation.

Hayashi et al. [12] reported our surgical results from second look operations. From 1995 to 2004, there were 6 patients (3 men and 3 women) who underwent this procedure. All patients were referred to our department after simple marginal

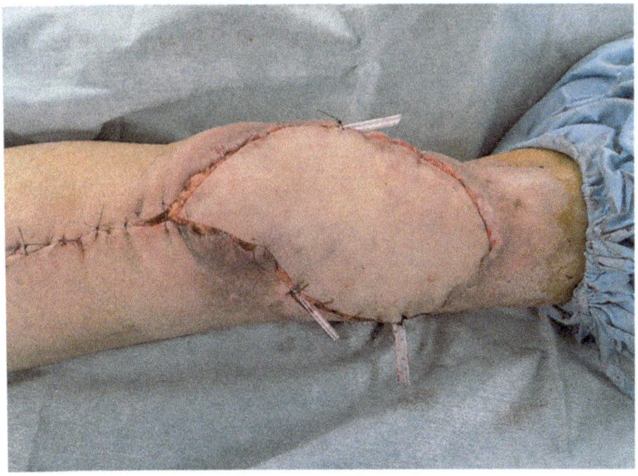

Fig. 21.13 Reconstruction with pedicled cutaneous flap from the anterolateral thigh after resection of tumor

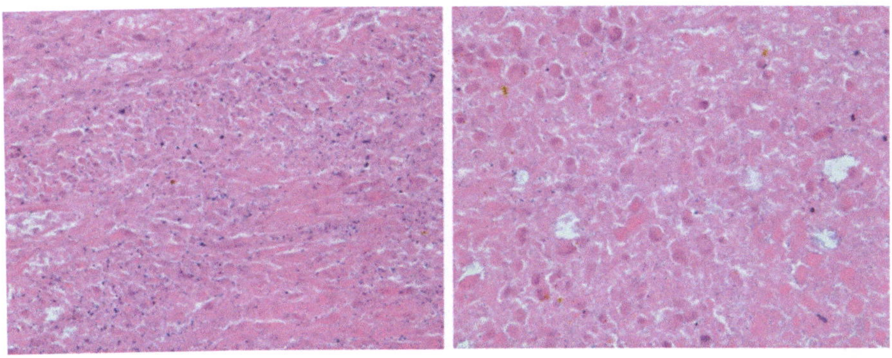

Fig. 21.14 Pathology of the resected specimen after RHC. Necrosis, degeneration treatment effects were observed throughout the tumor. No viable tumor was identified; thus, the tumor is considered to have had a complete response to the preoperative RHC therapy

resection performed by a previous physician and histologic diagnoses were obtained. Diagnoses included UPS (n = 3), sclerosing epithelioid fibrosarcoma (n = 1), myxoid liposarcoma (n = 1), and synovial cell sarcoma (n = 1). Before surgery, we performed RHC procedures as described above. Surgery was performed to excise scar tissue that was enhanced on preoperative MRI. Resected specimens were examined postoperatively under a microscope, evaluating the largest cut section.

In all 6 cases, no residual cancer cells were identified in the resected scar tissue; therefore, no additional wide excisions were performed. No major complications were observed during preoperative RHC, except for mild pain and minor burns. Burns were caused by hyperthermia in 3 patients. Dehiscence of the skin wound and delayed healing occurred in 3 patients. Preservation of adjacent structures was successful and postoperative function was 100 % by ISOLS/MSTS evaluation in

Table 21.3 Patients characteristics and clinical results of second-look operation

Cases	Age (year)	Histology	Location	Adjacent structure	Follow-up (year)	Oncological result
1	57	UPS	Forearm	Median nerve	17.6	CDF
2	54	Sclerosing epithelioid sarcoma	Forearm	Radial nerve	10.9	CDF
3	60	UPS	Forearm		10.0	CDF
4	49	Myxoid liposarcoma	Thigh	Femoral AVN	10.4	CDF
5	66	UPS	Forearm		8.2	CDF
6	39	Synovial sarcoma	Foot		8.1	CDF

all patients. During the follow-up period, there were no local recurrences identified, and all patients were alive at last follow-up. (Table 21.3)

References

1. Hattori M. Effect of hyperthermia on cultured malignant musculoskeletal tumor cells. Nagoya Med J. 1993;37:135–50.
2. Yonezawa M, Otsuka T, Matsui N, Tsuji H, Kato KH, Moriyama A, Kato T. Hyperthermia induces apoptosis in malignant fibrous histiocytoma cells in vitro. Int J Cancer. 1996;66:347–51.
3. Tsuji H. Experimental study of radio-hyperthermia for the malignant fibrous histiocytoma in vivo. The Journal of the Nagoya City University Medical Association. 2002;53:77–85 (Japanese).
4. Kal HB, Hahn GM. Kinetic responses of murine sarcoma cells to radiation and hyperthermia in vivo and in vitro. Cancer Res. 1976;36:1923–9.
5. Westermann AM, Wiedemann GJ, Jager E, Jager D, Katschinski DM, Knuth A, Vörde Sive Vörding PZ, Van Dijk JD, Finet J, Neumann A, Longo W, Bakhshandeh A, Tiggelaar CL, Gillis W, Bailey H, Peters SO, Robins HI; Systemic Hyperthermia Oncologic Working Group. Oncology.A systemic hyperthermia oncologic working group trial. Ifosfamide, carboplatin, and etoposide combined with 41.8 degrees C whole-body hyperthermia for metastatic soft tissue sarcoma. 2003;64:312–321.
6. Lehti PM, Moseley HS, Janoff K, Stevens K, Fletcher WS. Improved survival for soft tissue sarcoma of the extremities by regional hyperthermic perfusion, local excision and radiation therapy. Surg Gynecol Obstet. 1986;162:149–52.
7. Hill S, Fawcett WJ, Sheldon J, Soni N, Williams T, Thomas JM. Low-dose tumour necrosis factor alpha and melphalan in hyperthermic isolated limb perfusion. Br J Surg. 1993;80:995–7.
8. Grunhagen DJ, de Wilt JH, Graveland WJ, Verhoef C, van Geel AN, Eggermont AM. Outcome and prognostic factor analysis of 217 consecutive isolated limb perfusions with tumor necrosis factor-alpha and melphalan for limb-threatening soft tissue sarcoma. Cancer. 2006;106:1776–84.
9. Issels RD, Abdel-Rahman S, Wendtner C, Falk MH, Kurze V, Sauer H, Aydemir U, Hiddemann W. Neoadjuvant chemotherapy combined with regional hyperthermia (RHT) for locally advanced primary or recurrent high-risk adult soft-tissue sarcomas (STS) of adults: long-term results of a phase II study. Eur J Cancer. 2001;37:1599–608.

10. Issels RD, Lindner LH, Verweij J, Wust P, Reichardt P, Schem BC, Abdel-Rahman S, Daugaard S, Salat C, Wendtner CM, Vujaskovic Z, Wessalowski R, Jauch KW, Dürr HR, Ploner F, Baur-Melnyk A, Mansmann U, Hiddemann W, Blay JY, Hohenberger P; European Organization for Research and Treatment of Cancer Soft Tissue and Bone Sarcoma Group (EORTC-STBSG); European Society for Hyperthermic Oncology (ESHO). Neo-adjuvant chemotherapy alone or with regional hyperthermia for localized high-risk soft-tissue sarcoma: a randomized phase 3 multicenter study. Lancet Oncol. 2010;11:561–570.
11. Willett CG, Schiller AL, Suit HD, Mankin HJ, Rosenberg A. The histologic response of soft tissue sarcoma to radiation therapy. Cancer. 1987;60:1500–4.
12. Hayashi K, Yamada S, Inatani H, Okamoto H, Takeuchi A, Nishida H, Yamamoto N, Tsuchiya H, Otsuka T. Salvage method for unplanned excision of soft tissue sarcoma: long-term results of second-look surgery following radio-hyperthermo-chemotherapy. Anticancer Res. 2015;35:493–8.

Chapter 22
Combination by Hyperthermia and Chemotherapy: Lung Cancer

Takayuki Ohguri

Abstract In this chapter, the clinical results of chemotherapy with hyperthermia used for the treatment of lung cancer are reviewed. Some clinical results, including phase II clinical trials, demonstrated the feasibility and efficacy of systemic chemotherapy plus regional hyperthermia in patients with advanced non-small-cell lung cancer. Further evaluations via phase III, randomized clinical trials assessing chemotherapy with or without regional hyperthermia are warranted in these patients.

Keywords Hyperthermia • Chemotherapy • Lung Cancer • Pleural dissemination

22.1 Background

Although the combination of radiotherapy and hyperthermia has been the focus of most clinical attention, there is an equally strong rationale for combining hyperthermia with chemotherapy [1, 2]. In several experimental studies, moderate or even mild hyperthermia enhanced the *in vivo* cytotoxic effects of numerous chemotherapeutic agents, such as cisplatin, carboplatin, docetaxel, paclitaxel, irinotecan, oxaliplatin, and gemcitabine, thereby exhibiting thermal enhancement [2, 3].

In 2010, a large, randomized, phase III study of chemotherapy with or without regional hyperthermia for high-risk soft-tissue sarcoma showed that the addition of regional hyperthermia significantly improved survival rates [4]. However, clinical experience with a combination of chemotherapy and hyperthermia is still limited in patients with lung cancer. In this chapter, we review the reported results for hyperthermia in combination with chemotherapy for lung cancer.

Heating technique using an 8-MHz RF capacitive heating device for the tumors of the thoracic region is stated in this book at the chapter "Combination by Hyperthermia and Radiation (and Chemotherapy): Lung Cancer".

T. Ohguri (✉)
Department of Radiology, University of Occupational and Environmental Health,
1-1 Iseigaoka, Yahatanishi-ku, Kitakyushu 807-8555, Japan
e-mail: ogurieye@med.uoeh-u.ac.jp

S. Kokura et al. (eds.), *Hyperthermic Oncology from Bench to Bedside*,
DOI 10.1007/978-981-10-0719-4_22

22.2 Systemic Chemotherapy with Regional Hyperthermia for Lung Cancer

In 1993, Matsda et al. reported the multi-institutional analysis of cisplatin-based chemotherapy plus regional hyperthermia in patients with advanced lung cancer that could not be locally controlled by chemotherapy alone; an objective response was achieved in 3 (21 %) of 14 patients [5]. Jiang et al. conducted a Phase II trial of second-line docetaxel in conjunction with regional hyperthermia for advanced non-small-cell lung cancer (NSCLC) [6]. Hyperthermia was administered 1 h after chemotherapeutics (a regimen of docetaxel 40 mg/m^2 intravenous infusion on day1, 8 and 16, repeated every 4 weeks was chosen.) were given, and the total heating time was up to 60 min, twice a week. The regimen was well tolerated, and the median progression-free survival was 4 months.

In 2010, Shen et al. conducted a phase II randomized trial for the regimen of gemcitabine and cisplatin with or without regional hyperthermia in patients with advanced NSCLC [7]. Hyperthermia was administered 1 h after the chemotherapeutic drugs (a regimen of 1000 mg/m^2 intravenous gemcitabine on day 1 and day 8 and 80 mg/m^2 intravenous cisplatin on day 1 was performed.) were given, and the total heating time was up to 60 min, twice a week. The target temperature ranged from 39 °C to 42.5 °C; the intra-tumor temperature was directly measured for the chest wall tumors. Eighty patients were assigned to each of the two arms, and the combined regimens were well tolerated and safe. Significant advantages in the quality of life of patients with regional hyperthermia were recognized, although survival rates with or without hyperthermia were not different between the patients.

In 2003, Ohguri.et al reported a novel combination of systemic chemotherapy with carboplatin and paclitaxel (paclitaxel of 50 mg/m^2 and carboplatin of area under the curve of 1.0–1.5 for 3 out of 4 weeks), regional hyperthermia, and hyperbaric oxygen therapy to improve the clinical outcomes in NSCLC with multiple pulmonary metastases [8]. A favorable objective response rate of 64 % was demonstrated without severe toxicity. An interesting results in this study was that the re-introduction of carboplatin and paclitaxel caused an objective tumor response in 4 of 9 patients who had already received chemotherapy using carboplatin and paclitaxel (Fig. 22.1). In experimental studies, the use of hyperthermia with many chemotherapeutic drugs had the potential ability to reverse the drug resistance [9]. In an experimental study, the administration of hyperbaric oxygen therapy increased the effects of hyperthermic chemosensitization to carboplatin [10].

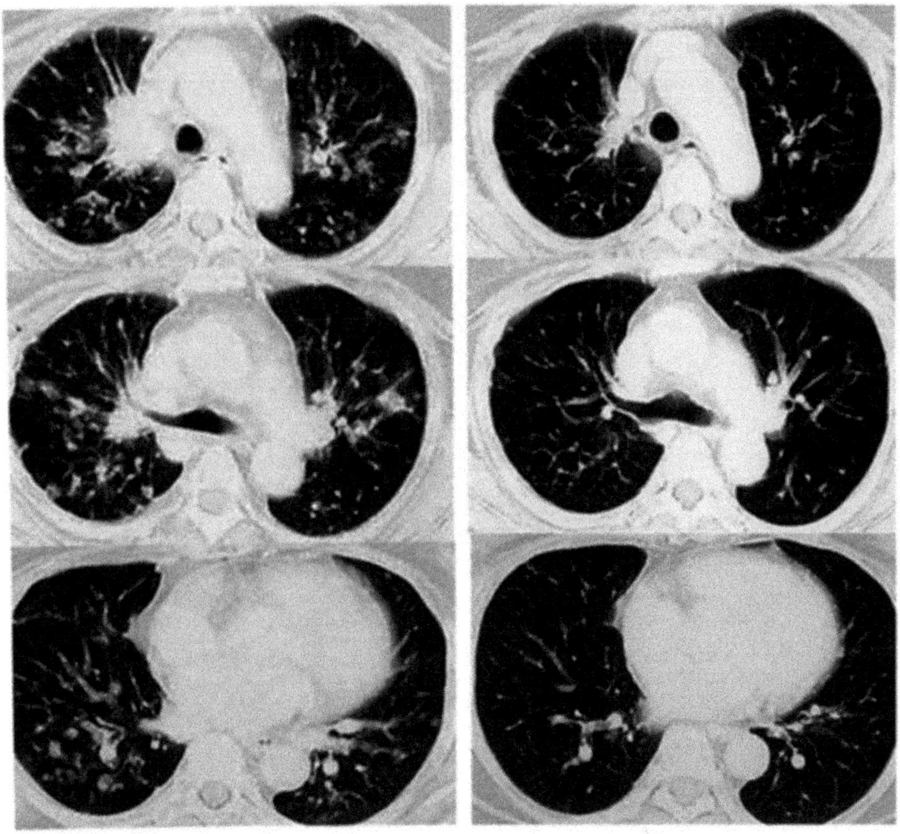

Fig. 22.1 A multiple chemotherapeutic drug-resistance case with non-small cell lung cancer of multiple pulmonary metastases was treated with the re-introduction of paclitaxel (50 mg/m^2) and carboplatin (area under the curve of 1.0) weekly for 3 out of 4 weeks combined with regional hyperthermia and hyperbaric oxygen therapy. Regional hyperthermia of the whole thoracic region was administered weekly during intravenous infusion of carboplatin, in addition, hyperbaric oxygen treatment was performed immediately after weekly chemo-hyperthermia. Complete response was achieved following 8 cycles of this combined treatment. Cited from [12]

22.3 Intrathoracic Chemotherapy Plus Regional Hyperthermia for Lung Cancer with Pleural Dissemination

The clinical outcomes for postoperative intrathoracic chemotherapy plus hyperthermia for lung cancer with pleural dissemination have been reported [11]. Regional hyperthermia using a radiofrequency device was applied through the chest wall to heat the pleura and pleural cavity immediately after the administration of 50–100 mg of cisplatin through a chest drainage tube (left in place at the time of surgery). Three courses of combined treatments were administered. The 5-year, local, relapse-free survival rates were promising at 76 %. The authors concluded that the combined treatments may be beneficial for regional disease control and for survival, particularly in patients without mediastinal lymph node metastasis.

References

1. Dewhirst MW, Viglianti BL, Lora-Michiels M, Hanson M, Hoopes PJ. Basic principles of thermal dosimetry and thermal thresholds for tissue damage from hyperthermia. Int J Hyperthermia. 2003;19(3):267–94. doi:10.1080/0265673031000119006. UDW7W1EPFQD1L140 [pii].
2. Ohguri T, Imada T, Korogi Y, Narisada H. Clinical results of systemic chemotherapy combined with regional hyperthermia. Therm Med. 2007;23(2):49–61.
3. Mohamed F, Marchettini P, Stuart OA, Urano M, Sugarbaker PH. Thermal enhancement of new chemotherapeutic agents at moderate hyperthermia. Ann Surg Oncol. 2003;10(4):463–8.
4. Issels RD, Lindner LH, Verweij J, Wust P, Reichardt P, Schem BC et al. Neo-adjuvant chemotherapy alone or with regional hyperthermia for localised high-risk soft-tissue sarcoma: a randomised phase 3 multicentre study. Lancet Oncol. 2010;11(6):561–70. doi:S1470-2045 (10)70071-1 [pii]10.1016/S1470-2045(10)70071-1.
5. Matsuda T, Shimoyama T, Ohno T, Maeta M, Takahashi M, Tsukiyama I, et al. Cancer treatment by hyperthermia, radiation and drugs. local hyperthermia combined with chemotherapy for solid tumours: analysis of joint research conducted at a number of institutions in Japan. London: Taylor and Francis; 1993.
6. Jiang Z, Yan W, Ming J, Yu Y. Docetaxel weekly regimen in conjunction with RF hyperthermia for pretreated locally advanced non-small cell lung cancer: a preliminary study. BMC Cancer. 2007;7:189. doi:1471-2407-7-189 [pii]10.1186/1471-2407-7-189.
7. Shen H, Li XD, Wu CP, Yin YM, Wang RS, Shu YQ. The regimen of gemcitabine and cisplatin combined with radio frequency hyperthermia for advanced non-small cell lung cancer: a phase II study. Int J Hyperthermia. 2011;27(1):27–32. doi:10.3109/02656736.2010.500645.
8. Ohguri T, Imada H, Narisada H, Yahara K, Morioka T, Nakano K et al. Systemic chemotherapy using paclitaxel and carboplatin plus regional hyperthermia and hyperbaric oxygen treatment for non-small cell lung cancer with multiple pulmonary metastases: preliminary results. Int J Hyperthermia. 2009;25(2):160–7. doi:910057552 [pii]10.1080/02656730802610357.
9. Jones EL, Samulski TV, Vujaskovic Z, Leonard RP, Dewhirst MW. Hyperthermia. Principles and practice of radiation oncology. 4th ed. Philadelphia: Lippincott Williams & Wilkins; 2003.
10. Ohguri T, Kunugita N, Yahara K, Imada H, Uemura H, Shinya N, et al. Efficacy of hyperbaric oxygen therapy combined with mild hyperthermia for improving the anti-tumour effects of carboplatin. Int J Hyperthermia. 2015;31(6):643–8. doi:10.3109/02656736.2015.1055832.
11. Kodama K, Doi O, Higashiyama M, Yokouchi H, Tatsuta M. Long-term results of postoperative intrathoracic chemo-thermotherapy for lung cancer with pleural dissemination. Cancer. 1993;72(2):426–31.
12. Ohguri T, Yahara K, Murakami M, Imada H, Terashima H, Korogi Y. Current status and future directions of hyperthermia at the University of Occupational and Environmental Health. Therm Med. 2010;26(3):87–96.

Chapter 23
Efficacy of Hyperthermia in Combination With Chemotherapy for Breast Cancer

Daigo Yamamoto, Chizuko Yamamoto, Satoru Iwase, and Hiroki Odagiri

Abstract *Background*: We previously reported that doxorubicin and cyclophosphamide (DC) followed by weekly paclitaxel is an active and manageable preoperative regimen for breast cancer patients (JBCRN-02 Trial).

Patients and Methods: We retrospectively investigated that the efficacy of hyperthermia in combination with chemotherapy for breast cancer. Four cycles of DC (doxorubicin: 60 mg/m^2 and cyclophosphamide: 600 mg/m^2) administered intravenously (i.v.) on day 1 every 21 days were followed by 12 cycles of paclitaxel i.v. (80 mg/m^2) every 7 days, prior to surgery. During chemotherapy, hyperthermia was performed using a Thermotron RF-8 capacitive heating device.

Results: 5 patients received hyperthermia during chemotherapy. The patient population was identified from a database of the Japan Breast Cancer Research Network. Clinical responses were rated as cPR in 4 patients (80 %), and cSD in 1 patient (20 %). Further, the cPR rate in ER/PgR– tumor was 100 %, while it in ER/PgR+ was 50 %. The value of Ki67 remarkably decreased in 4cases (80 %) and we observed significant association between Ki67 values and clinical response. In addition, 2 patients who developed grade 1 dermatitis during hyperthermia.

Conclusion: there is a possibility that high-Ki67 may become predictive factor in clinical response of neoadjuvant chemotherapy and hyperthermia.

Keywords Breast • Hyperthermia • Chemotherapy

D. Yamamoto (✉)
Department of Surgery, Kansai Medical University, Takii Hospital, 10-15 Fumizono-cho, Moriguchi, Osaka 570-8507, Japan

Internal medicine, Seiko Hospital, Osaka, Japan
e-mail: yamamotd@takii.kmu.ac.jp

C. Yamamoto
Internal medicine, Seiko Hospital, Osaka, Japan

S. Iwase
Department of Palliative Medicine, the institute of medical science, University of Tokyo Hospital, Tokyo, Japan

H. Odagiri
Department of Surgery, Hirosaki National Hospital, Aomori, Japan

S. Kokura et al. (eds.), *Hyperthermic Oncology from Bench to Bedside*,
DOI 10.1007/978-981-10-0719-4_23

23.1 Introduction

Neoadjuvant chemotherapy (NAC) has emerged as a promising step forward in the management of locally advanced breast cancer. When administered before surgery, chemotherapy may induce tumor shrinkage, facilitate surgery, and increase the breast-conserving surgery rate [1–3].

The effect of NAC is critical as a prognostic predictor for breast cancer, and tailor-made therapy can be based on that response. Anthracycline- and taxane-based regimens have been extensively studied in clinical trials and consequently are widely used. Although pCR of the tumor might be expected to represent a prognostic predictor after NAC, the low response rate remains problematic [4]. Therefore, a new and more effective strategy for the treatment of breast cancer is called for.

Hyperthermia, one of the oldest forms of cancer treatment involves selective heating of tumor tissues to temperatures ranging between 39 and 45 °C. Recent developments based on the thermoradiobiological rationale of hyperthermia indicate it to be a potent radio- and chemosensitizer. This has been further corroborated through positive clinical outcomes in various tumor sites using thermoradiotherapy or thermoradiochemotherapy approaches [5]. Our previous studies showed that hyperthermia accelerate the antitumor effects of radiation therapy and chemotherapy [6, 7]. However the mechanism is under investigation and a lot of clinical trials are ongoing. Recent clinical trials in terms with hyperthermia are registered in the ClinicalTrials.gov registry [8]. A total of 109 trials were identified in which hyperthermia was part of the treatment regimen. Of these, 49 trials (45 %) had hyperthermic intraperitoneal chemotherapy after cytoreductive surgery (HIPEC) as the primary intervention, and 14 other trials (13 %) were also testing some form of intraperitoneal hyperthermic chemoperfusion. Seven trials (6 %) were testing perfusion attempts to other locations (thoracic/pleural n = 4, limb n = 2, hepatic n = 1). Sixteen trials (15 %) were testing regional hyperthermia, 13 trials (12 %) whole body hyperthermia, seven trials (6 %) superficial hyperthermia and two trials (2 %) interstitial hyperthermia. One remaining trial tested laser hyperthermia.

Therefore, in this study we retrospectively examined the efficacy of hyperthermia in combination with chemotherapy for breast cancer and relationship between clinical responses of hyperthermia in combination with chemotherapy and Ki67 value.

23.2 Patients and Methods

23.2.1 A Multicenter Study

Forty-three patients received preoperative chemotherapy between April 2004 and March 2015 at four centers. Patients with bilateral, locally advanced, or metastatic disease were excluded. Among them, 5 patients received hyperthermia during chemotherapy. The patient population was identified from a database of the Japan Breast Cancer Research Network.

23.2.2 Treatment Plan

Preoperative chemotherapy was described previously [1]. In brief, 4 cycles of doxorubicin: 60 mg/m^2 and cyclophosphamide: 600 mg/m^2 (DC) administered intravenously (*i.v.*) on day 1 every 21 days were followed by 12 cycles of paclitaxel *i.v.* (80 mg/m^2) every 7 days, prior to surgery. The findings of physical examinations, tumor characteristics, initial dose of paclitaxel, number of treatment cycles, chemotherapy-related toxicities, and symptom severity were recorded every week.

In this subset analysis, during chemotherapy, hyperthermia was performed using a Thermotron RF-8 capacitive heating device (Yamamoto VINITA Co., Ltd., Japan) at a surface temperature of 38 °C–41 °C, for 50 min [6, 7]. The output power ranged from 1300–1400 W.

23.2.3 Assessment of Response to Therapy

A physical examination was performed and the performance status was assessed on day 1 of each course. Tumor assessment involved a physical examination before, during, and after every course and breast ultrasonography after 4 courses of DC regimen; the appearance of any new lesion was documented. The clinical response of bidimensionally measurable and assessable disease was classified as a complete response (CR), partial response (PR), stable disease (SD), or progressive disease (PD) according to WHO criteria. CR was defined as the disappearance of all clinical evidence of the tumor; PR was defined as a 50 % or more reduction in the sum of the products of measured lesions, or an estimated decrease in the tumor size of at least 50 %, without the appearance of new lesions; SD was defined as a decrease in the lesion size of less than 50 % for the sum of the products of measured lesions, or an estimated decrease of less than 50 % and increase of less than 25 %, without the appearance of new lesions. Any measured or estimated increase greater than 25 % or the appearance of new lesions was defined as PD. The clinical response was defined as the sum of CRs and PRs. Surgery was to be performed less than 4 weeks after the last chemotherapy course. *Histopathological examination* Pretreatment diagnosis was established by our pathologists using samples from core needle biopsy. The positive cell rates for ER/PgR and Ki-67 were determined by Immunohistochemistry [1].

23.3 Results

23.3.1 Patient Characteristics

Between April 2004 and March 2015, 43 patients were enrolled. The median age was 50 (range: 20–69) years. The majority of patients had T2 tumors. All the patients were evaluable regarding their response and toxicity. Clinical responses were rated as cCR in 9 patients (22 %), cPR in 25 patients (59 %), and cSD in 9 patients (19 %). On the contrary, 5 patients received hyperthermia during chemotherapy and clinical responses were rated as cPR in 4 patients (80 %), and cSD in 1 patients (20 %), while the cPR rate in ER/PgR– tumor was 100 %, while it in ER/PgR+ was 50 % (Table 23.1). Further we observed significant changes in Ki67 value after preoperative treatment. There were 4 cases where the Ki67 value decreased, 1case where Ki67 did not change (Fig. 23.1). Especially, cPR was seen in case 1–4 patient, while cSD was seen in case 5 patient.

In addition, we identified no severe toxicities and minor toxicities; 2 patients who developed grade 1 dermatitis during hyperthermia.

23.4 Discussion

In this study, we retrospectively examined the efficacy of hyperthermia in combination with chemotherapy for breast cancer patients. There was an additional effect of hyperthermia on chemotherapy in terms of the response, especially in ER/PgR negative tumor. In additions, the value of Ki67 remarkably decreased in 4cases (80 %) and we observed significant association between Ki67 values and clinical response. Therefore it may be possible to assess their predictive value for tumor response.

In pre-clinical studies it has been demonstrated that the rate of liposomal extravasation is enhanced 4–8 fold for temperatures in the target range of this trial [9]. In cats with soft tissue sarcomas, hyperthermia enhanced radio-labelled liposomal uptake by 4–16 fold compared to normothermia [10]. Hyperthermia also increases oxygen levels within the tumor, which is critical to the effectiveness of radiation and chemotherapy [11–14]. More recent studies [15, 16] showed that treatments of mild hyperthermia (41 °C–43 °C for 30–60 min) plus docetaxel induced G2/M cell cycle arrest in the MCF-7 and MDA-MB-453 cells. In additions, proteins in the MAPK pathway were expressed at higher levels in docetaxel-treated cells following mild hyperthermia than those in cells treated with docetaxel alone. HSP70 and Pgp expression levels were significantly increased following mild hyperthermia. It was concluded that treatments of mild hyperthermia plus docetaxel inhibited the proliferation of human breast cancer cells, promoted apoptosis of breast cancer cells, and produced synergistic antitumor effects. In a clinical setting, some studies [6, 17, 18] showed that neoadjuvant chemotherapy combined with

Table 23.1 Clinical responses

	CR	PR	SD	PD	RR(%)
Total (n = 5)		4	1	0	80.00 %
ER/PgR					
Negative (n = 4)		3	0	0	100.00 %
Positive (n = 1)		1	1	0	50.00 %

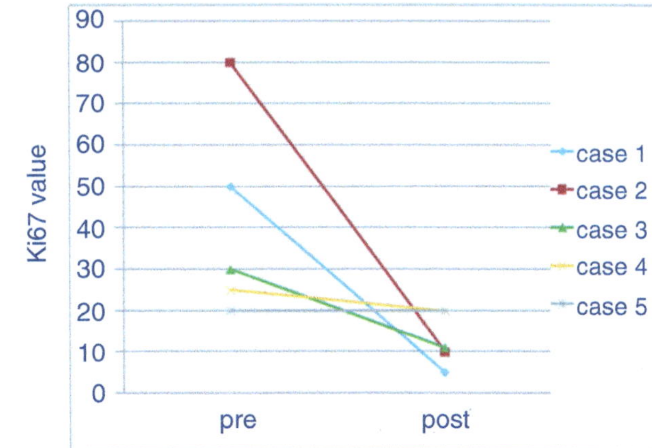

Fig. 23.1 The Ki67 values of pre-and post chemotherapy in 5 patients were shown. The Ki67 values in case1–4 patients decreased, while it in case5 patient did not change

hyperthermia is a feasible and well-tolerated treatment strategy in breast cancer patients.

However, mild hyperthermia remains modestly explored as method for improving the i.t. bioavailability of full intact mobs, such as Herceptin® [19]. The relationship between HER2 overexpression and the response to chemotherapy with trastuzumab is well-known. HER2 overexpression was suggested to be a predictor of the sensitivity to anthracycline chemotherapy [20]. Therefore hyperthermia and chemotherapy with trastuzumab are expected, although in this study, there was no HER2 positive case.

Therefore, hyperthermia combined with chemotherapy with trastuzumab may be useful for breast cancer treatment.

This study has a limitation and it was too small to cover all the characteristics of breast cancer. We are collecting more BC samples in order to increase our sample size and statistical power in future investigations. Especially, there is a possibility high-Ki67 may become predictive factor in clinical response.

The relationship between Ki67 value and the response to needs future investigation.

23.5 Conclusion

Hyperthermia and chemotherapy may be effective as preoperative adjuvant chemotherapy for Japanese women with breast cancer.

Conflict of Interest None.

References

1. Iwase S, Yamamoto D, Kuroda Y, Kawaguchi T, Kitamura K, Odagiri H, Teramoto S, Akazawa K, Nagumo Y. Phase II trial of preoperative chemotherapy for breast cancer: Japan Breast Cancer Research Network (JBCRN)-02 trial. Anticancer Res. 2011;31:1483–7.
2. Chaudry M, Lei X, Gonzalez-Angulo AM, Mittendorf EA, Valero V, Tripathy D, Hortobagyi GN, Chavez-MacGregor M. Recurrence and survival among breast cancer patients achieving a pathological complete response to neoadjuvant chemotherapy. Breast Cancer Res Treat. 2015;153:417–23.
3. Golshan M, Cirrincione CT, Sikov WM, Berry DA, Jasinski S, Weisberg TF, Somlo G, Hudis C, Winer E, Ollila DW. Impact of neoadjuvant chemotherapy in stage II–III triple negative breast cancer on eligibility for breast-conserving surgery and breast conservation rates: surgical results from CALGB 40603 (Alliance).; Alliance for clinical trials in oncology. Ann Surg. 2015;262:434–9.
4. Baulies S, Belin L, Mallon P, Senechal C, Pierga JY, Cottu P, Sablin MP, Sastre X, Asselain B, Rouzier R, Reyal F. Time-varying effect and long-term survival analysis in breast cancer patients treated with neoadjuvant chemotherapy. Br J Cancer. 2015;113:30–6.
5. Kouloulias V, Triantopoulou S, Uzunoglou N, Pistevou-Gompaki K, Barich A, Zygogianni A, Kyrgias G, Kardamakis D, Pectasidis D, Kouvaris J. Hyperthermia is now included in the NCCN clinical practice guidelines for breast cancer recurrences: an analysis of existing data.; Greek Society of Hyperthermic Oncology. Breast Care (Basel). 2015;10:109–16.
6. Yamamoto C, Yamamoto D, Tsubota Y, Sueoka N, Kawakami K, Yamamoto M. The synergistic effect of local microwave hyperthermia and chemotherapy for advanced or recurrent breast cancer. Gan To Kagaku Ryoho. 2014;41:1921–3.
7. Yamamoto D, Inui T, Tsubota Y, Sueoka N, Yamamoto C, Kuwana K, Yamamoto M. The utility of hyperthermia for local recurrence of breast cancer. World J Surg Oncol. 2012;10:201.
8. Cihoric N, Tsikkinis A, van Rhoon G, Crezee H, Aebersold DM, Bodis S, Beck M, Nadobny J, Budach V, Wust P, Ghadjar P. Hyperthermia-related clinical trials on cancer treatment within the clinicaltrials.gov registry. Int J Hyperthermia. 2015;31(6):609–14.
9. Kong G, Braun RD, Dewhirst MW. Hyperthermia enables tumor-specific nanoparticle delivery. Effect of particle size. Cancer Res. 2000;60:4440–5.
10. Matteucci ML, Anyarambhatla G, Rosner G, Azuma C, Fisher PE, Dewhirst MW, Needham D, Thrall DE. Hyperthermia increases accumulation of technetium-99 m-labeled liposomes in feline sarcomas. Clin Cancer Res. 2000;6:3748–55.
11. Griffin RJ, Okajima K, Barrios B, Song CW. Mild temperature hyperthermia combined with carbogen breathing increases tumor partial pressure of oxygen (pO2) and radio-sensitivity. Cancer Res. 1996;56:5590–3.
12. Song CW, Shakil A, Osborn JL, Iwata K. Tumour oxygenation is increased by hyperthermia at mild temperatures. Int J Hyperthermia. 1996;12:367–73.
13. Hehr T, Lamprecht U, Glocker S, Classen J, Paulsen F, Budach W, Bamberg M. Thermoradiotherapy for locally recurrent breast cancer with skin involvement. Int J Hyperthermia. 2001;17:291–301.

14. Vujaskovic Z, Poulson JM, Gaskin AA, Thrall DE, Page RL, Charles HC, MacFall JR, Brizel DM, Meyer RE, Prescott DM. Temperature-dependent changes in physiologic parameters of spontaneous canine soft tissue sarcomas after combined radiotherapy and hyperthermia treatment. Int J Radiat Oncol Biol Phys. 2000;46:179–85.
15. Lv F, Yu Y, Zhang B, Liang D, Li ZM, You WJ. Inhibitory effects of mild hyperthermia plus docetaxel therapy on ER(+/−) breast cancer cells and action mechanisms. Huazhong Univ Sci Technolog Med Sci. 2013;33(6):870–6.
16. Barnes JA, Dix DJ, Collins BW, Luft C, Allen JW. Expression of inducible Hsp70 enhances the proliferation of MCF-7 breast cancer cells and protects against the cytotoxic effects of hyperthermia. Cell Stress Chaperones. 2001;6:316–25.
17. Vujaskovic Z, Kim DW, Jones E, Lan L, McCall L, Dewhirst MW, Craciunescu O, Stauffer P, Liotcheva V, Betof A, Blackwell K. A phase I/II study of neoadjuvant liposomal doxorubicin, paclitaxel, and hyperthermia in locally advanced breast cancer. Int J Hyperthermia. 2010;26:514–21.
18. Craciunescu OI, Thrall DE, Vujaskovic Z, Dewhirst MW. Magnetic resonance imaging: a potential tool in assessing the addition of hyperthermia to neoadjuvant therapy in patients with locally advanced breast cancer. Int J Hyperthermia. 2010;26:625–37.
19. Escoffre JM, Deckers R, Sasaki N, Bos C, Moonen C. Mild hyperthermia influence on herceptin(®) properties. Radiol Oncol. 2015;49(1):41–9.
20. Penault-Llorca F, Abrial C, Mouret-Reynier MA, Raoelfils I, Durando X, Leheurteur M, Gimbergues P, Tortochaux J, Curé H, Chollet P. Achieving higher pathological complete response rates in HER-2-positive patients with induction chemotherapy without trastuzumab in operable breast cancer. Oncologist. 2007;12(4):390–6.

Chapter 24
Combination by Hyperthermia and Chemotherapy: Esophageal Cancer

Takayuki Ohguri

Abstract Chemotherapy has an important role in the treatment of esophageal cancer; e.g., neoadjuvant chemotherapy and concurrent chemoradiotherapy are used for locally advanced cases and for the systemic treatment of recurrent or metastatic disease. Local or regional hyperthermia can enhance the efficacy of systemic chemotherapy in specific areas. In this chapter, hyperthermic chemosensitization is outlined, and the clinical results of chemotherapy with hyperthermia used particularly for the treatment of esophageal cancer are presented.

Keywords Hyperthermia • Chemotherapy • Esophageal cancer

24.1 Hyperthermic Chemosensitization

Hyperthermia causes direct cytotoxicity in cancer and also acts as a radiation- and chemo-sensitizer. Although the combination of radiotherapy and hyperthermia has been the focus of much attention, an equally strong rationale for combining hyperthermia with chemotherapy is also recognized. The cytotoxic effects of numerous anti-cancer drugs, such as cisplatin, carboplatin, docetaxel, paclitaxel, irinotecan, oxaliplatin, and gemcitabine, were enhanced by hyperthermia *in vivo* even at mild temperatures such as 39–41 °C, and interactions were generally seen when the two treatments were administered in close sequence [1, 2].

Various mechanisms for the interaction between chemotherapy and hyperthermia have been proposed. These include increased drug uptake into cells, increased DNA damage, decreased DNA repair, reduced oxygen radical detoxification, and increased membrane damage. Interestingly, some experimental and clinical reports demonstrated that the addition of hyperthermia had the potential ability to reverse the chemotherapeutic drug resistance [3–6].

T. Ohguri (✉)
Department of Radiology, University of Occupational and Environmental Health, 1-1 Iseigaoka, Yahatanishi-ku, Kitakyushu 807-8555, Japan
e-mail: ogurieye@med.uoeh-u.ac.jp

S. Kokura et al. (eds.), *Hyperthermic Oncology from Bench to Bedside*,
DOI 10.1007/978-981-10-0719-4_24

Regional (e.g., pelvic, extremity, and specific organ) or local (e.g., interstitial and intracavitary local heating) hyperthermia can enhance the chemotherapeutic effect at various sites of the body. Regional hyperthermia tends to be less invasive than local hyperthermia. An approach of using systemic chemotherapy with regional hyperthermia should be increasingly utilized because it enhances the efficacy of systemic chemotherapeutic drugs in specific areas without increasing the toxicity. Heating technique for the tumors of the thoracic region is stated in this book at the chapter "Combination by Hyperthermia and Radiation (and Chemotherapy): Lung Cancer".

24.2 Clinical Results of Systemic Chemotherapy and Hyperthermia for Esophageal Cancer

In clinical trials, previous results for chemoradiotherapy have supported a standard of care in curable esophageal cancers [7]. Most failures of this combined modality are locoregional. The local recurrence rate at the first site was approximately 50 % [8]. There are several reports that endoscopic mucosal resection (EMR) could be a salvage option, although the application of salvage EMR is limited to an earlier superficial stage of the tumor [9]. Esophagectomy has a curative potential for the locoregional failure; however, surgical resection after definitive chemoradiotherapy is associated with high morbidity [10]. The treatment strategy is still not established. Systemic chemotherapy with regional hyperthermia may be an ideal treatment because hyperthermia can enhance the effects of systemic therapy for the recurrent or metastatic areas that had already been irradiated.

There are only a few clinical experiences of using a combination of systemic chemotherapy and hyperthermia [11]. Nishimura et al. reported the use of systemic chemotherapy combined with regional hyperthermia for residual or recurrent esophageal cancer after definitive chemoradiotherapy; an 8-MHz radiofrequency capacitive heating device was used, and patients received the combined chemotherapy of cisplatin/5-fluorouracil (cisplatin of 5 mg/m^2 on day 1 and day 15 and daily 5FU of 300 mg/m^2 per day), an oral fluoropyrimidine (daily 80 mg/m^2 for 4 weeks), or irinotecan (weekly 30 mg/m^2) [11]. Complete response was seen in 3 of 11 patients without severe toxicity. Thus, they concluded that the combined therapy was a feasible and potent salvage therapy for those patients if salvage surgery was not indicated.

In 2010, the valuable results of a large randomized phase III study for the addition of regional hyperthermia to systemic chemotherapy were released. 361 patients with localized high-risk soft-tissue sarcoma were enrolled [12]. The addition of regional hyperthermia to systemic chemotherapy significantly improved the local, progression-free, disease-free survival and the tumor response rate. These results encourage the future planning of clinical studies for systemic chemotherapy plus regional hyperthermia in patients with esophageal cancer.

References

1. Mohamed F, Marchettini P, Stuart OA, Urano M, Sugarbaker PH. Thermal enhancement of new chemotherapeutic agents at moderate hyperthermia. Ann Surg Oncol. 2003;10(4):463–8.
2. Jones EL, Samulski TV, Vujaskovic Z, Leonard RP, Dewhirst MW. Hyperthermia. Principles and practice of radiation oncology. 4th ed. Philadelphia: Lippincott Williams & Wilkins; 2003.
3. Hettinga JV, Konings AW, Kampinga HH. Reduction of cellular cisplatin resistance by hyperthermia – a review. Int J Hyperthermia. 1997;13(5):439–57.
4. Da Silva VF, Feeley M, Raaphorst GP. Hyperthermic potentiation of BCNU toxicity in BCNU-resistant human glioma cells. J Neurooncol. 1991;11(1):37–41.
5. Ohguri T, Imada H, Narisada H, Yahara K, Morioka T, Nakano K, et al. Systemic chemotherapy using paclitaxel and carboplatin plus regional hyperthermia and hyperbaric oxygen treatment for non-small cell lung cancer with multiple pulmonary metastases: preliminary results. Int J Hyperthermia. 2009;25(2):160–7. doi:910057552 [pii] 10.1080/02656730802610357.
6. Ohguri T, Imada T, Korogi Y, Narisada H. Clinical results of systemic chemotherapy combined with regional hyperthermia. Thermal Med. 2007;23(2):49–61.
7. Koshy M, Esiashvilli N, Landry JC, Thomas Jr CR, Matthews RH. Multiple management modalities in esophageal cancer: combined modality management approaches. Oncologist. 2004;9(2):147–59.
8. Roussel A, Haegele P, Paillot B, Gignoux M, Marinus A, Sahmoud T, et al. Results of the EORTC-GTCCG phase III trial of irradiation vs irradiation and CDDP in inoperable esophageal cancer. Proc Am Soc Clin Oncol. 1994;13:199.
9. Yano T, Muto M, Hattori S, Minashi K, Onozawa M, Nihei K, et al. Long-term results of salvage endoscopic mucosal resection in patients with local failure after definitive chemoradiotherapy for esophageal squamous cell carcinoma. Endoscopy. 2008;40(9):717–21. doi:10.1055/s-2008-1077480.
10. Swisher SG, Wynn P, Putnam JB, Mosheim MB, Correa AM, Komaki RR, et al. Salvage esophagectomy for recurrent tumors after definitive chemotherapy and radiotherapy. J Thorac Cardiovasc Surg. 2002;123(1):175–83. doi:S0022522302495116 [pii].
11. Nishimura S, Saeki H, Nakanoko T, Kasagi Y, Tsuda Y, Zaitsu Y, et al. Hyperthermia combined with chemotherapy for patients with residual or recurrent oesophageal cancer after definitive chemoradiotherapy. Anticancer Res. 2015;35(4):2299–303.
12. Issels RD, Lindner LH, Verweij J, Wust P, Reichardt P, Schem BC, et al. Neo-adjuvant chemotherapy alone or with regional hyperthermia for localised high-risk soft-tissue sarcoma: a randomised phase 3 multicentre study. Lancet Oncol. 2010;11(6):561–70. doi:S1470-2045(10)70071-1 [pii] 10.1016/S1470-2045(10)70071-1.

Chapter 25
Combination with Hyperthermia and Chemotherapy: Liver Cancer

Satoshi Kokura

Abstract In liver cancer, there are two kinds of liver cancer. One is a primary hepatocellular carcinoma and the other is metastatic liver cancer.

As for the out-of case of a surgery indication of these cancers, administrations into the hepatic artery of an anticancer agent, an arterial-embolism treatment (TACE), RFA, transient embolization by Degradable Starch Microspheres with anticancer agent or systemic chemotherapy are performed.

At present, liver resection and transplantation can improve the survival rate of HCC, but long waiting time due to the shortage of donor organs may result in tumor progression. Given these limitations, many nonsurgical methods have been proposed. However, in each monotherapy of these, the curative effect is restrictive and a multidisciplinary treatment is needed.

In these cases, the combined effect of hyperthermia is very high.

Keywords Chemoembolization • Systemic chemotherapy • Nanoparticle • Immunotherapy

25.1 Hepatocellular Carcinoma

Hepatocellular carcinoma (HCC) is the forth of male and sixth of female leading tumor in Japan, and it is estimated that its incidence will continue to rise in coming decades [1, 2]. Approximately 85 % of HCC is accompanied by cirrhosis, which works together with HCC leading to liver functional failure [3]. The pathogenesis for HCC is rather complicated since many risk factors are involved. The some risk factors are hepatitis B and aflatoxin in East Asia and Central Africa account for about 80 % of all cases [4–6], whereas hepatitis C and alcohol are risk factors in North America, Europe and Japan [7, 8]. The risk of carcinogenesis further increases with progression to hepatic cirrhosis in all liver disorders. Hepatic

S. Kokura (✉)
Kokura Lab. Faculty of Health and Medical Sciences, Kyoto Gakuen University, 18, Gotanda-cho, Yamanouchi, Ukyo-ku, Kyoto-shi, Kyoto 615-8577, Japan
e-mail: kokura@kyotogakuen.ac.jp

S. Kokura et al. (eds.), *Hyperthermic Oncology from Bench to Bedside*,
DOI 10.1007/978-981-10-0719-4_25

cirrhosis is an irreversible pathological change and inhibition of disease progression has previously been considered difficult.

At present, liver resection and transplantation can improve the survival rate of HCC, but long waiting time due to the shortage of donor organs may result in tumor progression. Given these limitations, many nonsurgical methods have been proposed.

25.2 Treatment Argorism of Hepatocellular Carcinoma

The Consensus-Based Clinical Practice Guidelines had proposed by the Japan Society of Hepatology (JSH) at 2010 version in Japan, and recently they have revised, updated (Fig. 25.1). No new treatments or molecular targeted agents have been developed for HCC since the 2010 JSH consensus-based treatment algorithm [9, 10] was adopted, so few changes were made to 2014 version. Although sorafenib is recommended for patients with minor portal vein invasion or portal invasion at the first portal branch (Vp1-3), the new algorithm reflects the consensus that it is not recommended for patients with portal invasion at the main portal branch (Vp4) due to the risk of hepatic failure. However, hepatic arterial infusion chemotherapy (HAIC) is still strongly recommended for patients with Vp4, and therefore recommendations regarding HAIC were left unchanged [11]. Moreover, because locoregional therapy for Child-Pugh C patients is now widely used and many

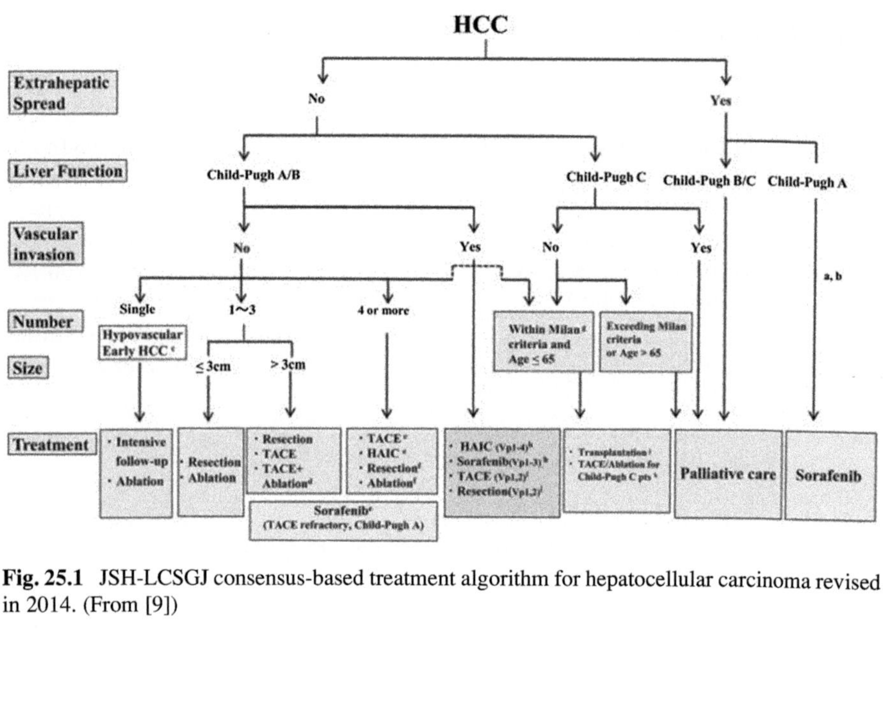

Fig. 25.1 JSH-LCSGJ consensus-based treatment algorithm for hepatocellular carcinoma revised in 2014. (From [9])

studies have reported its survival benefits, it is now described as a "well accepted treatment" rather than an "experimental treatment" in the revised algorithm.

25.3 Trans-Arterial Chemo-Embolozation (TACE)

Transarterial chemoembolization (TACE) is the current standard of care for patients with large or multinodular hepatocellular carcinoma (HCC). The rationale for TACE is that the intra-arterial injection of a chemotherapeutic drug such as doxorubicin or cisplatin followed by embolization of the blood vessel will result in a strong cytotoxic effect enhanced by ischemia [12]. The embolization end point is usually defined as stasis in the second- or third-order branches of the lobar hepatic artery and injection should be continued until near stasis is observed in the artery directly feeding the tumor. About the indication of TACE, patients should present a relatively well preserved liver function, defined as Child-Pugh (CP) $\leq$ B stage without ascites according to JSH-LCSG (Liver Cancer Study Group of Japan) guidelines [13].

Absolute contraindications to TACE are generally related to decompensated cirrhosis or impaired portal blood flow [14]. About the treatment schedule, current evidence suggests that one cycle of TACE may not be sufficient for effective treatment of intermediate-stage HCC. On the other hand, there is evidence suggesting that repeating TACE prolongs survival; however, current guidelines do not specify the criteria for treatment repetition. In particular, it should be noted that in bilobar tumors, the two hepatic lobes usually have to be treated in separate treatment sessions 2–4 weeks apart. However, there is evidence suggesting that the repetition of TACE with an aggressive schedule increases the incidence of adverse events [15]. Therefore, the experts in the field propose the on-demand repetition with longer intervals between treatments, rather than a regular predefined schedule [16, 17]. This has been recently confirmed by Terzi et al [18] in a series of 151 patients treated with on-demand conventional TACE (cTACE). In clinical practice, we found that tumor necrosis rate is low for patients with HCC received TACE, and the recurrence rate is high [19]. HCC received TACE could induce vascular endothelial growth factor which could promote vascular growth factor expression, increasing the risk of tumor recurrence and distant metastasis [20, 21]. And TACE has limitations, may not apply to all patients with liver function status, only the Child – Pugh grade A or B of the patients can be implemented. However, combined TACE and local hyperthermia is feasible and tolerable. We feel that this regimen would be a new promising modality in unresectable HCC. Further study is required to compare the therapeutic efficacy of this regimen to TACE alone.

25.4 Transient Embolization by Degradable Starch Microspheres (DSM)

We have used a new method of chemoembolization with degradable starch microspheres (DSM) for patients with malignant hepatic tumors. DSM, 40–45 μm in diameter, which are degraded by serum amylase, temporarily obstruct arterial blood flow at the arteriolar capillary bed. Experimental studies have demonstrated that such occlusion enhances the regional uptake and reduces systemic exposure to simultaneously administered arterial anticancer drugs. Transcatheter chemoembolization with DSM was performed in 44 cases (18 cases were received the combination with hyperthermia) of hepatocellular carcinoma. Doxorubicin mixed with DSM was injected into the patients with hepatocellular carcinoma through the proper hepatic artery. The therapeutic effect of this chemoembolization was evaluated by the change in tumor size measured by angiography or computed tomography. In hepatocellular carcinoma, tumor regression of over 50 % was observed in 11 of 26 patients. Elevated serum AFP level of more than 200 ng/ml was decreased in all 26 cases. Although half of the patients had transient pain within 2 h, no major side effects such as bone marrow suppression and hepatotoxicity were observed. Our results suggest that chemoembolization with DSM can be effectively used in the treatment of malignant hepatic tumors.

25.5 Hyperthermia Combined with DSM

Forty-eight cases with non-resectable hepatocellular carcinoma were examined. Chemoembolization using degradable starch microspheres (DSM) was performed in 26 cases. DSM, 40–45 μm in diameter, which are degraded by serum amylase, temporarily obstruct arterial blood flow at capillary bed. Doxorubicin mixed with DSM was injected into patients through the proper hepatic artery. Hyperthermia (8 MHz radiofrequency, Thermotron RF-8® which was made by Yamamoto VINITA Co., Ltd., Osaka, Japan) combined with chemoembolization was performed in 18 cases. Only hyperthermia treatment was performed 4 cases. In 8 cases treated by hyperthermia combined with chemoembolization, the intratumoral temperature was measured by a thermocouple thermometer during heating alone and heating after injection of DSM (Fig. 25.2). The therapeutic effect was evaluated by the change in tumor size measured by angiography or computed tomography. The efficacy of hyperthermia combined with chemoembolization was compared with that of chemoembolization alone. Intratumoral temperature was 1.0 ° C higher by heating after injection of DSM than by heating alone (Table 25.1). Partial response (tumor regression of over 50 %) was observed in 11 of 26 cases (42 %) with chemoembolization alone. Partial response was observed in 10 of 18 cases (67 %) with hyperthermia combined with chemoembolization (Table 25.2). One-year survival rate was 48 % in chemoembolization alone, against 88 % in

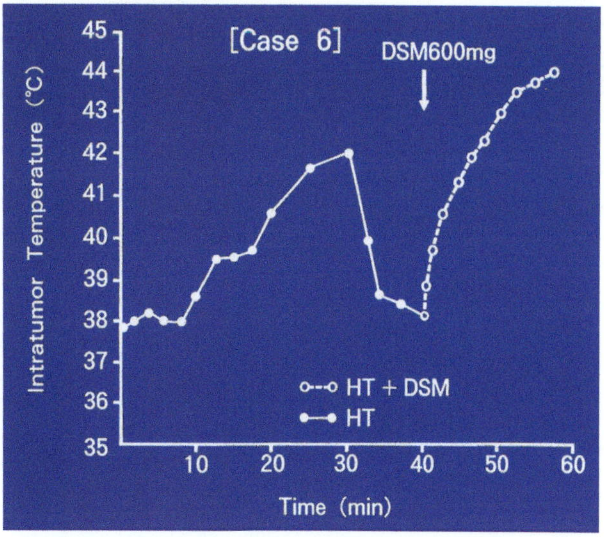

Fig. 25.2 A sequential change of the intratumor temperature during hyperthermia: If DSM is used, a heating efficiency become high

Table 25.1 Intratumor temperature of Hyperthermia alone and Hyperthermia with DSM

CASE	Intratumor maximal temperature	
	Hyperthermia alone	Hyperthermia with DSM
1	42.1 °C	43.0 °C
2	41.2 °C	42.0 °C
3	42.0 °C	42.0 °C
4	40.8 °C	42.8 °C
5	40.2 °C	40.6 °C
6	42.0 °C	43.9 °C
7	42.1 °C	43.1 °C
8	40.1 °C	41.1 °C
Mean	41.3 °C	42.3 °C

Table 25.2 Anti-tumor effect

Treatment method	Tumor reduction effect				Tumor regression of over 50 %
	CR	PR	NC	PD	
Hyperthermia alone	0	0	4	0	0 %(0/4)
Chemo-embolization	2	9	15	0	42 %(11/26)
Hyperthermia + Chemo-embolization	2	8	8	0	56 %(10/18)

hyperthermia combined with chemoembolization. And 2-year survival rate was 48 % in chemoembolization alone, against 68 % in hyperthermia combined with chemoembolization. (Fig.25.3) Our results suggest that hyperthermia combined

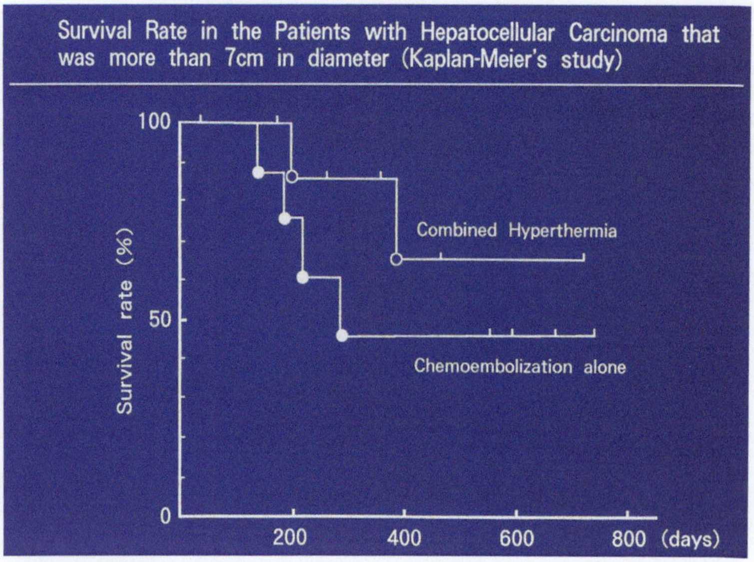

Fig. 25.3 Survival curve

with chemoembolization using DSM is effective in the treatment of hepatocellular carcinoma. Nagata Y et al. also investigated the effect of hyperthermia against for hepatocellular carcinoma. One hundred forteen patients with hepatocellular carcinoma treated between 1983 and 1995 underwent hyperthermia. Eight-megahertz RF capacitive heating equipment was used for the hyperthermia. Two opposing 25-cm electrodes were generally used for heating the liver tumors. Their standard protocol was to administer hyperthermia 40–50 min twice a week for a total of eight sessions. The liver tumor temperature was measured by microthermocouples when possible. Transcatheter arterial embolization, radiotherapy, immunotherapy, and chemotherapy were combined with hyperthermia treatment in accordance with each patient's liver function. The maximum tumor temperature, average tumor temperature, and minimum tumor temperature in the HCC were (mean +/− standard error) 41.2 +/− 0.2 ° C, 40.3 +/− 1.3 ° C, and 40.1 +/− 0.2 ° C, respectively. Of the 73 cases with HCC who were evaluated by computed tomography (CT), CR was achieved in 7 (10 %), PR in 15 (21 %), NR in 37 (51 %), and PD in 14 (19 %). The 1-year cumulative survival rate for HCC patients was 30.0 %, and the 5-year survival rate was 17.5 %. The sequelae of hyperthermia included focal fat necrosis in 20 patients (12 %), gastric ulceration in 4 (2 %), and liver necrosis in 1 (1 %). They concluded that Even though the thermometry results for liver tumors were not satisfactory, the treatment results are promising [22].

25.6 New Treatment Combined with Hyperthermia

Conventional strategies have failed to provide relevant survival benefit and are too toxic in the majority of patients who also have concomitant underlying hepatic dysfunction [23]. Challenges in advanced HCC treatment, therefore, exemplify the currently unmet need for novel, less toxic, and more effective strategies.

25.6.1 Nanoparticle + Hyperthermia

Cancer cell targeted drug delivery that minimizes toxicity and maximizes therapeutic benefit remains a fundamental challenge for successful cancer therapy. Steven A. Curly et al. investigated to devise a EGFR-targeted non-invasive chemo-thermo-therapy that can be readily translated to the clinic by taking advantage of recent advances in the understanding of (a) targeted drug delivery and (b) thermal therapy of cancers. The premise of such work is based on the synergistic interaction of conventional cytotoxic chemotherapeutic agents and hyperthermia [24]. Challenges in advanced HCC treatment, therefore, exemplify the currently unmet need for novel, less toxic, and more effective strategies. Prior *in vitro* and *in vivo* studies have demonstrated that this RF field-based nano-hyperthermia system is safe and effective in a rabbit model of liver cancer after direct intratumoral injection of single-walled carbon nanotubes, and demonstrated efficacy without toxicity after systemic delivery of EGFR-targeted gold nanoparticles (AuNPs) in a mouse model of pancreatic adenocarcinoma [25]. AuNPs are most extensively studied because they are already in clinical use and their physiochemical properties allow for a multitude of possible chemical linkages to targeting agents. For instance, Patra et al. successfully delivered gemcitabine non-covalently conjugated to AuNPs along with a targeting antibody (cetuximab, C225) against epidermal growth factor receptor (EGFR) in pancreatic adenocarcinoma cells [26]. According to these findings, I hypothesized that the nanoparticles bearing a targeting antibody and a chemotherapeutic agent could be systemically delivered to human HCC. Subsequent non-invasive RF induced-hyperthermia would then enhance tumor cell death because of the presence of drug and/or nanoparticles. Steven A. Curly et al. used cetuximab (C225) because surface EGFR-1 over-expression is associated with late stage disease and it is found in 66 % of human HCC tumors, making it a potential target [27]. It is also approved for use in other solid tumors.

25.6.2 Cytokine Induced Killer Cell + Hyperthermia

Adoptive cellular immunotherapy has been applied in the clinical treatment of advanced HCC due to the close relationship between the pathogenesis of HCC and the autoimmune system, and the persistence of pathogenic factors such as hepatitis and cirrhosis [28, 29]. Cellular immunotherapy includes several types of immunological cells, such as lymphokine-activated killer cells, tumor infiltrating lymphocytes and cytokine-induced killer (CIK) cells. CIK cells have been confirmed to have potential for immunotherapy against residual tumor cells. In recent years, several authors have reported that CIK cells were a heterogeneous population, and the major population to express both T cell marker monoclonal antibody and natural killer cell marker monoclonal antibody [30]. Cells with this phenotype are rare (1–5 %) in natural peripheral blood mononuclear cells. CIK cells are able to expand nearly 1000-fold when cultured in a cytokine cocktail, comprising interleukin (IL)-1α, interferon-γ (IFN-γ), IL-2 and mAbs against CD3, and have characteristics which are more effective in the treatment of tumors with a non-major histocompatibility complex (MHC)-restricted mechanism [31]. Thus, CIK cells may have some benefit in potential immunotherapeutic treatments in patients with HCC. Thus, CIK cells may have some benefit in potential immunotherapeutic treatments in patients with HCC. Hyperthermia treatment of tumors refers to tumor tissue, which is treated with a continuous direct current through two or more electrodes placed outside the tumor. Some researchers have suggested that hyperthermia can normalize cell growth, and accelerate cell division after inhibiting cell division when it becomes abnormally accelerated [30]. Hyperthermia is an alternative treatment for HCC patients and has a positive effect on tumors. I show here one clinical study, which was investigated the effect and tolerance of intraperitoneal perfusion of cytokine-induced killer (CIK) cells in combination with local radio frequency (RF) hyperthermia in patients with advanced primary hepatocellular carcinoma (HCC) [31]. Patients with advanced primary HCC were included in this study. CIK cells were perfused intraperitoneal twice a week, using 3.2×10^9–3.6×10^9 cells each session. Local RF hyperthermia was performed 2 h after intraperitoneal perfusion. Following an interval of 1 month, the next course of treatment was administered. Patients received treatment until disease progression. Tumor size, immune indices (CD3+, CD4+, CD3 + CD8+, CD3 + CD56+), alpha-fetoprotein (AFP) level, abdominal circumference and adverse events were recorded. Time to progression and overall survival (OS) were calculated. From June 2010 to July 2011, 31 patients diagnosed with advanced primary HCC received intraperitoneal perfusion of CIK cells in combination with local RF hyperthermia in their study. Patients received an average of 4.2 ± 0.6 treatment courses (range, 1–8 courses). Patients were followed up for 8.3 ± 0.7 months (range, 2–12 months). Following combination treatment, CD4+, CD3 + CD8+ and CD3 + CD56+ cells increased respectively. AFP decreased significantly and abdominal circumference decreased. The disease control rate was 67.7 %. The most common adverse events were low fever and slight abdominal erubescence, which resolved without treatment. The median time to progression was 6.1 month. The 3-, 6- and 9-months and 1-year survival rates were 93.5 %, 77.4 %, 41.9 % and 17.4 %, respectively. The median OS was 8.5 months. The most common causes of death in patients with advanced HCC are liver failure, ascites and obstruction. Control of the abdominal tumor is the principal goal of treatment, which can prolong survival and improve quality of life. Recently, intravenous infusion of CIK cells has been widely used in the treatment of HCC. However, the concentration of CIK cells is low in tumor tissues following intravenous infusion, and the anti-tumor effect is low. In this study, in order to achieve a high concentration of CIK cells in the abdomen, CIK cells were perfused intraperitoneally instead of intravenously. Intraperitoneal perfusion of CIK cells in combination with local RF hyperthermia can improve efficacy in clinical practice. The primary reason for this may be that hyperthermia accelerates the conjugation of CIK cells and tumor cells, which improves the reaction rate of CIK cells. Local RF hyperthermia after high-capacity perfusion of CIK cells improves the sensitivity of tumor cells, allowing CIK cells to permeate into tumor cells more effectively.

Local RF hyperthermia kills tumor cells by heating using physical energy and the temperature of tumor tissue rises to an effective treatment temperature. Studies have shown that high temperature can damage the membranes of mitochondria, lysosomes and endoplasmic reticulum and triggers the massive release of acid hydrolase from lysosomes, resulting in membranolysis, outflow of cytoplasm and death of cancer cells [32, 33]. In addition, some studies have reported that hyperthermia improves immune function by stimulating the development of the anti-tumor immune effect and resolving the inhibition of blocking factors in the immune system [34]. The core points behind this treatment were to administer intraperitoneal perfusion of CIK cells to improve immune function to kill the tumor, and local hyperthermia to enhance the concentration of CIK cells which directly affect liver tumor tissues and small lesions on the abdominal wall, which can damage cancer cells and effectively reduce abdominal dropsy. Local hyperthermia plays an important role in enhancing the concentration of CIK cells, which directly affect these parameters. In conclusion, intraperitoneal perfusion of CIK cells in combination with local RF hyperthermia is safe and effective for advanced HCC. More clinical trials with a large sample size are warranted to provide evidence for further applications. The addition of palliative chemotherapy or targeted therapy is worthy of further investigation.

25.7 Metastatic Liver Cancer (Mainly from Colon Cancer)

Metastatic disease to the liver is present at the time of diagnosis in 20 % of patients with colorectal cancer and develops in an additional 40 % over the course of their disease. In 30 % of patients, the liver is the only site of metastatic disease. Liver resection offers the best chance of cure for patients with liver metastases with 5-year survivals of up to 50 %. Unfortunately, only up to 25 % of eligible patients undergo resection because of co-morbid conditions. First line chemotherapy with oxaliplatin or irinotecan and the addition of a biologic agent have increased median survival time to 18–21 months [35]. However, after failing first line chemotherapy, the response rate to second line agents ranges from about 20–35 % [36, 37]. For these reasons, regional or liver-directed therapies are important for treatment of liver-predominant metastatic colorectal cancer.

Most effective liver-directed therapies are hyperthermia [22] or percutaneous radiofrequency ablation [38].

25.7.1 Chemoembolization by DSM or TACE

Since two decades transarterial chemoembolization (TACE) of liver metastases has been investigated in numerous studies. However, no standardized therapeutic procedure exists so far. Wasser K et al. [39] retrospectively investigated survival, response and side effects after TACE of liver metastases in 21 patients with colorectal cancer and results are compared with previous literature. As already shown in most previous studies, regardless of the used agents, also this investigation underlines the moderate therapeutic effect of TACE on colorectal liver metastases especially anti-tumor effect. So far, no significant survival benefit has been shown in the literature and the response rates are rather limited.

25.7.2 Hyperthermia Combined with Chemoembolization by DSM

We investigated anti-tumor effects of chemoembolization with degradable starch microspheres (DSM) combined with regional hyperthermia in patients with Hepatocellular carcinoma (48 cases) and in patients with metastatic liver cancer (25 cases). Chemoembolization with DSM was performed in 25 cases of metastatic liver cancer between 1985 and 1988. 3 cases of metastatic liver cancer were treated with the combination of hyperthermia. A catheter was placed in the proper hepatic artery via the trans-femoral approach. Mitomycin C mixed with DSM was injected every 2 or 3 weeks through the catheter. Thermotron RF-8® (which was made by Yamamoto VINITA Co., Ltd., Osaka, Japan), the heating device used in this study, is operated at 8 MHz radiofrequency. Hyperthermia treatment was applied twice a week. The therapeutic effect of this treatment was evaluated by the change in tumor size measured by angiography or computed tomography. In the patients with metastatic liver cancer, tumor regression over 33 % was observed in 65 % of the patients treated with chemoembolization alone and in 50 % of those with chemoembolization and HT. One year survival rate in the patients with metastatic liver cancer was 41 and 55 % in the treatment with chemoembolization alone, and chemoembolization and HT. These results suggest that chemoembolization using DSM combined with hyperthermia was markedly effective in the patients with metastatic liver cancer. Of course, the number of patients were very few. However, Nagata Y et al.[22] reported that more large study in 1997. Forty seven patients with metastatic liver cancers were treated between 1983 and 1995 combined or non combined hyperthermia. The 47 cases were metastatic liver cancer. Eight-megahertz RF capacitive heating equipment (Thermotron RF-8® which was made by Yamamoto VINITA Co., Ltd., Osaka, Japan) was used for the hyperthermia. Two opposing 25-cm electrodes were generally used for heating the liver tumors. Their standard protocol was to administer hyperthermia 40–50 min twice a week for a total of eight sessions. Transcatheter arterial embolization, radiotherapy, immunotherapy, and chemotherapy were combined with hyperthermia treatment in accordance with each patient's liver function. 38 patients who underwent more than four sessions of hyperthermia were evaluated in this study. Of the 47 cases involving metastatic liver cancer evaluated by CT, CR was achieved in 3 (7 %), PR in 17 (38 %), NR in 12 (27 %), and PD in 13 (29 %). The 1-year survival of metastatic liver cancer patients was 32.5 %, and the longest survival was 30 months. Nagata et al [22] concluded that the treatment results were promising. Further clinical trials of RF capacitive hyperthermia for the treatment of metastatic liver cancer should be encouraged.

25.7.3 *Systemic Chemotherapy Combined with Hyperthermia*

More than 50 % of patients with colorectal cancer develop metastasis, and about 20 % of patients have liver metastases at diagnosis. Surgery is currently the only curative treatment for metastases from colorectal cancer, but only 10–25 % of patients are suitable for resection. Even after curative surgery, 5-year survival rates are only 25–50 % in the most recent reports. There is no standard therapeutic approach after resection of metastases, and new strategies are needed to improve patient outcome. Adjuvant chemotherapy is a logical treatment for suspected micrometastases. Many previous studies have focused on intra-arterial chemotherapy, with or without systemic chemotherapy. This approach showed a 2-year survival benefit in only one randomized trial, in which intra-arterial and systemic chemotherapy was combined [40]. Intra-arterial chemotherapy requires highly trained surgical and medical oncologists, limiting its use in many centers. Recent schedules combining fluorouracil (FU) and oxaliplatin or irinotecan have considerably improved the results of systemic chemotherapy in the palliative treatment of patients with metastatic colorectal cancer, in terms of the response rate, progression-free survival, and overall survival. The use of new schedules, such as oxaliplatin, leucovorin (LV), and FU (FOLFOX), and irinotecan, LV, and FU (FOLFIRI), in addition to surgery, is thus a promising approach for resectable metastatic colorectal cancer. The use of monoclonal antibodies in first-line treatment of patients with colorectal cancer is justified in patients with the wild-type KRAS gene than in patients with KRAS mutations. Further trials of these drugs are warranted as second-line treatment for patients with KRAS wild-type tumors. As mentioned above, for the metastatic liver cancer, systemic chemotherapy is very common. But in this case, hyperthermia combined with chemotherapy is effective [41, 42]. The best timing of hyperthermia is same time treatment with chemotherapy. But if impossible, hyperthermia is performed immediately after systemic chemotherapy.

References

1. El-Serag HB, Mason AC. Rising incidence of hepatocellular carcinoma in the United States. N Engl J Med. 1999;340:745–50. doi:10.1056/NEJM199903113401001.
2. Bosch FX, Ribes J, Díaz M, Cléries R. Primary liver cancer: worldwide incidence and trends. Gastroenterology. 2004;127:S5–16. doi:10.1053/j.gastro.2004.09.011.
3. Nakayama H, Takayama T. Management before hepatectomy for hepatocellular carcinoma with cirrhosis. World J Hepatol. 2015;7(20):2292–302. doi:10.4254/wjh.v7.i20.2292.
4. Tejeda-Maldonado J, García-Juárez I, Aguirre-Valadez J, González-Aguirre A, Vilatobá-Chapa M, Armengol-Alonso A, Escobar-Penagos F, Torre A, Sánchez-Ávila JF, Carrillo-Pérez DL. Diagnosis and treatment of hepatocellular carcinoma: an update. World J Hepatol. 2015;7:362–76.
5. Forner A, Llovet JM, Bruix J. Hepatocellular carcinoma. Lancet. 2012;379:1245–55.
6. Trépo C, Chan HL, Lok A. Hepatitis B virus infection. Lancet. 2014;384:2053–63.
7. Webster DP, Klenerman P, Dusheiko GM. Hepatitis C. Lancet. 2015;385:1124–35.
8. Echeverría N, Moratorio G, Cristina J, Moreno P. Hepatitis C virus genetic variability and evolution. World J Hepatol. 2015;7:831–45.
9. Kudo M, Izumi N, Kokudo N, Matsui O, Sakamoto M, Nakashima O, Kojiro M, Makuuchi M. HCC expert panel of Japan society of hepatology management of hepatocellular carcinoma in Japan: consensus-based clinical practice guidelines proposed by the Japan society of hepatology (JSH) 2010 updated version. Dig Dis. 2011;29:339–64.
10. Arii S, Sata M, Sakamoto M, Shimada M, Kumada T, Shiina S, Yamashita T, Kokudo N, Tanaka M, Takayama T, Kudo M. Management of hepatocellular carcinoma: report of consensus meeting in the 45th annual meeting of the Japan society of hepatology (2009). Hepatol Res. 2010;40:667–85.
11. Kudo M. Treatment of advanced hepatocellular carcinoma with emphasis on hepatic arterial infusion chemotherapy and molecular targeted therapy. Liver Cancer. 2012;1:62–70.
12. Lencioni R, Petruzzi P, Crocetti L. Chemoembolization of hepatocellular carcinoma. Semin Intervent Radiol. 2013;30:3–11.
13. Masatoshi Kudo, Osamu Matsui, Namiki Izumi, Hiroko Iijima, Masumi Kadoya, Yasuharu Imai, Takuji Okusaka, Shiro Miyayama, Kaoru Tsuchiya, Kazuomi Ueshima, Atsushi Hiraoka, Masafumi Ikeda, Sadahisa Ogasawara, Tatsuya Yamashita, Tetsuya Minami, Koichiro Yamakado, and on behalf of the Liver Cancer Study Group of Japan. JSH consensus-based clinical practice guidelines for the management of hepatocellular carcinoma: 2014 Update by the Liver Cancer Study Group of Japan. Liver Cancer. 2014; 3(3–4): 458–468. doi:10.1159/000343875.
14. Bruix J, Sherman M. Management of hepatocellular carcinoma: an update. Hepatology. 2011;53:1020–2.

15. Ernst O, Sergent G, Mizrahi D, Delemazure O, Paris JC, L'Herminé C. Treatment of hepatocellular carcinoma by transcatheter arterial chemoembolization: comparison of planned periodic chemoembolization and chemoembolization based on tumor response. AJR Am J Roentgenol. 1999;172:59–64.
16. Raoul JL, Sangro B, Forner A, Mazzaferro V, Piscaglia F, Bolondi L, Lencioni R. Evolving strategies for the management of intermediate-stage hepatocellular carcinoma: available evidence and expert opinion on the use of transarterial chemoembolization. Cancer Treat Rev. 2011;37:212–20.
17. Bolondi L, Burroughs A, Dufour JF, Galle PR, Mazzaferro V, Piscaglia F, Raoul JL, Sangro B. Heterogeneity of patients with intermediate (BCLC B) Hepatocellular Carcinoma: proposal for a subclassification to facilitate treatment decisions. Semin Liver Dis. 2012;32:348–59.
18. Terzi E, Golfieri R, Piscaglia F, Galassi M, Dazzi A, Leoni S, Giampalma E, Renzulli M, Bolondi L. Response rate and clinical outcome of HCC after first and repeated cTACE performed "on demand". J Hepatol. 2012;57:1258–67.
19. Burrel M, Reig M, Forner A, Barrufet M, de Lope CR, Tremosini S, Ayuso C, Llovet JM, Real MI, Bruix J. Survival of patients with hepatocellular carcinoma treated by transarterial chemoembolization (TACE) using Drug Eluting Beads. Implications for clinical practice and trial design. J Hepatol. 2012;56(6):1330–5. doi:10.1016/j.jhep.2012.01.008. Epub 2012 Feb 5.
20. Llovet JM, Bru C, Bruix J. Prognosis of hepatocellular carcinoma: the BCLC staging classification. Semin Liver Dis. 1999;19:329–38.
21. Bruix J, Sherman M, Llovet JM, Beaugrand M, Lencioni R, Burroughs AK, Christensen E, Pagliaro L, Colombo M, Rodés J. Clinical management of hepatocellular carcinoma: conclusions of the Barcelona-2000 EASL conference. European Association for the study of the Liver. J Hepatol. 2001;35:421–30.
22. Nagata Y, Hiraoka M, Nishimura Y, Masunaga S, Mitumori M, Okuno Y, Fujishiro M, Kanamori S, Horii N, Akuta K, Sasai K, Abe M, Fukuda Y. Clinical results of radiofrequency hyperthermia for malignant liver tumors. Int J Radiat Oncol Biol Phys. 1997;38(2):359–65.
23. Llovet JM, Ricci S, Mazzaferro V, Hilgard P, Gane E, Blanc JF, de Oliveira AC, Santoro A, Raoul JL, Forner A, Schwartz M, Porta C, Zeuzem S, Bolondi L, Greten TF, Galle PR, Seitz JF, Borbath I, Häussinger D, Giannaris T, Shan M, Moscovici M, Voliotis D, Bruix J. SHARP investigators study group. Sorafenib in advanced hepatocellular carcinoma. N Engl J Med. 2008;359(4):378–90. doi:10.1056/NEJMoa0708857.
24. Mustafa Raoof MD, Corr SJ, Zhu C, Cisneros BT, Kaluarachchi WD, Phounsavath S, Wilson LJ, Curley SA. Gold nanoparticles and radiofrequency in experimental models for hepatocellular carcinoma. Nanomedicine. 2014;10(6):1121–30. doi:10.1016/j.nano.2014.03.004.
25. Gannon CJ, Cherukuri P, Yakobson BI, Cognet L, Kanzius JS, Kittrell C, et al. Carbon nanotube-enhanced thermal destruction of cancer cells in a noninvasive radiofrequency field. Cancer. 2007;110(12):2654–65.
26. Patra CR, Bhattacharya R, Wang EF, Katarya A, Lau JS, Dutta S, et al. Targeted delivery of gemcitabine to pancreatic adenocarcinoma using cetuximab as a targeting agent. Cancer Res. 2008;68(6):1970–8.
27. Buckley AF, Burgart LJ, Sahai V, Kakar S. Epidermal growth factor receptor expression and gene copy number in conventional hepatocellular carcinoma. Am J Clin Pathol. 2008;129 (2):245–51.
28. Thakur S, Singla A, Chawla Y, Rajwanshi A, Kalra N, Arora SK. Expansion of peripheral and intratumoral regulatory T-cells in hepatocellular carcinoma: a case-control study. Indian J Pathol Microbiol. 2011;54:448–53.
29. Gomez-Santos L, Luka Z, Wagner C, Fernandez-Alvarez S, Lu SC, Mato JM, Martinez-Chantar ML, Beraza N. Inhibition of natural killer cells protects the liver against acute injury in the absence of glycine N-methyltransferase. Hepatology. 2012;56:747–59.
30. Mesiano G, Todorovic M, Gammaitoni L, Leuci V, Giraudo Diego L, Carnevale-Schianca F, Fagioli F, Piacibello W, Aglietta M, Sangiolo D. Cytokine-induced killer (CIK) cells as

feasible and effective adoptive immunotherapy for the treatment of solid tumors. Expert Opin Biol Ther. 2012;12:673–84.

31. Kim HM, Kang JS, Lim J, Kim JY, Kim YJ, Lee SJ, Song S, Hong JT, Kim Y, Han SB. Antitumor activity of cytokine-induced killer cells in nude mouse xenograft model. Arch Pharm Res. 2009;32:781–7.
32. Mohamed F, Stuart OA, Glehen O, Urano M, Sugarbaker PH. Docetaxel and hyperthermia: factors that modify thermal enhancement. J Surg Oncol. 2004;88:14–20.
33. Xiao-Pu Wang, Meng Xu, Hong-Fei Gao, Jian-Fu Zhao, and Ke-Cheng Xu. Intraperitoneal perfusion of cytokine-induced killer cells with local hyperthermia for advanced hepatocellular carcinoma. World J Gastroenterol. 2013 May 21; 19(19): 2956–2962. Published online 2013 May 21. doi:10.3748/wjg.v19.i19.2956.
34. Ito A, Okamoto N, Yamaguchi M, Kawabe Y, Kamihira M. Heat-inducible transgene expression with transcriptional amplification mediated by a transactivator. Int J Hyperthermia. 2012;28:788–98.
35. Köhne CH, Lenz HJ. Chemotherapy with targeted agents for the treatment of metastatic colorectal cancer. Oncologist. 2009;14:478–88.
36. Tabernero J, Van Cutsem E, Lakomý R, et al. Aflibercept versus placebo in combination with fluorouracil, leucovorin and irinotecan in the treatment of previously treated metastatic colorectal cancer: prespecified subgroup analyses from the VELOUR trial. Eur J Cancer. 2014;50:320–31.
37. Peeters M, Price TJ, Cervantes A, et al. Final results from a randomized phase 3 study of FOLFIRI {+/−} panitumumab for second-line treatment of metastatic colorectal cancer. Ann Oncol. 2014;25:107–16.
38. de Baere T, Elias D, Dromain C, et al. Radiofrequency ablation of 100 hepatic metastases with a mean follow-up of more than 1 year. AJR Am J Roentgenol. 2000;175(6):1619–25.
39. Wasser K, Giebel F, Fischbach R, Tesch H, Landwehr P. Transarterial chemoembolization of liver metastases of colorectal carcinoma using degradable starch microspheres (Spherex): personal investigations and review of the literature. Radiologe. 2005;45(7):633–43.
40. Pernot S, Artru P, Mithieux F, Marsot J, Watkin E, Pellerin O, Lledo G, Chalabreysse P, Desramé J, Taieb J. Clin Res Hepatol Gastroenterol. Complete pathological response of unresectable liver metastases from colorectal cancer after trans-arterial chemoembolization with drug-eluting beads loaded with irinotecan (DEBIRI) and concomitant systemic FOLFOX: A case report from the FFCD 1201 trial. 2015 Jun 30. pii: S2210-7401(15) 00134–5. doi:10.1016/j.clinre.2015.06.004.
41. Angele MK, Albertsmeier M, Prix NJ, Hohenberger P, Abdel-Rahman S, Dieterle N, Schmidt M, Mansmann U, Bruns CJ, Issels RD, Jauch KW, Lindner LH. Effectiveness of regional hyperthermia with chemotherapy for high-risk retroperitoneal and abdominal soft-tissue sarcoma after complete surgical resection: a subgroup analysis of a randomized phase-III multicenter study. Ann Surg. 2014;260(5):749–54. doi:10.1097/SLA.0000000000000978. discussion 754–6.
42. Issels RD, Lindner LH, Verweij J, Wust P, Reichardt P, Schem BC, Abdel-Rahman S, Daugaard S, Salat C, Wendtner CM, Vujaskovic Z, Wessalowski R, Jauch KW, Dürr HR, Ploner F, Baur-Melnyk A, Mansmann U, Hiddemann W, Blay JY, Hohenberger P; European Organisation for Research and Treatment of Cancer Soft Tissue and Bone Sarcoma Group (EORTC-STBSG); European Society for Hyperthermic Oncology (ESHO). Neo-adjuvant chemotherapy alone or with regional hyperthermia for localised high-risk soft-tissue sarcoma: a randomised phase 3 multicentre study. Lancet Oncol 2010 Jun;11(6):561–70. doi:10.1016/S1470-2045(10)70071-1. Epub 2010 Apr 29.

Chapter 26
Hyperthermia Combined with Chemotherapy: Pancreatic Cancer

Takeshi Ishikawa

Abstract Pancreatic cancer has the highest mortality rate of all cancers in part because there are no diagnostic tools that allow for early detection. Treatment options are limited and mostly rely on chemotherapy alone or chemotherapy coupled with radiation for unresectable pancreatic cancer. These current treatments are insufficient and more effective, better tolerated regimens are required to improve the survival outcome of patients with pancreatic cancer. Hyperthermia as a treatment option has very strong biological and experimental rationale and in pancreatic cancer, the effects of chemotherapy combined with hyperthermia using several heating methods and technologies have been investigated in clinical settings. As such, non-invasive regional hyperthermia, whole body hyperthermia, and hyperthermic intraperitoneal chemotherapy (HIPEC) have been used to complement chemotherapy when treating pancreatic cancer patients. In Phase II clinical trials, non-invasive regional hyperthermia was shown to be effective in advanced pancreatic cancer when combined with chemotherapy; however, no pivotal phase III trials have been published to date. This chapter presents a comprehensive review of clinical studies in which hyperthermia is combined with chemotherapy to treat pancreatic cancer. Information relevant for clinical practice will be outlined, as well as studies that address the effects of hyperthermia on the quality of life of cancer patients.

Keywords Pancreatic cancer • Hyperthermia • Chemotherapy • Quality of life

26.1 Current Status of Pancreatic Cancer Treatment

Pancreatic cancer remains one of the most challenging malignancies to treat due to its aggressive nature and resistance to most available treatments. Prognosis in pancreatic cancer is generally dismal. Median survival for locally advanced disease is 6–10 months, while this falls to 3–6 months for metastatic disease; the overall

T. Ishikawa (✉)
Department of Gastroenterology & Hepatology, Graduate School of Medical Science, Kyoto Prefectural University of Medicine, Kyoto 602-8566, Japan
e-mail: iskw-t@koto.kpu-m.ac.jp

S. Kokura et al. (eds.), *Hyperthermic Oncology from Bench to Bedside*,
DOI 10.1007/978-981-10-0719-4_26

5-year survival rate is approximately 5 % [1, 2]. It represents the fifth frequent cause of cancer–related deaths in Japan and the fourth frequent in the United States. To make matters worse, the incidence of pancreatic cancer and its related death rate is on the increase. Tumor resection is the only curative treatment; sadly, approximately 80 % of patients are ineligible for surgery because of metastatic disease or unfavorable tumor location [2]. In patients not eligible for curative surgery, chemotherapy is used to extend or improve quality of life. In cases that are considered to be borderline resectable, neoadjuvant chemotherapy or chemoradiotherapy may be used before surgery to reduce the cancer to a level where surgery could be beneficial [3].

Gemcitabine was approved by the United States Food and Drug Administration (FDA) in 1997 after it improved survival in patients with advanced pancreatic cancer in a randomized trial [4]. Gemcitabine chemotherapy was the standard treatment for about a decade, but then a number of trials testing it in combination with other drugs failed to show significant survival advantages over gemcitabine monotherapy [5]. To date, randomized trial of two regimens-a combination of 5-FU, leucovorin, irinotecan, and oxaliplatin (FOLFIRINOX) [6] and gemcitabine plus protein-bound paclitaxel (nab-paclitaxel) [7]-have demonstrated significant prolongation of overall survival; the drawback is that these regimens are more toxic than gemcitabine monotherapy. At present, both FOLFIRINOX and gemcitabine with nab-paclitaxel are regarded as good options for patients able to tolerate the side-effects. However, despite the development of these combined chemotherapy regimens to improve survival, the treatment for pancreatic cancer in the palliative setting is still largely inadequate, and more effective, better tolerated regimens are required to improve the outcomes of patients with advanced pancreatic cancer.

26.1.1 Current Status of Hyperthermia Combined with Chemotherapy

Overall survival has not improved considerably in the last two decades and pancreatic cancer still remains a rapidly progressing disease with a very limited life expectancy. Therefore, it is critical to screen and evaluate complementary adjuvant treatments and hyperthermia may be one such option to prolong overall survival and improve quality of life [8].

Hypovascular tumors, such as pancreatic cancer, retain more heat than surrounding tissues therefore tumor temperature exceeds that of normal tissues. Thus, we consider pancreatic cancer a suitable target for hyperthermic treatment. Hyperthermia was shown to enhance the cytotoxicity of some chemotherapeutic agents by facilitating drug penetration into tissues and causing thermal destruction of cancer cells [9, 10]. Moreover, in a prior study, we demonstrated that hyperthermia enhances the cytotoxicity of gemcitabine, one of the key drugs for pancreatic cancer, by inhibiting gemcitabine-induced activation of nuclear factor kappa B (NF-κB) [10]. Considering these results, combining hyperthermia and chemotherapy could result in better treatment outcomes for pancreatic cancer patients. In this regard, there is an array of heating methods and technologies used in hyperthermia protocols and we will explore some of these options using published clinical data.

26.1.1.1 Non-invasive Deep Regional Hyperthermia

Regional hyperthermia seeks to increase the temperature of a tumor to an optimal range of 40–45 °C using an external heating device. The antenna system and capacitative system are two non-invasive deep regional heating technologies available delivers temperatures of between 39 and 44 °C [11]. Focused ultrasound (HIFU) is another method of non-invasive ablative heating and it is usually combined with an MRI for location control and temperature measurements [11].

The rationale for adding regional hyperthermia to chemotherapy grew out of a wealth of convincing basic research. The mechanisms of thermal enhancement when combined with anticancer agents include: (1) increased cellular uptake of anticancer agents [12], (2) increased oxygen radical production [13], (3) increased DNA damage and inhibition of repair [14], (4) to make resistant tumors responsive again to anticancer agents [10].

Although no randomized trial of hyperthermia combined with radiotherapy and/or chemotherapy have been conducted so far, several retrospective and prospective studies with small sample sizes have been done in which non-invasive deep regional hyperthermia was combined with chemotherapy and/or radiotherapy for advanced pancreatic cancer (Table 26.1). Hager et al reported a single arm prospective study in which patients with advanced, surgically partially resectable pancreatic cancer underwent a complex treatment regimen involving hyperthermia, hormone therapy (tamoxifen etc.), immunomodulating agents (thymus-peptides, mistletoe extract etc.), differentiating agents (high-dose vitamin A-E-palmitate, alphahydroxycalciferol etc.), and a mixed bacterial vaccine [15]. They reported that median OS was 10.8 months with a 5-year survival rate of 9 %. There were some limitations with the study design one being the lack of clear eligibility criteria. Some patients were not diagnosed histologically, and there was a lack of uniformity in the clinical stage of the patients (UICC stage 2–4). Most importantly, the treatment regimen used was not standard, it was rather unconventional, therefore the study is not reproducible. Thus, this study is of limited value in assessing the true effects of hyperthermia. We and other investigators conducted retrospective studies of regional hyperthermia combined with gemcitabine-based chemotherapy [16, 17]. In our study, seven unresectable pancreatic cancer patients received combination therapy with hyperthermia (capative system, once a week) and gemcitabine monotherapy and data were compared with seven historical controls treated with gemcitabine only at the same institution [16]. The median OS was 10.9 months, and it was longer than the historical controls (6.6 months) (Fig. 26.1). The disease control rate (DCR) was 57.1 %, whereas in gemcitabine alone group it

Table 26.1 Summary of studies on advanced pancreatic cancer treatments including non-invasive deep regional hyperthermia

Study	Type of study	Disease status	No. of patients	Hyperthermia method	Chemotherapy	Radiotherapy	Findings
Hager et al. [15]	Prospective, one arm phase 2	LAPC and MPC	46	Capative heating device	Enzyme therapy, Hormone therapy, Differentiating agents, Immune therapy, etc	–	Median OS: 10.8 months 1 year OS: 41 %
Ishikawa et al. [16]	Retrospective	LAPC and MPC	7	Once a week, capative heating device	GEM	–	Median OS: 10.9 months DCR: 57.1 %
Mueller-Huebenthal et al. [17]	Retrospective	LAPC and MPC	25	Twice a week	GEM-based	–	Median OS: 12.2 months 1 year OS: 51 % CR 1 pt, PR 7 pts
Zhang et al. [21]	Retrospective	LAPC	38	Twice a week	Tegafur-CF, GEM	Gamma-knife radiotherapy (3.0–4.5 Gy, 8–11fr)	1 year OS: 51.2 % 2 year OS: 26.5 % RR: 73.7 %
Maluta et al. [22]	Retrospective	LAPC	40	Twice a week, antenna heating device	GEM, GEM-CDDP, GEM-5FU, GEM-Oxa	30–66 Gy	Median OS: 15 month No increased CRT toxicity
Ishikawa et al. [18]	Prospective, one arm phase 2	LAPC and MPC	18	Once a week, capative heating device	GEM	–	Median OS: 8 months (17.7 months in LAPC) 1 year OS: 33 % DCR: 61.1 %
Tscoep-Lechner et al. [24]	Retrospective	GEM-refractory PC	23	Twice a week, antenna heating device	GEM-CDDP	–	Median OS: 12.9 months DCR: 50 %

LAPC locally advanced pancreatic cancer, *MPC* metastatic pancreatic cancer, *GEM* gemcitabine, *CF* calcium folinate, *Oxa* Oxaliplatin, *OS* overall survival, *CR* complete response, *PR* partial response, *DCR* disease control rate

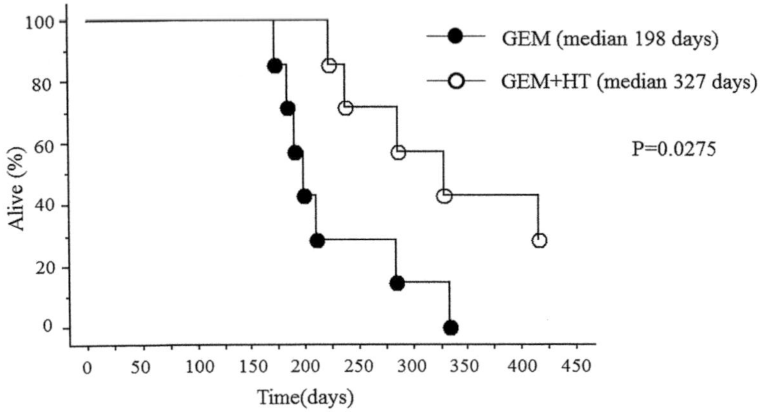

Fig. 26.1 Kaplan-Meier estimated overall survival (OS) in patients treated with gemcitabine plus hyperthermia and with gemcitabine alone. In this retrospective study, seven patients with unresectable pancreatic cancer received gemcitabine combined with hyperthermia. Data for these patients were compared with seven historical controls treated with gemcitabine alone at the same institution. Results showed that patients treated with gemcitabine plus hyperthermia had significantly better survival than the gemcitabine monotherapy group ($P = 0.0275$). From [16]

was 14.3 %[16]. In a German retrospective study, 25 unresectable pancreatic cancer patients received hyperthermia twice a week combined with gemcitabine-based standard chemotherapy [17]. Median OS was 12.2 months, and three patients showed complete response, while seven patients had a partial response (DCR = 65 %) [17]. These studies suggest that combining gemcitabine with regional hyperthermia may improve the survival of patients with advanced pancreatic cancer.

We proceeded to conduct a prospective phase II study to evaluate the efficacy and safety of regional hyperthermia combined with gemcitabine [18]. Eighteen unresectable pancreatic cancer patients (6 locally advanced and 12 metastatic) received hyperthermia with gemcitabine monotherapy at a dose of 1000 mg/m^2 on days 1, 8, 15 every 4 weeks. Regional capacitive hyperthermia was performed 1 day preceding or following gemcitabine administration. We based this schedule on in vitro study in which hyperthermia enhanced gemcitabine cytotoxicity, particularly when it was performed 24 h before and after gemcitabine treatment [10]. This resulted in a response rate of 11.1 %, DCR of 61.1 % and a median OS of 8 months. At 33.3 %, the 1-year survival rate exceeded the threshold required for the regimen to be considered effective. We also observed that locally advanced patients had a better outcome than metastatic patients (median OS of 17.7 months versus 5.2 months) (Fig. 26.2). The results in patients with locally advanced pancreatic cancer compare favourably with previous studies of 5-FU or gemcitabine with radiation, which reported a median OS of 8–10 months [19, 20].

Two retrospective studies done in China and Italy evaluated the efficacy of regional hyperthermia combined with gemcitabine-based chemotherapy and radiation in locally advanced pancreatic cancer patients [21, 22]. In the Chinese study,

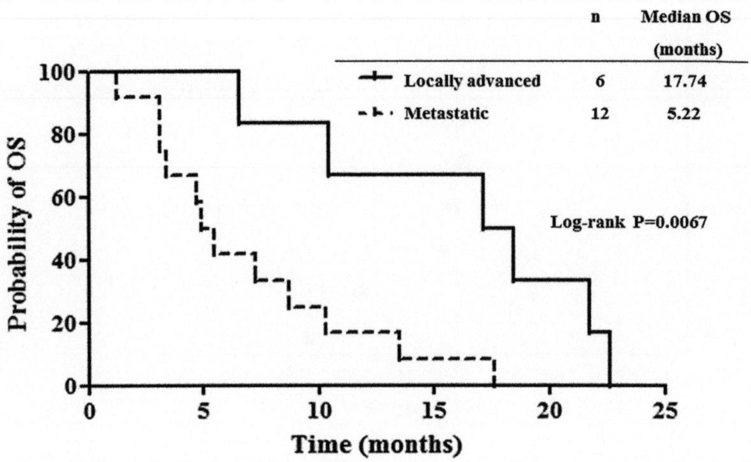

Fig. 26.2 Kaplan-Meier estimated overall survival (OS) in patients with locally advanced pancreatic cancer and patients with metastatic cancer. Survival curves differed significantly between patients with locally advanced pancreatic cancer and those with metastatic lesions when analyzed by log rank test (P = 0.0067). From [18]

75 patients underwent either gamma knife solo treatment or a combination therapy with gamma knife, hyperthermia and chemotherapy [21]. Patients in the combination group received hyperthermia twice a week simultaneously with chemotherapy using tegafur and calcium folinate or gemcitabine and CDDP. Patients in the gamma-knife combined with hyperthermia and chemotherapy group had higher 1 and 2 year survival rates (51.2 and 26.5 %) than patients in the gamma knife solo group (45.2 and 17.6 %) [21]. The Italian study showed a median OS of 15 months in patients with locally advanced pancreatic cancer treated with a combination of radiation therapy with regional hyperthermia (twice a week) and chemotherapy (gemcitabine alone or in association with either oxaliplatin, CDDP or 5FU) [22]. We have recently demonstrated in an *in vitro* study that hyperthermia suppressed TGF-β-induced epithelial-to-mesenchymal transition (EMT), which is crucial in cancer invasion and metastasis [23]. Since regional hyperthermia combined with chemotherapy or chemoradiotherapy can improve the prognosis of locally advanced pancreatic cancer, we believe that EMT inhibition may be one of the mechanisms involved.

A study done in Germany evaluated the efficacy of chemotherapy combined with regional hyperthermia [24] as a second-line treatment for gemcitabine-refractory advanced pancreatic cancer. In that retrospective study, patients received gemcitabine and CDDP combined with regional radiative hyperthermia on days 2 and 4 each week for 4 months. The DCR for patients with available CT scans was 50 % and the median time to progression was 4.3 months. That study is still active with phase II clinical trials being conducted on this treatment protocol in gemcitabine-refractory pancreatic patients.

The European Society for Hyperthermic Oncology (ESHO) HEAT phase III trial (Hyperthermia European Adjuvant Trial) is an ongoing study to evaluate gemcitabine solo treatment versus combining gemcitabine and CDDP with regional hyperthermia in R0/R1 resected pancreatic cancer patients. This is the first large-scaled, randomized trial for regional hyperthermia in pancreatic cancer, and as such we are looking forward to the results.

26.1.1.2 Superficial Heating (Intraoperative)

While local intraoperative hyperthermia for pancreatic cancer is rarely performed, intraoperative superficial hyperthermia combined with radiotherapy has shown promising results without systemic toxicity [25, 26]. For example, Kouloulias et al. demonstrated a potential advantage of combining intraoperative hyperthermia, chemotherapy and postoperative radiotherapy as a palliative treatment in patients with unresectable pancreatic cancer [27]. In that study, ten patients with unresectable pancreatic cancer received multiple sessions of chemotherapy plus radiotherapy (45Gy), and a single session of intraoperative hyperthermia using a waveguide-type applicator (433 MHz) during bypass surgery. Their findings showed a significant decrease in serum values of both CEA and CA19-9 after combined therapy and there was a significant positive impact of thermal parameters on overall survival. Moreover, the study suggested that a combined therapy that includes hyperthermia could reduce pain and improve the quality of life of patients [27]. The study population was relative small in that study so we believe that a randomized trial is needed to further assess the impact of this procedure on survival and palliation.

26.1.1.3 Hyperthermic Intraperitoneal Chemotherapy (HIPEC)

Peritoneal carcinomatosis is known to negatively impact survival, and both conventional surgery and systemic chemotherapy exert little effect on survival of these patients. It was reported that hyperthermic intraperitoneal chemotherapy (HIPEC) mostly combined with cytoreductive surgery resulted in significant improvements in survival, morbidity and mortality in patients with peritoneal carcinomatosis [28]. Although no trials specific to pancreatic cancer have been published for HIPEC, there is strong evidence pointing to the superior efficacy of HIPEC over conventional treatments in the management of peritoneal carcinoma arising from gastric and colorectal cancer [28–30]. Additionally, pharmacologic data suggests that a benefit of intraoperative gemcitabine monotherapy administered intraperitoneally with hyperthermia is that the exposure of peritoneal surfaces to intraperitoneal gemcitabine is approximately 200–500 times the exposure that occurs within the plasma [31]. For these reasons we believe that HIPEC could possibly exert superior efficacy and the rationale for using it to treat pancreatic cancer patients is quite strong.

26.1.1.4 Whole Body Hyperthermia

There are few clinical reports available on whole body hyperthermia with chemotherapy for pancreatic cancer. In one such report, Bakshandeh-Bath et al. demonstrated the feasibility of hyperthermia in 13 advanced pancreatic patients treated with carboplatin and gemcitabine combined with 41.8 °C whole body hyperthermia [32]. Partial response was achieved by 23 % of the patients and 38 % achieved stable disease; the median survival was 11.4 months. They concluded that whole body hyperthermia combined with chemotherapy may lead to clinical benefits for some advanced pancreatic cancer patients; however, since no trial was performed the results were merely case-based.

26.2 Hyperthermia and Quality of Life

Existing data is heterogeneous and so far there are no well-designed large-scale trials for pancreatic cancer, however, evidence suggests that at a minimum quality of life is improved by supplementing treatment with hyperthermia. Hager et al demonstrated that hyperthermia combined with various treatments such as anti-hormonal therapy and immunomodulating agents improved quality of life in most pancreatic cancer patients treated [15]. Pain was reduced in 56 % of patients and appetite improved in 48 % of patients [15]. For patients with peritoneal metastases, a recent systematic review suggested that hyperthermic intraperitoneal chemotherapy (HIPEC) with cytoreductive surgery could preserve or even improve patients' quality of life [33]. No patients with pancreatic cancer were included in this study, however, other tumor sites including colorectal, gastric and ovarian cancer were represented. Since HIPEC combined with cytoreductive surgery has been proven to positively affect the quality of life of patients with various cancers, I believe it be a promising strategy that can be applied to pancreatic cancer. As such, a step in the right direction would be to create a well-designed prospective study to clarify the significance of hyperthermia on the quality of life of patients with advanced pancreatic cancer.

26.3 Conclusion

An important advantage of hyperthermia combined with chemotherapy is that unlike radiation therapy, it does not exacerbate adverse events, in particular, the hematotoxicity of cytotoxic agents. In recent years, the use of more intensive chemotherapy regimens such as FOLFIRINOX and gemcitabine plus nab-paclitaxel improved prognosis of patients with advanced pancreatic cancer. On the other hand, these regimens are suboptimal as they are more toxic; more than

20 % of patients in Japan who received FOLFIRINOX experienced febrile neutropenia. In analyzing the clinical data available on treatments that combined hyperthermia and chemotherapy, I anticipate that this combination will not exacerbate the adverse events of chemotherapy but will instead enhance the anti-tumor effects in pancreatic cancer patients. As a result, it is necessary to further assess the efficacy of hyperthermia combined with these new intensive regimens in future clinical trials.

Hyperthermia has been proven to modulate various cells of the innate and adaptive immune system directly or indirectly [34–36]. As is the case for other cancers, immune therapies such as cancer vaccines and immune checkpoint inhibitors are being developed for pancreatic cancer [37, 38]. Hyperthermia is reported to modify the tumor microenvironment thereby preventing the development of immunological tolerance. Since hyperthermia is expected to enhance anti-tumor effects of immune therapies, we propose that clinical trials should be done to assess the impact of using hyperthermia as a supplement to various cancer immune therapies.

Hyperthermia is a potent sensitizer for radiotherapy, chemotherapy, and immune therapies, each of which has significantly advanced in this decade. A multimodal treatment approach that relies on hyperthermia combined with state-of-the-art radiation, chemotherapy, and immune therapy can result in great clinical benefits for pancreatic cancer patients. We believe that, in the years to come, a stronger focus on the research and development of these novel yet safe modals is what is required for a treatment breakthrough to emerge that ultimately results in longer survival and higher quality of life for pancreatic cancer patients.

References

1. Siegel R, DeSantis C, Virgo K, Stein K, Mariotto A, Smith T, et al. Cancer treatment and survivorship statistics. CA Cancer J Clin. 2012;62(4):220–41. doi:10.3322/caac.21149.
2. Siegel R, Naishadham D, Jemal A. Cancer statistics. CA Cancer J Clin. 2013;63(1):11–30. doi:10.3322/caac.21166.
3. Heinemann V, Haas M, Boeck S. Neoadjuvant treatment of borderline resectable and non-resectable pancreatic cancer. Ann Oncol. 2013;24(10):2484–92. doi:10.1093/annonc/mdt239.
4. Burris 3rd HA, Moore MJ, Andersen J, Green MR, Rothenberg ML, Modiano MR, et al. Improvements in survival and clinical benefit with gemcitabine as first-line therapy for patients with advanced pancreas cancer: a randomized trial. J Clin Oncol Off J Am Soc Clin Oncol. 1997;15(6):2403–13.
5. Di Marco M, Di Cicilia R, Macchini M, Nobili E, Vecchiarelli S, Brandi G, et al. Metastatic pancreatic cancer: is gemcitabine still the best standard treatment? (Review). Oncol Rep. 2010;23(5):1183–92.
6. Conroy T, Desseigne F, Ychou M, Bouche O, Guimbaud R, Becouarn Y, et al. FOLFIRINOX versus gemcitabine for metastatic pancreatic cancer. N Engl J Med. 2011;364(19):1817–25. doi:10.1056/NEJMoa1011923.
7. Von Hoff DD, Ervin T, Arena FP, Chiorean EG, Infante J, Moore M, et al. Increased survival in pancreatic cancer with nab-paclitaxel plus gemcitabine. N Engl J Med. 2013;369(18):1691–703. doi:10.1056/NEJMoa1304369.

8. Roesch M, Mueller-Huebenthal B. Review: the role of hyperthermia in treating pancreatic tumors. Indian J Surg Oncol. 2015;6(1):75–81. doi:10.1007/s13193-014-0316-5.
9. Engelhardt R. Hyperthermia and drugs. Recent results in cancer research Fortschritte der Krebsforschung. Progres dans les recherches sur le cancer. 1987;104:136–203.
10. Adachi S, Kokura S, Okayama T, Ishikawa T, Takagi T, Handa O, et al. Effect of hyperthermia combined with gemcitabine on apoptotic cell death in cultured human pancreatic cancer cell lines. Int J Hyperthermia. 2009;25(3):210–19. doi:10.1080/02656730802657036.
11. Soares PI, Ferreira IM, Igreja RA, Novo CM, Borges JP. Application of hyperthermia for cancer treatment: recent patents review. Recent Pat Anticancer Drug Discov. 2012;7(1):64–73.
12. Maeta M, Sawata T, Kaibara N. Effects of hyperthermia on the metabolism of 5-fluorouracil in vitro. Int J Hyperthermia. 1993;9(1):105–13.
13. Hirano H, Tabuchi Y, Kondo T, Zhao QL, Ogawa R, Cui ZG, et al. Analysis of gene expression in apoptosis of human lymphoma U937 cells induced by heat shock and the effects of alpha-phenyl N-tert-butylnitrone (PBN) and its derivatives. Apoptosis. 2005;10(2):331–40. doi:10.1007/s10495-005-0807-5.
14. Eppink B, Krawczyk PM, Stap J, Kanaar R. Hyperthermia-induced DNA repair deficiency suggests novel therapeutic anti-cancer strategies. Int J Hyperthermia. 2012;28(6):509–17. doi:10.3109/02656736.2012.695427.
15. Hager ED, Migeod F, Koomagi R, Schrittwieser G, Krautgartner I. Multimodal complementary therapy for advanced pancreas carcinoma; translation of publication of Deutsche. Zeitschrift fur Onkologie. 2006;38:100–7.
16. Ishikawa T, Kokura S, Oyamada H, Inui T, Okita M, Isozaki Y, et al. Effects of a sequential combination of hyperthermia and gemcitabine in the treatment of advanced unresectable pancreatic cancer: a retrospective study. Thermal Med. 2008;24(4):131–9.
17. Mueller-Huebenthal B. Hyperthermia and gemcitabine for the treatment of advanced inoperable pancreatic cancer: a retrospective analysis of 25 cases. The 6th binneal conference of Indian Association of Hyperthermia Oncology IAHOM; Bangalore 2010.
18. Ishikawa T, Kokura S, Sakamoto N, Ando T, Imamoto E, Hattori T, et al. Phase II trial of combined regional hyperthermia and gemcitabine for locally advanced or metastatic pancreatic cancer. Int J Hyperthermia. 2012;28(7):597–604. doi:10.3109/02656736.2012.695428.
19. Ishii H, Okada S, Tokuuye K, Nose H, Okusaka T, Yoshimori M, et al. Protracted 5-fluorouracil infusion with concurrent radiotherapy as a treatment for locally advanced pancreatic carcinoma. Cancer. 1997;79(8):1516–20.
20. Philip PA, Mooney M, Jaffe D, Eckhardt G, Moore M, Meropol N, et al. Consensus report of the national cancer institute clinical trials planning meeting on pancreas cancer treatment. J Clin Oncol Off J Am Soc Clin Oncol. 2009;27(33):5660–9. doi:10.1200/JCO.2009.21.9022.
21. Zhang LP, Nie Q, Kang JB, Wang B, Cai CL, Li JG et al. [Efficacy of whole body gamma-knife radiotherapy combined with thermochemotherapy on locally advanced pancreatic cancer]. Ai zheng = Aizheng = Chin J Cancer 2008;27(11):1204–7.
22. Maluta S, Schaffer M, Pioli F, Dall'oglio S, Pasetto S, Schaffer PM, et al. Regional hyperthermia combined with chemoradiotherapy in primary or recurrent locally advanced pancreatic cancer: an open-label comparative cohort trial. Strahlentherapie und Onkologie: Organ der Deutschen Rontgengesellschaft [et al]. 2011;187(10):619–25. doi:10.1007/s00066-011-2226-6.
23. Kimura-Tsuchiya R, Ishikawa T, Kokura S, Mizushima K, Adachi S, Okajima M, et al. The inhibitory effect of heat treatment against epithelial-mesenchymal transition (EMT) in human pancreatic adenocarcinoma cell lines. J Clin Biochem Nutr. 2014;55(1):56–61. doi:10.3164/jcbn.14-8.
24. Tschoep-Lechner KE, Milani V, Berger F, Dieterle N, Abdel-Rahman S, Salat C, et al. Gemcitabine and cisplatin combined with regional hyperthermia as second-line treatment in patients with gemcitabine-refractory advanced pancreatic cancer. Int J Hyperthermia. 2013;29(1):8–16. doi:10.3109/02656736.2012.740764.
25. Ashayeri E, Bonney G, DeWitty RL, Goldson AL, Leffall LD, Thomas JN. Preliminary survivorship report on combined intraoperative radiation and hyperthermia treatments for unresectable pancreatic adenocarcinoma. J Natl Med Assoc. 1993;85(1):36–40.

26. Ryan TP, Colacchio TA, Douple EB, Strohbehn JW, Coughlin CT. Techniques for intraoperative hyperthermia with ultrasound: the Dartmouth experience with 19 patients. Int J Hyperther. 1992;8(4):407–21.
27. Kouloulias VE, Nikita KS, Kouvaris JR, Golematis BC, Uzunoglu NK, Mystakidou K, et al. Intraoperative hyperthermia and chemoradiotherapy for inoperable pancreatic carcinoma. Eur J Cancer Care. 2002;11(2):100–7.
28. Loggie BW, Fleming RA, McQuellon RP, Russell GB, Geisinger KR. Cytoreductive surgery with intraperitoneal hyperthermic chemotherapy for disseminated peritoneal cancer of gastrointestinal origin. Am Surg. 2000;66(6):561–8.
29. Mi DH, Li Z, Yang KH, Cao N, Lethaby A, Tian JH, et al. Surgery combined with intraoperative hyperthermic intraperitoneal chemotherapy (IHIC) for gastric cancer: a systematic review and meta-analysis of randomised controlled trials. Int J Hyperthermia. 2013;29 (2):156–67. doi:10.3109/02656736.2013.768359.
30. Esquivel J, Sticca R, Sugarbaker P, Levine E, Yan TD, Alexander R, et al. Cytoreductive surgery and hyperthermic intraperitoneal chemotherapy in the management of peritoneal surface malignancies of colonic origin: a consensus statement. Society of Surgical Oncology. Ann Surg Oncol. 2007;14(1):128–33. doi:10.1245/s10434-006-9185-7.
31. Sugarbaker PH, Stuart OA, Bijelic L. Intraperitoneal gemcitabine chemotherapy treatment for patients with resected pancreatic cancer: rationale and report of early data. Int J Surg Oncol. 2011;2011:161862. doi:10.1155/2011/161862.
32. Bakshandeh-Bath A, Stoltz AS, Homann N, Wagner T, Stolting S, Peters SO. Preclinical and clinical aspects of carboplatin and gemcitabine combined with whole-body hyperthermia for pancreatic adenocarcinoma. Anticancer Res. 2009;29(8):3069–77.
33. Seretis C, Youssef H. Quality of life after cytoreductive surgery and intraoperative hyperthermic intraperitoneal chemotherapy for peritoneal surface malignancies: a systematic review. Eur J Surg Oncol. 2014;40(12):1605–13. doi:10.1016/j.ejso.2014.08.477.
34. Yanase M, Shinkai M, Honda H, Wakabayashi T, Yoshida J, Kobayashi T. Antitumor immunity induction by intracellular hyperthermia using magnetite cationic liposomes. Jpn J Cancer Res Gann. 1998;89(7):775–82.
35. Ito A, Shinkai M, Honda H, Wakabayashi T, Yoshida J, Kobayashi T. Augmentation of MHC class I antigen presentation via heat shock protein expression by hyperthermia. Cancer Immunol Immunother CII. 2001;50(10):515–22.
36. Skitzki JJ, Repasky EA, Evans SS. Hyperthermia as an immunotherapy strategy for cancer. Curr Opin Investig Drugs. 2009;10(6):550–8.
37. Uram JN, Le DT. Current advances in immunotherapy for pancreatic cancer. Curr Probl Cancer. 2013;37(5):273–9. doi:10.1016/j.currproblcancer.2013.10.004.
38. Di Caro G, Castino GF, Bergomas F, Cortese N, Chiriva-Internati M, Grizzi F, et al. Immune-based therapies in pancreatic and colorectal cancers and biomarkers of responsiveness. Expert Rev Anticancer Ther. 2014;14(10):1219–28. doi:10.1586/14737140.2014.947277.

Chapter 27
Effect of Hyperthermia in Combination with Chemotherapy and Radiation Therapy for Treatment of Urological Cancer: Prostate Cancer and Urothelial Cancer

Kosuke Ueda, Fumiko Maeda, and Mayumi Oota

Abstract This report describes the clinical result of hyperthermia for urological cancer; first, the effects of regional hyperthermia combined with low-dose anticancer drugs for the treatment of 7 high-grade prostate cancer patients with Gleason scores of 8 or higher.

The combination therapy resulted in three cases in CR and four cases in PR based on the PSA level. One patient with Stage D2 died (duration of survival 132 months), but six survived. The duration of survival ranged from 9 to 156 months with 20 months of median survival time. No severe side effects were observed in any of the seven patients, so they were able to undergo treatment without discontinuation.

Thus, the treatment was proven to be effective for high-grade prostate cancer, contributing to long-term survival of the patients.

In addition, regional hyperthermia combined with low-dose anticancer drugs (CDDP, gemcitabine and MMC) was similar to prostate cancer, administered to urothelial cancer patients (5 patients with renal pelvis cancer, 3 patients with ureter cancer and 10 patients with bladder cancer).

The treatment resulted in CR: 7 cases, PR: 3 cases, NC: 1 case, and PD: 7 cases with 8 patients deceased (duration of survival: 9–91 months, median 24 months). The rest of the patients are still alive.

Thus the result indicates that the treatment is able to prolong survival of urothelial cancer patients, similar to prostate cancer patients.

Keywords Prostate cancer • Urothelial cancer • Low-dose chemotherapy • Regional hyperthermia

K. Ueda (✉) • F. Maeda • M. Oota
Department of Urology, Hachiya Orthopedic Hospital, Nagoya Prostate Center, 2-4 Suemoridori, Chikusaku, Nagoya, Aichi 464-0821, Japan
e-mail: uro-ueda@hachiya.or.jp

S. Kokura et al. (eds.), *Hyperthermic Oncology from Bench to Bedside*,
DOI 10.1007/978-981-10-0719-4_27

27.1 Prostate Cancer

27.1.1 Introduction

Prostate cancer is one of the critical diseases in men, and in recent years, prostate cancer patients have been increasing with westernization of eating habits. Castration-refractory prostate cancer in particular is refractory to various treatments, resulting in a poor prognosis. The combined treatment of docetaxel and steroid drugs, or a newly developed anticancer agent is administered to these patients. But they have not offered long-term survival, despite good clinical results in the short term [1].

Meanwhile, hyperthermia treatment for prostate cancer is not commonly administered. In the urology field, only thermotherapy using HIFU (high-intensity focused ultrasaound) is performed for patients with localized prostate cancer [2].

Then our hospital administered combined treatment of hyperthermia and low-dose anticancer drug, reporting good clinical results [3, 4].

In addition to the studies work mentioned above, we here report our long-term clinical results from the combination treatment of hyperthermia and low-dose anticancer drugs after radiation therapy for prostate cancer.

27.1.2 Patients and Methods

From April, 2000 to July, 2015, seven patients were enrolled. All of the patients who developed high grade prostate cancer with more than a Gleason score of 8 have been treated in the hospital.

Of seven patients (ages: 53–82, mean: 69), five were classified as Gleason score 8, one patient as Gleason score 9 and one patient as Gleason score 10.

Their cancer stages were categorized as follows.

Six patients were categorized as T3, and one patient as T4. Five patients were also classified as N0 and two patients as N1. As for M category, five patients were classified as M0 and two patients as M1 (stage D2).

The treatment used cisplatin (CDDP) or docetaxel as an anticancer drug. Patients were intravenously given 5 mg of drugs each diluted in 500 ml saline 30 min before hyperthermia treatment.

Basically, CDDP and docetaxel were alternately administered weekly, for a total of 10 times, (3 cases CDDP 10 times, docetaxel 10 times, 4 cases docetaxel 10 times and CDDP 10 times). Patients underwent 20 rounds of chemotherapy. After confirming less than the normal PSA level, the treatment was continued on a monthly basis as a maintenance chemotherapy with CDDP and docetaxel alternately administered a total of 10 times.

Hyperthermia with an 8 MHz-RF (pros-8) was delivered to the pelvic region using capacitive heating with two electrodes for 40–50 min at 600–1000 radiofrequency watts. The treatment consisted of 10 cycles. The total hyperthermia treatment varied from 1 to 13 cycles (mean: 3 cycles).

The efficacy of treatment was evaluated by the rate of decline in the PSA level because 5 cases of local invasive prostate cancer did not show any evaluable lesion (MRI failed to confirm any lesion in all 5 cases) and in 2 cases of progressive cancer, bone metastases were the only evaluable lesions. The partial response was defined as more than 50 % of the reduction and the complete response as more than 90 % reduction in PSA level. The efficacy of treatment was assessed by the duration of survival.

27.1.3 Response to Treatment

The clinical results showed a complete response in 3 cases (42.8 %) and a partial response in 4 cases. The duration of survival after the start of treatment ranged from 9 to 156 months. One patient with stage D2 cancer died, but the other six are still alive. No severe side effects were observed, although three patients suffered mild general fatigue. The patients were able to continue the treatment.

27.1.4 Discussion and Conclusion

Castration-resistant prostate cancer has a poor prognosis, and docetaxel and corticosteroid drug are commonly used for the treatment. But the clinical results are only temporary [5].

Recently newly developed anticancer drugs such as Enzalutamide and Abiraterone acetate are administered patients, but compared with the control group, the treatment is only able to extend the duration of survival for just a few months [6, 7].

In addition, it also produces multiple severe side effects with death resulting from the severe side effects reported in Japan.

Therefore, our developed treatments with as few side effects as possible, enable patients to receive long-term treatment with a durable response. Hyperthermia treatment combined with anticancer drugs or immune cell therapy in particular is able to increase the efficacy of treatment [8].

Especially, the anticancer drug regimen is very important. Takahashi reported that low-dose administration of anticancer drugs would inhibit metastases without severe side effects when the drugs are given quite regularly like a metronome [9]. Hyperthermia is also proved to enhance the permeability of anticancer drugs to tumor cells [10].

Based on these results, we have developed a combination treatment of hyperthermia and low-dose anticancer drugs and have treated castration-resistant prostate cancer for more than 10 years [4].

In this study, we examined the clinical results from the combined therapy of regional hyperthermia and low-dose anticancer drug after radiation therapy for patients with high-grade prostate cancer whose Gleason score was more than 8. The clinical results showed a complete response in 3 cases (42.8 %) and a partial response in 4 cases. The duration of survival after the start of treatment ranged from 9 to 156 months (median: 20 months). Six of seven patients are still alive, although one patient with progressive cancer succumbed. But the patients survived 132 months, proving the efficacy of treatment.

All of five patients with local invasive prostate cancer have maintained a PSA level of less than 0.1 ng/ml, so they are likely to have long-time survival.

Many researchers have reported efficacy of combined treatment of radiation therapy and hyperthermia for the treatment [11].

Few researchers, however, have studied the clinical result of hyperthermia after radiation therapy. While radiation therapy is proved to be effective for peritumoral areas due to the oxygen effect, hyperthermia is useful to treat hypoxic tumors with poor blood flow. So hyperthermia treatment after radiation therapy medically makes sense.

Patients undergo a blood chemical test to check side effects. They did not show any decline in the number of white blood cells and platelets but revealed the tendency to increase lymphocytes after treatment. Three patients temporarily suffered from mild fatigue after treatment of 1 day, so the treatment could be continued. One of two patients with stage D2 cancer survived 11 years (132 months) and received outpatient treatment just before death.

Therefore, the combined treatment with low-dose chemotherapy and hyperthermia after radiation therapy is useful even if patients have bone metastases. Hyperthermia is reportedly not only effective for tumor cells but also improves immunity via HSP expression. So together with localized curative effects, the treatment has proved useful to improve the general condition of patients, leading to long-term survival.

27.2 Urothelial Cancer (Renal Pelvis Cancer, Ureter Cancer and Bladder Cancer)

27.2.1 Introduction

Cockett et al. first administered combination therapy with irradiation and hot water irrigation in a dog model [12], and Hisazumi et al. administered the combination therapy of hyperthermia with intra-vesical perfusion of peplomycin and hot water irrigation for bladder carcinoma [13].

The treatment resulted in 22.2 % response rate (CR + PR), but was not useful for invasive bladder carcinoma. Therefore, for invasive carcinoma, systemic chemotherapy (M-VAC = methotrexate, vinblastine, adriamysin and cisplatin) or GC (gemcitabine and cisplatin) was performed. But the patients developed severe side effects and long-term survival was not possible [14, 15].

Therefore, we began to perform our treatment combining regional hyperthermia using an 8 MHz-radiofrequency hyperthermic machine (Yamamoto VINITA Co., Ltd., Osaka, Japan) and intravenous chemotherapy (cisplatin or gemcitabine) for the treatment of invasive urothelial carcinomas 10 years ago. The result was already described [16].

Then we report the result of combined treatment with low-dose chemotherapy and regional hyperthermia for progressive urothelial carcinomas (renal pelvis carcinoma, ureter carcinoma and bladder carcinoma) for 10 years.

27.2.2 Patient and Methods

The study enrolled 18 patients with progressive urothelial cancer in the hospital for 11 years. Five patients had renal pelvis cancer, three had ureter cancer and ten had bladder cancer. Nine patients were classified as T2 (5 cases as T2a and 4 cases as T2b), eight patients as T3 and one patient as T4 according to UCC-TNM classification. As for pretreatment, six patients underwent total nephro-ureterectomy, six patients trans-urethral resection of the bladder tumor, one patient a partial cystectomy, one patient a total cystectomy, two patients received radiation therapy and two did not undergo any surgical operation.

27.2.3 Methods

All of 18 patients received combination of hyperthermia and low-dose anticancer drug using cisplatin (CDDP), gemcitabine (GEM),and mytomycin C (MMC).

The dosage amount of CDDP, GEM and MMC was 5 mg, 100 mg and 0.1 mg respectively, for each treatment. CDDP or MMC dissolved in 500 ml of saline or electrolytic solution was intravenously given to patients 30 min before hyperthermia.

GEM dissolved in 200 ml of saline was intravenously given to patients after hyperthermia. The patients underwent treatment once weekly. One cycle of treatment consisted of ten treatments. The clinical results were confirmed after each cycle of treatment. When the positive results were confirmed, patients underwent similar treatment once monthly as maintenance therapy. When the response was not confirmed, drugs were changed and hyperthermia was performed once weekly. The total cycle of hyperthermia treatment ranged from one to 99 (mean: 3).

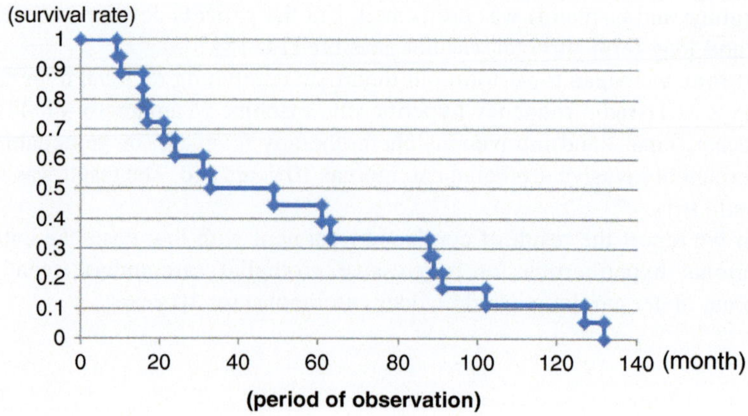

Fig. 27.1 Survival curve of urothelial cancer patients

27.2.4 *Results*

The treatment responses were as follows: 7 patients in CR (complete response), 3 patients in PR (partial response), one patient in NC (no change), and 7 patients in PD (progressive disease). A total of 10 cases (CR + PR) showed a response, so the response rate was 55.5 %. Ten patients remained alive, (survival duration: 17–-132 months; median: 61 months). Eight patients died (survival duration: 9–-91 months; median: 24 months). One patient died of pneumonia and seven patients of cancer (Fig. 27.1). Five patients temporarily developed minor general fatigue, but the treatment could continue.

27.2.5 *Discussion*

Progressive urothelial carcinoma tends to recur and is refractory even if the patients undergo operation. The patients with bladder cancer in particular often undergo transurethral resection (TUR) as an initial treatment but the cancer is highly recurrent.

Intravesical instillation therapy with bacillus Calmette-Guérin is effective to prevent recurrence [17], but the cancer frequently recurs. It progresses to invasive cancer after repeating TUR. Then total cystectomy is performed.

When the cancer reappears regardless of the treatment mentioned above, systemic chemotherapy is administered but the treatment has serious side effects and negatively affects the general status of the patient. Thus a long survival is not expected [13, 14].

Our hospital has developed a treatment without side effects for a good clinical result and long survival duration [16]. 18 cases of urothelial cancer treated in our institution for 11 years were enrolled in this study.

The treatment resulted in 10 cases of CR and PR out of these 18 cases (response rate: 55.5 %).

Ten patients are still alive (duration of survival: 17–132 months median: 61 months) but eight patients died. One of the eight patients died of pneumonia caused by a complication from a fall. The remaining 7 patients eventually died of cancer (duration of survival: 9–91 months; median: 24 months). While many of the surviving patients were stage T2, 3 patients underwent hyperthermia therapy after radiation.

Thus, similar to prostate cancer, the local control is important even for progressive carcinomas, so hyperthermia after radiation therapy is very effective. As mentioned in the article on prostate cancer above, metronomic low-dose chemotherapy with anticancer drug is able to inhibit metastasis and generate a synergy effect with hyperthermia [9]. In addition, it improves immunity through HSP, and accordingly patients are able to enhance their QOL [8]. The treatment contributes to longer-term survival.

27.3 Conclusion

In summary, the hyperthermia treatment combined with low-dose anticancer drug proved to be effective for high-grade prostate cancer and progressive urothelial carcinomas without serious side effects, leading to a better prognosis.

References

1. Nakai Y, Nishimura K, Nakayama M, Uemura M, Takayama H, Nonomura N, Tsujimura A. Weekly, low-dose docetaxel combined with estramastine for Japanese castration-resistant prostate cancer: its efficacy and safety profile compared with tri-weekly standard-dose treatment. Int J Clin Oncol. 2014;19:165–72.
2. Uchida T, Nagata Y, Hyodo T, Saito T, Irie A. Treatment of localized prostate cancer using high-intensity focused ultrasound. BJU Int. 2006;97:56–61.
3. Ueda K, Aoki Y, Hachiya Y. Combination therapy combining low-dose cisplatin and regional hyperthermia for treatment of progressive hormone refractory prostatic carcinoma. Thermal Med. 2007;23:95–100.
4. Ueda K, Maeda F, Hasegawa T, Ito Y: Castration-resistant prostate cancer treated with low -dose chemotherapy and regional hyperthermia for obtaining long survival. Thermal Med 2014;30(Suppl):153.
5. Petrylak DP, Tangan CM, Hussan MHA, Primo NL, Jeffer AJ, Mary ET, Patrick AB, Donna B, Carol M, Manish K, Mitchell CB, Eric JS, Derec R, David C. Docetaxel and estramustine compared with mitoxantrone and prednisone for advanced refractory prostate cancer. N Engl J Med. 2004;351:1513–20.

6. Scher HI, Fizazi K, Saad R, Taplin ME, Sternberg CN, et al. Increased survival with Enzalutamide in prostate cancer after chemotherapy. N Engl J Med. 2012;367:1187–97.
7. Ryan CJ, Smith MR, Fizazi S, Saad F, Sternberg CN, Miller K, Logothetis CJ, et al.: Abiraterone acetate plus prednisone versus placebo plus prednisone in chemotherapy-naïve men with metastatic castration-resistant prostate cancer (COU-AA-302): final overall survival analysis of a randomised, double-blind, placebo-controlled phase 3 study. doi.org/10.1016/S1470-2045(14)71205-7.
8. Miyata K, Hasegawa T, Maeda K, Ando S, Amano S, Fukuyama A, et al. The activation of immunological activity and anti-tumor effects by mild hyperthermia. Jpn J Hyperthermic Oncol. 2005;21:33–45.
9. Takahashi Y, Nishioka K. Survival without tumor shrinkage: reevaluation of survival gain by cytostatic effect of chemotherapy. J Natl Cancer Inst. 1995;87:1262–3.
10. Ono H, Ando S, Suzuki T, Monzen H, Amano M, Terai K, Takahashi T, Hasegawa T.: The drug uptake in the tumor with mild hyperthermia treatment in combination with chemotherapy in vivo. Jpn J Hyperthermic Oncol 2006;22:23–33.
11. Abe M, Hiraoka H, Takahashi M, Egawa S, Matsuda T, Onoyama Y, Morita K, Kakehi M, et al. Multi-institutional studies on hyperthermia using 8-MHz radiofrequency capacitive heating device (Thermotron RF-8) in combination with radiation for cancer therapy. Cancer. 1986;58:1589–95.
12. Cockett ATK, Kazman M, Nakamura R, Fingerhut A, Stein JJ. Enhancement of regional bladder megavoltage irradiation in bladder cancer using local bladder hyperthermia. J Urol. 1967;97:1034–9.
13. Nakajima K, Hisazumi H, Misaki T, Sugata T, Koshida K, et al. Hyperthermic perfusion therapy using Peplomysin for bladder cancer. Acta Urol Jpn. 1985;31:1559–63.
14. Saxman SB, Propert KJ, Einhorn LH, Crawford ED, Tnnock I, et al. Long-term follow up of a phase 3 intergroup study of cisplatin alone or in combination with methotrexate, vinblastine and doxorubicin in patients with metastatic urothelial carcinoma: a cooperative group study. J Clin Oncol. 1997;15:2564–9.
15. Gunlusoy B, Arslan M, Vardar E, Degirmenci T, Kara C, Ceylan Y, Kozacioglu Z. The efficacy and toxicity of gemcitabine and cisplatin chemotherapy in advanced/metastatic bladder urothelial carcinoma. Actas Urol Esp. 2012;36:515–20.
16. Ueda K, Maeda F, Ito Y. Combined treatment with low-dose chemotherapy and regional hyperthermia for progressive urothelial cancer. Themal Med. 2011;27:109–12.
17. Morales A, Nickel JC, Wilson JWL. Dose response of bacillus Calmete-Guerin in the treatment of superficial bladder cancer. J Urol. 1992;157:1256–8.

Chapter 28
Combination by Hyperthermia and Surgery

Takayuki Asao, Hiroyuki Kuwano, Takeo Takahashi, and Takashi Nakano

Abstract Hyperthermia combined with radiation and/or chemotherapy has been used as a preoperative adjuvant therapy in patients with esophageal and rectal cancer. The intraluminal heating method with radiofrequency or microwave was employed first, and then replaced by the external heating method as new equipment was developed. For the treatment of deep-seated tumors, optimizing and standardizing the heating operation is important for improving the effects of radiation and chemotherapy. Pathological CR was seen in 20–30 % of patients treated with preoperative hyperthermo-chemoradiation therapy for esophageal and rectal caner. An anal-preserving operation for tumors near the anus could be possible in patients with pathological CR. In addition to the prevention of local recurrence, improvement of survival has been expected; however, an effective adjuvant chemotherapy that can reduce distant metastasis has not been established and warrants further clinical trials.

In conclusion, preoperative hyperthermia combined with radiation and/or chemotherapy with surgical procedures following may benefit patients with locally advanced cancer by decreasing local recurrence and increasing functional preservation.

Keywords Rectal cancer • Anal preservation surgery • RF hyperthermia

---

T. Asao (✉)
Big Data Center for Integrative Analysis, Gunma University Initiative for Advance Research (GIAR), 3-39-22 Showa-machi, Maebashi, Gunma 371-8511, Japan
e-mail: asao@gunma-u.ac.jp

H. Kuwano
Department of General Surgical Science, Gunma University, 3-39-22 Showa-machi, Maebashi, Gunma 371-8511, Japan

T. Takahashi
Department of Radiology, Saitama Medical Center, Saitama Medical School, 1981 Kamodatsujido-cho, Kawagoe, Saitama 350-8550, Japan

T. Nakano
Department of Radiology and Radiation Oncology, Gunma University, 3-39-22 Showa-machi, Maebashi, Gunma 371-8511, Japan

S. Kokura et al. (eds.), *Hyperthermic Oncology from Bench to Bedside*,
DOI 10.1007/978-981-10-0719-4_28

28.1 Neoadjuvant Therapy with Hyperthermia Before Surgery

Preoperative adjuvant therapy with hyperthermia has been reported in patients with esophageal cancer, lower rectal cancer, and sarcoma. The primary aim of treatment before surgery is to prevent local recurrence after resection; therefore, the therapy is indicated only in patients with esophageal cancer and lower rectal cancer in the field of surgery because these organs are not covered with serosa, and tumor cells can easily invade the surrounding tissues. These scattering cells cause local recurrence after "curative" surgical resection. Surgeons have intended for hyperthermia to enhance the efficacy of chemotherapy and radiation therapy. In experimental studies, the effect of hyperthermia alone is seen at 40–42 °C; however, clinical heating to a high temperature is impossible in deep-seated organs such as the esophagus and rectum. Practically, the type of hyperthermia utilized before surgery is so-called "mild hyperthermia," and most of these therapies include concomitant chemotherapy and/or radiation therapy with hyperthermia. The role of hyperthermia is expected to be a radiation or chemo sensitizer. The great advantage of hyperthermia for an adjuvant therapy before surgery is that it is tolerable for a majority of patients without severe toxicity [1], and it is preferred as a preoperative treatment to reduce complications after surgery.

28.1.1 Development of Hyperthermia for Adjuvant Preoperative Therapy

One of the most serious problems of preoperative hyperthermia is the difficulty in heating deep-seated tumors that are surgically resectable. Several methods have been developed to heat tumors properly and safely. As a general method for preoperative hyperthermia, intraluminal heating has been replaced by external heating as effective external heating equipment has been developed. First, an intraluminal microwave device with radiation therapy for rectal cancer was used [2–4]. Intraluminal radio frequency capacitive [5, 6] and extracorporeal radiofrequency equipment [7, 8] followed the microwave devices. Ohno et al. reported a 30 % pathologic CR rate in rectal cancer patients treated with an intraluminal radiofrequency capacitive device combined with chemoradiation [6]; unfortunately, the endo-RF system is no longer available. Recently, 8 MHz of radiofrequency equipment (Thermotron RF-8, Yamamoto VINITA Co., Ltd., Osaka, Japan) was employed for deep hyperthermia in Japan [7]. In Europe, the annular phased-array system was typically used for patients with rectal cancer [8, 9]. In most trials, external deep hyperthermia was commonly performed once a week for approximately 1 h. Intraoperative hyperthermia has been reported immediately after amputation or resection of the rectum with hot saline with a cytotoxic agent in the pelvic cavity [10].

There have been several reports of preoperative hyperthermia of sarcoma [11–13]; however, they are not described in this chapter.

28.2 Preoperative Adjuvant Therapy with Hyperthermia for Esophageal Cancer

In 1994, a randomized prospective study was undertaken to determine the effect of hyperthermia in patients with resectable squamous cell carcinoma of the thoracic esophagus [14–16]. In these reports, the pathological CR rate was 25 % in the hyperthermo-chemoradiation group, while only 5.9 % in the chemoradiation group. Moreover, the 3-year survival rate was reported to be 50.4 % in the group with hyperthermia and 24.2 % in the group without hyperthermia. The same survival benefit was also reported in another study [17, 18]. A phase II study of chemoradiation with external hyperthermia using an AMC phased array of four 70-MHz antennas showed that 66 % of patient underwent R0 operations, and pathologic CR was seen in 19 % of patients [19].

28.3 Chemoradiation Therapy with Hyperthermia for Rectal Cancer

Chemoradiation therapy before surgery for locally advanced rectal cancer is a standard treatment when the lesion is located in the lower part of the rectum. Large-scale randomized trials for advanced rectal cancer were conducted and, as a result, preoperative therapy combined with long-term radiation and concomitant chemotherapy became the standard protocol [20].

The additional effect of hyperthermia on unresectable or recurrent colorectal cancer has been reported [21, 22]. Recently, a phase II trial showed that 92.2 % of R0 resections were achieved in patients with locally advanced fixed T4 rectal cancer using chemoradiation with local hyperthermia [23]. The same effect of hyperthermia when combined with the chemoradiation therapy before surgery was expected in patients with resectable rectal cancer [24]. The aim of preoperative treatment was the prevention of local recurrence after surgical resection the first time; however, anal preservation and the prolongation of overall survival after surgery are also expected today [25].

The typical length of time between the end of preoperative treatment and surgery is 6–8 weeks, considering the prolonged effects of chemoradiation and the time needed for normal tissue to recover from radiation. In our retrospective study of cases with short waiting times, a correlation can be seen [26], and there was no case with CR in which surgery was performed within 4 weeks of preoperative chemoradiation therapy's conclusion.

28.3.1 Modified Protocol of Hyperthermo-Chemoradiation Therapy

In Fig. 28.1, our modified protocol of hyperthermo-chemoradiation therapy (HCRT) for lower advanced rectal cancer is shown. After evaluating the rectal cancer with pelvic MRI, PET-CT, CT, endoscopy, and pathological examination of the rectal tumor with biopsy specimens, chemoradiation therapy combined with RF hyperthermia is started. The total dose of radiation is 50 Gy (2 Gy × 25) with Intensity Modulated Radiation Therapy (IMRT) for 5 weeks. Oral administration of capecitabine is synchronized with radiation therapy for 5 days per week, 1700 mg/m^2/day. This dose of capecitabine is relatively small, as it is expected to enhance the patient's sensitivity to radiation. In our original protocol of hyperthermo-chemoradiation with RF heating (HCRT), constant infusion of 5-FU only at night was employed in anticipation of effecting chronotherapy [27], and, since 2011, 5-FU infusion has been replaced by oral administration of capecitabine, a treatment available on an outpatient basis [28, 29]. Clinical trials have also demonstrated that adding oxaliplatin to radiation as an intensive concurrent chemotherapy did not improve surgical outcomes [30].

In our study, hyperthermia was performed using Thermotron RF-8 (Yamamoto VINITA Co., Ltd., Osaka, Japan), which can heat the deep parts of the body extracorporeally with an RF antenna. RF hyperthermia was administrated 5 times for 5 weeks with 50 min irradiation on the same day to start the radiation cycle. The method used to control the energy output of the equipment is described in 28.3.5.

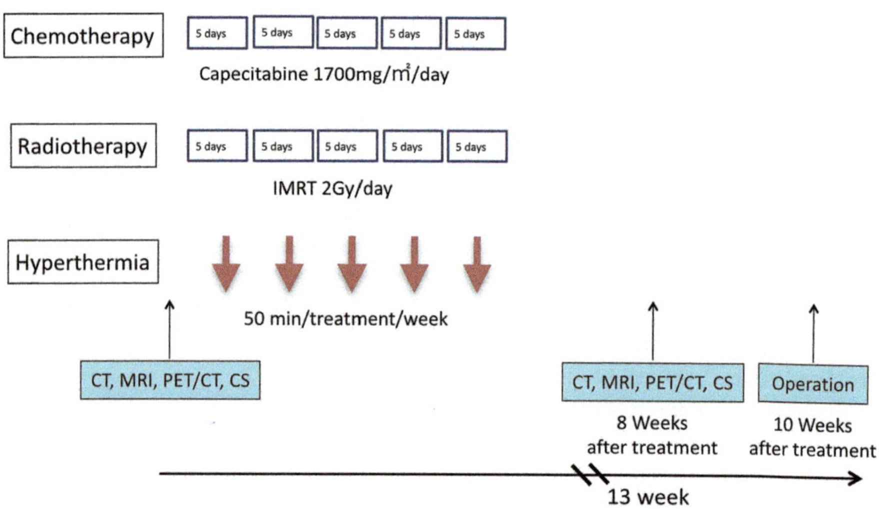

Fig. 28.1 HCRT treatment schedule. Preoperative treatment for locally advanced low rectal cancer is performed for 5 weeks, and surgery is carried out with a 10-week interval

Patients underwent surgical resection after HCRT. The interval between HCRT and surgery was 8–10 weeks. The waiting time before surgery was decided in accordance with our retrospective study [26].

28.3.2 Radiation Field for Lower Rectal Cancer

A short-term radiation schedule with whole pelvic field radiation seemed to result in severe complications [30]. The EORTC 22921 trial, in which a narrow radiation field was used, showed good response results and a low rate of complications [31]. We also changed the radiation field from the conventional whole pelvic field to a narrow field (Fig. 28.2) when the total dose of radiation was increased from 40 Gy to 50 Gy.

Total mesorectal excision (TME) is a standard operating procedure in which precise and sharp dissection is undertaken around the integral mesentery of the hindgut [32]. Figure 28.3 shows the dissection line around mid rectum. During the operation after preoperative chemo-radiation therapy, TME is employed; the mesorectum, which is the pathway of lymph-node metastasis along superior rectal artery, is resected surgically. Therefore, radiation therapy at the level of S1-2 might be wasted because metastatic lesions in the upper part of whole pelvic field radiation are resected by TME. Moreover, local recurrence common after anterior rectal resection for lower rectal cancer is seen near the tumor bed while recurrence in the upper part of the presacral space seldom occurs. [33] There is no theoretical reason for whole pelvic field radiation in preoperative treatment for lower rectal cancer, since we recommend limitation to a narrow field of radiation focused on presacral adjustment of the tumor and lateral lymph nodes near the rectum. The narrow field focused on the necessary site may contribute to a reduction in the radiation and surgical complications, especially by decreasing the small intestine's exposure to radiation.

28.3.3 Indications of HCRT for Rectal Cancer

Preoperative chemoradiation therapy is performed in patients with advanced lower rectal cancer; cT3, T4, and/or positive lymph nodes; and without distant metastasis by pretreatment examinations including PET-CT and MRI in accordance with guidelines. As a phase II clinical trial, the case of cT2 is included in the indication criteria when the lesion is located near the puborectal muscle, an important component of the anal sphincter, and abdominoperineal resection is needed for resection with no residual lesion (R0).

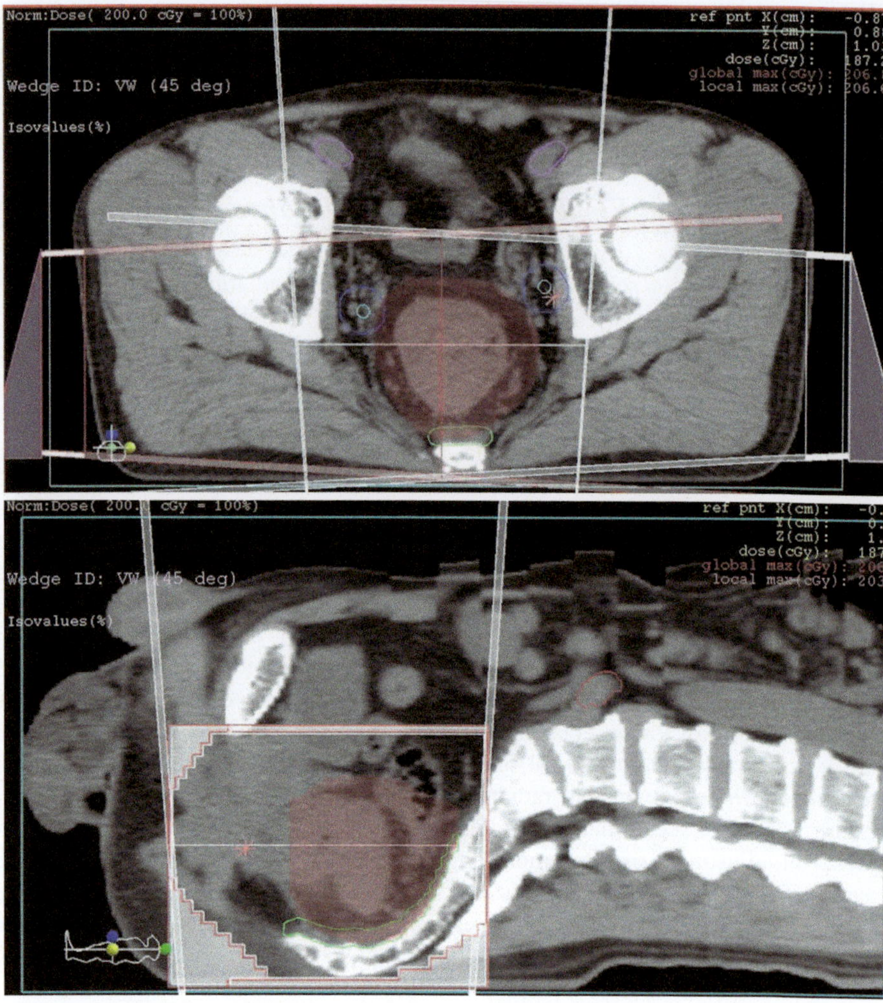

Fig. 28.2 Typical planning CT for HCRT. The *upper* margin of the radiation field is located at the S3 level. A localized field that includes the rectal tumor, anus, and lateral lymph nodes is employed

28.3.4 CR Rate and the Survival Benefit of HCRT for Rectal Cancer

In the pretreatment of rectal cancer, approximately 20 % of the pathologic complete response (pCR) rate was achieved by HCRT for rectal cancer [27, 34]. Hyperthermia seemed to increase the response of both T and N stages as compared to chemoradiation without hyperthermia in the retrospective study [25]. The same pCR rate (20.0 %) was seen in our clinical trial using RF hyperthermia with chemoradiation therapy as mentioned in this chapter.

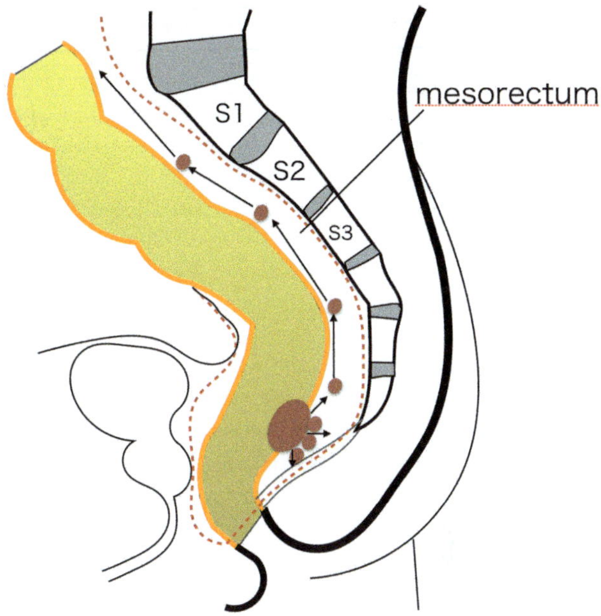

Fig. 28.3 Superior pathway of cancer spreading from a low rectal tumor. The tumor is spreading along the superior rectal artery, since the lymph node metastasis is located in the mesorectum. These metastatic lesions are removed by total mesorectal excision (TME). The resection line of TME is shown with a broken *red line* (- - - -). As for preoperative radiation therapy for lower rectal tumors, the superior margin of the radiation field at the S3 level is reasonable for avoiding intestinal irradiation

The Dutch Colorectal Cancer Group reported that short-term radiation therapy without chemotherapy had no survival benefit as compared to TME alone [35]. Since survival is mainly determined by distant metastases, efforts to increase survival benefits should be directed toward preventing systemic disease. To improve overall survival as well as local control, more intensive chemotherapy as a neoadjuvant or adjuvant therapy after operation will be needed in HCRT [36].

28.3.5 Standardization of RF Hyperthermia of the Rectum

Heating in deep hyperthermia could depend on the energy balance in the tumor, that is, the inflow of heat provided from the hyperthermia equipment and the outflow removed by the body's cooling mechanism. There has been a trial that measured temperature using MRI, but it is not yet available clinically [37]. Measuring the temperature of a patient's deep-site tumor is not practical, since it is difficult to estimate whether the heat energy received during hyperthermia is optimal or not. The thermal dose is not the same for all tumors, even when using the same heating procedure for the same duration with the same output setting of the equipment. Another problem of hyperthermia is that no standardization for treatment with reference points has been established, such as the recommended mg/m^2 for chemotherapy or Gy for radiation therapy. Standardization parameters that can increase the efficacy of the treatment are required for hyperthermia to be approved as a standard treatment method in oncology.

Moreover, radio-frequency (RF) hyperthermia has a fundamental flaw, the hot spot phenomenon, which does not allow continuation of the treatment without lowering the output. RF output, in practice, is started from 300 W and increased by the operator to 1200 W until output-limiting symptoms occur. As these symptoms are seen, the operator decreases the output level, increasing it again when the pain disappears, that is, the energy output is roughly adjusted by the operator from patient to patient according to operator's experience.

Shoji et al. [38] reported from a retrospective evaluation that patients with output-limiting symptoms showed greater thickness of the abdominal wall, fat areas of the intestinal organs, and total fat area than did those without output-limiting symptoms. They proposed a new heating control rule for radio-frequency hyperthermia in consideration of a patient's thickness of fat of the abdominal wall, visceral fat area, and total fat area. An output control method was proposed, in which the tolerance energy output is decided in the first session, and the same tolerant output is used as the starting point of heating after the first session. Figure 28.4. shows the energy output of the conventional heating method (above) and the new output control (below). Patients can receive the maximum level of heating energy without complications by using this output-control method.

Further study is desired to establish the optimal thermal treatment for each patient. In other words, the personalization of hyperthermia using standardized parameters is needed.

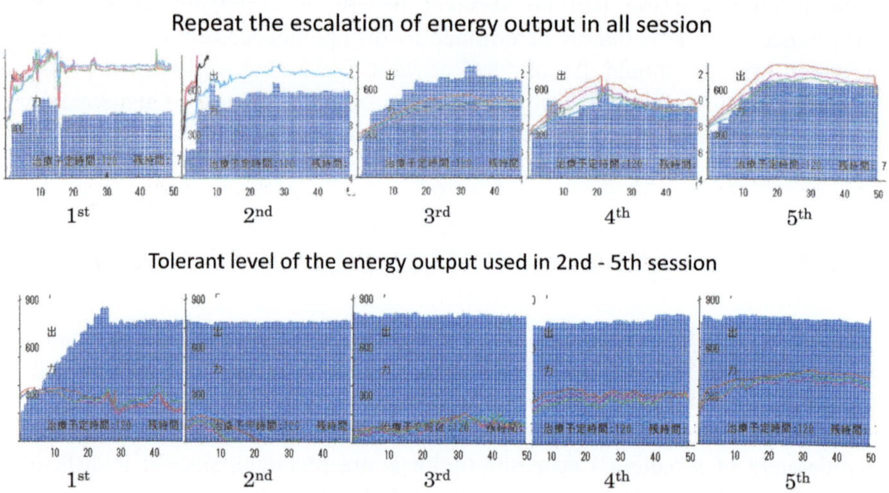

Fig. 28.4 The new method of controlling the energy output level during hyperthermia. In the conventional output control (*above*), escalation of the energy output is repeated in 5 sessions of hyperthermia. The tolerance level is determined in the first session and is used as the starting output level for the second to fifth treatment sessions (*below*)

28.3.6 Surgical Strategy for Patients with Clinical CR After HCRT

The strategy we devised for advanced lower rectal cancer was altered histologically as shown in Table 28.1, in which the sphincter-preserving rate for advanced lower rectal cancer is also shown. The rate of sphincter preservation was increased according to the development of treatment. Introduction of intersphincteric resection (ISR) for patients with clinical CR in 2006 augmented the sphincter preservation rate to 70 %. An increased CR rate could permit increased indications for sphincter-preserving surgery, including local excision. Schroeder et al. showed that hyperthermia combined with chemoradiation increased the rates of complete pathological response and of sphincter-sparing surgery [39]. Additionally, it is possible that local resection of primary lesion will be allowed in the selected cases of anal or low-lying rectal cancer (Fig. 28.5.). The feasibility of anal-preserving surgery mainly depends on the local response to preoperative treatment. Therefore, increased efficacy of hyperthermia in adjuvant therapy will be beneficial for patients with rectal cancer near the anus. Oncological safety and indications will be clarified in future clinical studies [40].

Besides the functional preservation of the anus, improvement of survival is an important goal of preoperative treatment for patients with advanced lower rectal cancer. Overall survival is dependent on the control of distant metastatic lesions. Radiation and hyperthermia are the modality of local treatment, as well as surgery. Therefore, pre- or post-operative chemotherapy may be needed to improve survival after surgery. Breugom et al. reported that adjuvant fluorouracil-based chemotherapy did not improve overall survival, disease-free survival, or distant recurrences [41, 42]. However, they also indicate that the adjuvant chemotherapy might benefit patients with tumors 10–15 cm from the anal verge, in terms of disease-free survival and distant recurrence. In the phase II randomized trial, improvement of

Table 28.1 Increase of anal preservation rate along with improvement of strategy with preoperative therapy and surgery for advanced low rectal cancer

Year	Preoperative therapy	Surgical procedures	Rate of anal preservation
1991-	Radiotherapy	TME with lateral lymph-node dissection	33.3 %
1996-	Radiotherapy with hyperthermia	TME alone	50 %
2002-	Radiotherapy (40 Gy), Hyperthermia, 5-FU civ. at night	TME alone	71 %
2006-	Radiotherapy (50 Gy), Hyperthermia, 5-FU civ at night	TME or ISR	90.2 %
2011-	Radiotherapy (50 Gy), Hyperthermia, Capecitabine p.o.	TME or ISR	77 %

General Surgical Science, and Radiation Oncology, Graduate School of Medicine, Gunma University
Department of Radiology and Radiation, Graduate School of Medicine, Gunma University

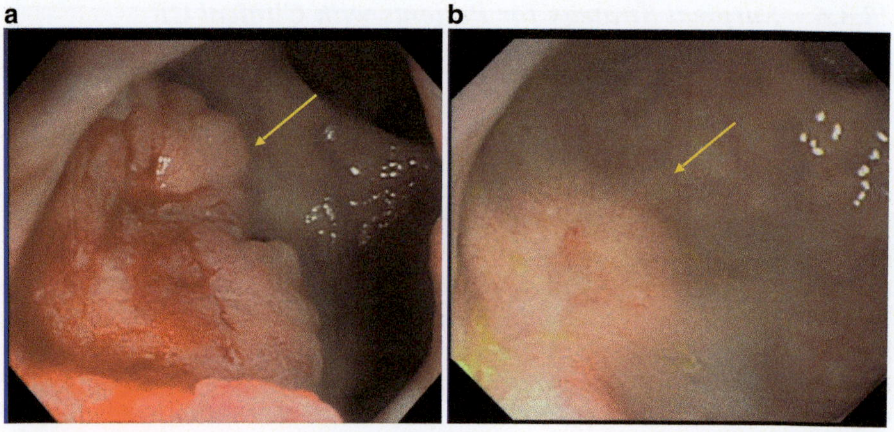

Fig. 28.5 Endoscopic findings before and after HCRT. (**a**) Endoscopic examination before HCRT shows locally advanced rectal cancer. The anal canal was involved at the distal margin of the tumor. (**b**) The tumor has disappeared and appeared as a scar just before the operation. Pathological CR is proven by the specimen of local excision with a partial external sphincter muscle

disease-free survival was observed in the group of adjuvant FOLFOX as compared with fluorouracil-based chemotherapy in patients with locally advanced rectal cancer after preoperative chemoradiotherapy and total mesorectal excision [43]. This finding needs to be further investigated in a randomized controlled trial to obtain the benefits of chemotherapy.

28.4 Prediction of Response After HCRT for Rectal Cancer

Several randomized trials indicated that the long-term preoperative treatment and an interval time of more than 8 weeks before the surgical procedure are recommended for advanced rectal cancer [20]. Neoadjuvant treatment over a long period has induced investigators to search for a good prediction marker that can estimate the response to preoperative treatment beforehand [44–46].

Regarding HCRT response, Ebara et al. [46] retrospectively investigated the biopsy specimens from primary lesions of rectal cancer before preoperative treatment and reported that L-type amino-acid transporter 1 (LAT1) expression can predict the response to preoperative HCRT in advanced rectal cancer. In this study, LAT1-negative tumors had an 81.8 % probability of a good response and an 18.2 % probability of a poor response. Shioya et al. [45] also reported immunostaining of hypoxia-inducible factor 1α (HIF-1α) using biopsy specimens as a survival marker for rectal cancer patients treated with HCRT. They showed that the overall 3-year survival rate in the HIF-1α-negative group was 85.2 %, which was significantly better than 60.6 % in the HIF-1α-positive group.

In the future, personalized medicine for preoperative treatment might be possible based on these markers using immunostaining on biopsy specimens. Further studies using a large number of samples to validate and reconfirm these findings are warranted.

28.5 Hyperthermia from the Bench to the Bedside

As mentioned above, HIF-1α [45], a hypoxic marker, was shown to be a prognostic marker for rectal cancer patients after surgical resection with preoperative HCRT. The tumor microenvironment is commonly hypoxic, poorly perfused, and exhibits abnormally high interstitial fluid pressure. Such a microenvironment can significantly reduce the efficacy of chemo and radiation therapies. Recently, cancer stem cells have also shown resistance potential to chemo and radiation therapy. These dormant cells in the hypoxic microenvironment may cause tumor residue after chemo and/or radiation therapy in the primary lesion and distant metastasis.

In the mouse xenograft model, hyperthermia can reduce the interstitial fluid pressure of tumors and enhance the effect of radiation [47]. Yamauchi et al. [48] showed that cluster cells, aggressive cancer cells, are resistant to chemotherapy as compared with the same cells under the monolayer culture condition and, conversely, that cluster formation can induce hyperthermia sensitivity (Fig. 28.6). In this study, using a 40-μm pore filter, researchers selected a cell cluster of small aggressive cancer cells to avoid necrosis in the center of a large spheroid.

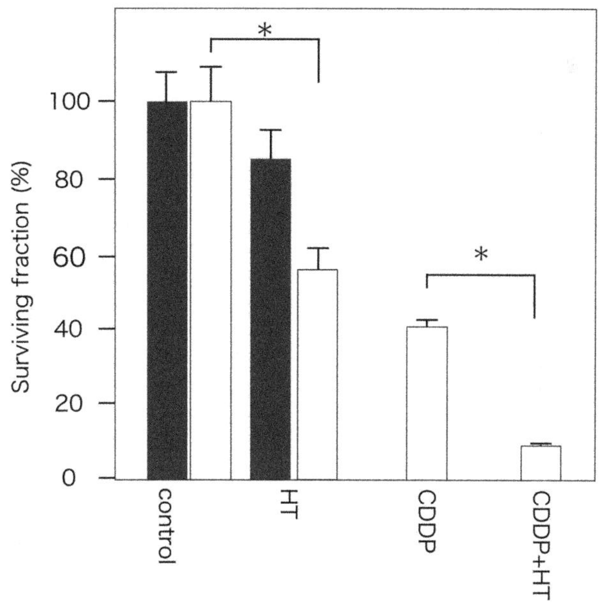

Fig. 28.6 Cluster cells in vitro are more sensitive to hyperthermia than are monolayer cells. Cluster cells (□) were more sensitive to hyperthermia than were monolayer cells (■). However, cluster formation induced cis-diamino-dichloroplatinum (CDDP) resistance. No monolayer cells formed colonies after single CDDP and CDDP combined with hyperthermia

Interestingly, they showed that hyperthermia sensitivity reverts to that under the monolayer condition after dispersion to single cells; the change in sensitivity is reversible.

These results from the "bench" support the significance of combining hyperthermia with chemo and radiation therapy at the "bedside," as hyperthermia has the potential to treat cancer cells in the hypoxic microenvironment that are resistant to radiation and chemotherapy.

References

1. Triantopoulou S, Efstathopoulos E, Platoni K, Uzunoglou N, Kelekis N, Kouloulias V. Radiotherapy in conjunction with superficial and intracavitary hyperthermia for the treatment of solid tumors: survival and thermal parameters. Clin Transl Oncol. 2012;15:95–105.
2. You QS, Wang RZ, Suen GQ, Yan FC, Gao YJ, Cui SR, et al. Combination preoperative radiation and endocavitary hyperthermia for rectal cancer: long-term results of 44 patients. Int J Hyperthermia. 1993;9:19–24.
3. Berdov BA, Menteshashvili GZ. Thermoradiotherapy of patients with locally advanced carcinoma of the rectum. Int J Hyperthermia. 1990;6:881–90.
4. Mori M, Sugimachi K, Matsuda H, Ohno S, Inoue T, Nagamatsu M, et al. Preoperative hyperthermochemoradiotherapy for patients with rectal cancer. Dis Colon Rectum. 1989;32:316–22.
5. Ichikawa D, Yamaguchi T, Yoshioka Y, Sawai K, Takahashi T. Prognostic evaluation of preoperative combined treatment for advanced cancer in the lower rectum with radiation, intraluminal hyperthermia, and 5-fluorouracil suppository. Am J Surg. 1996;171:346–50.
6. Ohno S, Tomoda M, Tomisaki S, Kitamura K, Mori M, Maehara Y, et al. Improved surgical results after combining preoperative hyperthermia with chemotherapy and radiotherapy for patients with carcinoma of the rectum. Dis Colon Rectum. 1997;40:401–6.
7. Furuta K, Konishi F, Kanazawa K, Saito K, Sugawara T. Synergistic effects of hyperthermia in preoperative radiochemotherapy for rectal carcinoma. Dis Colon Rectum. 1997;40:1303–12.
8. Rau B, Wust P, Hohenberger P, Löffel J, Hünerbein M, Below C, et al. Preoperative hyperthermia combined with radiochemotherapy in locally advancedrectal cancer: a phase II clinical trial. Ann Surg. 1998;227:380–9.
9. Rau B, Wust P, Tilly W, Gellermann J, Harder C, Riess H, et al. Preoperative radiochemotherapy in locally advanced or recurrent rectal cancer: regional radiofrequency hyperthermia correlates with clinical parameters. Int J Radiat Oncol Biol Phys. 2000;48:381–91.
10. Takahashi M, Fujimoto S, Kobayashi K, Mutou T, Kure M, Masaoka H, et al. Clinical outcome of intraoperative pelvic hyperthermochemotherapy for patients with Dukes' C rectal cancer. Int J Hyperthermia. 1994;10:749–54.
11. Issels RD, Lindner LH, Verweij J, Wust P, Reichardt P, Schem BC, et al. Neo-adjuvant chemotherapy alone or with regional hyperthermia for localised high-risk soft-tissue sarcoma: a randomised phase 3 multicentre study. Lancet Oncol. 2010;11:561–70.
12. Wendtner CM, Abdel-Rahman S, Krych M, Baumert J, Lindner LH, Baur A, et al. Response to neoadjuvant chemotherapy combined with regional hyperthermia predicts longterm survival for adult patients with retroperitoneal and visceral high-risk soft tissue sarcomas. J Clin Oncol. 2002;20:3156–64.
13. ESMO/European Sarcoma Network Working Group. Soft tissue and visceral sarcomas: ESMO Clinical Practice Guidelines for diagnosis, treatment and follow-up. Ann Oncol. 2012;23 Suppl 7:vii92–9.

14. Sugimachi K, Kuwano H, Ide H, Toge T, Saku M, Oshiumi Y. Chemotherapy combined with or without hyperthermia for patients with oesophageal carcinoma: a prospective randomized trial. Int J Hyperthermia. 1994;10:485–93.
15. Kitamura K, Kuwano H, Watanabe M, Nozoe T, Yasuda M, Sumiyoshi K, Saku M, Sugimachi K. Prospective randomised study of hyperthermia combined with chemoradiotherapy for esophageal carcinoma. J Surg Oncol. 1995;60:55–8.
16. Saeki H, Kawaguchi H, Kitamura K, Ohno S, Sugimachi K. Recent advances in preoperative hyperthermochemoradiotherapy for patients with esophageal cancer. J Surg Oncol. 1998;69:224–9.
17. Wang J, Li D, Chen N. Intracavitary microwave hyperthermia combined with external irradiation in the treatment of esophageal cancer. Zhonghua Zhong Liu Za Zhi. 1996;18:51–4.
18. Ohga T, Kimura Y, Futatsugi M, Miyazaki M, Saeki H, Nozoe T. Surgical and oncological advances in the treatment of esophageal cancer. Surgery. 2002;131(1 Suppl):S28–34.
19. Hulshof MC, Van Haaren PM, Van Lanschot JJ, Richel DJ, Fockens P, Oldenborg S, et al. Preoperative chemoradiation combined with regional hyperthermia for patients with resectable esophageal cancer. Int J Hyperthermia. 2009;25:79–85.
20. Ortholan C, Francois E, Thomas O, Benchimol D, Baulieux J, Bosset JF, et al. Role of radiotherapy with surgery for T3 and resectable T4 rectal cancer:evidence from randomized trials. Dis Colon Rectum. 2006;49:302–10.
21. Nishimura Y, Hiraoka M, Akuta K, Jo S, Nagata Y, Masunaga S, et al. Hyperthermia combined with radiation therapy for primarily unresectable and recurrent colorectal cancer. Int J Radiat Oncol Biol Phys. 1992;23:759–68.
22. van der Zee J, González González D, van Rhoon GC, van Dijk JD, van Putten WL, Hart AA. Comparison of radiotherapy alone with radiotherapy plus hyperthermia in locally advanced pelvic tumours: a prospective, randomised, multicentre trial. Dutch Deep Hyperthermia Group. Lancet. 2000;355:1119–25.
23. Barsukov YA, Gordeyev SS, Tkachev SI, Fedyanin MY, Perevoshikov AG. Phase II study of concomitant chemoradiotherapy with local hyperthermia and metronidazole for locally advanced fixed rectal cancer. Colorectal Dis. 2013;15:1107–14.
24. Ohno S, Sumiyoshi Y, Mori M, Sugimachi K. Hyperthermia for rectal cancer. Surgery. 2002;131(1 Suppl):S121–7.
25. Kang MK, Kim MS, Kim JH. Clinical outcomes of mild hyperthermia for locally advanced rectal cancer treated with preoperative radiochemotherapy. Int J Hyperthermia. 2011;27 (5):482–90.
26. Kato T, Fujii T, Ide M, Takada T, Sutoh T, et al. Effect of long interval between hyperthermochemoradiation therapy and surgery for rectal cancer on apoptosis, proliferation and tumor response. Anticancer Res. 2014;34(6):3141–6.
27. Asao T, Sakurai H, Harashima K, Yamaguchi S, Tsutsumi S, Nonaka T, et al. The synchronization of chemotherapy to circadian rhythms and irradiation in pre-operative chemoradiation therapy with hyperthermia for local advanced rectal cancer. Int J Hyperthermia. 2006;22:399–406.
28. Tsutsumi S, Tabe Y, Fujii T, Yamaguchi S, Suto T, Yajima R, et al. Tumor response and negative distal resection margins of rectal cancer after hyperthermochemoradiation therapy. Anticancer Res. 2011;31:3963–7.
29. O'Connell MJ, Colangelo LH, Beart RW, Petrelli NJ, Allegra CJ, Sharif S, et al. Capecitabine and oxaliplatin in the preoperative multimodality treatment of rectal cancer: surgical end points from National Surgical Adjuvant Breast and Bowel Project trial R-04. J Clin Oncol. 2014;32:1927–34.
30. Peeters KC, van de Velde CJ, Leer JW, Martijn H, Junggeburt JM, Kranenbarg EK, et al. Late side effects of short-course preoperative radiotherapy combined with total mesorectal excision for rectal cancer: increased bowel dysfunction in irradiated patients--a Dutch colorectal cancer group study. J Clin Oncol. 2005;23:6199–206.
31. Bosset JF, Calais G, Mineur L, Maingon P, Radosevic-Jelic L, Daban A, et al. Enhanced tumorocidal effect of chemotherapy with preoperative radiotherapy for rectal cancer:preliminary results--EORTC 22921. J Clin Oncol. 2005;23:5620–7.

32. MacFarlane JK, Ryall RD, Heald RJ. Mesorectal excision for rectal cancer. Lancet. 1993;341:457–60.
33. Joye I, Haustermans K. Clinical target volume delineation for rectal cancer radiation therapy: time for updated guidelines? Int J Radiat Oncol Biol Phys. 2015;91:690–1.
34. Maluta S, Romano M, Dall'oglio S, Genna M, Oliani C, Pioli F, et al. Regional hyperthermia added to intensified preoperative chemo-radiation in locally advanced adenocarcinoma of middle and lower rectum. Int J Hyperthermia. 2010;26:108–17.
35. Peeters KC, Marijnen CA, Nagtegaal ID, Kranenbarg EK, Putter H, Wiggers T, et al. The TME trial after a median follow-up of 6 years: increased local control but no survival benefit in irradiated patients with resectable rectal carcinoma. Ann Surg. 2007;246:693–701.
36. Bosset JF, Calais G, Mineur L, Maingon P, Stojanovic-Rundic S, Bensadoun RJ, et al. EORTC Radiation Oncology Group. Fluorouracil-based adjuvant chemotherapy after preoperative chemoradiotherapy in rectal cancer: long-term results of the EORTC 22921 randomised study. Lancet Oncol. 2014;15(2):184–90.
37. Gellermann J, Wlodarczyk W, Feussner A, Fähling H, Nadobny J, Hildebrandt B, et al. Methods and potentials of magnetic resonance imaging for monitoring radiofrequency hyperthermia in a hybrid system. Int J Hyperthermia. 2005;21:497–513.
38. Shoji H, Motegi M, Osawa K, Okonogi N, Okazaki A, Andou Y, et al. A novel strategy of radiofrequency hyperthermia (neothermia) in combination with preoperative chemoradiotherapy for the treatment of advanced rectal cancer: a pilot study. Cancer Med. 2015;4:834–43.
39. Schroeder C, Gani C, Lamprecht U, von Weyhern CH, Weinmann M, Bamberg M, et al. Pathological complete response and sphincter-sparing surgery after neoadjuvant radiochemotherapy with regional hyperthermia for locally advanced rectal cancer compared with radiochemotherapy alone. Int J Hyperthermia. 2012;28:707–14.
40. Ng AK, Recht A, Busse PM. Sphincter preservation therapy for distal rectal carcinoma: a review. Cancer. 1997;79:671–83.
41. Breugom AJ, Swets M, Bosset JF, Collette L, Sainato A, Cionini L, et al. Adjuvant chemotherapy after preoperative (chemo)radiotherapy and surgery for patients with rectal cancer: a systematic review and meta-analysis of individual patient data. Lancet Oncol. 2015;16:200–7.
42. Bosset JF, Calais G, Mineur L, Maingon P, Stojanovic-Rundic S, Bensadoun RJ, et al. Fluorouracil-based adjuvant chemotherapy after preoperative chemoradiotherapy in rectal cancer: long-term results of the EORTC 22921 randomised study. Lancet Oncol. 2014;15:184–90.
43. Hong YS, Nam BH, Kim KP, Kim JE, Park SJ, Park YS, et al. Oxaliplatin, fluorouracil, and leucovorin versus fluorouracil and leucovorin as adjuvant chemotherapy for locally advanced rectal cancer after preoperative chemoradiotherapy (ADORE): an open-label, multicentre, phase 2, randomized controlled trial. Lancet Oncol. 2014;15:1245–53.
44. Sato H, Suzuki Y, Ide M, Katoh T, Noda SE, Ando K, et al. HLA class I expression and its alteration by preoperative hyperthermo-chemoradiotherapy in patients with rectal cancer. PLoS One. 2014;9(9):e108122.
45. Shioya M, Takahashi T, Ishikawa H, Sakurai H, Ebara T, Suzuki Y, et al. Expression of hypoxia-inducible factor 1α predicts clinical outcome after preoperative hyperthermo-chemoradiotherapy for locally advanced rectal cancer. J Radiat Res. 2011;52(6):821–7.
46. Ebara T, Kaira K, Saito J, Shioya M, Asao T, Takahashi T, et al. L-type amino-acid transporter 1 expression predicts the response to preoperative hyperthermo-chemoradiotherapy for advanced rectal cancer. Anticancer Res. 2010;30(10):4223–7.
47. Winslow TB, Eranki A, Ullas S, Singh AK, Repasky EA, Sen A. A pilot study of the effects of mild systemic heating on human head and neck tumour xenografts: analysis of tumour perfusion, interstitial fluid pressure, hypoxia and efficacy of radiation therapy. Int J Hyperthermia. 2015;31:693–701.
48. Yamauchi H, Asao T, Yazawa S, Kigure W, Morita H, Kobayashi T, et al. Cell to cell interaction in clusters enhances thermosensitivity in HT29 human colon cancer cells. Hepatogastroenterology. 2011;58:1958–62.

Chapter 29
Combining Hyperthermia and Immunotherapy: NK Therapy and Hyperthermia

Hiroshi Terunuma, Noriyuki Nishino, Tsuyoshi Abe, Akiko Yoshimura, Atsushi Toki, Tatsuaki Ishiguro, Xuewen Deng, and Atsushi Terunuma

Abstract The NK cell is a major player in anti-cancer immunity, and hyperthermia augments host immunity against cancer. In cancer patients, NK cell activity is often suppressed. Therefore, when treating cancer patients with hyperthermia, it is desirable to restore the NK cell activity by co-administering a large number of highly activated NK cells that are prepared using a sophisticated cell culture strategy. This chapter describes *ex vivo*-expanded NK cells used in immunotherapy, the effects of hyperthermia on NK activity, and clinical applications of the combinatory therapy with hyperthermia and NK cells.

Keywords NK cell • NK cell therapy • Immunotherapy • Hyperthermia • Fever-range hyperthermia • Clinical application

---

H. Terunuma (✉)
Tokyo Clinic, 1F Shin-Otemachi Building, 2-2-1 Otemachi Chiyoda-ku, Tokyo 100-0004, Japan

Southern Tohoku General Hospital, 7-115 Yatsuyamada, Koriyama 963-8563, Japan

Biotherapy Institute of Japan, 2-4-8 Edagawa Koutou-ku, Tokyo 135-0051, Japan
e-mail: terunuma@bij-net.com

N. Nishino • T. Abe
Southern Tohoku General Hospital, 7-115 Yatsuyamada, Koriyama 963-8563, Japan

A. Yoshimura • A. Toki • T. Ishiguro
Tokyo Clinic, 1F Shin-Otemachi Building, 2-2-1 Otemachi Chiyoda-ku, Tokyo 100-0004, Japan

X. Deng
Biotherapy Institute of Japan, 2-4-8 Edagawa Koutou-ku, Tokyo 135-0051, Japan

A. Terunuma
Tokyo Clinic, 1F Shin-Otemachi Building, 2-2-1 Otemachi Chiyoda-ku, Tokyo 100-0004, Japan

Biotherapy Institute of Japan, 2-4-8 Edagawa Koutou-ku, Tokyo 135-0051, Japan

S. Kokura et al. (eds.), *Hyperthermic Oncology from Bench to Bedside*,
DOI 10.1007/978-981-10-0719-4_29

29.1 Introduction

As described in Chap. 12, hyperthermia enhances immune responses and potentiates cancer treatment. The positive effects of hyperthermia on anti-cancer immune responses can be achieved when the host immune system is functioning adequately. In cancer patients, however, the number and activity of immune cells are often decreased due to immunosuppression by the cancer as well as by therapeutic modalities such as chemotherapy and radiation. Under such circumstances, hyperthermia should be combined with a means of immunopotentiation in order to increase the activity of anti-cancer cytotoxic immune cells and exert synergistic effects on cancer treatment.

In this chapter, we will discuss immunotherapy that uses cultured NK cells, as well as its utility in combinatory therapy with hyperthermia.

29.2 Immunotherapy Using NK Cells

NK cells have been used in cancer treatment for the past 30 years. In early clinical studies, it was a challenge to expand NK cells in cell culture while maintaining their strong cytotoxic activity. It is only in recent years that new technologies of NK cell culture have been developed to generate a large number of NK cells of potentiated cytotoxicity.

To date, several culture methods of NK cells from peripheral blood mononuclear cells (PBMCs) have been reported [1–5]. Most of them use allogeneic cancer cells or genetically modified cells as feeder cells to be co-cultured with PBMCs, which contain NK cells [2–5]. Such an approach, however, comes with the technical challenge of constantly keeping the feeder cells in good condition, as well as with the hurdle of obtaining formal approvals for clinical application. In contrast, we have developed an easy-to-use cell culture kit for *ex vivo* expansion of NK cells that does not require feeder cells or NK cell purification from PBMCs [6, 7]. The kit allows NK cells to increase in number for more than 4 weeks. By using the kit, the number of NK cells can sometimes expand more than several thousand-fold from PBMCs of healthy individuals, and 400-fold on average from PBMCs of patients with advanced cancer [1].

Most of the NK cells expanded with the kit demonstrate the cell surface phenotype of CD3-negative and CD56-bright when analyzed by flow cytometry (Fig. 29.1) [1, 8]. Unlike a CD56-bright NK cell subset in the peripheral blood, the *ex vivo*-expanded CD3-negative CD56-bright NK cells are active not only in cytokine production but also in cytotoxicity [1, 8, 9]. The cytotoxicity of the *ex vivo*-expanded NK cells is several times greater than that of CD56-dim NK cells in the peripheral blood [1, 8, 9]. This is explained, at least partly, by the following observations on the *ex vivo*-expanded NK cells: enhanced expression of activating NK receptors and increased secretion of IFN-γ upon stimulation

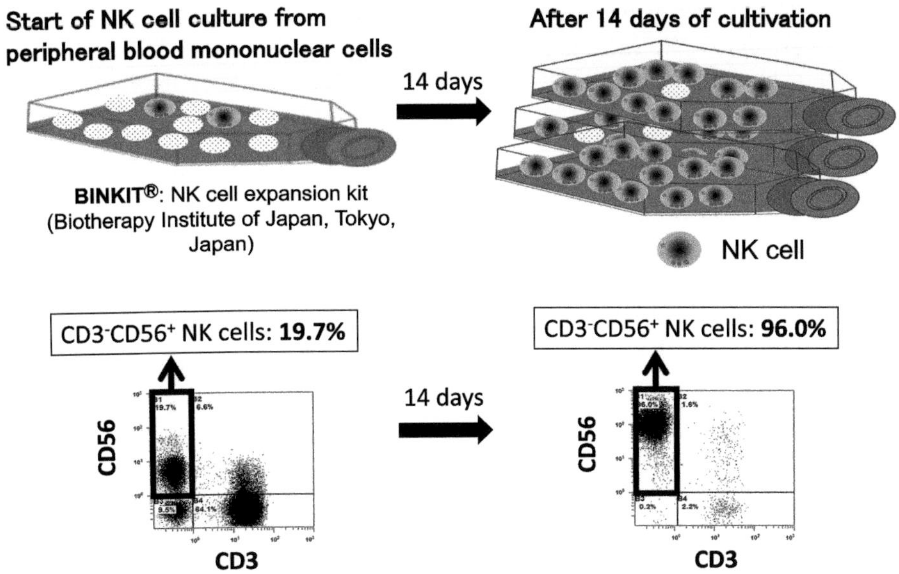

Fig. 29.1 A safe and simple NK cell expansion method. Based on [10]

[1, 8]. They also show a significantly higher cytotoxicity than uncultured NK cells against colon and uterine cancer cells that express a cancer stem cell marker CD133 [1, 9]. Furthermore, they potentiate antibody-dependent cell-mediated cytotoxicity (ADCC) when used with antibody drugs [1]. This is achieved through CD16 Fc receptors on the surfaces of NK cells, which recognize antibodies bound to the surfaces of target cells.

Additionally, the superiority of *ex vivo*-expanded NK cells over activated T cells, conventionally used for cancer immunotherapy, is evident. We have confirmed that the *ex vivo*-expanded NK cells produce more cytotoxic molecules (e.g., perforin and granzyme) and cytokines, and are more potent in cytotoxicity and ADCC compared to αβT cells or γδT cells [6, 9].

When the *ex vivo*-expanded NK cells are administered intravenously, a rise in the number of NK cells in the peripheral blood exceeds the number of administered NK cells [8, 10]. Moreover, the numbers of T and B cells in the peripheral blood are increased even if those types of cells have not been administered [8]. Cytotoxicity of PBMCs is also increased and sustained for at least 10 days after the intravenous administration [8, 10].

Adverse events associated with the intravenous administration of *ex vivo*-expanded NK cells are rare. Between 2010 and 2014, a total of 16,885 treatments were conducted using the *ex vivo*-expanded NK cells, and the associated adverse events were as follows: coughing and shortness of breath in four treatments (0.024 %); exacerbations of preexisting autoimmune diseases in three treatments (0.018 %). There were several cases of chills and low-grade fever, but there were no serious adverse events, and the treatment can therefore be considered safe.

The results indicate that NK cell therapy can be safely conducted using the *ex vivo*-expanded NK cells to increase the number of activated NK cells in the peripheral blood. This is the foundation of combinatory therapy with hyperthermia, which is expected to help the NK cells by providing microenvironments for enhanced immune reactions as described below.

29.3 Mechanisms of Enhanced Anti-Cancer Cytotoxicity in Combinatory Therapy with *Ex Vivo*-Expanded NK Cells and Fever-Range Hyperthermia

As discussed in Chap. 12, the effector function of NK cells is regulated by a balance between activating and inhibitory NK-receptor signaling pathways. Because the expression of MHC class I antigens frequently becomes reduced or lost when cells become malignant or metastatic (Table 12.1 in Chap. 12), the MHC class I-negative malignant cells cannot be destroyed by CTLs but can be destroyed by NK cells due to the attenuated inhibitory signals. In *ex vivo*-expanded NK cells, the expression of activating NK receptors, such as NKG2D and NKp44, is elevated, whereas that of inhibitory NK receptors, such as CD159a, shows no elevation [1]. On the other hand, fever-range hyperthermia on cancer cells stimulates the expression of an NK activation ligand MICA, whereas the expression of an NK inhibitory ligand MHC class I remains unchanged (Fig. 12.2 in Chap. 12) [1, 9, 11]. Collectively, *ex vivo*-expanded NK cells, which possess high numbers of activating NK receptors, receive strong activation signals from NK activation ligands expressed in abundance on fever-range hyperthermia-treated cancer cells, leading to aggressive attacks on cancer cells.

Hyperthermia also induces heat shock proteins (HSPs) in cancer cells, and enhances antigen-specific cytotoxic T lymphocyte (CTL) induction via efficient cross-presentation by dendritic cells (DCs) [12–14], which is also positively affected by Th1-type cytokines, such as IFN-γ, produced by *ex vivo*-expanded NK cells.

Furthermore, hyperthermia reinforces immunotherapy by removing Tregs so that NK cells, DCs, and CTLs are released from immunosuppression [15, 16].

As described above, the combination of NK cell therapy and hyperthermia is a promising approach to targeting cancer cells by potentiating both innate immunity (via enhanced NK cell cytotoxicity) and acquired immunity (via efficient CTL induction).

29.4 Clinical Application of Combinatory Therapy with Hyperthermia and *Ex Vivo*-Expanded NK Cells

Although several clinical trials of NK cell therapy have been done [17–21], no clinical trials on the combination of NK cell therapy with hyperthermia have been reported. Combinatory therapy using hyperthermia and *ex vivo*-expanded NK cells has been used in our clinic [22]. To the combinatory therapy, we sometimes add low-dose chemotherapy, which has been reported safe and effective when used with hyperthermia [23, 24].

Frequent and repeated use of low-dose chemotherapy has been proposed as metronomic chemotherapy, which is intended to suppress Treg and restore anti-cancer immune responses [25]. Additionally, by employing DNA-damaging agents (e.g., cisplatin and gemcitabine), NK cytotoxicity can be augmented via enhanced expression of NKG2D ligands, such as MICA/B, on cancer cells [26, 27].

Here we present a case of advanced cancer treated with the combination of hyperthermia, immunotherapy, and low-dose chemotherapy. The patient was an 85-year-old male patient with pancreatic cancer and metastatic liver tumors. He was treated with hyperthermia delivered using an 8-MHz capacitive heating device (Thermotron RF8 by Yamamoto VINITA Co., Ltd., Osaka, Japan) once a week, immunotherapy with NK cells and DC once every other week, and was given 600 mg of gemcitabine once every other week. Three months into the treatment, the primary lesion in the pancreas shrunk and the liver tumors disappeared (Fig. 29.2). Generally, the median survival time for stage IV pancreatic cancer patients under the standard treatment including chemotherapy is 7.5 months, and the 1-year survival rate is less than 20 % (Fig. 29.3a) [28]. However, when treated with a combination of hyperthermia, chemotherapy, and *ex vivo*-expanded NK cells (sometimes with DCs as well) with or without radiation treatment, the median survival time for stage IV pancreatic cancer patients is 13 months and the 1-year survival rate is more than 50 % (Fig. 29.3b and c).

As we reported previously, the combinatory therapy with hyperthermia and *ex vivo*-expanded NK cells has also been used as a palliative treatment [9, 22]. Patients with advanced cancer refractory to conventional therapies were treated with hyperthermia, *ex vivo*-expanded NK cells with or without cultured T cells and DCs, and sometimes low-dose chemotherapy. Hyperthermia was delivered using an 8-MHz capacitive heating device once every 1–3 weeks, and *ex vivo*-expanded NK cells were administered once every 3–4 weeks. Six months after the initial treatment, a quarter of the patients showed complete or partial responses, and another quarter had long stable diseases. Of note, the combinatory therapy can be performed at outpatient clinics, even on patients with progressive diseases, all the while maintaining their quality of life. The preliminary results suggest that the combinatory therapy with hyperthermia and *ex vivo*-expanded NK cells is an effective approach without severe adverse events to treating advanced cancer patients who are refractory to conventional therapies.

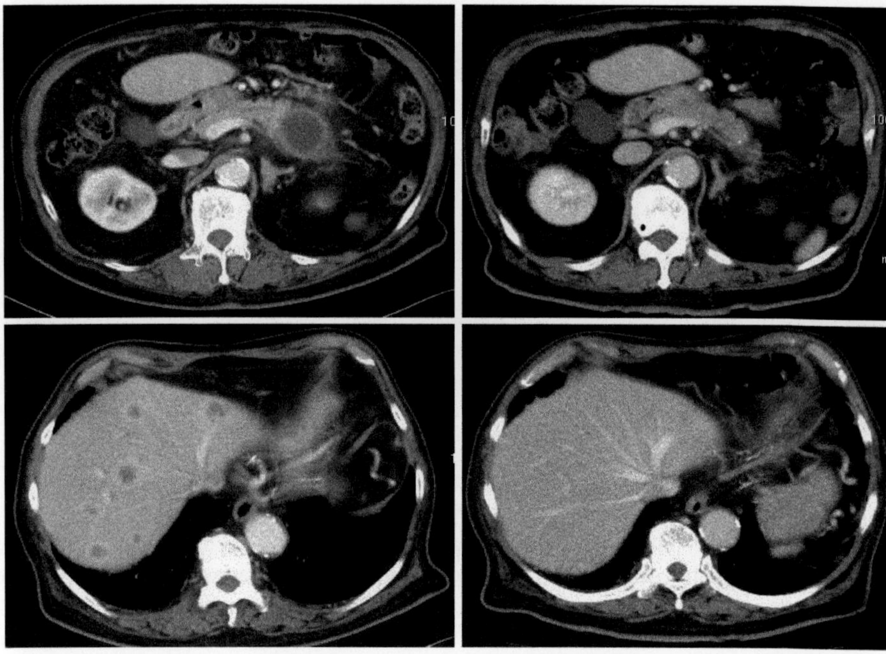

Fig. 29.2 Anti-cancer effects of autologous *ex vivo*-expanded NK cells. Images of an 85-year-old male patient with pancreatic cancer and multiple liver metastases. He received hyperthermia delivered using an 8-MHz capacitive heating device (Thermotron RF8 by Yamamoto VINITA Co., Ltd., Osaka, Japan) once a week, 600 mg of gemcitabine at 2-week intervals, and immunotherapy with *ex vivo*-expanded NK cells (7.9×10^9 cells/injection), *ex vivo*-expanded T cells (4.0×10^9 cells/injection), and dendritic cells (1.3×10^7 cells/injection) injected five times at 2-week intervals after receiving informed consent from the patient. Three months into the treatment, the primary lesion in the pancreas shrunk and the liver tumors disappeared.

Importantly, NK cell therapy may prevent both cancer onset and relapse. While it is difficult to evaluate the immune potency against cancer, NK activity can be assessed by measuring the cytotoxicity of PBMCs against cultured cancer cells using a relatively simple assay [1, 29]. In a study of 3625 healthy individuals, cancer incidence through 11 years of follow-up has been reported to be higher by 1.7-fold in individuals with low NK activity compared to those with medium to high NK activity [29]. We have also reported that NK activity is significantly lower in patients with breast cancer, either at the initial onset or relapse, compared to individuals without cancer [30].

NK cell therapy and hyperthermia can also be used during the perioperative period. NK activity has been shown to decline after surgery and can persist for weeks [31]. The decreased NK activity can increase the risk of cancer development, metastasis, or recurrence [29–32]. Therefore, it would be reasonable to treat

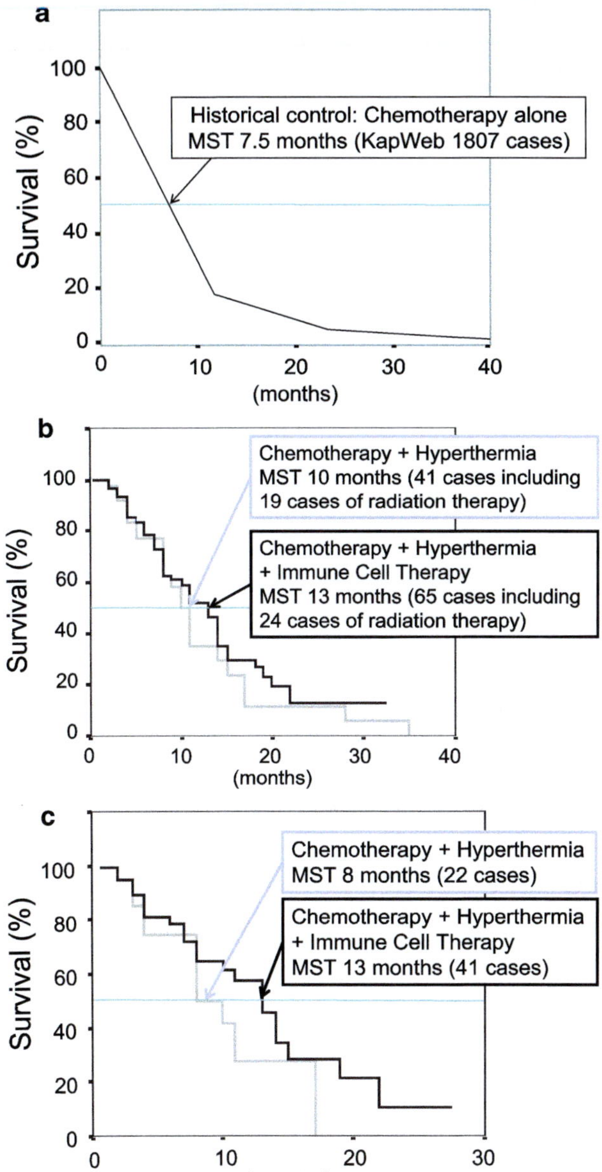

Fig. 29.3 A retrospective analysis of survival in patients with stage IV pancreatic cancer who were treated with hyperthermia and *ex vivo*-expanded NK cells. The survival of patients was compared between three patient subgroups: (i) our patients treated with hyperthermia without immune cell therapy; (ii) our patients treated with hyperthermia and immune cell therapy; (iii) 1807 patients in a data set based on the treatment outcomes of 32 Japanese hospitals in 2005 concerning cancer survival rates at a Japanese association of clinical cancer centers who received neither hyperthermia nor immune cell therapy and were instead treated with conventional

perioperative patients with *ex vivo*-expanded NK cells and hyperthermia, both of which potentiate NK activity, so that the risks of initial onset, metastasis, and recurrence can be reduced.

29.5 Conclusion

Hyperthermia modulates immune cells and sets the stage for enhancing anti-cancer immune reactions. By adding to it a large number of highly activated NK cells, which are major players in one's innate immunity against cancer, effective anti-cancer immunotherapy can be achieved. More clinical studies on the combinatory therapy shall tell us which types of cancer are best suited for the therapy in order to sustain patient lives and a good quality of life.

References

1. Deng X, Terunuma H, Nieda M, et al. Synergistic cytotoxicity of *ex vivo* expanded natural killer cells in combination with monoclonal antibody drugs against cancer cells. Int Immunopharmacol. 2012;14:593–605.
2. Berg M, Lundqvist A, McCoy P, et al. Clinical-grade *ex vivo*-expanded human natural killer cells upregulate activating receptors and death receptor ligands and have enhanced cytolytic activity against tumor cells. Cytotherapy. 2009;11:341–55.
3. Voskens CJ, Watanabe R, Rollins S, et al. *Ex-vivo* expanded human NK cells express activating receptors that mediate cytotoxicity of allogeneic and autologous cancer cell lines by direct recognition and antibody directed cellular cytotoxicity. J Exp Clin Cancer Res. 2010;29:134.
4. Lapteva N, Durett AG, Sun J, et al. Large-scale *ex vivo* expansion and characterization of natural killer cells for clinical applications. Cytotherapy. 2012;14:1131–43.
5. Fujisaki H, Kakuda H, Shimasaki N, et al. Expansion of highly cytotoxic human natural killer cells for cancer cell therapy. Cancer Res. 2009;69:4010–17.
6. Deng X, Terunuma H, Terunuma A, et al. *Ex vivo*-expanded natural killer cells kill cancer cells more effectively than *ex vivo*-expanded γδT cells or αβT cells. Int Immunopharmacol. 2014;22:486–91.
7. Dewan MZ, Terunuma H, Toi M, et al. Potential role of natural killer cells in controlling growth and infiltration of AIDS-associated primary effusion lymphoma cells. Cancer Sci. 2006;97:1381–7.

Fig. 29.3 (continued) therapies, mostly chemotherapy alone (**a**) [28]. All of our 106 patients received hyperthermia (**b**). Out of the 106 patients, 63 patients did not receive radiation treatment (**c**). Informed consent was obtained from all patients, and 65 out of 106 patients (41 out of the 63 patients without radiation therapy) received NK cell therapy. In addition to the NK cell therapy, some patients also received T cell therapy and dendritic cell therapy. *MST:* median survival time

8. Terunuma H, Deng X, Dewan Z, et al. Potential role of NK cells in the induction of immune responses: implications for NK cell-based immunotherapy for cancers and viral infections. Int Rev Immunol. 2008;27:93–110.
9. Terunuma H, Deng X, Toki A, et al. Effects of hyperthermia on the host immune system: from NK cell-based science to clinical application. Thermal Med. 2012;28:1–9.
10. Deng X, Ashiba K, Terunuma H, et al. Injection of *ex vivo* expanded NK cells activates NK cells in blood. Med Sci Dig. 2015;41:318–21 [Japanese].
11. Ostberg JR, Dayanc BE, Yuan M, et al. Enhancement of natural killer (NK) cell cytotoxicity by fever-range thermal stress is dependent on NKG2D function and is associated with plasma membrane NKG2D clustering and increased expression of MICA on target cells. J Leukoc Biol. 2007;82:1322–31.
12. Srivastava PK, Udono H, Blachere NE, Li Z. Heat shock proteins transfer peptides during antigen processing and CTL priming. Immunogenetics. 1994;39:93–8.
13. Binder RJ, Srivastava PK. Peptides chaperoned by heat-shock proteins are a necessary and sufficient source of antigen in the cross-priming of CD8+ T cells. Nat Immunol. 2005;6:593–9.
14. Takahashi T. Effects of hyperthermia and modification of thermosensitivity. Thermal Med. 2007;23:171–9.
15. Guo J, Zhu J, Sheng X, et al. Intratumoral injection of dendritic cells in combination with local hyperthermia induces systemic antitumor effect in patients with advanced melanoma. Int J Cancer. 2007;120:2418–25.
16. Terunuma H, Wada A, Deng X, et al. Mild hyperthermia modulates the relative frequency of lymphocyte cell subpopulations: an increase in a cytolytic NK cell subset and a decrease in a regulatory T cell subset. Thermal Med. 2007;23:41–7.
17. Krause SW, Gastpar R, Andreesen R, et al. Treatment of colon and lung cancer patients with ex vivo heat shock protein 70-peptide-activated, autologous natural killer cells: a clinical phase I trial. Clin Cancer Res. 2004;10:3699–707.
18. Iliopoulou EG, Kountourakis P, Karamouzis MV, et al. A phase I trial of adoptive transfer of allogeneic natural killer cells in patients with advanced non-small cell lung cancer. Cancer Immunol Immunother. 2010;59:1781–9.
19. Rizzieri DA, Storms R, Chen DF, et al. Natural killer cell-enriched donor lymphocyte infusions from A 3-6/6 HLA matched family member following nonmyeloablative allogeneic stem cell transplantation. Biol Blood Marrow Transpl. 2010;16:1107–14.
20. Curti A, Ruggeri L, D'Addio A, et al. Successful transfer of alloreactive haploidentical KIR ligand-mismatched natural killer cells after infusion in elderly high risk acute myeloid leukemia patients. Blood. 2011;118:3273–9.
21. Geller MA, Cooley S, Judson PL, et al. A phase II study of allogeneic natural killer cell therapy to treat patients with recurrent ovarian and breast cancer. Cytotherapy. 2011;13:98–107.
22. Terunuma H, Deng X, Nishino N, et al. NK cell-based autologous immune enhancement therapy (AIET) for cancer. J Stem Cells Regen Med. 2013;9:9–13.
23. Ueda K, Aoki Y, Hachiya Y. Combination therapy combining low dose cisplatin and regional hyperthermia for treatment of progressive hormone refractory prostatic carcinoma. Thermal Med. 2007;23:95–100.
24. Ueda K, Maeda F, Ito Y. Combined treatment with low dose chemotherapy and regional hyperthermia for progressive urothelial cancer. Thermal Med. 2011;27:109–12.
25. Pasquie E, Kavallaris M, Andre N. Metronomic chemotherapy: new rationale for new directions. Nat Rev Clin Oncol. 2010;7:455–65.
26. Gasser S, Orsulic S, Brown EJ, Raulet DH. The DNA damage pathway regulates innate immune system ligands for the NKG2D receptor. Nature. 2005;436:1186–90.
27. Xu X, Rao GS, Groh V, et al. Major histocompatibility complex class I-related chain A/B (MICA/B) expression in tumor tissue and serum of pancreatic cancer: role of uric acid accumulation in gemcitabine-induced MICA/B expression. BMC Cancer. 2011;11:194.
28. Survival statistics of Japanese association of Clinical Cancer Centers: survival rate census. https://kapweb.chiba-cancer-registry.org. Accessed 1 Dec 2015.

29. Imai K, Matsuyama S, Miyake S, et al. Natural cytotoxic activity of peripheral-blood lymphocytes and cancer incidence: an 11-year follow-up study of a general population. Lancet. 2000;356:1795–9.
30. Dewan MZ, Takada M, Terunuma H, et al. Natural killer activity of peripheral-blood mononuclear cells in breast cancer patients. Biomed Pharmacother. 2009;63:703–6.
31. Pollock RE, Lotzová E, Stanford SD. Mechanism of surgical stress impairment of human perioperative natural killer cell cytotoxicity. Arch Surg. 1991;126:338–42.
32. Tai LH, Zhang J, Scott KJ, et al. Perioperative influenza vaccination reduces postoperative metastatic disease by reversing surgery-induced dysfunction in natural killer cells. Clin Cancer Res. 2013;19:5104–15.

Chapter 30
Combination by Hyperthermia and Immunotherapy: DC Therapy and Hyperthermia

Tsutomu Takeda and Takashi Takeda

Abstract We treated 2322 patients with advanced or recurrent cancer using immunotherapy or hyperthermia (HT) (between July 2005 and March 2015). HT was used in 2235 cancer patients, cytokine-activated T cells (CAT) in 1256 patients, and dendritic cells (DC) in 1447 patients. Response to therapy was evaluated in 2217 patients. Clinical benefit case (complete response (CR), partial response (PR) or 6-month stable disease (SD)) was noted in 350 patients, including 35 who showed CR. The rate of beneficial effects from DC increased from 7.1 to 19.1 % of patients with the addition of HT. The rate of beneficial effects from DC + CAT increased from 15.4 to 21.6 % of patients who also received HT. Maximal response rate (21.6 %) was thus noted in patients who received combination therapy comprising CAT + DC + HT. DC with HT had beneficial effects on multiple metastases observed in multiple important solid organs in breast cancer cases.

Keywords Immunotherapy • Hyperthermia • Dendritic cells • Cancer

30.1 Introduction

Some basic studies have shown that hyperthermia (HT) restores and reinforces the immune system that has been suppressed by advanced cancer [1–4]. In addition, we investigated injection of dendritic cells (DC) into breast cancer masses at Osaka University in 1992, revealing subsequent degeneration and necrosis in cancer tissues [5]. Clinically, we performed immunotherapy for numerous cancer patients without HT until 2005, achieving partial response (PR) or even complete response (CR) in a few cases. After 2005, we performed immunotherapy with HT for many cancer patients, obtaining not only PR, but also relatively frequent CR at the Osaka Cancer Immuno-chemotherapy Center over 10 years. We found that HT enhanced

---

T. Takeda (✉) • T. Takeda
Osaka Cancer Immuno-chemotherapy Center, Kyohrinkai, 1-1-44, Tenmabashi, Kita-ku, Osaka 530-0042, Japan
e-mail: tom0221tom@spice.ocn.ne.jp

S. Kokura et al. (eds.), *Hyperthermic Oncology from Bench to Bedside*,
DOI 10.1007/978-981-10-0719-4_30

immunotherapy in clinical cases. DC therapies combined with HT offer more beneficial effects, particularly in cases with multiple metastases in multiple important solid organs.

30.2 Patients

We treated 2322 patients with advanced or recurrent cancer using immunotherapy or HT between July 2005 and March 2015. HT was applied using a Thermotron RF-8 heating device (Yamamoto VINITA Co., Ltd., Osaka, Japan) in 2235 cancer patients, cytokine-activated T cells (CAT) were administered in 1256 patients, and DC in 1447 patients. CAT were prepared using anti-CD3-activated peripheral lymphocytes and injected by intravenous drip. DC were usually injected into the tumor or regional lymph nodes.

Treatment comprised the combination of CAT + DC + HT in 825 patients, DC + HT in 567 patients, CAT + HT in 333 patients, and HT alone in 467 patients (Fig. 30.1).

The percentages of patients receiving chemotherapy and radiotherapy did not differ significantly between therapy groups.

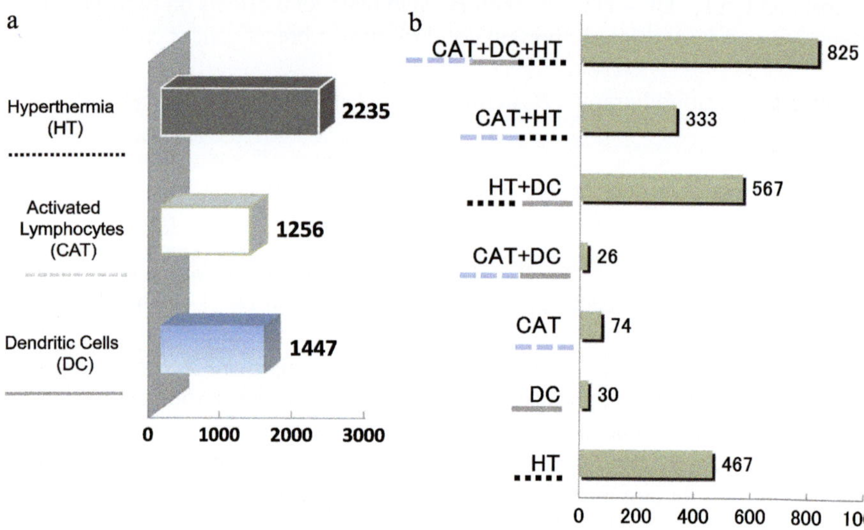

Fig. 30.1 Total treatment 2322 cases for advanced or recurrent cancers by hyperthermia or immunotherapy (between July 2005 and March 2015). (**a**) Classification by the kind of treatment. (**b**) Classification by the combination therapy

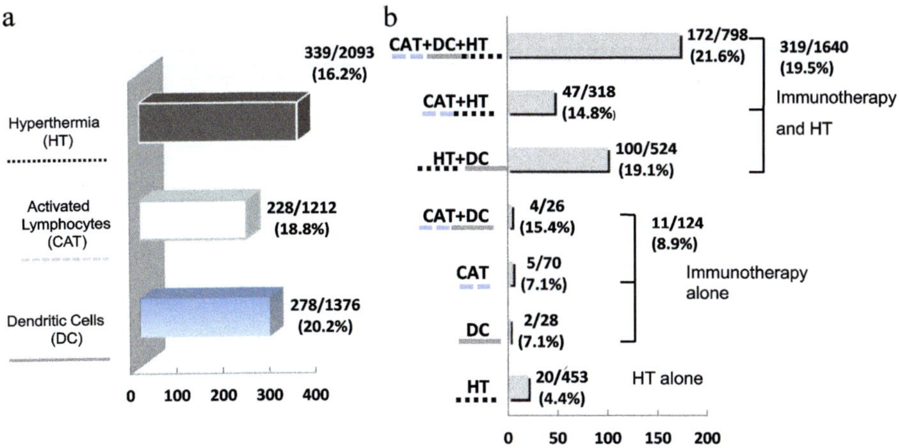

Fig. 30.2 Number and percentages of clinical benefit cases that showed CR, PR and long SD. (**a**) Classification by the kind of treatment. (**b**) Classification by the combination therapy. A total of 350 patients (15.7 %) among 2217 evaluated cases showed such response. (between July 2005 and March 2015)

30.3 Response to Therapy

The effects of treatment were judged according to the guidelines for response evaluation criteria in solid tumors (RECIST), such as CR, PR, stable disease (SD) and progressive disease (PD).

Response to therapy was evaluated in 2217 patients. Clinical benefit case (CR, PR or 6-month SD) was shown in 350 patients, including 35 who showed CR. The beneficial effect of DC increased from 7.1 to 19.1 % in patients who also received HT. The beneficial effect of DC + CAT increased from 15.4 to 21.6 % in patients who also received HT. The highest response rate (21.6 %) was thus noted in patients who received the combination therapy of CAT + DC + HT. However, response to HT alone was only 4.0 % (Fig. 30.2).

30.4 Recurrent Form and Effects

We analyzed effective cases according to the form of recurrence in several kinds of cancer. HT alone was effective for skin and regional lymphocytes. CAT was effective for liver or lung metastases. DC with HT had beneficial effects on multiple metastases observed in multiple important solid organs. Detailed data for breast cancer cases are shown in Table 30.1.

Table 30.1 Clinical data of 33 clinical benefit cases of evaluated advanced or recurrent breast cancer cases including 5 CR cases (2005/7 ~ 2013/11). (CR cases: ◎)

			Treatment					Regions of recurrence and metastases									Stage					
	NO.	name	HT	CAT	DC	drug	Effect◎: CR	LN	Bone	Liver	Lung	Dist-LN	Pleura, Peritoneum	Skin	Brain, Spine	others	IIIb, c	IV	HR	Her 2	Survival time	Prognosis
HT alone	1	MY1	○			○	○	○									○		+	-	2y10m	alive
	2	KM	○				○	○						○				○	+	-	1y1m	unknown
	3	SA	○				○	○						○				○	+	1+	3y6m	alive
CAT alone	4	MF1		○		○	○				○		○					○	+	-	1y0m	alive
HT and IT	5	MK	○	○	○	○	◎	○	○				○					○	+	3+	6y0m	alive
	6	NR	○		○		◎	○	○	○			○					○	-	-	1y11m	alive
	7	HM	○	○	○		○	○	○		○							○	+	-	1y6m	alive
	8	MF2	○	○	○	○	○		○					○				○	+	-	7y0m	alive
	9	OK1	○	○		○	○	○	○	○	○		○	○	○			○	+	-	3y2m	dead
	10	NI	○	○			○		○	○	○			○	○			○	+	-	1y5m	dead
	11	KE1	○	○	○	○	○	○		○	○	○						○	+	-	4y1m	alive
	12	OK2	○	○	○	○	○	○	○		○	○	○	○		Eye, Ovary		○	+	2+	4y4m	alive
	13	MM1	○	○	○	○	○	○	○									○	+	-	1y2m	unknown
	14	KY1	○	○	○	○	○	○	○	○		○	○		○			○	-	3+	3y0m	alive
	15	MT	○	○		○	○	○	○									○	+	-	2y8m	dead
	16	KE2	○		○		◎						○	○				○	+	-	0y9m	alive
	17	YF	○	○	○		◎	○			○		○					○	-	-	1y9m	alive
	18	MY3	○	○	○		○	○		○								○	+	-	0y6m	unknown
	19	TM	○	○	○	○	○									Contra-Breast	○		-	-	1y3m	unknown
	20	US	○	○	○		○	○	○		○		○					○	+	1+	3y5m	alive
	21	HK	○	○	○		○		○	○								○	+	-	1y7m	alive

	22	FY	○		○	○	○		○	○			○					○	+	-	1y5m	alive
	23	ON	○	○	○	○	○		○		○							○	+	2+	3y1m	alive
	24	MM2	○		○	○	○					○		○	○	Contra-Breast		○	+	3+	1y0m	alive
	25	IN	○		○	○	○				○							○	-	-	1y0m	alive
	26	KT	○		○	○	○				○		○					○	+	2+	1y0m	alive
	27	YM1	○	○	○	○	○	○	○		○		○	○		Contra-Breast		○	-	2+	2y2m	dead
	28	YM2	○	○	○	○	○	○	○		○		○		○			○	+	-	1y3m	unknown
	29	IT	○		○		○	○	○									○	+	1+	0y8m	unknown
	30	KY2	○	○	○		○	○	○	○	○		○					○	+	2+	0y8m	unknown
	31	MU	○		○		○				○							○	+	1+	1y1m	dead
	32	MY2	○	○	○	○	○		○									○	+	-	1y0m	alive
	33	NM	○	○		○	◉	○									○		+	1+	1y10m	unknown

HT hyperthermia, *CAT* activated lymphocyte, *DC* dendritic cell, *IT* immunotherapy (CAT and DC), *LN* regional lymph node, *Dist-LN* distant lymph node, *HR* hormonal receptor

30.5 Case Report

(Case 1: NR, No. 6 in Table 30.1) A 60-year-old woman was diagnosed with right breast cancer. Right mastectomy was performed in 2005, followed by chemotherapy. Immunohistochemical staining showed Er-, PgR-, Her2- triple-negative cancer. Five years after surgery, multiple bone and lymph node metastases were identified. Because her last chemotherapy had caused severe side effects, she declined chemotherapy this time. She was referred to our clinic in 2010, where treatment included weekly HT and injection of immature and peptide vaccine DC cells into the cervical lymph nodes four times. After these DC therapies, all lymph node and bone metastases disappeared (Fig. 30.3). This case was considered to have achieved clinical CR.

(Case 2: KE1, No. 11 in Table 30.1) A 60-year-old woman was diagnosed with right breast cancer and right mastectomy was performed in 1990, followed by endocrine therapy. Immunohistochemical staining showed Er+, PgR+, Her2-. Six years after surgery, metastases to the lung, liver and mediastinal lymph nodes were identified. New chemotherapy failed to induce a response. She was referred to our clinic in 2009, where treatment included weekly HT, and injection of immature and peptide vaccine DC cells into the cervical lymph nodes eight times. After these DC therapies, liver and mediastinal lymph node metastases disappeared and lung metastases were reduced in size (Fig. 30.4). This case was considered to have achieved clinical PR.

(Case 3: OT) A 63-year-old man was diagnosed with rectal cancer. Anterior resection was performed in 2009. Soon after surgery, abscess was identified in the

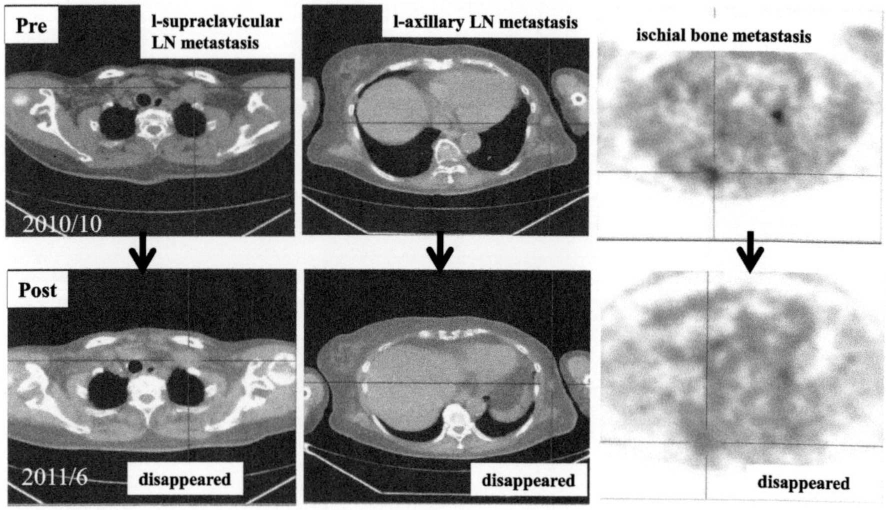

Fig. 30.3 (Case 1: NR, No.6 of Table 30.1). A 60-year-old female patient; right breast cancer with bone and lymph node metastases. CT and PET taken before and after the combination therapy of hyperthermia and DC. Note the disappearance of multiple lymph node and bone metastases

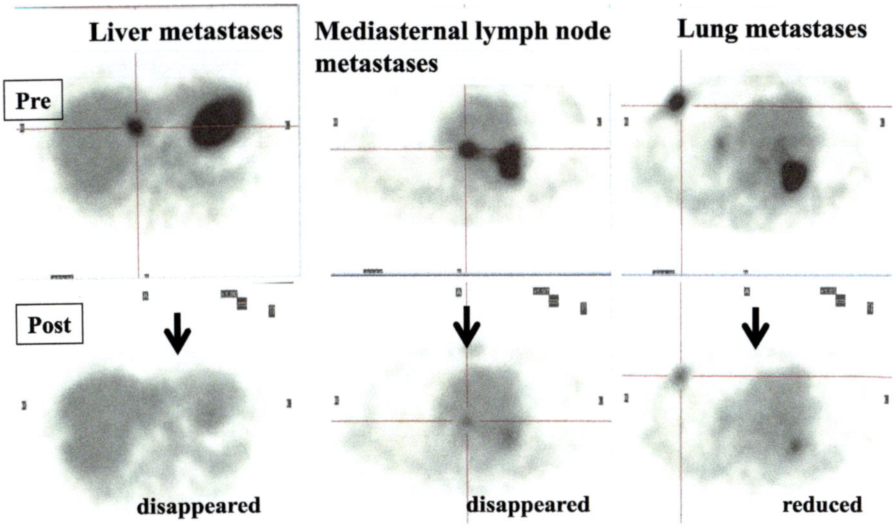

Fig. 30.4 (Case 2: KE1, No.11 of Table 30.1). A 60-year-old female patient; right breast cancer with liver, lung and lymph node metastases. PET/CT taken before and after the combination therapy of hyperthermia and DC. Note the disappearance of liver and lymph node metastases and reduction of lung metastases

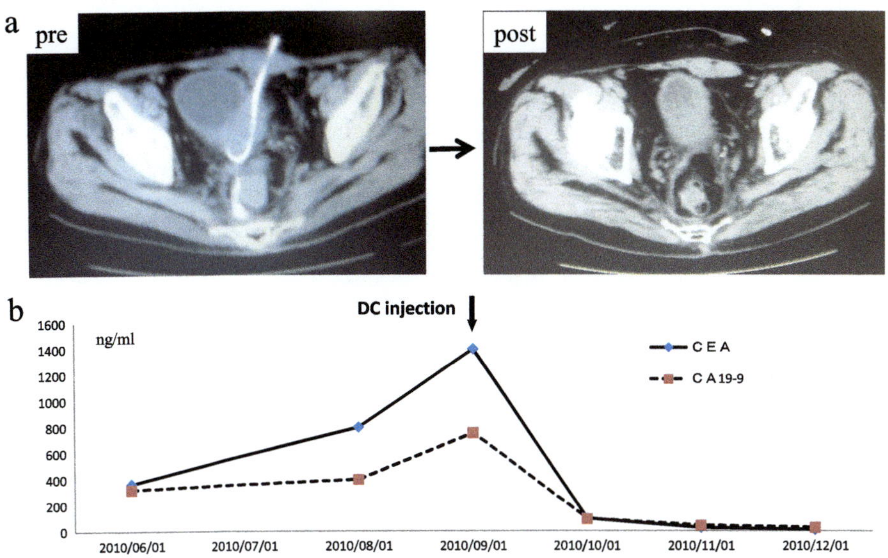

Fig. 30.5 (Case 3: OT). A 63-year-old male patient; rectal cancer with abscess and liver and lung metastases. (**a**) CT taken before and after the combination therapy of hyperthermia and DC. Note the disappearance of abscess in pelvic cavity. (**b**) The level of CEA and CA19-9 decreased to the normal range after the combination therapy of hyperthermia and DC

pelvic cavity, contraindicating chemotherapy. Liver and lung metastases were identified. HT was performed for 3 months, but metastases increased in size and tumor marker levels rose. He was referred to our clinic in July 2010. DC cells were injected into the left cervical lymph nodes four times with HT. Four months later, both the abscess and all metastases had disappeared, and all tumor marker levels had normalized (Fig. 30.5a and b). This case was considered to have achieved clinical CR.

(Case 4: OR) A 45-year-old woman was diagnosed with colon cancer in Australia. Left hemicolectomy was performed in 2003, followed by chemotherapy.

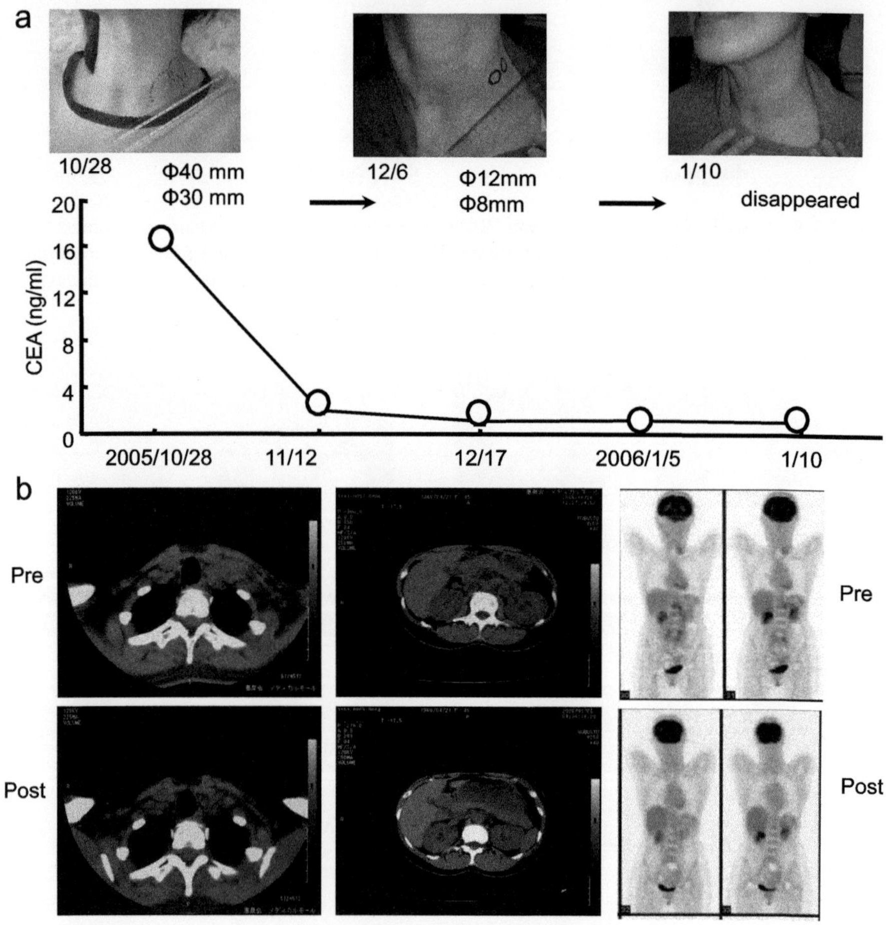

Fig. 30.6 (Case 4: OR). A 45-year-old female patient; colon cancer with multiple lymph node metastases. (**a**) Cervical metastatic tumors disappeared and the level of CEA decreased to the normal range after the combination therapy of hyperthermia, injection of DC and CAT. (**b**) CT and PET taken before and after the combination therapy of hyperthermia, injection of DC and CAT. Note the disappearance of multiple metastases in the cervical and abdominal regions following the treatment

Two years after surgery, multiple lymph nodes were identified in the paraaortic and left cervical regions. Chemotherapy with various chemotherapeutic agents failed to induce a response and the tumors increased in size. She was referred to our clinic in October 2005, where treatment included weekly HT (from 2005/11/7 to 2006/1/5) and injection of immature DCs into the left cervical lymph nodes twice (2005/11/1, 12/12), followed by intravenous infusion of CAT seven times (from 2005/11/7 to 2006/1/5). PET/CT conducted 3 months later showed the disappearance of all lymph nodes, as well as the normalization of all tumor marker levels (Fig. 30.6a and b). This case was considered to have achieved clinical CR.

30.6 Conclusion

Cancer patients are increasing year by year and new treatments are constantly being developed, but we have yet to gain control over cancer and the various remedies remain troubled by adverse effects. For cancer patients, minimizing the side effects while optimizing results is an important purpose.

HT alone does not show attractive results for cancer patients, but does appear to reinforce many therapeutic options. There are few side effects and many effects in DC therapy, but good results are not achieved anytime. DC therapy combined by HT offer a solid contribution to the multidisciplinary treatment of cancer patients.

References

1. Ito A, Shinkai M, Honda H, et al. Augmentation of MHC class I antigen presentation via heat shock ptotein expression by hyperthermia. Cancer Immunol Immunother. 2001;50:515–22.
2. Takahashi T, Mitsuhashi N, Sakurai H, et al. Modification of tumor-asociated antigen expression on human lung cancer cells by hyperthermia and cytokine. Anticancer Res. 1995;15:2601–6.
3. Ostberg JR, Kabingu E, Repasky EA. Thermal regulation of dendritic cell activation and migration from skin explants. Int J Hyperthermia. 2003;19:520–33.
4. Takeda T, Takahashi T, Yamamoto I, et al. Hyperthermia enhances immunotherapy in cancer patients: clinical and experimental analyses. Thermal Med. 2012;28:11–6.
5. Takeda T, Kobayashi T, Monden T, et al. The effect of local immunotherapy for breast cancer using a mixture of OK-432 and fibrinogen supplemented with activated macrophages. Biotherapy. 1994;7:47–54.

Chapter 31
Effects of Fever-Range Hyperthermia on T Cell-Mediated Immunity: Possible Combination of Hyperthermia and T Cell-Based Cancer Immunotherapy

Keishi Tanigawa, Yusuke Ito, and Yasunobu Kobayashi

Abstract Elevated body temperature has been thought to play an important role in the regulation of immune responses, and accumulating evidence in thermal medicine indicates that hyperthermia could be a useful combination therapy to enhance the efficacy of cancer immunotherapy. However, the intrinsic effects of elevated body temperature on the immune system are poorly understood, particularly in humans. Future clinical studies are expected to elucidate the practical utility of hyperthermia, and in particular fever-range whole-body hyperthermia in combination with T-cell and/or DC-based cancer immune cell therapy.

Keywords T cell • Fever-range hyperthermia • Cancer immune cell therapy

31.1 Introduction

Fever is regarded as one of the body's defense systems against infectious pathogens, and the concept that an increase in body temperature plays a beneficial role in immune defense has been well accepted [1]. Nonetheless, fever itself is a complex physiological response [2], and the intrinsic effects of elevated body temperature on the immune system are still under investigation. Indeed, a number of studies have investigated the effects of thermal stimuli on immunocompetent cells involved in both innate and adaptive immune function, and functional changes of immune cells including T cells, natural killer (NK) cells, and lymphokine-activated killer cells were examined in various experimental conditions. These earlier *in vitro* studies revealed that the temperature in an experiment is critical, and that around the fever range (*e.g.* 39–40.5 °C) and not the "heat shock" temperature (*e.g.* 41–43 °C) is the most favorable for immune responses [3].

K. Tanigawa (✉) • Y. Ito • Y. Kobayashi
Bio-Thera Clinic, 5-6-12 Shinjuku, Shinjuku-ku, Tokyo 1600022, Japan
e-mail: tanigawa@jb-t.co.jp

S. Kokura et al. (eds.), *Hyperthermic Oncology from Bench to Bedside*,
DOI 10.1007/978-981-10-0719-4_31

Here we summarize the effects of thermal stimuli on T cell responses, and in particular those in the so-called fever-range or physiological thermal stress on T cell responses. Also, the possible combination of fever-range thermal therapy with T cell-based immunotherapy is discussed.

31.2 Induction of T Cell Mediated Immune Responses

Figure 31.1 shows some pivotal events that are involved in the induction of successful T cell mediated immune responses and those in the immune effector phase. To activate and produce antigen specific effector T cells, they first need to encounter antigen presenting cells in secondary lymphoid organs such as draining lymph nodes and Peyer's patch. Dendritic cells (DC) are the only antigen presenting cells that are capable of initiating T cell-mediated immune responses [4]. These professional antigen-presenting cells usually reside in peripheral tissues and play the role of a kind of sentinel. When DC capture antigens including foreign antigens, self-antigens, and occasionally some fragments of tumor cells that were generated in consequence of the interaction of tumor and innate immune cells, DC leave peripheral tissue, invade into lymph vessels, migrate to secondary lymphoid organs such as lymph nodes and Peyer's patches, and finally get to the paracortical T cell area in the these secondary lymphoid organs. In the meantime, circulating lymphocytes, in particular CD62L and CCR7 positive cells continuously patrol the paracortical T cell areas in secondary lymphoid organs in order to search for their cognate antigens. These T cells in secondary lymphoid organs can encounter cognitive antigens when those antigen-loading DCs arrived at paracortical T cell-area in secondary lymphoid organs. Eventually DC present carrying antigens to

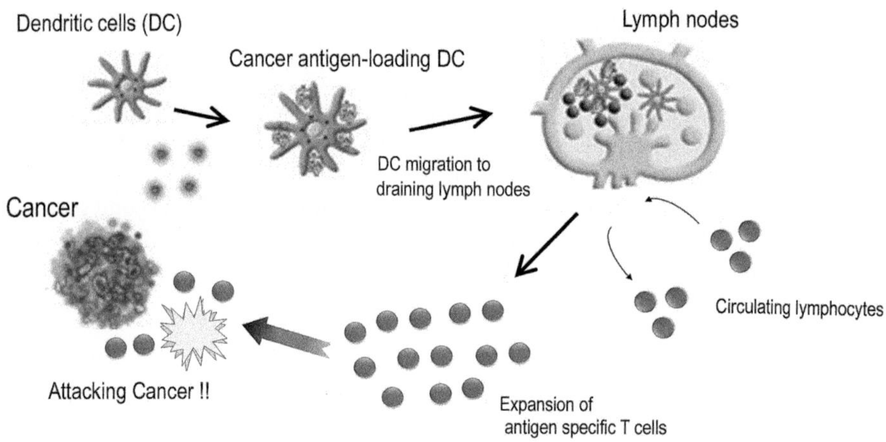

Fig. 31.1 T cell-mediated tumor immunogenicity

antigen specific T cells, and this is the very beginning of the induction phase of T cell mediated immune response.

It was shown *ex vivo* that this migration capacity of DC is increased by fever-range thermal treatment [5]. Using the mice ear skin explant culture system, Ostberg et al. tried to examine the effect of thermal stimuli on the migration capacity of epidermal Langerhans cells (LC); one of the best-characterized tissue-resident DC reside in the epidermis of the skin or mucosal tissues [6]. In this experiment system, spontaneous migration of LC from the epidermis to the dermis was observed when the ear skins were split into dorsal and ventral halves and cultured dermal side down in the culture medium. Using this system they found that short time culture (6.5 h) of skin explants at 40 °C significantly increased LC migration out of the epidermis. It was also shown that migrating LCs displayed a more matured phenotype, and higher stimulating capacity in a mixed lymphocyte reaction compared with controls. These results indicated that mild thermal stimuli enhanced DC activation and function, and thereby stimulated induction of T cell mediated immune responses.

After encounter the specific antigen, T cells are induced to proliferate and differentiate into effector T cells. After the elicitation phase, these effector cells invade target tissues and attach to target cells such as tumor cells as illustrated in Fig. 31.1. In the next section, we will see how thermal stress stimulates T cell function by each process described above.

31.3 Effects of Fever-Range Whole-Body Hyperthermia on T Cell Responses in Mice

A number of studies in mice system indicate that fever-range thermal stimuli increase the variety of T cell responses *in vitro*, *ex vivo* and *in vivo*. In some experiments, T cells were heated at fever-range *in vitro*, and in some of the *ex vivo* and *in vivo* studies, mice were placed in an environment chamber at approximately 39 °C for 6 h or more in order to increase and maintain their core body temperature in the fever range. Some representative studies are discussed in detail below.

31.3.1 Effects on T Cell Homing to Secondary Lymphoid Organs

To encounter antigens carried from peripheral tissues by DC, homing of circulating T cells to secondary lymphoid organs is an indispensable step. Some specific phenotypes of circulating cells such as CD62L positive cells are capable of access to high endothelial venules (HEVs) in lymph nodes and Peyer's patches, and they then extravasate across HEV into lymph nodes and Peyer's patches.

Wang et al. first showed that fever-range thermal stimuli *in vitro* increased in L-selectin–dependent adhesion of human lymphocytes to HEVs in lymph nodes and Peyer's patches [7]. They also showed that fever-range heat treatment stimulates L-selectin-dependent homing of murine splenocytes to peripheral lymph nodes and Peyer's patches in mouse systems. In addition, Evans et al. showed that fever-range whole-body heat treatment in mice stimulated CD62L and $\alpha 4\beta 7$ integrin-dependent adhesion of lymphocytes to HEVs [8]. Also, Chen et al. showed that fever-range thermal stress increased endothelial expression of intercellular adhesion molecule 1 and CCL21 chemokine, both of which are involved in enhanced trafficking of lymphocyte across HEVs [9]. These results suggested that hyperthermia, in particular whole-body fever-range hyperthermia, enhanced homing of naïve phenotype T cell to secondary lymphoid organs where they encounter cognate antigen-loading DC, and thereby accelerate the induction of T cell-mediated immune response.

From the results of these investigations, Chen and Evans et al. suggested a new perspective on the role of HEVs in immune defense system: "the 'HEV axis' as a thermally sensitive alert system that heightens immune surveillance during inflammation by amplifying lymphocyte trafficking to lymphoid organs" [9].

31.3.2 Effects on T Cell Differentiation into Effector Cells

Differentiation of T cells into effector cells is the next step, and it was also shown to be enhanced by fever-range hyperthermia. Mace et al. showed that temporary exposure of naïve T cells ($CD62L^{hi}$, $CD44^{lo}$) to 39.5 °C *in vitro* resulted in enhanced differentiation of these cells into effector phenotype cells ($CD62L^{lo}$, $CD44^{hi}$) compared to cells exposed at 33 and 37 °C [10]. This increased differentiation of naïve T cells was also observed following whole–body hyperthermia (39.5 °C, 6 h) *in vivo*. They showed that the increase in membrane fluidity of T cells resulted in various molecular events to facilitate effective interaction of T cells with antigen presenting cells. Indeed, an increase in the cluster formation of the $GM1^{+}$ CD-microdomain in $CD8^{+}$ T cells, clustering $TCR\beta$ and the CD8 co-receptor, and enhanced conjugate formation between T cells and antigen-presenting cells were observed by thermal stimuli. These results suggest that a heat stress-induced increase in membrane fluidity is one of the primary events, and it subsequently triggers a cascade of various molecular events that eventually make T cells crosstalk more rapidly and efficiently with antigen-presenting cells.

31.3.3 Effects on T Cell Infiltration in Tumor Tissue

After induction and expansion of antigen specific T cells, one of the critical steps for immune effector responses is efficient arrival and invasion of effector cells at the target site.

Using a hapten-induced contact hypersensitivity model in mice, Ostberg et al. examined the effects of fever-range whole-body hyperthermia on the elicitation of contact hypersensitivity in mice ear skins [11]. They found that application of fever-range whole-body hyperthermia resulted in enhanced ear swelling responses when haptens were applied to the ear skin of hapten-sensitized mice. Histochemical analysis of the site of inflammation indicated significant enlargement of blood vessels and increased numbers of lymphocytes per ear in the blood vessels of the ear, suggesting that application of fever-range whole-body hyperthermia during the elicitation phase of contact hypersensitivity enhances the homing of antigen-specific T cells to the inflammatory site.

Regarding anti-tumor immune response, Fisher et al. indicated that fever-range whole-body hyperthermia in tumor-bearing mice (core temperature elevated to approximately 39.5 °C for 6 h) induced a marked increase in intratumoral infiltration of CD8 positive T cells [12]. Cytofluorography experiments demonstrated that fever-range whole-body hyperthermia increased endogenous CD8 positive T cell localization in tumors, and 90–95 % of these cells were primarily effector/memory phenotype ($CD44^{hi}$). Interestingly they also showed that fever-range whole-body hyperthermia lowered the number of tumor-infiltrating regulatory T cells ($CD4^{+}$, $CD25^{+}$, $FoxP3^{+}$), and thus substantially increased the overall effector/regulatory cell ratio of T cells in tumor tissues. By contrast, no significant changes were observed in tumor-infiltrating macrophages ($CD11b^{+}$, $Ly6C^{-}$, $Ly6G^{-}$), polymorphonuclear cells ($CD11^{b}$, $Ly6G^{+}$, $Ly6C^{-}$), and both granulocytic ($CD11b^{+}$, $Ly6G^{+}$, $Ly6C^{lo}$) and monocytic ($CD11b^{+}$, $Ly6C^{hi}$, $Ly6G^{-}$) myeloid derived suppressor cells. For the elicitation of successful immune responses, for instance immunity against tumors, understanding how effector cell invasion into tumor tissue and access to tumor cells is a practical issue to improve and strengthen anti-tumor immune response. The results of Fisher et al. suggested that fever-range whole body hyperthermia provides many favorable effects in the elicitation phase of anti-tumor immune response as well.

31.3.4 Effects on Effector T Cell Cytokine Production and Their Cytotoxicity

The effects of increased temperature, in particular fever-range thermal stress, on antigen-specific T cell functions were reported by Mace et al. [13]. They examined whether and how fever-range thermal stimuli increase tumor-antigen specific T cell functions; interferon gamma (IFN-γ) production and cytotoxicity against target cells.

They found that IFN-γ production in antigen-specific effector CD8+ T cell was temperature sensitive, and effector T cells cultured at 39.5 °C (for 2–6 h) *in vitro* showed significantly more IFN-γ production compared with those cultured at 33 °C and 37 °C. Similar to these effects, effector T cells cultured at 39.5 °C for 6 h displayed enhanced cytotoxicity against target tumor cells. These results indicated that antigen-dependent T cell activity is enhanced following fever-range hyperthermia, and these effects could be expected in patients being treated with thermal therapies.

31.4 Effects of Fever-Range Whole Body Hyperthermia on T Cell Responses in Humans

To date, only a few studies have examined the effects of whole-body fever-range hyperthermia on T-cell-mediated immunity in humans. Kappel et al. showed that whole-body heat treatment using a hot water bath (39.5 °C) led to transient changes in the number of circulating T cells in healthy volunteers [14]. Similarly, Kraybill et al. showed that transient decreases in the number of circulating lymphocytes in cancer patients were induced by fever-range whole-body heat treatment [15]. In addition, it was shown that whole-body hyperthermia in cancer patients provided as an adjunct to chemotherapy induced T-cell activation, such as an increase in the percentage of CD69-expressing $CD4^+$ T cells and elevations of intracellular IFN-γ and tumor-necrosis factor-α in $CD8^+$ T cells [16]. The results in these studies suggest that heat treatment, particularly that in the fever-range, has some impact on T cells in humans as well. To better understand the functional changes in heat treatment-induced T cells, we designed an experiment to examine whether fever-range whole-body hyperthermia stimulates T-cell function, and in particular cytokine production in humans [17]. Details of the study are given in the following section.

31.4.1 Effects of Fever-Range Whole-Body Hyperthermia on T Cell Cytokine Production in Human

To examine whether fever-range whole-body hyperthermia stimulates T-cell function, we used the Heckel ht3000 (Heckel Medizintechnik GmbH), an infrared irradiation device specifically designed for whole-body hyperthermia. Healthy volunteers were exposed to infrared until the rectal temperature reached 38.5 °C. They were then wrapped in a leather tent for 60 min to maintain the rectal temperature above 38.5 °C. Under this heating protocol, it typically took approximately 60 min for the rectal temperature to reach 38.5 °C, and the rectal temperature still increased gradually even after terminating heating. This elevated rectal temperature was maintained at approximately 39 °C during the 60-min heat-retention phase, in which the volunteer was covered with a leather tent. After removal of the leather tent, the elevated temperature gradually decreased and returned to the normal level approximately 2 h after the end of the heat-retention phase. In consideration of these temperature changes, we obtained blood samples four times: before the treatment, immediately after the 60 min heat-retention phase at which the rectal temperature was elevated maximally, and also 24 h and 48 h after the treatment, at which times the rectal temperature had completely returned to the normal level. Peripheral blood mononuclear cells (PBMC) were then prepared and cytokine production from T cells was examined.

First, to induce non-specific cytokines productions from T cells, PBMCs prepared from each donor were cultured in the presence of mAbs against CD3 and CD28 which provide both primary and co-stimulatory signals with resting T cells that enable them to produce a variety of cytokines. Although the production levels of IFN-γ and interleukin-2 (IL-2) in response to mAbs against CD3 and CD28 considerably varied from donor to donor, we found significant increases in the production of both IFN-γ (approx. 1.9-fold) and IL-2 (approx. 1.6-fold) in cells prepared immediately after heat treatment compared with those prepared before treatment. Interestingly, T cells in PBMCs prepared 24 h after the heat treatment still produced elevated amounts of IFN-γ (approx. 2.0-fold) and IL-2 (approx. 1.6-fold) compared with those prepared before treatment. These results indicate that the ability of T cells to produce IFN-γ and IL-2 was enhanced by whole body heat treatment, and also that the heat-induced enhancement of cytokine production appeared to be a long-lasting effect. In contrast, the production levels of both cytokines from T cells in PBMCs prepared 48 h after the treatment returned to approximately the same levels as before treatment.

We also examined the effects of heat treatment on antigen-specific T cell-responses. Two common recall antigens, tuberculin purified protein derivative (PPD) and CEF-viral peptide, were selected, and the effects of heat treatment on IFN-γ production in T cells in response to these common recall antigens were examined. To do this, PBMCs were co-cultured with PPD-pulsed or CEF viral peptides-pulsed autologous monocyte-derived DCs, and IFN-γ production in the supernatants of these cultures was measured for 24 h. A foreign antigen, keyhole limpet hemocyanin (KLH), was used as the negative control.

As in the case for stimulation with anti-CD3/CD28 mAbs, the production level of IFN-γ induced by PPD-loading or CEF viral peptides-loading DCs had a wide range. For instance, before heat treatment, the range of PPD-induced IFN-γ production was from 774 to 12,479 pg/ml, with a mean of 5812 ± 4235 pg/ml, while that of CEF viral peptides ranged from 334 to 3938 pg/ml, with a mean of 1610 ± 1462 pg/ml. Also, for both antigens, a marked increase in IFN-γ production was observed in cells prepared immediately after and 24 h heat treatment of PBMC co-cultured with antigen-loading DCs. Indeed, the fold increases in PPD-induced IFN-γ production were approximately 1.6 and 1.8 immediately after and 24 h after the treatment, respectively, and in the case of CEF viral peptides-induced IFN-γ production, they were approximately 1.9-fold and 2.2-fold immediately after and 24 h after the treatment, respectively, as compared to the control (before treatment). These results indicate that heat treatment stimulates both PPD-specific and viral peptide-specific T-cell responses, and this effect of an increase in body temperature lasted for at least 24 h, even after the heat treatment. Compared with the responses to these recall antigens, the foreign antigen, KLH, did not induce IFN-γ production from PBMCs of any donors at any time point of preparation, indicating that enhanced cytokine production induced by whole-body heat treatment is an antigen-specific response.

31.4.2 Molecules Involved in the Enhanced T Cells Function Induced by Fever-Range Hyperthermia

It is still difficult to estimate what effect an increase in body temperature has on T-cell activation. It is also difficult to estimate how this enhanced effect lasts even 24 h after the treatment in humans. The molecular events involved in enhanced T-cell responsiveness should be investigated to answer these questions.

Since synthesis of heat-shock proteins (HSPs) was shown to increase with elevated body temperature in fever-range whole-body hyperthermia [18], their function should apparently be investigated. Regarding T cell function, it has been shown that HSPs are involved in antigen presentation and cross-presentation in DCs by delivering chaperoned antigenic peptides to MHC class I molecules, thereby inducing antigen-specific T-cell activation [19]. However, the direct effects of HSPs on T cells, particularly the effects on cytokine production by T cells, have not been clarified yet. Breloer et al. showed that recombinant human HSP60 stimulates antigen-dependent T-cell activation and IFN-γ production [20]. In contrast, using highly purified recombinant proteins, Wang et al. showed that recombinant HSP60 and HSP70 had no effects on the release of cytokines by murine and human T cells, including IFN-γ and IL-2 [21]. In this sense, the effects of HSPs on T cells are contradictory, and for this reason their function must continue to be investigated carefully in order to clarify whether and how HSPs are involved in T-cell responsiveness during heat stress.

Also, we have to consider the possible involvement of stress-related hormones such as cortisol and catecholamines. Indeed, in humans, a previous study on fever-range whole-body hyperthermia in healthy volunteers showed that the concentrations of serum cortisol, plasma epinephrine and norepinephrine increased significantly during elevated body temperature [14]. Cortisol, a major glucocorticoid (GC) hormone, is the primary end product of the hypothalamic-pituitary-adrenal axis in the central nervous system, and T cells could be its target for the action via GC receptors [22]. Epinephrine and norepinephrine, the end products in activation of the sympatho-adrenomedullary axis in the sympathetic nervous system, are also reported to modify T-cell function via β2-adrenergic receptors [23]. Generally, the effects of these neuroendocrine products are regarded as immunosuppressive [24]; however, their immune-enhancing effects such as enhanced contact hypersensitivity response were also reported [25, 26]. Further studies are needed to elucidate the precise mechanisms involved in the effects of whole-body heating on T-cell activation, and to achieve this the systemic effects of these neuroendocrine products should be taken into account in future studies.

31.5 Fever-Range Hyperthermia Could Be a Combination Therapy to Enhance T Cell-Based Immune Therapy

Based on the results in mice and humans indicating the immune-stimulating effects of hyperthermia on a variety of T cell functions, one can anticipate the possible application of heat treatment in combination with T-cell-based therapy. Particularly, the combination of fever-range whole-body hyperthermia with T-cell-based cancer immunotherapy appears to be a practical combination to enhance anti-tumor immunity in cancer patients.

Over the last two decades, T-cell-based therapy has been considered to be one of the most promising fields of investigation in clinical cancer research. It includes adoptive transfer using tumor-infiltrating T cells [27] or gene-modified T cells such as chimeric antigen receptor-modified T cells [28]. In Japan, CD3-activated autologous T cells are prevalent in many private clinics [29] and those in combination with tumor antigen-pulsed DC vaccination [30] are also provided to cancer patients in some clinic. These treatments have been reported to elicit durable tumor regression and/or prolonged overall survival in several types of cancer.

We introduced a whole-body hyperthermia device (the heckel HT3000) several years ago, and have been studying how to combine it with cancer immuno-cell therapy; adoptive transfer of activated T cells and DC-based vaccination. Patients with various solid tumors were vaccinated once a week with antigen-pulsed DC vaccine prepared from autologous monocyte-derived DC. These DC were pre-treated with tumor antigenic peptide as well as KLH as a vaccine adjuvant, and were injected intradermally near an inguinal nodal region. Autologous CD3-activated T cells were also infused in each vaccination period. Some patients simultaneously received whole-body hyperthermia in combination with DC-vaccination and CD3-activated T cells infusion. For whole-body hyperthermia, patients were exposed to infrared A until the rectal temperature reached 38.5 °C. They were then wrapped in a synthetic leather tent for 60 min in order to maintain the core body temperature above 38.5 °C. To assess the induction of immune responses in patients who received antigen-pulsed DC-vaccination, we examined the onset of skin reaction at the vaccination site. The onset of this DTH-like skin reaction indicates antigen specific T-cell responses (against tumor antigens and/or against KLH), and we thus examined how much vaccine treatment was required in

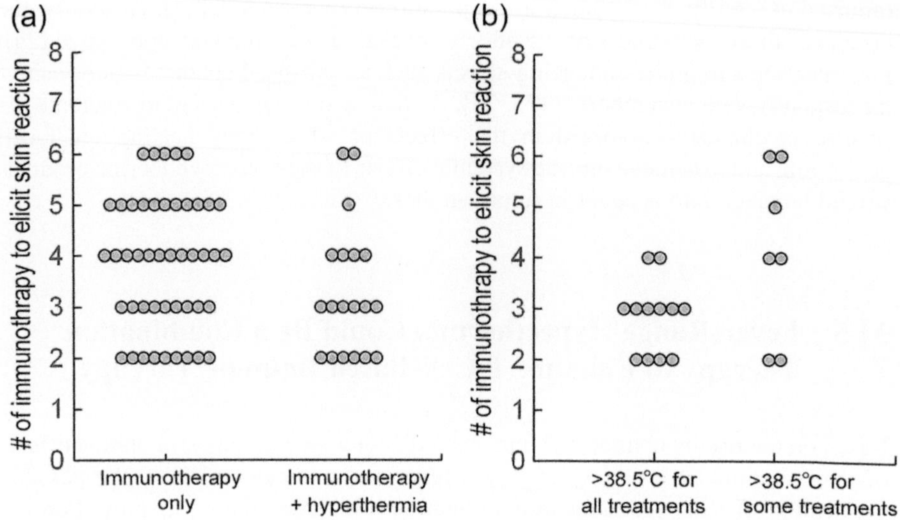

Fig. 31.2 Number of the treatment of immunotherapy required to elicit DTH-like skin reaction. (**a**) Comparison between patients who received immunotherapy alone or in combination with hyperthermia, (**b**) Comparison of patients' core body temperatures during hyperthermia

each patients whose skin reaction sizes after 48 h of DC-vaccination became consistently larger than 1.5 cm in diameter.

As shown in Fig. 31.2a, the average number of vaccinations to induce skin reaction was 3.87 in patients who received immune cell therapy alone. By contrast, in patients who received both immune cell therapy and whole-body hyperthermia, induction of skin reaction appeared to be slightly accelerated, and the average number of vaccine treatments was 3.32. Moreover, the onset of DTH-like skin reaction of patients whose core body temperature was successfully elevated above 38.5 °C in every treatment appeared to be accelerated, compared to those of patients in whom body core temperature elevation failed on some occasions (Fig. 31.2b). The results in these preliminary analyses indicated that fever-range whole-body heat treatment appeared to make a positive impact on the induction of T cell-mediated immune responses when it was provided with DC vaccine and activated T cell infusion, suggesting that it could be a practical combination therapy to enhance the efficacy of cancer immune cell therapy.

31.6 Conclusion

As discussed above, evidence has accumulated to indicate the immune-stimulatory effects of heat treatment and its possible application in combination with cancer immunotherapy. The temperature is critical; temperature around the fever range (*e.g.* 39–40.5 °C) appears to be the most favorable for immune responses in humans. Future clinical studies are expected to elucidate the practical utility of T-cell and/or DC-based cancer immunotherapy provided in combination with fever-range hyperthermia.

References

1. Zhang HG, Mehta K, Cohen P, Guha C. Hyperthermia on immune regulation: a temperature's story. Cancer Lett. 2008;271:191–204.
2. Mackowiak PA. Concepts of fever. Arch Intern Med. 1998;158:1870–81.
3. Baronzio GF, Della Seta R, D'Amico M, Baronzio A, Freitas I, Forzenigo G, et al. Effects of local and whole body hyperthermia on immunity. In: Baronzio GF, Hager ED, editors. Gerorgetown: Landes BioScience; 2006, p. 247–75.
4. Banchereau J, Briere F, Caux C, Davoust J, Lebecque S, Liu YJ, et al. Immunobiology of dendritic cells. Annu Rev Immunol. 2000;18:767–811.
5. Ostberg JR, Kabingu E, Repasky EA. Thermal regulation of dendritic cell activation and migration from skin explants. Int J Hyperthermia. 2003;19:520–33.
6. Steinman RM. The dendritic cell system and its role in immunogenicity. Annu Rev Immunol. 1991;9:271–96.
7. Wang WC, Goldman LM, Schleider DM, Appenheimer MM, Subjeck JR, Repasky EA, et al. Fever-range hyperthermia enhances L-selectin-dependent adhesion of lymphocytes to vascular endothelium. J Immunol. 1998;160:961–9.
8. Evans SS, Wang WC, Bain MD, Burd R, Ostberg JR, Repasky EA. Fever-range hyperthermia dynamically regulates lymphocyte delivery to high endothelial venules. Blood. 2001;97:2727–33.
9. Chen Q, Fisher DT, Clancy KA, Gauguet JM, Wang WC, Unger E, et al. Fever-range thermal stress promotes lymphocyte trafficking across high endothelial venules via an interleukin 6 trans-signaling mechanism. Nat Immunol. 2006;7:1299–308.
10. Mace TA, Zhong L, Kilpatrick C, Zynda E, Lee CT, Capitano M, et al. Differentiation of CD8 + T cells into effector cells is enhanced by physiological range hyperthermia. J Leukoc Biol. 2011;90:951–62.
11. Ostberg JR, Gellin C, Patel R, Repasky EA. Regulatory potential of fever-range whole body hyperthermia on langerhans cells and lymphocytes in an antigen-dependent cellular immune response. J Immunol. 2001;167:2666–70.
12. Fisher DT, Chen Q, Skitzki JJ, Muhitch JB, Zhou L, Appenheimer MM, et al. IL-6 trans-signaling licenses mouse and human tumor microvascular gateways for trafficking of cytotoxic T cells. J Clin Invest. 2011;121:3846–59.
13. Mace TA, Zhong L, Kokolus KM, Repasky EA. Effector CD8+ T cell IFN-γ production and cytotoxicity are enhanced by mild hyperthermia. Int J Hyperthermia. 2012;28:9–18.
14. Kappel M, Stadeager C, Tvede N, Galbo H, Klarlund PB. Effects of *in vitro* hyperthermia on natural killer cell activity, *in vitro* proliferative responses and blood mononuclear cell subpopulations. Clin Exp Immunol. 1991;84:175–80.
15. Kraybill WG, Olenki T, Evans SS, Ostberg JR, O'Leary KA, Gibbs JF, et al. A phase I study of fever-range whole body hyperthermia (FR-WBH) in patients with advanced solid tumours: correlation with mouse models. Int J Hyperthermia. 2002;18:253–66.
16. Atanackovic D, Nierhaus A, Neumeier M, Hossfeld DK, Hegewisch-Becker S. 41.8 °C whole body hyperthermia as an adjunct to chemotherapy induces prolonged T cell activation in patients with various malignant diseases. Cancer Immunol Immunother. 2002;51:603–13.
17. Kobayashi Y, Ito Y, Ostapenko VV, Sakai M, Matsushita N, Imai K, et al. Fever-range whole-body heat treatment stimulates antigen-specific T-cell responses in humans. Immunol Lett. 2014;162:256–61.

18. Ostberg JR, Kaplan KC, Repasky EA. Induction of stress proteins in a panel of mouse tissues by fever-range whole body hyperthermia. Int J Hyperthermia. 2002;18:552–62.
19. Tsan MF, Gao B. Heat shock protein and immune system. J Leukoc Biol. 2009;85:905–10.
20. Breloer M, Dorner B, Moré SM, Roderian T, Fleischer B, von Bonin A. Heat shock proteins as "danger signals": eukaryotic Hsp60 enhances and accelerates antigen-specific IFN-γ production in T cells. Eur J Immunol. 2001;31:2051–9.
21. Wang Y, Gao B, Tsan MF. Induction of cytokines by heat shock proteins and concanavalin a in murine splenocytes. Cytokine. 2005;32:149–54.
22. Herold MJ, McPherson KG, Reichardt HM. Glucocorticoids in T cell apoptosis and function. Cell Mol Life Sci. 2006;63:60–72.
23. Elenkov IJ, Wilder RL, Chrousos GP, Vizi ES. The sympathetic nerve – an integrative interface between two supersystems: the brain and the immune system. Pharmacol Rev. 2002;52:595–638.
24. Padgett DA, Glaser R. How stress influences the immune response. Trends Immunol. 2003;24:444–8.
25. Dhabhar FS, McEwen BS. Enhancing versus suppressive effects of stress hormones on skin immune function. Proc Natl Acad Sci U S A. 1999;96:109–64.
26. Dhabhar FS. Effects of stress on immune function: the good, the bad, and the beautiful. Immunol Res. 2014;58:193–210.
27. Rosenberg SA, Restifo NP, Young JC, Morgan RA, Dudley ME. Adoptive cell transfer: a clinical path to effective cancer immunotherapy. Nat Rev Cancer. 2008;8:299–308.
28. Fedorov VD, Sadelain M, Kloss CC. Novel approaches to enhance the specificity and safety of engineered T cells. Cancer J. 2014;20:160–5.
29. Takayama T, Skine T, Makuuchi M, Yamasaki S, Kosuge T, Yamamoto J, et al. Adoptive immunotherapy to lower postsurgical recurrence rates of hepatocellular carcinoma: a randomized trial. Lancet. 2000;356:802–7.
30. Shimizu K, Kotera Y, Aruga A, Takeshita N, Takasaki K, Yamamoto M. Clinical utilization of postoperative dendritic cell vaccine plus activated T-cell transfer in patients with intrahepatic cholangiocarcinoma. J Hepatobiliary Pancreat Sci. 2012;19:171–8.

Chapter 32
Combination of Hyperthermia and Immunotherapy: Hyperthermia and Naïve T-cell Therapy

Satoshi Kokura

Abstract Since many of the patients who receive adoptive immunotherapy also receive chemotherapy and/or radiation therapy, evaluation of the curative effect of adoptive immunotherapy itself is difficult. To overcome this, I perform immune monitoring to examine how much individual immunotherapies contribute to the curative effects. As a result, I found that higher whole blood levels of Th1 cytokines prodicing potential may be used to predict better clinical responses. In addition, I showed that the naïve T-cell adoptive therapy, which we developed, has superior clinical efficacy when compared to other immunotherapies. This chapter also describes the results of our clinical trial. In addition, I describe the application of hyperthermia, which can influence the immune-response of a tumor tissue. Basic experiments showed that hyperthermia could be applied to strengthen a patient's immune system. Next, we performed naïve T-cell adoptive immunotherapy combined with hyperthermia. Finally, I will describe a case of advanced gastric cancer that showed complete remission after treatment with naïve T-cell adoptive immunotherapy combined with hyperthermia.

Keywords Adoptive T cell transfer therapy • Naïve T cell • Clinical trial

32.1 Immune Monitoring of Adoptive Cell Transfer (ACT) Immunotherapy

In our study, we demonstrated that adoptive T cell therapy had a strong potential to enhance whole blood production of the Th1 cytokine, interferon (IFN)-γ, in advanced pancreatic cancer patients in whom immunosuppression and immune escape mechanisms may be present [1]. Immune function was tested using venous blood obtained from patients prior to the administration of cultured cells (baseline)

S. Kokura (✉)
Kokura Lab. Faculty of Health and Medical Sciences, Kyoto Gakuen University, 18, Gotanda-cho, Yamanouchi, Ukyo-ku, Kyoto-shi, Kyoto 615-8577, Japan
e-mail: kokura@kyotogakuen.ac.jp

S. Kokura et al. (eds.), *Hyperthermic Oncology from Bench to Bedside*,
DOI 10.1007/978-981-10-0719-4_32

and during the follow up, which occurred 4 weeks after the 2nd and 6th cultured cell infusion. Methods for quantifying IFN-α production in whole human blood have been described previously [2]. Other cytokines were also quantified according to procedures described previously [3, 4].

A multivariate Cox proportional hazards analysis identified the increase of whole blood IFN- γ levels after adoptive T cell therapy as a predictive factor of the clinical response to the therapy; this factor was independent of the baseline IFN-γ levels and the performance status of the patients. Moreover, in agreement with previous studies [5–8], the number of peripheral regulatory T cells (Tregs) at baseline was an independent prognostic marker of overall survival (OS).

It is generally accepted that CD-8 T cell responses are an important indicator of the effectiveness of cancer immunotherapies, such as cancer vaccines. Immunological assays, such as the enzyme-linked immunospot (ELISPOT) assay or the human leukocyte antigen (HLA)-peptide multimer assay, are very efficient at detecting antigen-specific CD-8 T cells. Subsequently, they are frequently used to detect immunizing antigen-specific T cells after immunotherapy for validating surrogate endpoints that correlate with clinical outcomes. However, with a few exceptions, there have been poor correlations between the magnitude of the immunizing antigen-specific T-cell responses measured by these immunological assays and improvements in clinical outcomes [9, 10]. The most obvious explanation for this phenomenon is that the immune-monitoring assays measured irrelevant T-cell responses that had no effect on tumor destruction. Another possible explanation is that the assays were unable to detect how the T cells responded to different tumor-rejection antigens other than the immunizing antigen; this could result from immunotherapy-induced tumor lysis and endogenous priming with new tumor-derived antigens. Some recent immunotherapy trials have provided supporting evidence that the immunological phenomenon of determinant antigen or epitope spreading may play an important role in clinical responses to immunotherapy [9, 11, 12]. Butterfield et al. reported that new post-vaccine responses to non-vaccine antigens were observed in clinical responders, while non-responders did not display reactivity to epitopes other than those used for the vaccination [9, 12]. Although the ELISPOT and HLA-peptide multimer assays are quite efficient at detecting antigen-specific effector T cells, these assays are limited in their ability to assess post-immunotherapy responses to new tumor-derived antigens. To accurately predict clinical responses after immunotherapy, it is necessary to apply more comprehensive and functional assays in addition to these antigen-specific assays.

It is important to use structure-based assays, such as the HLA-peptide multimer assay, to dissect the antitumor responses and predict clinical outcomes after immunotherapy; it is also equally important to analyze the function of effector cells [13]. Analysis of the antigen-driven expression of cytokines can be used to determine the functional property of effector T cells. IFN-γ expression has been shown to be highly reproducible and sensitive enough to detect effector T-cell function [14]. Recently, June et al. have demonstrated that the clinical response of patients treated with genetically engineered T cells was accompanied by a delayed increase in the serum levels of cytokines, i.e., interleukin (IL)-6, IFN-γ, IL-8, and IL-10 [15, 16]. In their study, IFN-γ peaked on day 17 after adoptive cell therapy, when it showed the largest relative change when compared with baseline levels. This result in the study by June et al. showing that IFN-γ plays an important role in adoptive T-cell therapy emphasizes the relevance of our results.

The whole blood IFN-γ production assay used in our study has several advantages. Uno et al. reported that the value of IFN-γ production in whole blood often differs from that in peripheral blood mononuclear cells (PBMC). Serum factors, such as IL-10, can affect IFN-γ production. As IFN-γ production values that are determined from isolated PBMC would lack information about the influence of serum on the cells, our whole blood method is likely a more comprehensive tool for evaluating the actual immune responses of patients [17, 18]. Moreover, whole blood IFN-γ production assays are already available for immune monitoring of multiple cancer immunotherapies against unknown tumor antigens.

One of the limitations of our study is whether the increase in whole blood IFN-γ production truly reflected treatment-induced immune responses. Thus, when the immunogenic antigens are known, antigen-specific assays, such as ELISPOT or HLA-peptide multimer assays, should be used. This would allow researchers to assess the relationship between the assay results and whole blood IFN-γ levels to validate the reliability of the assay as a surrogate marker of clinical outcome. By combining whole blood IFN-γ assays and antigen-specific immunological assays, we believe it may be possible to construct a novel assay system for monitoring the immune response of patients to cancer immunotherapy. We demonstrated that adoptive T-cell therapy increased the whole blood level of the Th1 cytokine, IFN-γ, in advanced pancreatic cancer patients, who were in a strong immunosuppressive state. The increase in whole blood IFN-γ levels after adoptive T-cell therapy was shown to be independently related to OS in a multivariate Cox proportional hazards analysis. These findings indicated that the assay for whole blood IFN-γ levels offers promise as a convenient and efficient method for evaluating clinical responses to cancer immunotherapies.

32.2 Clinical Trial of Naïve T Cell Therapy

T-cell differentiation is a progressive process characterized by phenotypic and functional changes. Previous mouse studies and clinical trials have demonstrated that naïve and early differentiated T cells may be the optimal population for ACT-based immunotherapies due to their *in vivo* persistence, high proliferative potential, receptiveness to homeostatic and costimulatory signals, homing to secondary lymphoid tissues, and ability to secrete IL-2 [19–22]. In addition to this, Yu et al. have reported that a recombinant human fibronectin fragment, FN-CH296, acting with anti-cluster of differentiation (CD) 3 induces naïve and early differentiated T-cell proliferation. These evidences suggest that FN-CH296/anti-CD3 stimulation may be an efficient way to generate a large number of naïve and early differentiated T cells suitable for ACT with higher and longer persistence *in vivo* [23]. We built on these prior results by assessing the safety and efficacy of FN-CH296/anti-CD3-stimulated T-cell therapy, which we called naïve T-cell therapy, in patients with advanced cancer. The primary objective of this Phase I clinical trial was to assess the safety and adverse-event profiles of FN-CH296-stimulated T-cell therapy in patients with advanced cancer. Our secondary objective was to assess the antitumor activity of ACT therapy. This study was conducted with a standard 3 + 3 Phase I design that investigated the dose-limiting toxicities (DLTs) occurring over a 28-day period after the second infusion of cultured lymphocytes. We used an accelerated titration design to assess the safety of the number of adoptive lymphocytes at 1×10^9 (cohort 1), 3×10^9 (cohort 2), and 9×10^9 (cohort 3) lymphocytes per person. Written informed consent was obtained from all patients. This study is registered in the UMIN Clinical Trials Registry with the identifier UMIN000001835. Patients enrolled in this trial fulfilled the following eligibility criteria: histologically or cytologically confirmed esophageal cancer, gastric cancer, colorectal cancer, pancreatic cancer, biliary tract cancer, or non-small lung cancer; residual disease after standard treatment with no other curative treatment options available; a life expectancy of at least 3 months; and at least 4 weeks since their last chemotherapy or radiation therapy. Immune function was tested using venous blood obtained from the patients prior to the administration of cultured cells (baseline) and during the follow up, which occurred 4 weeks after the 2nd and 6th cultured cell infusion. Cytokine levels in the samples were measured with a Bio-Plex Multiplex Cytokine Array System (Bio-Rad Laboratories, CA, USA) according to the manufacturer's instructions. The Multiplex Th1/Th2 Bead Kit (Bio-Rad Laboratories) was used to measure the following cytokines: IL-2, IL-4, IL-5, IL-10, IL-12 (p70), IL-13, tumor necrosis factor (TNF)-α, IFN-γ, and granulocyte-monocyte colony-stimulating factor (GM-CSF). Data acquisition and analysis were conducted with the Bio-Plex Manager Software, version 5.0. Since the levels of cytokines without PHA-P stimulation were extremely low (data not shown), we only assessed those with PHA-P stimulation in this study. The median age was 60 years (range, 43 to 75 years) and all patients had an ECOG performance score of 0 or 1. Three patients had colorectal cancer, two patients had bile duct cancer, and one patient each had pancreatic cancer, gastric cancer, hepatocellular carcinoma, or lung cancer. All patients underwent two infusions of cultured cells, and eight patients, who wished to continue treatment, completed four further T-cell infusions. Non-hematological events, including fatigue and anorexia, were observed, but no ACT-related severe or unexpected toxicities were observed. As for the clinical outcomes, of the nine patients, one (11.1%) exhibited complete remission (CR), one (11.1%) had partial remission (PR), four (44.4%) had stable disease (SD) and three (33.3%) had progressive disease (PD). The response rate was 22.2% and the disease control rate (DCR = CR + PR + SD) was 66.7%. To determine the immune responses in patients receiving novel ACT, whole blood cytokine assays were performed using venous blood obtained from the patients. The levels of IFN-**γ** treatment; in particular, the levels of IFN-**γ**, IL-12, and GM-CSF increased by a multiple of 10 or more times.

Next, we evaluated whole blood cytokine levels according to objective tumor response. The levels of IFN-γ, IL-2, IL-12, and GM-CSF in patients with PR/CR increased by more than two times after the treatment, while the levels in patients with SD or PD decreased or did not change significantly after the treatment. The results presented in this study indicated that FN-CH296-stimulated T-cell therapy was very well tolerated in our patient sample. No significant ACT-related grade 3 or 4 adverse events occurred in any patient and the ACT resulted in tumor regression in two patients: CR in one patient with rectal cancer and PR in another with bile duct cancer. The response rate was 22.2% and the DCR was 66.7%. Despite the small size of the sample, these results are promising. Previous studies have demonstrated that IFN-γ plays an important role in cancer immunotherapy and IFN-γ expression in T cells is considered to be highly correlated with therapeutic success. We have also demonstrated that the assay of whole blood IFN-γlevels was an efficient method for evaluating clinical response to cancer immunotherapies [24, 25]. In this study, the whole blood IFN-γ levels in cohort 3 increased by up to 10 or more times after six infusions of cultured cells. Whole blood IFN-γ levels in PR or CR cases increased by more than two times, whereas those in PD or SD cases showed no significant increase after the treatment. The finding in our prior study [20] showing that the increase in whole blood IFN-γ levels after ACT was independently related to overall survival in cancer patients emphasizes the relevance of the results gained in this study. In addition to IFN-γ, the levels of other Th1 cytokines, such as IL-2, IL-12, and TNF-α, also increased in the patients of cohort 3 and in patients with PR/CR; in contrast, the levels of Th2 cytokines, such as IL-4, IL-5, and IL-13, decreased after the treatment. In addition, the whole blood GM-CSF level increased in the patients of cohort 3 and in patients who experienced PR/CR. It was previously reported that cell-mediated immunity is preferentially activated by Th1 cytokines, whereas Th2 cytokines suppress cell-mediated immunity [26]. GM-CSF is a pleiotropic cytokine that stimulates dendritic cells (DCs) and promotes the uptake of tumor antigens by DCs, leading to T-cell cross-priming and activation of the immune system against specific antigens [27]. Consequently, based on the results from the immune monitoring performed in this present study, we are inclined to believe that fibronectin CH296-stimulated T-cell therapy may exert anti-tumor effects by activating cell-mediated immunity.

32.3 The Effect of Hyperthermia on the Immune System

It has been shown that hyperthermia in the range of 39°C to 44°C can arrest cancer cell proliferation and kill cells *in vitro* and *in vivo*. The effects are dependent on two factors, namely, the temperature and duration of hyperthermia, which are collectively referred to as thermal dose [28]. When cancer cells are heated, the heat alters the characteristics of the membrane [29–31], proteins, which are the most sensitive biomolecules [28], and DNA repair mechanisms. A positive correlation between thermal dose and cytotoxicity has been known historically [32, 33], so it is believed that inducing hyperthermia by increasing the heating temperature results in an increased anti-tumor effect.

Recently, clinical local hyperthermia is used either for ablation purposes as an alternative to surgery or in combination with chemotherapy and/or radiation therapy to enhance the tumoricidal effects of those therapies [34]. It has been reported that many clinical cases treated with hyperthermia at the temperature of 39°C to 45°C showed unexpected benefits, including improvement of anti-tumor immunity [35]. As it has become apparent that activating the immune system is crucial to successfully treating metastatic cancer, the potential of boosting anti-tumor immunity by heating tumors has become a growing area of cancer research. In this section, several mechanisms by which heating tumors can elicit anti-tumor immune responses, including tumor cell damage, tumor surface molecule changes, production of heat-shock proteins (HSPs), and the direct effects on immune cells, are discussed.

It was becoming clear that different mechanisms of immune activation occur at different temperatures within the range of 39°C to 45°C. Hyperthermia can increase the visibility of tumor cells to the immune system. It was reported that heating tumor cells *in vitro* at 39.5°C for 6 h increased the surface expression of major histocompatibility complex (MHC) class I polypeptide-related sequence A (MICA), an NKG2D ligand, but not MHC class I, making the cells more sensitive to lysis by natural killer (NK) cells [36], and that tumor cells heated *in vitro* at 43°C for 30 min had increased surface MHC class I levels [37], which allows better recognition by $CD8^+$ T cells.

It is well known that hyperthermia can induce HSPs in heated tumor tissue. These HSPs also contribute to hyperthermia-induced anti-tumor immunity. HSPs are usually divided into subgroups based on their molecular size. Among them, Hsp70 is the most recognized HSP that can stimulate the immune system. Hsp70 has an epitope that is recognized by NK cells and stimulates NK cell proliferation and cytolytic activities [38–40]. Hsp70 is also released by heat-stressed cancer cells and directly binds to toll-like receptor (TLR) 2 and TLR4 on antigen-presenting cells (APCs), such as DCs, to activate cytokine production and antigen uptake by the APCs [41–44]. HSPs can transfer potential tumor antigens to APCs [45]. It is reported that those APCs are able to cross-present tumor antigens from HSP complexes to $CD8^+$ T cells via MHC class I and thus elicit tumor-specific $CD8^+$ T-cell responses [46, 47].

When tumors are treated with local hyperthermia, tumor-infiltrating immune cells will also be exposed to heat. Fever is a natural immune stimulatory mechanism associated with fighting infections, so it is not surprising that fever-level hyperthermia has stimulatory effects on leukocytes. According to *in vitro* study results, heating human NK cells at 39.5°C for 6 h does not alter the surface expression level of NKG2D, but it does result in NKG2D clustering, as seen on NK cells activated with IL-2, leading to better lysis activity [36]. The results also showed that heating killer $CD8^+$ T cells *in vitro* at 39.5°C for 6 h enhances their antigen-specific IFN-γ production and target tumor cell-killing ability [48]. Almost all relevant literature agrees that activation of immune cells by direct heating is only observed in the fever-range of temperatures, and not at higher temperatures above 41°C [49].

32.4 Combination Therapy with Hyperthermia and Naïve T Cell Therapy

It is likely that hyperthermia amplifies the effects of other immunotherapies, such as checkpoint blockade and adoptive T-cell therapies, as indicated by some studies [50]. Hyperthermia cannot always be applied uniformly when a cancer tissue is heated. Therefore, the actual immunological modification resulting from hyperthermia may differ depending on the intratumoral environment. Although uniform heating would be ideal, there are many factors, such as the blood flow inside of a tumor, which may prevent uniform heating.

Despite these challenges, we performed combined treatment of hyperthermia with the naïve T-cell adoptive immunotherapy that we developed. Similar to chemotherapy or radiotherapy patients, the patients treated with hyperthermia and naïve T-cell adoptive immunotherapy are often also treated with other therapies. However, the response rate, DCR, and overall survival period have yet to be sufficiently investigated to determine the validity of this combined therapy. Immune monitoring was performed in patients treated with naïve T-cell adoptive immunotherapy alone or with the naïve T-cell adoptive immunotherapy combined with hyperthermia. The methods of immune monitoring were the same as those mentioned above for the clinical trial on the naïve T-cell adoptive immunotherapy.

Whole blood cytokine assays were performed using venous blood obtained from the patients. We investigated 85 cases that were treated with the naïve T-cell adoptive immunotherapy alone, and 158 cases that were treated with the naïve T-cell adoptive immunotherapy combined with hyperthermia. First, we compared the carcinomas between the two groups. Figure 32.1 shows the levels of each cytokine; these were calculated as a ratio of the value measured 14 days after the 4th cultured cell infusion (10 weeks after the start of the naïve T-cell adoptive immunotherapy) to the baseline level measured before starting the naïve T-cell adoptive immunotherapy. Although the levels of IFN-γ, TNF-α, GM-CSF, and IL-4 were higher in the combination group when compared to the naïve T-cell adoptive immunotherapy alone group, the differences were not statistically significant.

Next, we examined a hepatocellular carcinoma and a metastatic liver cancer. The levels of IFN-α, IFN-γ, TNF-α, and IL-5 were higher in the combination group when compared to those of the naïve T-cell adoptive immunotherapy alone group (Fig. 32.2). The levels of Th2 cytokines also increased slightly, likely in response to the increase in Th1 cytokines resulting from the naïve T-cell adoptive immunotherapy. When hypothermia was applied, the levels of each cytokine increased, although a similar pattern to that seen with the naïve T-cell adoptive immunotherapy alone was maintained. As mentioned above, the Phase I trial of our naïve T-cell

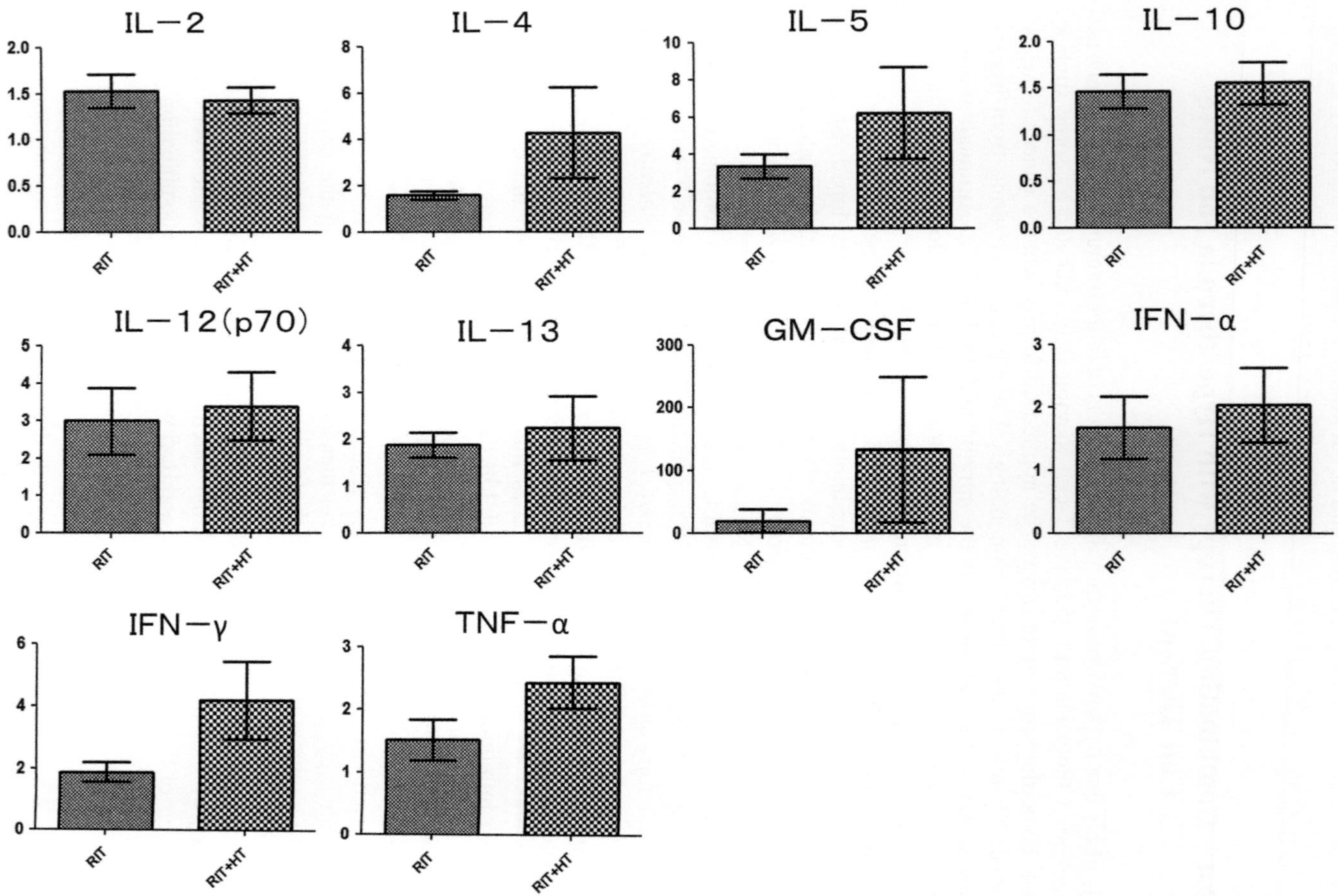

Fig. 32.1 Immune monitoring of cytokine levels in all cases. RIT: naïve T-cell adoptive immunotherapy, HT: hyperthermia (RIT alone: 85 cases, RIT + HT: 158 cases)

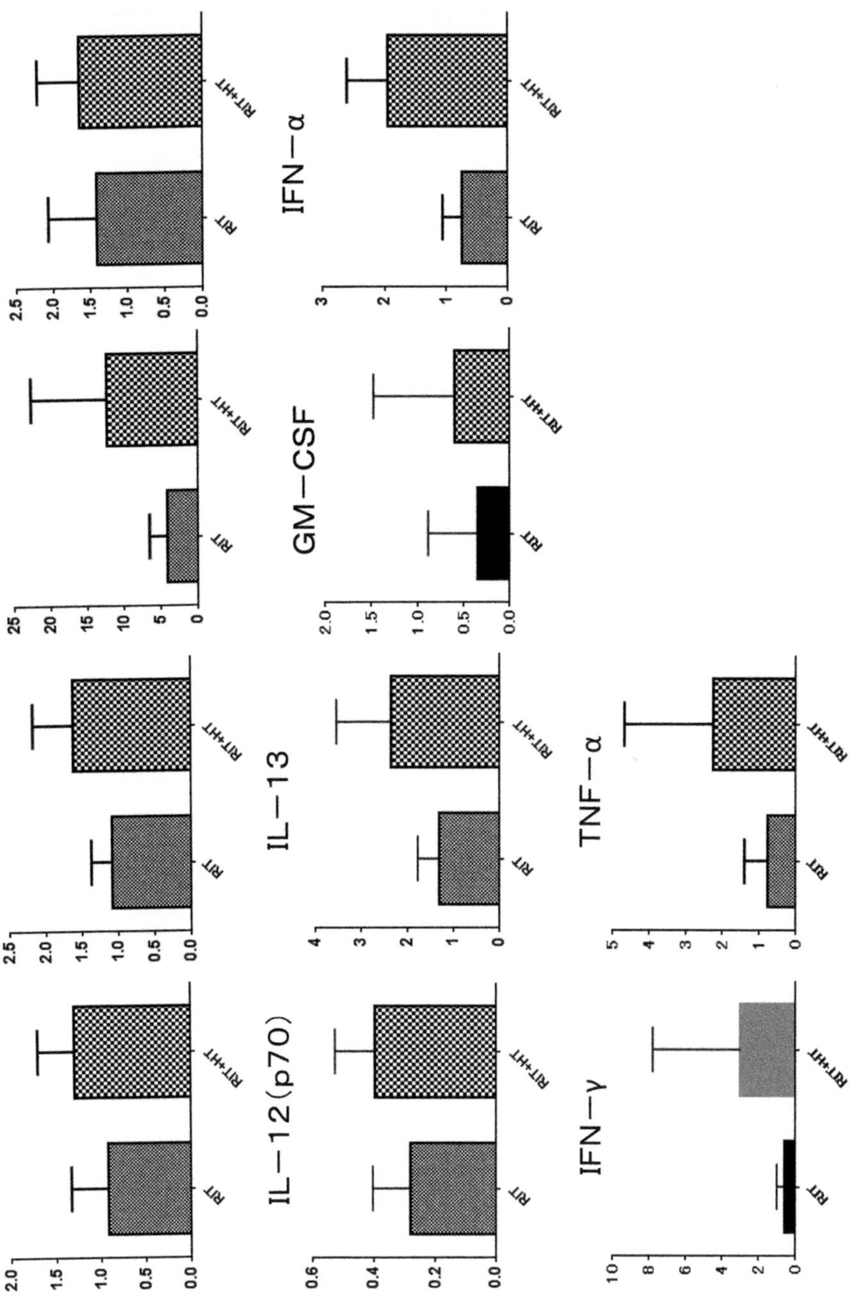

Fig. 32.2 Immune monitoring of cytokine levels in liver cancer cases. RIT: naïve T-cell adoptive immunotherapy, HT: hyperthermia (RIT alone: 6 cases, RIT +HT: 16 cases)

adoptive immunotherapy showed that the levels of IFN-γ, IL-2, IL-12, and GM-CSF in patients with PR/CR increased by more than two times after the treatment, while the levels in patients with SD or PD decreased or did not change significantly after the treatment. Therefore, using hyperthermia together with naïve T-cell adoptive immunotherapy may contribute to a further improvement in tumor reduction. Effects on the overall survival period are presently under investigation.

32.5 Case Presentation

The case of a patient with advanced gastric cancer, who rejected standard chemotherapy and received combined hyperthermia and naïve T cell transfer therapy is described herein. The patient was a 62-year-old man. He experienced abdominal discomfort and went to a clinic in his neighborhood. In that clinic, he underwent gastric endoscopy and computed tomography (CT). The results showed advanced gastric cancer with liver metastases and ascites. From a cytological examination of the ascites, the patient was diagnosed with peritonitis carcinomatosa. At first, he agreed to receive titanium silicate (TS)-1 + cisplatin combination therapy, which is a standard chemotherapy, and he received TS-1 for 1 week. However, he experienced very severe adverse effects and refused further chemotherapy. Thereafter, he received the naïve T cell transfer therapy once every 2 weeks and the hyperthermia therapy once a week from August 3rd, 2013, to March 17th, 2014. We performed gastric endoscopy and CT again on April 23rd, 2014. Surprisingly, the advanced gastric cancer and the ascites had completely disappeared. The liver metastasis was almost completely cured (Figs.32.3, 32.4). Two years after the initial diagnosis, the

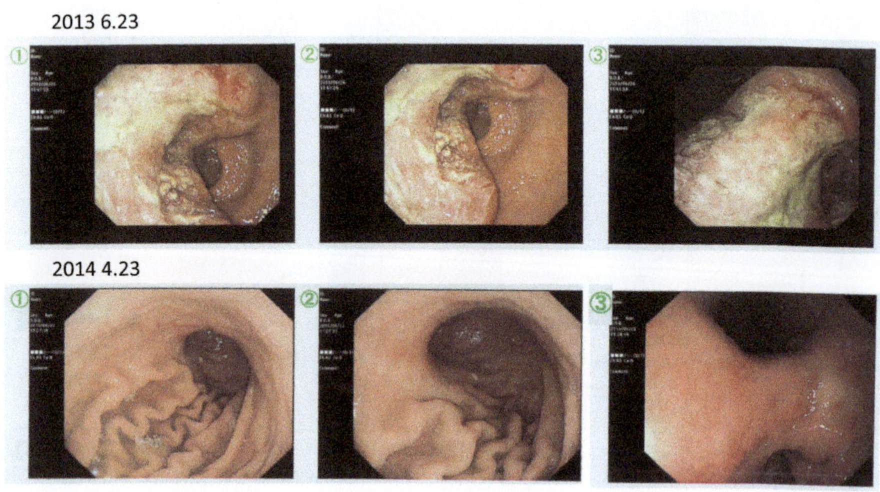

Fig. 32.3 Gastric endoscopy image of the patient before and after 8 months of treatment with hyperthermia + naïve T-cell adoptive immunotherapy

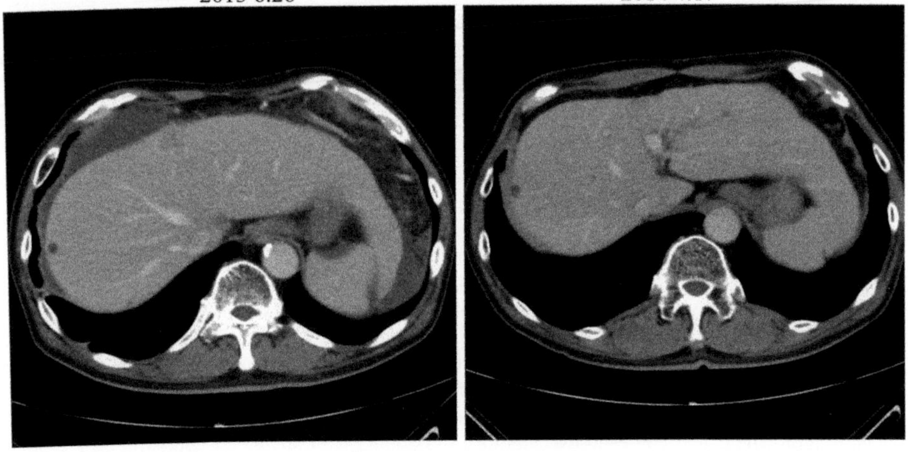

Fig. 32.4 Computed tomography image of the patient before and after 8 months of treatment with hyperthermia + naïve T-cell adoptive immunotherapy

advanced gastric cancer, ascites, and liver metastasis remained undetectable as of April, 2015.

References

1. Liyanage UK, Goedegebuure PS, Moore TT, et al. Increased prevalence of regulatory T cells (Treg) is induced by pancreas adenocarcinoma. J Immunother. 2006;29:416–24.
2. Uno K, Nakano K, Maruo N, Onodera H, Mata H, et al. Determination of interferon-alpha-producing capacity in whole blood cultures from patients with various diseases and from healthy persons. J Interferon Cytokine Res. 1996;16:911–8.
3. Ishikawa T, Kokura S, Sakamoto N, Matsumoto T, Funaki J, Adachi S, et al. Adoptive cellular therapy enhances the helper T cell response and reduces the number of regulatory T cells. Exp Ther Med. 2011;2:737–43.
4. Ishikawa T, Kokura S, Sakamoto N, Okajima M, Matsuyama T, et al. Relationship between circulating cytokine levels and physical or psychological functioning in patients with advanced cancer. Clin Biochem. 2012;45:207–11.
5. Liyanage UK, Moore TT, Joo HG, et al. Prevalence of regulatory T cells is increased in peripheral blood and tumor microenvironment of patients with pancreas or breast adenocarcinoma. J Immunol. 2002;169:2756–61.
6. Curiel TJ, Coukos G, Zou L, et al. Specific recruitment of regulatory T cells in ovarian carcinoma fosters immune privilege and predicts reduced survival. Nat Med. 2004;10:942–9.
7. Sasada T, Kimura M, Yoshida Y, et al. CD41CD251 regulatory T cells in patients with gastrointestinal malignancies: possible involvement of regulatory T cells in disease progression. Cancer. 2003;98:1089–99.
8. Kono K, Kawaida H, Takahashi A, et al. CD4(1)CD25high regulatory T cells increase with tumor stage in patients with gastric and esophageal cancers. Cancer Immunol Immunother. 2006;55:1064–71.

9. Butterfield LH, Ribas A, Dissette VB, et al. Determinant spreading associated with clinical response in dendritic cell-based immunotherapy for malignant melanoma. Clin Cancer Res. 2003;9:998–1008.
10. Ribas A, Timmerman JM, Butterfield LH, et al. Determinant spreading and tumor responses after peptide-based cancer immunotherapy. Trends Immunol. 2003;24:58–61.
11. Ranieri E, Kierstead LS, Zarour H, et al. Dendritic cell/peptide cancer vaccines: clinical responsiveness and epitope spreading. Immunol Invest. 2000;29:121–5.
12. Butterfield LH, Comin-Anduix B, Vujanovic L, et al. Adenovirus MART-1-engineered autologous dendritic cell vaccine for metastatic melanoma. J Immunother. 2008;31:294–309.
13. Yee C, Greenberg P. Modulating T-cell immunity to tumours: new strategies for monitoring T-cell responses. Nat Rev Cancer. 2002;2:409–19.
14. Kammula US, Lee KH, Riker AI, et al. Functional analysis of antigen-specific T lymphocytes by serial measurement of gene expression in peripheral blood mononuclear cells and tumor specimens. J Immunol. 1999;163:6867–75.
15. Porter DL, Levine BL, Kalos M, et al. Chimeric antigen receptor-modified T cells in chronic lymphoid leukemia. N Engl J Med. 2011;365:725–33.
16. Kalos M, Levine BL, Porter DL, et al. T cells with chimeric antigen receptors have potent antitumor effects and can establish memory in patients with advanced leukemia. Sci Transl Med. 2011;3:95ra73.
17. Uno K, Setoguchi J, Tanigawa M, et al. Differential interleukin 12 responsiveness for interferon gamma production in advanced stages of cancer patients correlates with performance status. Clin Cancer Res. 1998;4:2425–32.
18. Miyata H, Uno K, Ono T, et al. Low density lipoprotein apheresis ameliorates interferon-gamma production in patients with nephrotic syndrome. Ther Apher Dial. 2012;16:189–94.
19. Gattinoni L, Klebanoff CA, Palmer DC, Wrzesinski C, Kerstann K, et al. Acquisition of full effector function in vitro paradoxically impairs the in vivo antitumor efficacy of adoptively transferred CD8+ T cells. J Clin Invest. 2005;115:1616–26.
20. Robbins PF, Dudley ME, Wunderlich J, El-Gamil M, Li YF, et al. Cutting edge: persistence of transferred lymphocyte clonotypes correlates with cancer regression in patients receiving cell transfer therapy. J Immunol. 2004;173:7125–30.
21. Zhou J, Shen X, Huang J, Hodes RJ, Rosenberg SA, et al. Telomere length of transferred lymphocytes correlates with in vivo persistence and tumor regression in melanoma patients receiving cell transfer therapy. J Immunol. 2005;175:7046–52.
22. Huang J, Khong HT, Dudley ME, El-Gamil M, Li YF, et al. Survival, persistence, and progressive differentiation of adoptively transferred tumor- reactive T cells associated with tumor regression. J Immunother. 2005;28:258–67.
23. Yu SS, Nukaya I, Enoki T, Chatani E, Kato A, et al. In vivo persistence of genetically modified T cells generated ex vivo using the fibronectin CH296 stimulation method. Cancer Gene Ther. 2008;15:508–16.
24. Ishikawa T, Kokura S, Sakamoto N, Matsumoto T, Funaki J, Adachi S, et al. Adoptive cellular therapy enhances the helper T cell response and reduces the number of regulatory T cells. Exp Ther Med. 2011;2:737–43.
25. Ishikawa T, Kokura S, Sakamoto N, Okayama T, Endo M, et al. Whole blood interferon-gamma levels predict the therapeutic effects of adoptive T-cell therapy in patients with advanced pancreatic cancer. Int J Cancer. 2013;133:1119–26.
26. Romagnani S. Human TH1 and TH2 subsets: doubt no more. Immunol Today. 1991;12:256–7.
27. Metcalf D, Begley CG, Johnson GR, Nicola NA, Vadas MA, et al. Biologic properties in vitro of a recombinant human granulocyte-macrophage colony-stimulating factor. Blood. 1986;67:37–45.
28. Hildebrandt B, Wust P, Ahlers O, Dieing A, Sreenivasa G, Kerner T, et al. The cellular and molecular basis of hyperthermia. Crit Rev Oncol Hematol. 2002;43:33–56.
29. Vidair CA, Dewey WC. Evaluation of a role for intracellular Na+, K+, Ca2+, and Mg2+ in hyperthermic cell killing. Radiat Res. 1986;105:187–200.

30. Stevenson MA, Calderwood SK, Hahn GM. Effect of hyperthermia (45 °C) on calcium flux in Chinese hamster ovary HA-1 fibroblasts and its potential role in cytotoxicity and heat resistance. Cancer Res. 1987;47:3712–7.
31. Malhotra A, Heynen ML, Lepock JR. Role of extracellular calcium in the hyperthermic killing of CHL V79 cells. Radiat Res. 1987;112:478–89.
32. Sapareto SA, Dewey WC. Thermal dose determination in cancer therapy. Int J Radiat Oncol Biol Phys. 1984;10:787–800.
33. Lepock JR. Cellular effects of hyperthermia: relevance to the minimum dose for thermal damage. Int J Hyperthermia. 2003;19:252–66.
34. van der Zee J. Heating the patient: a promising approach? Ann Oncol. 2002;13:1173–84.
35. Frey B, Weiss EM, Rubner Y, Wunderlich R, Ott OJ, Sauer R, et al. Old and new facts about hyperthermia-induced modulations of the immune system. Int J Hyperthermia. 2012;28:528–42.
36. Ostberg JR, Dayanc BE, Yuan M, Oflazoglu E, Repasky EA. Enhancement of natural killer (NK) cell cytotoxicity by fever-range thermal stress is dependent on NKG2D function and is associated with plasma membrane NKG2D clustering and increased expression of MICA on target cells. J Leukoc Biol. 2007;82:1322–31.
37. Ito A, Tanaka K, Kondo K, Shinkai M, Honda H, Matsumoto K, et al. Tumor regression by combined immunotherapy and hyperthermia using magnetic nanoparticles in an experimental subcutaneous murine melanoma. Cancer Sci. 2003;94:308–13.
38. Multhoff G, Botzler C, Jennen L, Schmidt J, Ellwart J, Issels R. Heat shock protein 72 on tumor cells: a recognition structure for natural killer cells. J Immunol. 1997;158:4341–50.
39. Botzler C, Li G, Issels RD, Multhoff G. Definition of extracellular localized epitopes of Hsp70 involved in an NK immune response. Cell Stress Chaperones. 1998;3:6–11.
40. Roigas J, Wallen ES, Loening SA, Moseley PL. Heat shock protein (Hsp72) surface expression enhances the lysis of a human renal cell carcinoma by IL-2 stimulated NK cells. Adv Exp Med Biol. 1998;451:225–9.
41. Vabulas RM, Ahmad-Nejad P, Ghose S, Kirschning CJ, Issels RD, Wagner H. Hsp70 as endogenous stimulus of the Toll/interleukin-1 receptor signal pathway. J Biol Chem. 2002;277:15107–12.
42. Asea A, Rehli M, Kabingu E, Boch JA, Bare O, Auron PE, et al. Novel signal transduction pathway utilized by extracellular Hsp70: role of toll-like receptor (TLR) 2 and TLR4. J Biol Chem. 2002;277:15028–34.
43. Todryk S, Melcher AA, Hardwick N, Linardakis E, Bateman A, Colombo MP, et al. Heat shock protein 70 induced during tumor cell killing induces Th1 cytokines and targets immature dendritic cell precursors to enhance antigen uptake. J Immunol. 1999;163:1398–408.
44. Noessner E, Gastpar R, Milani V, Brandl A, Hutzler PJ, Kuppner MC, et al. Tumor-derived heat shock protein 70 peptide complexes are cross-presented by human dendritic cells. J Immunol. 2002;169:5424–32.
45. Suzue K, Zhou X, Eisen HN, Young RA. Heat shock fusion proteins as vehicles for antigen delivery into the major histocompatibility complex class I presentation pathway. Proc Natl Acad Sci U S A. 1997;94:13146–51.
46. Suto R, Srivastava PK. A mechanism for the specific immunogenicity of heat shock protein-chaperoned peptides. Science. 1995;269:1585–8.
47. Binder RJ, Srivastava PK. Peptides chaperoned by heat-shock proteins are a necessary and sufficient source of antigen in the cross-priming of CD8+ T cells. Nat Immunol. 2005;6:593–9.
48. Mace TA, Zhong L, Kokolus KM, Repasky EA. Effector CD8+ T cell IFN-gamma production and cytotoxicity are enhanced by mild hyperthermia. Int J Hyperthermia. 2012;28:9–18.
49. Ostberg JR, Repasky EA. Emerging evidence indicates that physiologically relevant thermal stress regulates dendritic cell function. Cancer Immunol Immunother. 2006;55:292–8.
50. Bear AS, Kennedy LC, Young JK, Perna SK, Mattos Almeida JP, Lin AY, et al. Elimination of metastatic melanoma using gold nanoshell-enabled photothermal therapy and adoptive T cell transfer. PLoS One. 2013;8:e69073.

Chapter 33
High-Temperature Hyperthermic Intraperitoneal Chemotherapy (H-HIPEC) with Cytoreductive Surgery for Patients with Peritoneal Metastases of Colorectal Cancer and Appendiceal Pseudomyxoma

Kanji Katayama, Mitsuhiro Morikawa, Kenji Koneri, Makoto Murakami, Yasuo Hirono, Takanori Goi, and Akio Yamaguchi

Abstract Procedures of our H-HIPEC and cytoreductive surgery (CRS) for patients with peritoneal metastases from colorectal cancer and appendiceal pseudomyxoma (PMP) are reported.

Procedure: After resection of the tumor and regional lymph nodes and CRS for peritoneal metastases, without performing systemic peritonectomy, H-HIPEC was performed against the remaining superficial malignancy. Saline containing cisplatin, mitomycin C and etoposide were heated to 56 °C and were poured into the abdomen and stirred carefully by hand. Intraabdominal temperature was monitored and the thermal dose (TD) was calculated the equivalent time at 43 °C. HIPEC lasted until the TD reached to 25 min.

Results: Among all patients with peritoneal metastases from colon cancer, the H-HIPEC group had a 28.5 % 5-year survival rate, with a MST of 20.5 months, whereas those of the non HIPEC were only 5.4 % and 8.3 months ($p = 0.01$). The 5-year survival of patients with PMP treated by CRS and H-HIPEC was 82.5 %. Good quality of life and improved prognosis ware obtained by H-HIPEC without massive resection of the peritoneum or other organs, although it requires a high-level postoperative general management technique.

Conclusions: H-HIPEC after CRS is an effective and safe therapy for peritoneal metastases from colorectal cancer and PMP. During HIPEC, we heat the whole serosa over 43 °C and evaluate treatment by "Thermal Dose 43 °C". Intensive care with sufficient infusion is essential to avoid renal damage after HIPEC.

K. Katayama (✉)
Cancer Care Promotion Center, Medical School Hospital, University of Fukui, 23 Matsuoka-Shimoaitsuki, Eiheiji-cho, Fukui 910-1193, Japan
e-mail: kanji@u-fukui.ac.jp

M. Morikawa • K. Koneri • M. Murakami • Y. Hirono • T. Goi • A. Yamaguchi
First Department of Surgery, University of Fukui, School of Medicine, Fukui, Japan

S. Kokura et al. (eds.), *Hyperthermic Oncology from Bench to Bedside*,
DOI 10.1007/978-981-10-0719-4_33

Keywords Peritoneal • Metastases • Colon cancer • Pseudomyxoma peritonei • HIPEC • Thermal dose

33.1 Introduction

Treatment for colorectal cancer has recently made great advancements, especially in the field of chemotherapy; development of molecular target drugs allows patients who have unresectable metastases to achieve long-term survival. On the other hand, the peritoneum is one of the major metastatic organs of colorectal cancer, involving 13 % of cases [1]. Peritoneal metastases have a worse prognosis than metastases to other organs [2]. In addition, peritoneal metastases bring on the carcinomatous ascites that press into the abdominal cavity and markedly decreases the patient's quality of life (QOL) because of gastrointestinal tract obstruction and undernutrition. Tumor resection at operation is the most effective treatment for peritoneal metastases from colorectal cancer [3], while systemic chemotherapy offers less benefit than for other metastases. However, complete resection is difficult in cases of peritoneal dissemination. Therefore, there is no standard treatment for peritoneal metastases at present.

Pseudomyxoma peritonei (PMP) is derived mostly from rupture of an appendiceal or ovarian cystadenoma or cystadenocarcinoma [4, 5], producing a large quantity of mucinous substance [6]. Though the malignancy level is relatively low, radical resection is difficult and the prognosis is poor [7]. Systemic chemotherapy offers almost no benefit to patients with PMP, and after repeated tumor excisions over many years, the patients die in pain. Sugarbaker [6], Yan [7], and Bijelic [8] reported that HIPEC treatment was the most effective approach after cytoreduction surgery (CRS) by performing possible peritonectomy. However, it has been reported that the complication rate, morbidity, and mortality caused by CRS followed by HIPEC were very high [9]. In most institutions, HIPEC with a relatively low temperature of around 41–42 °C has been effectively performed after subtotal peritonectomy involving massive resection of the gastrointestinal tract [10, 11].

In this study, primary foci were resected, and CRS was performed for peritoneal metastases greater than 5 mm in size (non-peritonectomy), and HIPEC was conducted in the relatively high temperature zone of around 43–44 °C (high-temperature HIPEC; H-HIPEC) for peritoneal metastases of colorectal cancer and appendiceal PMP. In this paper, the precise procedures of our H-HIPEC treatment and the response to CRS and H-HIPEC in patients with peritoneal metastases of colorectal cancer and appendiceal PMP are reported.

33.2 Methods

33.2.1 Ethics

The procedures of this study received ethical approval from the institutional committee responsible for human experimentation at the University of Fukui, and all those involved gave their informed consent.

33.2.2 H-HIPEC Procedure

After resection of the primary foci with appropriate dissection of regional lymph nodes and CRS for peritoneal metastases greater than 5 mm and re-construction, H-HIPEC was performed. We did not perform systemic peritonectomy or ablation of other organs, but we treated the remaining superficial malignancy by chemo-hyperthermia.

A Thompson retractor was used, and the abdominal wall was slung up to maintain sufficient perfusion space (Fig. 33.1) [12–14]. An outflow tube was placed into the pouch of Douglas. The retractor was pulled fully upwards and fixed tightly to a vertical stand. This procedure allowed the abdominal cavity to extend wide enough and allow perfusate to spread throughout the peritoneal cavity

Four liters of saline containing 100 mg/body of cisplatin (CDDP), 20 mg/body of mitomycin C (MMC), and 200 mg/body of etoposide were heated to 56 °C in the reservoir in a water bath. Two liters of the infusate were poured into the abdomen

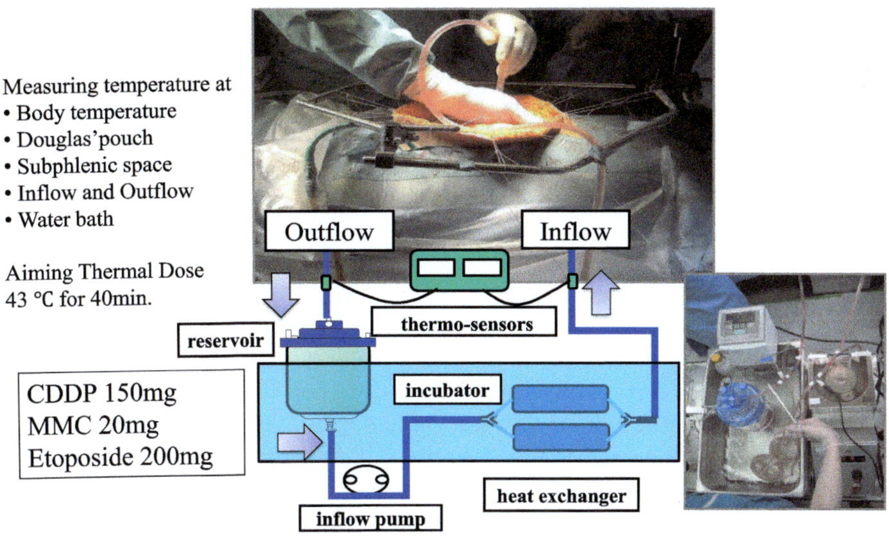

Fig. 33.1 H-HIPEC with newly developed disposable circuit

and stirred carefully for about 5 min so that no part of the abdomen was excessively heated. An additional two liters of the infusate were heated in the water bath and pumped into the circulation between the abdominal cavity and a reservoir at ~500 ml per minute. The newly developed disposable circulating circuit was used. This circuit consisted of a reservoir tank and a heat exchanger and inflow and outflow tubes. Thermometers that can measure inflow and outflow temperatures were included in the tubes.

It is important to heat the whole abdominal cavity evenly and continually during H-H by stirring by hand IPEC. Temperatures in the abdomen were measured using a 22-gauge thin thermocouple thermometer at the serosal surface in the subphrenic space and the pouch of Douglas, and the temperature of the infusate was also measured in the inflow tube, outflow tube, and water bath. The temperature of the peritoneal surface was maintained at approximately 43 °C by controlling the temperature of the water bath and the speed of the pump. The thermal dose [15, 16] (TD) obtained during treatment was calculated simultaneously during H-HIPEC and expressed in terms of equivalent time at 43 °C (Table 33.1). HIPEC was performed until the TD reached 25 min, which was actually obtained in around 40 to 50 min. During the treatment, a clear sterilized cover made for intra-operative fluoroscopy was attached to the retractor, with stirring by the arms through its slits, and then the cover was continuously aspirated with an aspirator to prevent transpiration of the anticancer drug.

33.2.3 Post Treatment Intensive Care

It is very important to manage the water balance of the patient after H-HIPEC to avoid renal failure. Because heating with a high temperature injures the whole peritoneum, intensive circulation care similar to the care after a severe systemic burn is required. The volume of the infusion depends on the size of the patient's body and the TD obtained and calculated during the treatment. The maintenance infusion was infused at about 35 ml/body weight (kg), including the necessary electrolytes, glucose, and amino acids. It was necessary to infuse additional bicarbonate Ringer's solution (or acetic acid Ringer's solution) at 300 to 500 ml/h within the first 24 h after HIPEC. The average total volume of infusion on the 1st day was around 12,000 ml in the case of a 50 kg patient (Fig. 33.2). It is very important to regulate the volume of infusion using various data showing the circulatory system's state, such as the urine flow per hour, central venous pressure, peripheral venous resistance, cardiac output, and stroke volume variation (SVV). After HIPEC, the systemic vascular resistance level was about half by the isolation of so-called burn substances from the burned tissue. Therefore, it was necessary to infuse a large quantity of extracellular fluid to fill the doubled vascular bed. A Vigileo monitor (Edwards Lifesciences Co. Irvine, CA. U.S.A.) was used to monitor the CC and SVV.

Table 33.1 The substitution calculating formula of the Thermal Dose 43 °C (TD43) numerical table

	Pre	1	3	5	10	15	20	25	30	35	40	45	50	55	60	Total TD	MEAN	S.D.	VAR
Douglas' pouch temp.(°C)	37	42	44	43	43	42	42	43	43	43	43						42.7	0.55	0.30
TD43.0 (min)		0.3	3.7	2.5	2.5	2.2	1.3	2.9	2.5	3.8	3.8	0	0	0	0	25.40			
Lt subphlenic temp.(°C)	37	42	43	43	43	43	44	42	42	43	43						42.76	0.50	0.25
TD43.0 (min)		0.3	1.3	2	4.4	3.8	8.7	1.6	1.9	2.5	5.7	0	0	0	0	32.28			
Body temp. (Esophag. ;°C)	36	36	36	37	38	38	39	39	39	39	39								
Body temp. (U.B;°C)	36	36	38	39	39	39	40	40	40	40	40								
Inflow temp.(°C)		50	50	48	47	46	49	48	49	49	49								
Outflow temp.(°C)		40	41	41	40	40	40	41	41	40	40								
Pump flow (L/min)		0.6	0.8	0.8	0.6	0.4	0.4	0.6	0.5	0.6	0.6								
Water bath temp.(°C)		60	60	59	58	58	58	59	59	59	59								

C3: =IF(C2 > =43,2^(C2-43),4^-(43-C2))
D3: =IF(D2 > =43,(D1-C1)*2^(D2-43),(D1-C1)*4^-(43-D2))
Copy similarly from E3 to P3
P3: =IF(P2 > =43,(P1-O1)*2^(P2-43),(P1-O1)*4^-(43-P2))
Q3: =SUM(C3:P3)
R2: =AVERAGE(C2:N2)
S2: =STDEV(C2:N2)
T2: =VAR(C2:P2)
Copy line 2 and 3 to line 4 and 5
TD 43 is calculated with a PC every 5 min and multiply. We aim to obtain over 40 min in therapeutic HIPEC

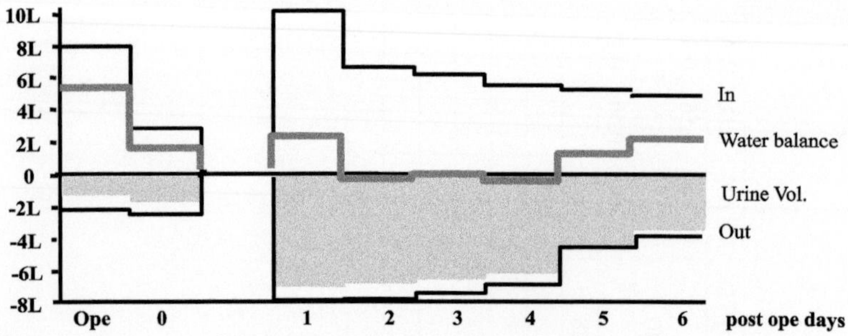

Fig. 33.2 Post H-HIPEC infusion therapy and water balance in a case. The volume of infusion follows treatment of the broad burn and amounts over 10 l in a day

33.3 Patients

33.3.1 Colorectal Cancer

Thirty-eight patients underwent resection of the primary tumor and CRS for peritoneal metastases greater than 5 mm (systemic peritonectomy was not performed) and H-HIPEC at the Department of Gastrointestinal Surgery, University of Fukui between 1984 and 2013 (Table 33.2). According to the Japanese classification of colorectal carcinoma version 8 [17], the peritoneal metastases were grossly graded as follows: P1 for metastases into adjacent localized peritoneum; P2 for slight metastases into remote peritoneum; and P3 for marked metastases into remote peritoneum.

There were 20 P1/P2 patients and 18 P3 patients. Among the P1/P2 patients, 7 underwent primary tumor resection and H-HIPEC, followed by guideline-prescribed postoperative systemic chemotherapy at other hospitals. Thirteen patients suffered recurrence of treatment-resistant peritoneal metastases, for which HIPEC was consequently performed at our hospital.

33.3.2 PMP

H-HIPEC was performed for 13 cases of appendiceal-origin PMP. Nine of the 13 patients were female and four were male, ranging from 40 to 73 years of age (Table 33.3). Ronnett pathologically classified PMP into three categories as follows [18]: Disseminated Peritoneal Adenomucinosis (DPAM), with a relatively good prognosis and accounting for more than half of cases in Western countries; Peritoneal Mucinous carcinomatosis (PMCA), with a poor prognosis, which is generally said to be equivalent to that of peritoneal metastases from colon cancer; and Intermediate type (PMCA-I), which is intermediate between the two previous types.

Table 33.2 Patient characteristics who received CRS + HIPEC

	HIPEC (n = 38)	non HIPEC (n = 64)
Age(years)	55.8	67.6
Sex(n)		
Male	19	45
Female	19	19
Location(n)		
Cecum	9	12
Ascending	7	12
Transverse	4	14
Descending	3	1
Sigmoid	8	13
Rectum	7	12
Pathology(n)		
Well/Moderately differentiated	25	37
Others	12	27
Peritoneal metastasis stage(n)		
P1	11	17
P2	9	21
P3	18	26

Table 33.3 Prognosis after HIPEC, by Ronnett's classification

Case	Patho.type	Recurrence	Survival month
65 F	DPAM	none(HIPEC × 2)	175 M alive
73 F	DPAM	none	63 M alive
58 M	DPAM	none	55 M alive
71 F	DPAM	none(HIPEC × 2)	43 M alive
70 F	DPAM	none	34 M alive
61 F	DPAM	none	14 M alive
53 F	PMCA-I	none(HIPEC × 2)	60 M alive
67 F	PMCA-I	none	11 M alive
66 M	PMCA	yes	97 M alive
59 M	PMCA	yes	36 M dead
59 M	PMCA	none	43 M alive
71 F	PMCA	yes	38 M alive
40 F	PMCA	yes	20 M dead
63 M	PMCA	none	36 M alive
46 F	PMCA	yes	19 M alive
63 M	PMCA	yes	11 M alive

Pathological diagnosis from the resected specimens showed that six patients had DPAM, two of the female patients had PMCA-I, and the other eight patients had PMCA. In all PMCA cases, multilayered tumor cells and a partially infiltrated area were observed. In three recurrence cases, tumor re-resection and HIPEC were repeated.

33.3.3 Criteria for H-HIPEC Therapy

The criteria for H-HIPEC therapy were as follows: (1) histopathological findings confirmed primary colorectal cancer with peritoneal metastases or histopathologically proven PMP in which the appendiceal tumor was clear on imaging; (2) resection of colorectal cancer or appendiceal cyst adenocarcinoma with extended (D3) lymph node dissection and CRS for peritoneal metastases greater than 5 mm (Japanese Society for Cancer of Colon and Rectum, 2009); (3) when hematogenous metastases were confirmed, they were curatively resected (except peritoneal dissemination); (4) an Eastern Cooperative Oncology Group performance status (PS) of 0 or 1; (5) function of the major organs was maintained (Renal, Liver, Pulmonary, Heart, Circulation, Bone marrow etc.); (6) under 75 years old; (7) provided written, informed consent; and (8) all patients were followed-up for recurrence at regular intervals for 5 years, underwent chest X-ray, computed tomography, and colonoscopy, etc.

33.3.4 Statistical Analyses

Survival time was estimated using the Kaplan-Meier method. The outcomes of the two groups were compared by the log-rank test, with $p < 0.05$ considered significant.

33.4 Results

33.4.1 Heating Effect

The intraperitoneal space was sufficiently heated, and 25 min of "TD 43 °C" was achieved in the left sub-phrenic space and in the pouch of Douglas in all cases. Figure 33.3 shows the monitored temperature of an average case during H-HIPEC treatment.

33.4.2 Characteristics of Patients with Peritoneal Metastases from Colorectal Cancer

There were no significant differences in age, sex, tumor location, histopathological findings, and peritoneal metastasis grade between the groups (Table 33.1).

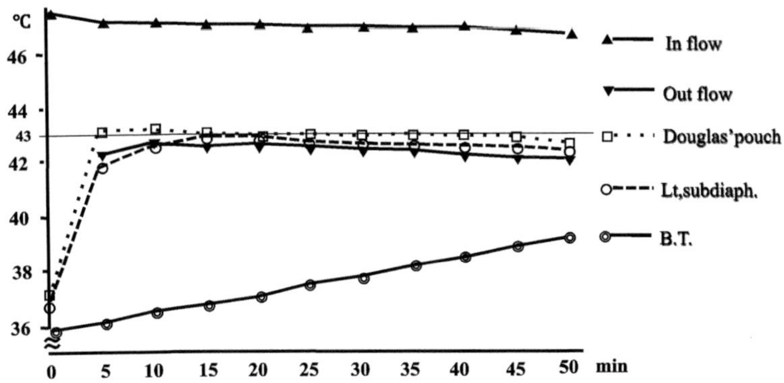

Fig. 33.3 Temperature monitoring during H-HIPEC

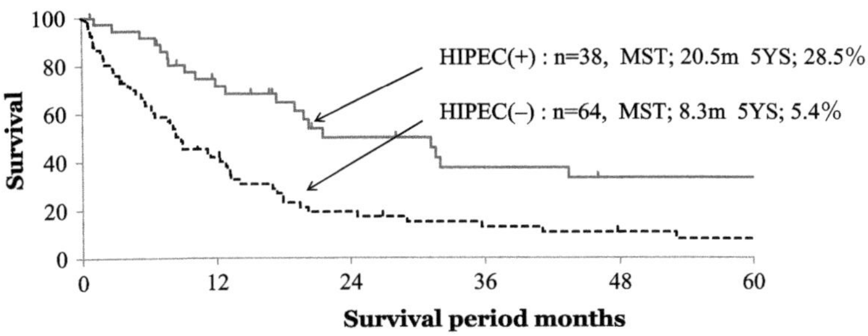

Fig. 33.4 Kaplan-Meier estimate of overall survival curves of all patients peritoneal metastases from colorectal cancer received CRS and H-HIPEC treatment. Five years survival rate was 28.5 % by the HIPEC treatment. HIPEC(+)cases survived significantly longer ($p = 0.01$ Generalized Wilcoxon test)

33.4.3 Survival Rate of H-HIPEC with Peritoneal Metastases of Colorectal Cancer

Among all peritoneal metastases patients, the H-HIPEC group had a 28.5 % 5-year survival rate, with a median survival time (MST) of 20.5 months (Fig. 33.4), whereas the 5-year survival and MST of the no HIPEC group were only 5.4 % and 8.3 months, respectively. There was a significant difference in the two survival curves ($p = 0.01$). The overall survival rate of each peritoneal metastasis grade, P1/P2 and P3, was then estimated. In the P1/P2 grade group, the HIPEC group had a 32.4 % 5-year survival rate (Fig. 33.5). In the P3 grade group, the HIPEC group had a 21.6 % 5-year survival rate, with an MST of 19.0 months (Fig. 33.6). Both the P1/P2 and P3 groups that received H-HIPEC survived significantly longer than those groups that did not receive H-HIPEC.

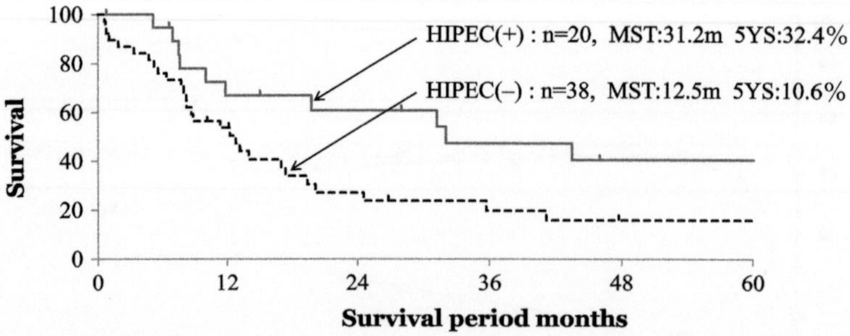

Fig. 33.5 Kaplan-Meier estimate of overall survival curves of p1 + p2 patients with peritoneal metastases from colorectal cancer received CRS and H-HIPEC treatment. Five years survival rate was 31.2 % by the HIPEC treatment. HIPEC(+)cases survived significantly longer (p = 0.001 Generalized Wilcoxon test)

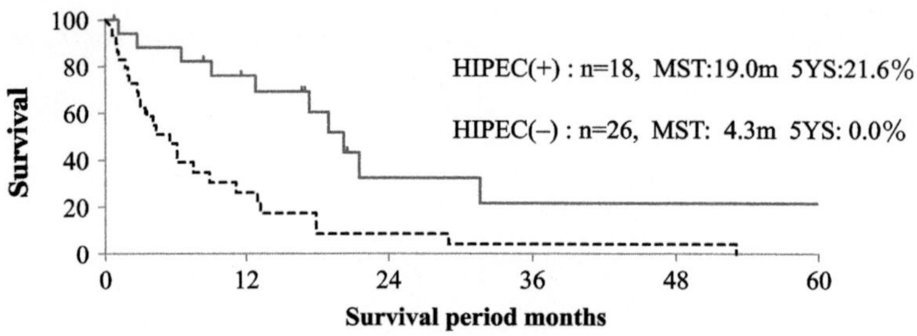

Fig. 33.6 Kaplan-Meier estimate of overall survival curves of p3 patients with peritoneal metastases from colorectal cancer received CRS and H-HIPEC treatment. Five years survival rate was 21.6 % by the HIPEC treatment. HIPEC(+) though no patient could Survive over 5 years without H-HIPEC treatment (p = 0.01 Generalized Wilcoxon test)

33.4.4 Survival Rate of H-HIPEC with PMP

Figure 33.7 shows the overall survival curve of all PMP patients who received H-HIPEC treatment. The 5-year survival rate of all patients treated by CRS and H-HIPEC was 82.5 %. Figure 33.8 shows the overall survival curve of PMCA patients with H-HIPEC. The 5-year survival rate of PMCA patients treated by CRS and HIPEC was 66.7 %. Concerning the treatment of PMP, we considered that good quality of life and improved prognosis may be obtained by performing H-HIPEC over 43 °C without massive resection of the peritoneum or alimentary tract, although it requires a high-level postoperative general management technique.

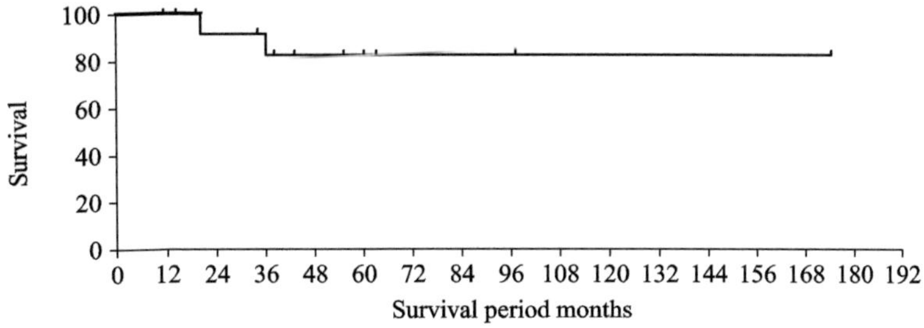

Fig. 33.7 Kaplan-Meier estimate of overall survival curves of all PMP patients with H-HIPEC treatment. Five years survival rate of all patients was 82.5 % by CRS and HIPEC treatment

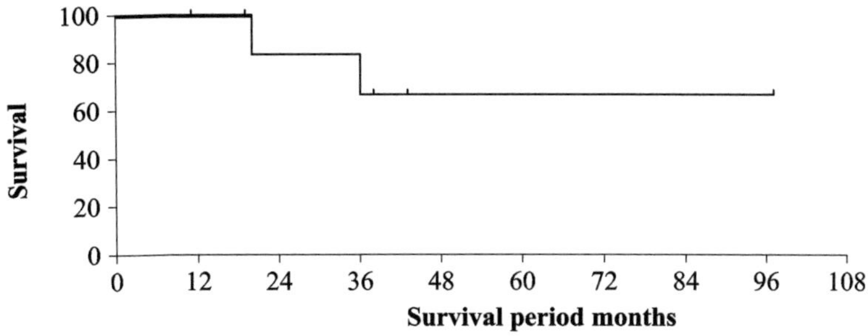

Fig. 33.8 Kaplan-Meier estimate of overall survival curves of PMCA PMP patients with H-HIPEC. Five years survival rate of PMCA was 66.7 % by CRS and HIPEC treatment

33.4.5 Complications Caused by HIPEC

The complications caused by HIPEC were investigated. Among the 54 H-HIPEC procedures for 51 patients who underwent HIPEC from 1984 to 2013, there were 5 renal disorders, and 3 patients developed anastomotic leakage (Table 33.4). Only one patient developed acute renal failure that needed artificial dialysis, but but recovered before long. Treatment was not associated with any mortality.

33.5 Discussion

In recent years, with technological developments in surgery and advances in anti-cancer drugs and molecular targeted drugs for chemotherapy [19, 20], outcomes for unresectable advanced colorectal cancer have improved. However, there are no definitive treatments for peritoneal metastases, there are no reported large-scale trials, and the outcomes have not improved. Generally, systemic chemotherapy is performed for peritoneal metastases of colorectal cancer, but it is not effective compared with hematogenous metastases. The lack of effectiveness of

Table 33.4 Complications after HIPEC

Complications	HIPEC (n = 38) colo-rectal Ca	HIPEC (n = 19) PMP	Total (n = 57)
Heart disease n (%)	0 (0)	0 (0)	0 (0)
Pulmonary complication n (%)	1 (2.6)	1 (5.3)	2 (3.5)
Renal disease n (%)	4 (10.5)	1 (5.3)	5 (8.8)
Ileus n (%)	0 (0)	0 (0)	0 (0)
Anastomotic stenosis n (%)	0 (0)	0 (0)	0 (0)
Leakage n (%)	2 (5.2)	1 (5.3)	3 (5.3)
Pelvic abscess n (%)	0 (0)	0 (0)	0 (0)
Wound infection n (%)	0 (0)	0 (0)	0 (0)
Morbidity	6 (1.6)	3 (1.6)	10 (1.8)
Mortarity	0 (0)	0 (0)	0 (0)

chemotherapy is likely due to the poor blood flow of the peritoneum, which decreases the concentrations of anticancer drugs. This has been called the blood-peritoneal barrier [21].

Sugarbaker [6], Yan [7], and Bijelic [8] reported that HIPEC treatment was the most effective after cytoreduction by total or subtotal peritonectomy. However, a very high rate of complications and serious problems has been reported [9]. In this study, the effectiveness of HIPEC was investigated after performing resection of the primary tumor and CRS for peritoneal metastases greater than 5 mm (non-peritonectomy) for peritoneal metastases of colorectal cancer.

Koppe et al [22] reported that the median survival rate of patients with peritoneal metastases treated with systemic chemotherapy was 5.2–12.6 months. However, the treatments used in this study showed better outcomes than seen in patients with systemic chemotherapy; this was because complete tumor resection of the peritoneal dissemination was impossible, and the anticancer drugs of systemic chemotherapy could not infiltrate into peritoneal tumor tissue at an effective concentration. Whereas HIPEC treatment facilitates direct contact between anticancer drugs and peritoneal metastases smaller than 5 mm, it also allows infiltration of the drugs at a high concentration because the electrolyte barrier of peritoneal tumors might be broken by the hyperthermic effect [23].

HIPEC treatment seems to be one of the effective treatments for all grades of peritoneal metastases of colorectal cancer, including PMP from appendiceal cystic tumors. Smeenk et al [11] reported that the median survivals of PMCA and PMCA-I after treatment were 13 and 24 months, respectively, with no 5-year survival in PMCA. Though the present data are derived from fewer cases compared with their data, survival of the present cases was comparatively satisfactory. Concerning the treatment of PMP, we considered that a good quality of life and an improved prognosis may be obtained by performing H-HIPEC with a temperature over 43 °C without massive resection of the peritoneum or alimentary tract, though it requires high-level postoperative general management technique.

In addition, some patients have shown an improvement in gastrointestinal obstruction because of decreased carcinomatous ascites and so-called 'jelly belly'. It is thought that H-HIPEC may contribute to a major improvement in patients' quality of life.

In many western institutions, HIPEC with a relatively low temperature around 41–42 °C has been effectively performed after subtotal peritonectomy involving massive resection of the gastrointestinal tract as a positive treatment [10, 11, 24]. However, in many cases, subtotal peritonectomy needs to be combined with total gastrectomy, subtotal colectomy, and massive resection of the small intestine, which is likely to result in a markedly decreased quality of life for the patient postoperatively.

As postoperative considerations, the abdominal cavity developed a widespread burn after HIPEC that caused increased blood vessel permeability and an increased circulating vascular floor [25, 26]. Furthermore, CDDP, which has renal toxicity, was used [27]. Thus, there is a need to perform intensive infusion control to avoid renal dysfunction due to acute tubular damage. Four cases of renal dysfunction were seen from 1984 to 2004, when HIPEC was started. Then, from 1995, intensive respiratory and circulatory management was performed in patients under sedation in the intensive care unit on a respirator after HIPEC to maintain renal blood flow. This postoperative management decreased some complications caused by HIPEC, and it was found that perioperative care could be performed more safely.

Recently, some molecules have been investigated as biomarkers of sensitivity to anticancer drugs [28]. Fujishima et al. [29] reported that the effectiveness of one of them in HIPEC was seen to decrease when MUC2 protein expression, which is in the MUC family, was seen in colorectal cancer cells. The MUC family is related to secretion of mucin, which plays a large role in the protective mechanism of the gastrointestinal tract [30]. They stated that MUC2 protein protect colorectal cancer cells against hyperthermic sensitivity by decreasing perviousness of anticancer drugs. Thus, they suggested that MUC2 was useful as an indicator to assess the effectiveness of HIPEC in colorectal cancer.

HIPEC was more effective in treating colorectal cancer patients with newly diagnosed P1 or P2 peritoneal metastases if it was carried out prior to systemic chemotherapy, instead of being used to treat peritoneal metastatic lesions resistant to systemic chemotherapy. This may be because genetic changes in the metastases make them less sensitive to HIPEC, since it is generally accepted that chemotherapy can alter the genomes of cancer cells. Another possible reason is that repetitive laparotomy was required, especially for patients with PMP, resulting in some adhesions left in place, disrupting the even spread of the warm liquid throughout all parts of the peritoneum.

33.6 Conclusion

H-HIPEC is an effective and safe therapy after resection of the primary tumor and CRS for peritoneal metastases greater than 5 mm (non-peritonectomy).

For treating peritoneal metastases from colorectal cancer and PMP, H-HIPEC is effective only for cases in which the primary lesion and lymph nodes have been removed.

During HIPEC, it is desirable to heat the whole serosa to 43 °C and evaluate treatment using "Thermal Dose 43 °C". Intensive care with sufficient infusion volumes, as in the treatment of widespread burns, is essential to avoid renal damage after HIPEC.

References

1. Janye DG, Fook S, Loi C, Seow-Choen F. Peritoneal carcinomatosis from colorectal cancer. Br J Surg. 2002;89(12):1545–50.
2. Rempone B, Schiavone B, Martino A, Confuorto G. Current role of hyperthermic intraperitoneal chemotherapy in the treatment of peritoneal carcinomatosis from colorectal cancer. World J Gastroenterol. 2010;16(11):1299–302.
3. Glehen O, Kwiatkowski F, Sugarbaker PH, Elias D, Levine EA, De Simone M, Barone R, Yonemura Y, Cavaliere F, Quenet F, Gutman M, Tentes AA, Lorimier G, Bernard JL, Bereder JM, Porcheron J, Gomez-Portilla A, Shen P, Deraco M, Rat P. Cytoreductive surgery combined with perioperative intraperitoneal chemotherapy for management of peritoneal carcinomatosis from colorectal cancer: a multi-institutional study. J Clin Oncol. 2004;22 (16):3284–92.
4. Dixit A, Robertson JHP, Mudan SS, et al. Appendiceal mucoceles and pseudomyxoma peritonei. World J Gastroenterol. 2007;13(16):2381–4.
5. Miner TJ, Shia J, Jaques DP, et al. Long-term survival following treatment of pseudomyxoma peritonei. An anlysis of surgical therapy. Ann Surg. 2005;241(2):300–8.
6. Sugarbaker PH. Peritonectomy procedures. Ann Surg. 1995;221(1):29–42.
7. Yan TD, Black D, Savady R, Sugarbaker PH. Systematic review on the efficacy of cytoreductive surgery combined with perioperative intraperitoneal chemotherapy for peritoneal carcinomatosis from colorectal carcinoma. J Clin Oncol. 2006;24(24):4011–9.
8. Bijelic L, Yan TD, Sugarbaker PH. Failure analysis of recurrent disease following complete cytoreduction and perioperative intraperitoneal chemotherapy in patients with peritoneal carcinomatosis from colorectal cancer. Ann Surg Oncol. 2007;14(8):2281–8.
9. Younan R, Kusamura S, Baratti D, Cloutier AS, Deraco M. Morbidity, toxicity, and mortality classification systems in the local regional treatment of peritoneal surface malignancy. J Surg Oncol. 2008;98(4):253–7.
10. Sugarbaker PH. Cytoreductive surgery and peri-operative intraperitoneal chemotherapy as a curative approach to pseudomyxoma peritonei syndrome. Eur J Surg Oncol. 2001;27:239–43.
11. Smeenk RM, Verwaal VJ, Antonini N, et al. Survival analysis of pseudomyxoma peritonei patients treated by cytoreductive surgery and hyperthermic intraperitoneal chemotherapy. Ann Surg. 2007;245:104–9.
12. Katayama K, Yamaguchi A, Murakami M, Koneri K, Nagano H, Honda K, Hirono Y, Goi T, Iida A, Ito H. Chemo-hyperthermic peritoneal perfusion (CHPP) for appendiceal pseudomyxoma peritonei. Int J Clin Oncol. 2009;14(2):120–4.

13. Katayama K, Note M, Fujita H, et al. Continuous hyperthermic peritoneal perfusion using peritoneal cavity expander. Hypertherm Oncol. 1988;1:450–1.
14. Sugarbaker PH. Peritoneal surface oncology: review of a personal experience with colorectal and appendiceal malignancy. Tech Coloproctol. 2005;9(2):95–103.
15. Sapareto SA, Dewey WC. Thermal dose determination in cancer therapy. Program and abstracts. The 2nd annual meeting. North American Hyperthermia Group. 1982 27.
16. Dewey WC, Hopwood LE, Sapareto SA, Gerweck LE. Cellular response to combinations of hyperthermia and radiation. Radiology. 1997;123(2):463–74.
17. Japanese Society for Cancer of Colon and Rectum. Japanese classification of colorectal carcinoma. 8th ed. Tokyo: Kanehara Publishers; 2009.
18. Ronnett BM, Zahn CM, Kurman RJ, et al. Disseminated peritoneal adenomucinous and peritoneal mucinous carcinomatosis: a clinicopathologic analysis of 109 cases with emphasis on distinguishing pathologic features, site of origin, prognosis, and relationship to "pseudomyxoma peritonei". Ann J Surg Pathol. 1995;19:1390–408.
19. Goldberg RM, Sargent DJ, Morton RF, Fuchs CS, Ramanathan RK, Williamson SK, Findlay BP, Pitot HC, Alberts SR. A randomized controlled trial of fluorouracil plus leucovorin, irinotecan, and oxaliplatin combinations in patients with previously untreated metastatic colorectal cancer. J Clin Oncol. 2004;22(1):23–30.
20. Tol J, Koopman M, Rodenburg CJ, Cats A, Creemers GJ, Schrama JG, Erdkamp FL, Vos AH, Mol L, Antonini NF, Punt CJ. A randomised phase III study on capecitabine, oxaliplatin and bevacizumab with or without cetuximab in first-line advanced colorectal cancer, the CAIRO2 study of the Dutch Colorectal Cancer Group (DCCG). An interim analysis of toxicity. Ann Oncol. 2008;19(4):734–8.
21. Sugarbaker PH, Stuart OA, Vidal-Jove J, Pessagno AM, DeBruijn EA. Pharmacokinetics of the peritoneal-plasma barrier after systemic mitomycin C administration. Cancer Treat Res. 1996;82:41–52.
22. Koppe MJ, Boerman OC, Oyen WJ, Bleichrodt RP. Peritoneal carcinomatosis of colorectal origin: incidence and current treatment strategies. Ann Surg. 2006;243(2):212–22.
23. Los GL, van Vugt MJ, Pinedo HM. Response of peritoneal solid tumours after intraperitoneal chemohyperthermia treatment with cisplatin or carboplatin. Br J Cancer. 1994;69(2):235–41.
24. Glehen O, Mithieux F, Osinsky D, et al. Surgery combined with peritonectomy procedures and intraperitoneal chemotherapy in abdominal cancers with peritoneal carcinomatosis; a phase II study. J Clin Oncol. 2003;21(5):799–806.
25. Sugarbaker PH, Stuart OA, Vidal-Jove J, Pessagno AM, DeBruijn EA. Studies of the peritonea-plasma barrier after systemic mitomycin C administration. Reg Cancer Treat. 1993;4:188–94.
26. Katayama K, Makimoto K, Maeda H, Takeuchi K, Fujita T, Kobayashi T. Continuous Hyperthermic Peritoneal Perfusion (CHPP)-Thermal dose during CHPP and response to treatment. Hypertherm Oncology Jpn. 1997;97:50–3.
27. Marie HH, Prasad D. Cisplatin nephrotoxicity:molecular mechanisms. Cancer Ther. 2003;1:47–61.
28. Giampieri R, Scartozzi M, Del Prete M, Maccaroni E, Bittoni A, Faloppi L, Bianconi M, Cecchini L, Cascinu S. Molecular biomarkers of residence to anti-EGFR treatment in metastatic colorectal cancer, from classical to innovation. Crit Rev Oncol Hematol. 2013;88 (2):272–83.
29. Fujishima Y, Goi T, Kimura Y, Hirono Y, Katayama K, Yamaguchi A. MUC2 protein expression status is useful in assessing the effects of hyperthermic intraperitoneal chemotherapy for peritoneal dissemination of colon cancer. Int J Oncol. 2012;40(4):960–4.
30. Knowles MR, Boucher RC. Mucus clearance as a primary innate defense mechanism for mammalian airways. J Clin Invest. 2002;109(5):571–7.

Chapter 34
A New Comprehensive Treatment for Peritoneal Metastases Using Cytoreductive Surgery Combined with Hyperthermic Intraperitoneal Chemoperfusion

Yutaka Yonemura, Akiyoshi Mizumoto, Haruaki Ishibashi, Kazuyoshi Takeshita, Oliver Glehen, and Paul H. Sugarbaker

Abstract A game-changing therapy for peritoneal metastasis (PM), now called "comprehensive treatment" was first established in the late 1990s. The treatment consists of aggressive cytoreductive surgery (CRS) combined with perioperative intraperitoneal/systemic chemotherapy. PM is considered as local disease, and the rationale behind the treatment is to remove macroscopic tumors and eradicate residual micrometastasis using perioperative chemotherapy (POC). Comprehensive treatment consists of laparoscopic evaluation of the tumor load, POC, and CRS. POC includes six procedures including laparoscopic hyperthermic intraperitoneal chemotherapy (LHIPEC), neoadjuvant intraperitoneal/systemic chemotherapy (NIPS), hyperthermic intraoperative intraperitoneal chemotherapy (HIPEC),

Y. Yonemura (✉)
Regional Cancer Therapies, Peritoneal metastasis center, NPO to support peritoneal surface malignancy treatment, Kishiada Tokushukai Hospital and Kusatsu general Hospital, 510, Fukushima-Cho, Simogyou-Ku, Kyoto City, Kyoto, Japan, Zip:600-8189

Kishiwada Tokushukai Hospital, Kishiwada, Japan

Kusatsu general hospital, Kusatsu, Japan
e-mail: y.yonemura@coda.ocn.ne.jp

A. Mizumoto
Kusatsu general hospital, Kusatsu, Japan

H. Ishibashi • K. Takeshita
Kishiwada Tokushukai Hospital, Kishiwada, Japan

O. Glehen
Dēpartement de Chirurgie Gēnerale, Centre Hospitalier Lyon-Sud Hospices Civils de Lyon, Universitē Lyon, 69495 Lyon, France

P.H. Sugarbaker
Center of Gastrointestinal Malignancies, Program in Peritoneal Surface Malignancies, MedStar Washington Hospital Center, Washington, DC 20010, USA

S. Kokura et al. (eds.), *Hyperthermic Oncology from Bench to Bedside*,
DOI 10.1007/978-981-10-0719-4_34

extensive intraoperative peritoneal lavage (EIPL), early postoperative intraperitoneal chemotherapy (EPIC), and late postoperative systemic chemotherapy. In a study of gastric cancer patients with P0/Cy1 status that employed comprehensive treatment, POC was confirmed to eradicate intraperitoneal micrometastasis. For clinical standardization of HIPEC, the concept of the thermal dose should be introduced. One thermal dose is equivalent to 30-min treatment at 43 °C. In gastric cancer, one thermal dose of HIPEC has been shown to reduce the peritoneal cancer index of 3.5, and changed the peritoneal cytology from positive to negative in 70 % of patients.

In gastric cancer, ovarian cancer, colorectal cancer, mesothelioma and pseudomyxoma peritonei, meta-analyses of randomized controlled studies demonstrated that HIPEC significantly improved survival after CRS.

These results suggest that comprehensive treatment is a breakthrough that improves the prognosis of patients with PM.

Keywords Peritoneal metastasis • HIPEC • Peritonectomy • Comprehensive treatment

34.1 Introduction

Until the early 1990s, peritoneal metastasis (PM) had been considered a terminal disease, and patients with PM were treated by palliative surgery or systemic chemotherapy. However, in the late 1990s, a paradigm shift occurred [1–4]. PM is considered a local disease, confined to the peritoneal cavity, and combination of cytoreductive surgery (CRS) to remove macroscopic detectable PM and perioperative chemotherapy (POC) to eradicate micrometastases became a new and successful therapy that is now termed "comprehensive treatment". In comprehensive treatment, a patient is treated with neoadjuvant chemotherapy, CRS, hyperthermic intraperitoneal chemotherapy (HIPEC), early postoperative intraperitoneal chemotherapy (EPIC), and postoperative chemotherapy. After the introduction of comprehensive treatment, postoperative survival significantly improved compared with that after the standard surgery or systemic chemotherapy alone. Additionally, the treatment is considered a curative approach for selected patients who have limited PM due to colorectal and gastric cancer. As a result, the National Institute of Health and Clinical Excellence (NICE) declared that HIPEC followed by CRS is the state-of-the art treatment for colorectal cancer PM, confined in one or two peritoneal sectors with curative intent (IPG331. http://www.nice.org.uk/guidance/IPG331. accessed 12 August 2013) [5].

In this chapter, the principles and latest treatment results using comprehensive treatment are presented.

34.2 Outline of the Comprehensive Treatment for PM

As shown in Fig. 34.1, the comprehensive treatment comprises laparoscopic evaluation of the tumor load, perioperative chemotherapy (POC) and CRS. POC includes six procedures, including laparoscopic hyperthermic intraperitoneal chemotherapy (LHIPEC), neoadjuvant intraperitoneal/systemic chemotherapy (NIPS), hyperthermic intraoperative intraperitoneal chemotherapy (HIPEC), extensive intraoperative peritoneal lavage (EIPL), early postoperative intraperitoneal chemotherapy (EPIC) and late postoperative systemic chemotherapy [2, 3].

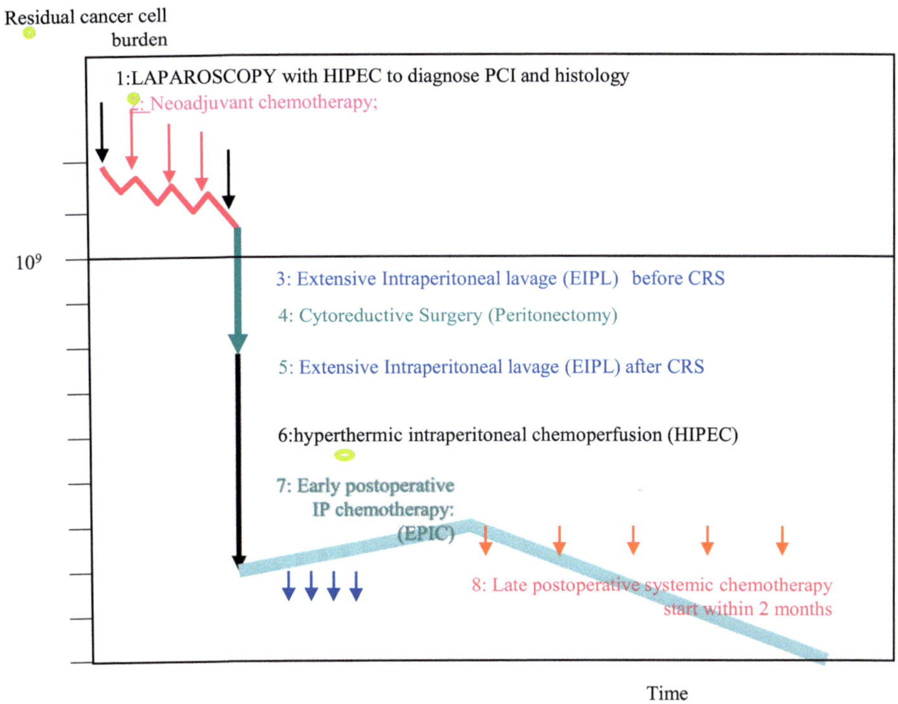

Fig. 34.1 Schedule of a comprehensive treatment for patients with PM. The comprehensive treatment consists of eight procedures, which are laparoscopic evaluation of the tumor load, perioperative chemotherapy (POC) and CRS. POC includes laparoscopic hyperthermic intraperitoneal chemotherapy (LHIPEC), neoadjuvant intraperitoneal/systemic chemotherapy (NIPS), hyperthermic intraoperative intraperitoneal chemotherapy (HIPEC), extensive intraoperative peritoneal lavage (EIPL), early postoperative intraperitoneal chemotherapy (EPIC) and late postoperative systemic chemotherapy

34.2.1 *Peritoneal Cancer Index (PCI) and Completeness of Cytoreduction Score (CC Score)*

The peritoneal cancer index (PCI) is an international classification of PM, which was approved by experts at the 5th international meeting on peritoneal surface malignancy treatment in Milan, 2006 [6, 7]. PCI reflects not only PM distribution but also PM volume and is the most useful classification method for quantitative estimation of PM. The abdomen is divided into 13 compartments. The tumor involvement in each compartment is macroscopically evaluated by lesion size, using scores ranging from 0 to 3, and the tumor load in the abdominal cavity is evaluated by PCI scores ranging from 0 to 39 (Fig. 34.2) [6, 7]. After undergoing CRS, PCI score is considered the most important of several prognostic factors for patients with PM in many different primary sites (Figs. 34.3, 34.4, 34.5, 34.6). PCI cutoff levels for good and bad prognosis have been reported by several researchers. Table 34.1 shows the PCI cutoff levels for patients with gastric cancer, colorectal cancer, pseudomyxoma peritonei, and ovarian cancer, who were treated at our hospitals. In PM from gastric cancer, each of three papers proposed different cutoff levels (8, 12, and 20) [2, 8, 9]. Coccolini studied the cutoff levels by conducting a meta-analyses of 20 studies and reported that the PCI cutoff level for favorable prognosis is PCI 12 [10].

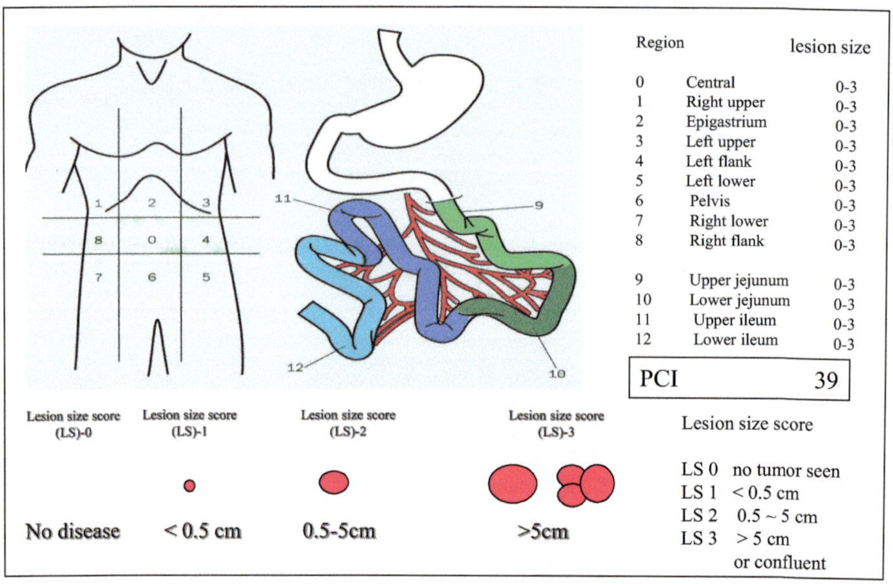

Fig. 34.2 Peritoneal Cancer Index (PCI). PCI was approved at 5th International Workshop on Peritoneal Surface Malignancy, Milan (published in J Surg Oncol. 2008;98:228–231. The abdomen is divided into 13 compartments. The tumor involvement in each compartment is macroscopically evaluated by lesion size scores from 0 to 3, and the tumor load in the abdominal cavity is evaluated by PCI scores from 0 to 39

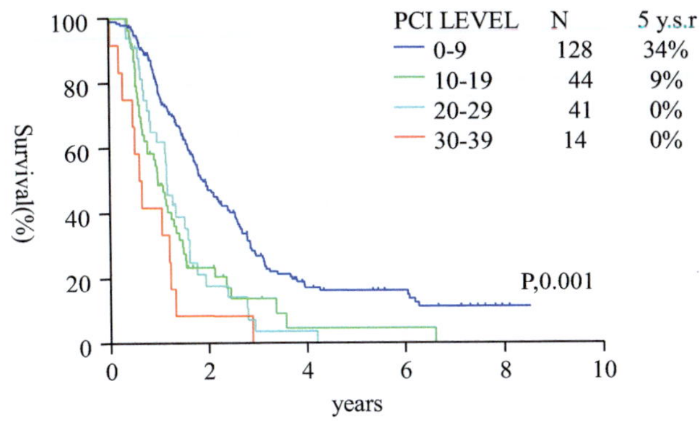

Fig. 34.3 PCI scores as important prognostic indicator after CRS plus HIPEC for 227 patients with PM from gastric cancer

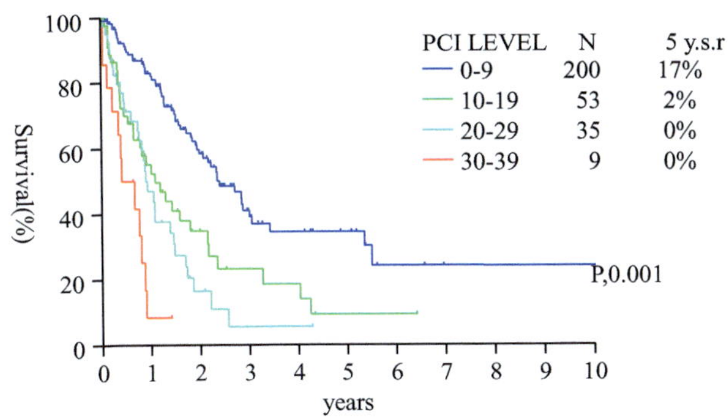

Fig. 34.4 PCI scores as important prognostic indicator after CRS plus 297 HIPEC for patients with PM from colorectal cancer

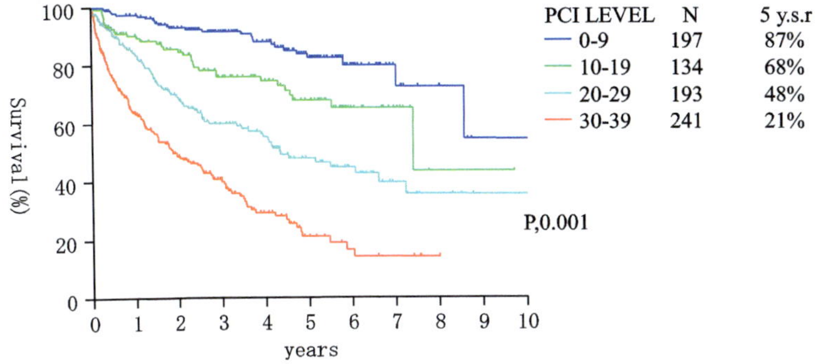

Fig. 34.5 PCI scores as important prognostic indicator after CRS plus HIPEC for 865 patients with pseudomuxoma peritonei

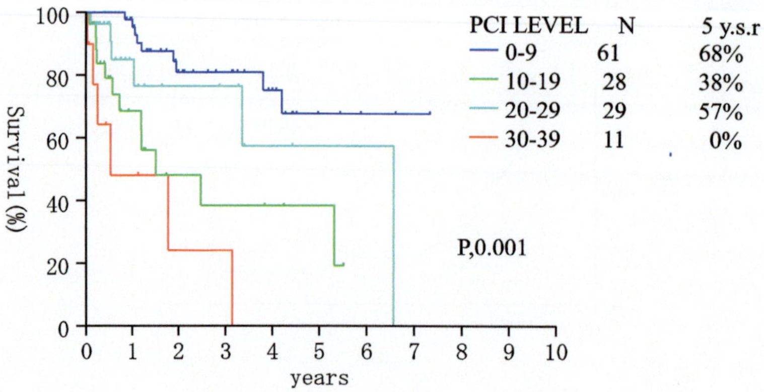

Fig. 34.6 PCI scores as important prognostic indicator after CRS plus HIPEC for 129 patients with ovarian cancer or primary peritoneal carcinomatosis

Table 34.1 PCI cutoff levels and 5-year survival rates in patients with peritoneal metastases, treated at Peritoneal Surface Malignancy Treatment Centers belonging to Japanese NPO to support peritoneal surface malignancy treatment

	PCI cut-off level	No of cases	5-year survival	X^2	P
Gastric cancer	6>=	231	15 %	62.9	P<0.0001
	7=<	241	2 %		
Colorectal cancer	15>=	161	29 %	61.3	P<0.0001
	16=<	77	0 %		
Pseudomyxoma peritonei	27>=	577	68 %	167.6	P<0.0001
	28=<	269	22 %		
Ovarian cancer or PPC#	20>=	88	61 %	9.7	P=0.0018
	21=<	42	39 %		

PPC = primary peritoneal carcinomatosis

In colorectal cancer, PCI cutoff levels have been reported to range from 5 to20 [2, 5, 11, 12]. The most stringent selection criterion performing the comprehensive treatment is a PCI cutoff level of 5 [12], and NICE recommends that the PCI cutoff level should be 6 [5]. In contrast, Huang and Yan et al. proposed a PCI cutoff level of 20 for a significantly favorable prognosis after CRS [12, 13], Our study confirmed that a PCI score less than 16 was associated with 5-year survival rate of 29 %, while that 16 or greater was associated with a 0 % survival rate (Table 34.1) [2].

Pseudomyxoma peritonei (PMP) with noninvasive malignant behavior has an excellent rate of survival (94 %) at 20 years, if the PCI is less than 20 [14]. In our experience, the statistically significant PCI cutoff level for favorable prognosis in patients with PMP was 27 [2].

In stage III or IV ovarian cancer, Cotte claims that a 'PCI less than 12 resulted in a significantly more favorable prognosis [15]. In our experience, the 5-year survival rate of 126 stage III and IV ovarian cancer patients was 61 % those with a PCI of < 20, and 39 % in those with a PCI of > 21 (Table 34.1). Survival rates of patients with PCI scores greater than cutoff level are not improved by aggressive CRS. All authors reported that patients with PCI scores greater than the cutoff level should not be treated with aggressive CRS because the postoperative mortality and morbidity after aggressive CRS and HIPEC are still higher than those after standard CRS.

The completeness of cytoreduction (CC) score is now used to assess the residual tumor load after CRS [16]. CC0 means complete removal of macroscopic peritoneal nodules. CC1, CC2, and CC3 scores indicate that the residual tumor diameters are < 2.5 mm, between 2.5 mm and 25 mm, and > 25 mm, respectively. CC1, CC2 and CC3 indicate incomplete cytoreduction. CC scores are independent prognostic factors for patients undergoing CRS [1, 3].

Survival after complete cytoreduction is significantly better than after incomplete cytoreduction in patients with PM from gastric, colorectal, ovarian cancer and PMP (Figs. 34.7, 34.8, 34.9, 34.10). Different from gastric and colorectal cancer, in

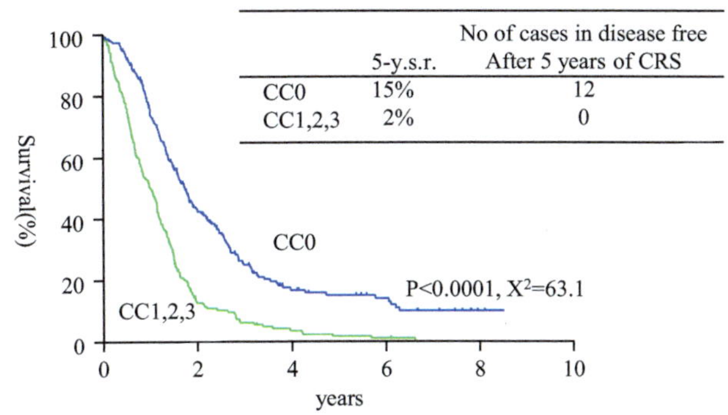

Fig. 34.7 Survival curves of 472 patients with PM from gastric cancer. Survival of patients received CC-0 resection (N = 243) was significantly better than the survival of those underwent CC1,2,3 (incomplete cytoreduction, N = 229)

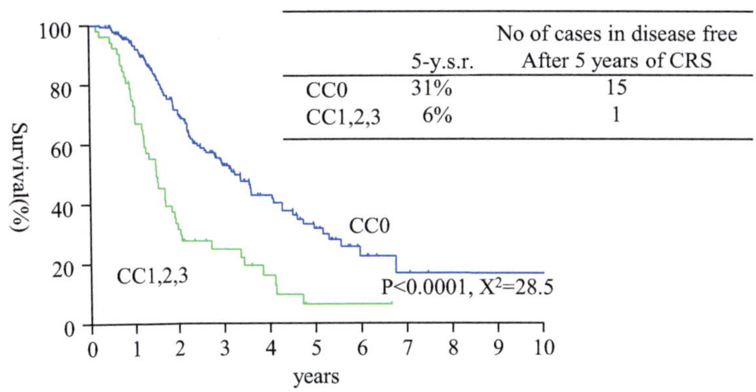

Fig. 34.8 Survival curves of 235 patients with PM from colorectal cancer. Survival of patients received CC-0 resection (complete cytoreduction, N = 183) was significantly better than the survival of those underwent CC1,2,3 (incomplete cytoreduction, N = 52)

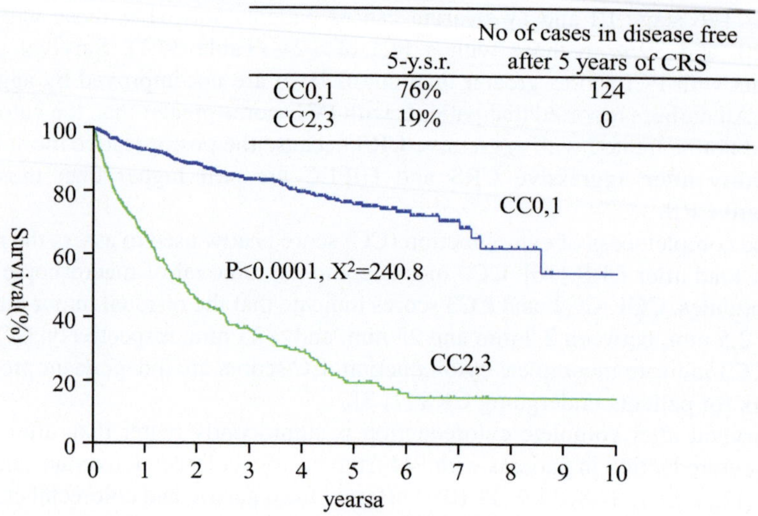

Fig. 34.9 Survival curves of 804 patients with pseudomuxoma peritonei Survival of patients received CC0,1 resection (N = 552) was significantly better than the survival of those underwent CC2,3 (N = 252)

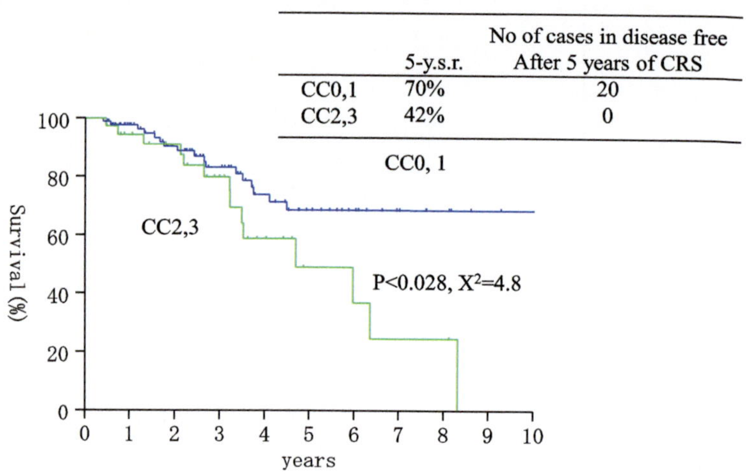

Fig. 34.10 Survival curves of 127 patients with PM from ovarian cancer or primary peritoneal carcinomatosis. Survival of patients received CC-0,1 resection (N = 99) was significantly better than the survival of those underwent CC2,3 (N = 28)

patients with PM from ovarian cancer, cytoreduction down to nodules less than 2.5 mm in diameter is expected to result in long-term survival (Figs. 34.9, 34.10). Ovarian cancer is known to be remarkably more sensitive to chemotherapy than gastric or colorectal cancer. Complete response to perioperative chemotherapies is frequently experienced even though tumor burdens in those with CC1 or CC2 remain after CRS. For ovarian cancer patients, CRS is indicated for patients who are expected to undergo CC2 resection.

34.2.2 Laparoscopic Diagnosis of PCI

Accuracy of computed tomography (CT) or positron emission tomography (PET) for the diagnosis of PCI is considered to be low when the diameter of PM is less than 10 mm [17, 18]. In particular, small nodules on the small bowel mesentery are difficult to assess. For preoperative diagnosis of PCI in gastric cancer, Hong et al. reported that PM could not be diagnosed in 39 of 137 patients (28.5 %) with PM [19]. Accordingly, there is no imaging modality that can sufficiently predict PCI. By the late 2000s, diagnostic laparoscopy was being performed to determine accurate PCI, followed by histologic and cytologic analysis [20]. Laparoscopy is a highly accurate diagnostic modality for determining PCI. Valle et al. reported that PCI is almost the same when evaluated by preoperative laparoscopy or at laparotomy [20].

34.2.3 Laparoscopic Hyperemic Intraperitoneal Chemoperfusion

Valle et al. also reported that CC0 was achievable in less than 30 % of gastric cancer patients who had not been treated with neoadjuvant chemotherapy (NAC) [20]. NAC may reduce PCI levels below the cutoff levels, which is one of its aims, along with eradicating micrometastases and free peritoneal floating cells before CRS.

Neoadjuvant laparoscopic HIPEC (LHIPEC) was developed to reduce PCI and preserve intact peritoneal sectors as much as possible. Just after laparoscopic diagnosis, extensive intraoperative peritoneal lavage (EIPL) using physiological saline delivered by a HIPEC machine was performed to wash away non-adherent peritoneal cancer cells [21]. Laparoscopic HIPEC is done using 4 L of saline with anticancer drugs heated to 42–43 °C [22].

34.2.4 Neoadjuvant Chemotherapy (Bidirectional Intraperitoneal and Systemic Chemotherapy)

Systemic chemotherapy has little effect on PM because the blood-peritoneal barrier (BPB) [23] impedes the penetration of drugs from the systemic circulation into the peritoneal cavity.

In gastric cancer patients with PM, Hong et al. reported that all patients treated with only systemic chemotherapy died within 5 years [19]. Furthermore, Franko et al. reported that systemic chemotherapy for colorectal cancer patients with PM was not beneficial for survival [24]. Additionally, Kerscher et al. reported that the survival rates for colorectal cancer patients with PM when treated with modern systemic chemotherapy like FOLFOX or FOLFIRI were not significantly different from those when patients were treated using old-fashioned chemotherapy of 5-fluorouracil (5-FU) [25]. Accordingly, they proposed that colorectal cancer patients with PM should be treated with therapies other than systemic chemotherapy [25].

In contrast, intraperitoneal (IP) chemotherapy concentrates drugs loco-regionally in the peritoneal cavity to a greater extent than does systemic chemotherapy [26, 27]. Compared with systemic chemotherapy, IP chemotherapy has been reported to improve both response rates and survival in gastric cancer, colorectal cancer and ovarian cancer [28–31].

Yonemura et al. reported a new protocol named neoadjuvant intraperitoneal/systemic chemotherapy (NIPS), using IP docetaxel and cisplatin (CDDP) combined with oral administration of S1 [28].

More recently, a new bidirectional intraperitoneal and systemic induction chemotherapy (BISIC) method was reported [32]. In this treatment, oral S1 is administered for 14 days at a dose of 60 mg/m^2/day, followed by 7 days rest. Docetaxel (30 mg/m^2) and CDDP (30 mg/m^2) are administered by IP infusion on day 1, and the same dose of docetaxel and CDDP are administered intravenously on day 8. Administration of docetaxel and CDDP from both sides of the peritoneal surface enable a wider area of subperitoneal tissue to be treated than does IP administration alone. Ishigami et al. reported a new BISIC method using systemic and IP paclitaxel combined with S1 [31]. The overall response rate and 1-year overall survival rate were 56 % and 78 %, respectively.

Grade-3, -4, -5 toxicity rates during NIPS and BISIC were lower than after systemic chemotherapies [28–32].

In a systematic review and meta-analysis of treatment for patients with gastric cancer with PM, IP chemotherapy combined with CRS was associated with significant improvement in overall survival [33].

Thus, these indicate that NIPS and BISIC should be utilized to reduce PCI, eradicate free peritoneal floating cancer cells, and pathological response before patients undergo CRS.

34.2.5 Extensive Intraoperative Peritoneal Lavage (EIPL)

Extensive intraoperative peritoneal lavage (EIPL) using physiological saline is performed to wash away non-adherent peritoneal cancer cells and mucinous materials in cases of PMP [21]. To perform EIPL, 1 L of saline is administered into the peritoneal cavity, and then completely aspirated using a suction tube. This procedure is repeated 10 times. EIPL is performed before and after CRS. Kuramoto et al. performed a randomized study in gastric cancer patients with positive peritoneal wash cytology. Patients were allocated into either CRS + EIPL + intraperitoneal administration of CDDP (EIPL Group) or CRS + intraperitoneal administration of CDDP (non-EIPL Group). Patients in the EIPL Group survived significantly longer than those in non-EIPL Group.

34.2.6 *Cytoreductive Surgery After NIPS or BISIC*

CRS and HIPEC are performed 3 to 4 weeks after the last cycle of neoadjuvant chemotherapy. During the operation, EIPL is performed before and after CRS to remove free peritoneal floating cancer cells as well as tumor cells spillover from lymphatic vessels and blood vessels damaged during CRS. The final goal of CRS is to establish complete cytoreduction of the primary tumor, metastases in regional lymph nodes, and peritoneal surface.

In addition, HIPEC is done to eradicate residual micrometatases, which remain on the peritoneal cavity after CRS and EIPL. In gastric and colorectal cancer, CRS is indicated for patients with PCI below cutoff levels and who are supposed to undergo complete resection (CC0). In contrast, CC1 or CC2 resection may be recommended even in patients with higher PCI than the cut-off levels because chemosensitivity against ovarian cancer is high.

In the patients with low-grade appendiceal or ovarian mucinous neoplasm who are supposed to receive incomplete cytoreduction due to PCI greater than cutoff levels, primary tumor and metastatic tumors, which produce mucinous materials, should be removed as much as possible. Even after incomplete cytoreduction, quality of life will be improved [14].

34.2.7 *Early Postoperative Intraperitoneal Chemotherapy (EPIC)*

EPIC begins just after CRS and HIPEC when the residual cancer load in the peritoneal cavity is minimum. Immediately after CRS and HIPEC is the best time for eradicating the residual invisible cancer cells. Yu et al. reported that EPIC (intraperitoneal administration of 5- 5FU on postoperative days 1–4 after gastrectomy for advanced gastric cancers) was effective for the prevention of peritoneal recurrence [34]. For colorectal cancer with PM, the median survival (32 months) of patients treated with CRS followed by EPIC (5-FU 550 mg/m2 from days 1–6, and repeated monthly for 8 total courses) was significantly longer than that of the control group (14 months, CRS + systemic chemotherapy) [35]. Cashin et al. reported that the overall survival (OS) of patients treated with CRS + EPIC was 25 months and the 5-year survival rate was 18 %. In contrast, OS and 5-year survival rates after CRS + HIPEC were significantly better than those after CRS + EPIC [36].

These results indicate that CRS combined with HIPEC and EPIC may improve survival of patients with PM.

In contrast, Cao C et al. performed a systematic review and meta-analysis from 47 papers concerning CRS with perioperative intraperitoneal chemotherapy for PM with colorectal origin [37]. However, the pooled data did not show a significant improvement in OS for patients treated by CRS and EPIC [37]. In colorectal cancer patients with PM, Lam et al. reported that the 3-year OS and recurrence-free survival rates were 50 % and 21 % for CRS + HIPEC + EPIC and 46 % and 6 %, respectively for CRS + HIPEC. HIPEC + EPIC patients experienced more grade III/IV complications (42.6 % vs 19.6 %) [38]. Thus, the prognostic significance of EPIC needs to be further evaluated.

34.2.8 Postoperative Systemic Chemotherapy After CRS and Intraoperative Intraperitoneal Chemotherapy

In gastric cancer patients with intraperitoneal free cancer cells but no macroscopic PM, postoperative systemic chemotherapy using S1 has been reported to improve the prognosis of patients after gastrectomy [39, 40]. Figure 34.11 shows survival curves of gastric cancer-patients with positive cytology and no macroscopic PM (P0,Cy1) after treatment of CRS plus perioperative intraperitoneal chemotherapy of BISIC/HIPEC. Five-year survival was 54 %. This result indicates that intraperitoneal micrometastasis can be cured with perioperative intraperitoneal chemotherapy.

Glehen et al. performed a retrospective international multicenter study to evaluate the prognostic indicators of colorectal cancer patients with PM. They found that the independent prognostic factors were completeness of CRS, treatment with a second procedure, PCI, lymph node metastasis, age, tumor differentiation, liver metastasis and postoperative adjuvant chemotherapy. However, sex, preoperative chemotherapy, and perioperative intraperitoneal chemotherapy were not significant prognostic factors [41]. Median survival time of patients treated with postoperative systemic chemotherapy was significantly longer than that of patients without postoperative systemic chemotherapy (25.2 months vs. 15.6 months) [41]. After complete cytoreduction, postoperative systemic chemotherapy improves survival.

34.3 Effects of HIPEC

Hyperthermia enhances the cytotoxicity on cancer cells when combined with certain chemotherapeutic agents. Accordingly, many clinical trials for HIPEC have been reported, but protocols have not been consistent. The temperatures

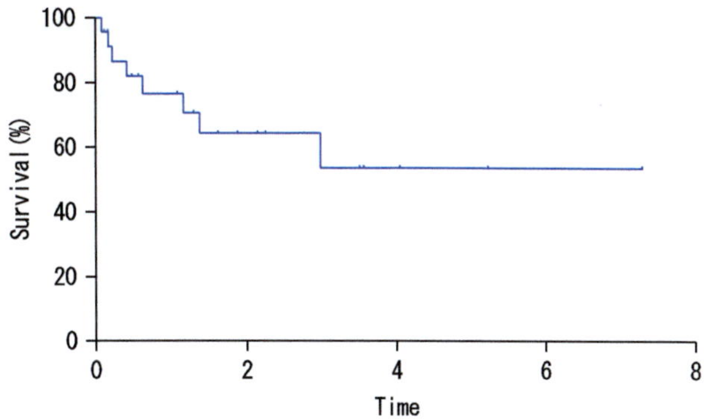

Fig. 34.11 Survival curves of patients with positive peritoneal wash cytology but no macroscopic metastasis treated by comprehensive treatment (N = 24). Five-year survival rate as 54 %

used varied from 41 to 43 °C, as did the duration of treatment and chemotherapeutic drugs have also varied.

The relationship that temperature and exposure time have on cytotoxicity during hyperthermic treatment has been reported to be exponential [42]. This relationship can be stated as : one degree increase in temperature above 43 °C requires two-fold decrease in time for the same effect at 43 °C. In contrast, one degree decrease below 43 °C needs three- to four-fold increase in time for the same effects at 43 °C. When the treatment temperature is higher than 43 °C, cells are killed according to the time-dependent and exponential manner. If the cells treated below 43 °C, cells survive by thermo-tolerance, which is induced by the production of heat-shock protein [41]. Accordingly, a thermal dose that is clinically relevant to the cytotoxic effect is needed to the standardize HIPEC. In HIPEC, Yonemura et al. reported that in HIPEC one thermal dose is equivalent to treatment for 30 min at 43 °C [43]. If the temperature is 42 °C, the treatment time should be prolonged 90–120 min to obtain a cytotoxic effect equivalent to that of treatment for 30 min at 43 °C. In contrast, 15 min is enough when the temperature is 44 °C. However, when temperature is higher than 44 °C, intestinal necrosis or anastomotic insufficiency may develop. Shimizu et al. reported that intraperitoneal hyperthermia in rats treated for 30 min at up to 44.0 °C for 30 min. had no adverse effects on the healing of how intestinal anastomosis healed [44]. However, intraperitoneal hyperthermia for 30 min at 45 °C resulted in 90 % mortality.

From these results, HIPEC at 43 °C to 43.5 °C and a treatment time of 40 min is recommended. During HIPEC, even temperatures between 43 °C–43.5 °C should be maintained at all peritoneal surfaces by stirring the heated saline by hand.

34.3.1 Direct Effects of HIPEC

Many studies have shown the enhanced cytotoxic effects of chemotherapy when it is done in conjunction with hyperthermia. In clinical HIPEC, treatment time is limited and the intraperitoneal temperature cannot be always controlled at 43 °C–

43.5 °C. Accordingly, drugs that have synergistic effects with hyperthermia are used for HIPEC. So far, however, the direct effects of HIPEC on PM have only been reported in one paper [22]. Yonemura et al. studied changes in PCI score 1 month after laparoscopic HIPEC. Fifty gastric cancer patients with PM received laparoscopic diagnosis of PCI, HIPEC was performed at 43–43.5 C for 60 min adding 3 L of saline plus 12.5 mg/m^2 of Mitomycin C with Cisplatin (50 mg/m^2). A second session of LHIPEC (2nd LHIPEC) was conducted 1 month after the first session of LHIPEC (1st LHIPEC) in all cases. There were no intraoperative complications or mortality after LHIPEC. Four (8 %) patients developed mild Grade II azotemia. Ascites were completely abolished or decreased in 22 of 34 (64.7 %) patients and peritoneal cytology changed from positive to be negative in 14 of 20 (70 %) patients at the 2nd LHIPEC. A complete response was observed in 6 (12 %) patients, and PCIs were significantly reduced from the 1st LHIPEC to the 2nd LHIPEC (14.3 ± 10.2 vs. 10.8 ± 10.5 at $p < 0.05$). Furthermore, lesion size scores for small bowel mesentery decreased from the 1st to 2nd LHIPEC (6.56 ± 2.92 vs. 5.25 ± 3.78, $P = 0.016$). These results indicate that diagnostic and therapeutic laparoscopy can be performed safely in patients with PM of GC. Laparoscopic HIPEC can be applied as a neoadjuvant treatment in order to reduce tumor burden and to improve disease control until complete managements can be achieved in patients with PM of GC.

34.3.2 Methods of HIPEC and Effects on Survival After CRS Plus HIPEC

To maximize the advantage of HIPEC, several HIPEC devices have been developed. Essentially, they are capable of delivering constant hyperthermia using a closed continuous circuit with a pump, a heater, a heat exchanger, and a real-time temperature monitoring. During HIPEC, temperature probes are placed at different sites of the circuit and the peritoneal cavity: heat exchanger, heat generator, and temperature probes on the pelvic cavity and both spaces under diaphragm. The intra-abdominal temperature range during HIPEC should be maintained between 41.5 °C–43 °C .

The meta-analyses of treatment results for PM from various primary sites indicate that HIPEC + CRS improve survival improvement compared with CRS + systemic chemotherapy [33]. According, HIPEC is considered a state-of-the art method for prevention and treatment of PM.

34.3.2.1 Gastric Cancer

In gastric cancer treatments, there has been only one prospective randomized trial to confirm the effect of HIPEC [45]. Yang et al. reported the results of RCT for gastric cancer patients with PM that consisted of those with CRS + HIPEC and those with CRS alone. Medications were MMC and CDDP, and treatment was 60 min at 42 C. Patients in the CRS + HIPEC group survived significantly longer than those in the CRS alone group. The authors concluded that HIPEC using MMC and CDDP might contribute to an improved survival rate. Table 34.2 shows the methods of HIPEC for gastric cancer. The most commonly used drugs were MMC and CDDP. Recently, docetaxel and CDDP have been used based on the results after LHIPEC [32].

Coccolini et al. performed a meta-analysis on eight RCTs to verify the effects of HIPEC on survival [33]. They found that HIPEC significantly improved the 1-, 2- and 3-year survival rates after CRS with HIPEC. Additionally, HIPEC had positive effects on the prevention of peritoneal recurrence after curative resection of gastric cancer with serosal invasion. Multivariate analysis showed that independent factor for a better prognosis was complete cytoreduction, PCI below cutoff level, histologic responder after neoadjuvant chemotherapy, and HIPEC [2].

However, morbidity rate may increase with HIPEC. Development of complications has been shown to be an independent prognostic factor for poor survival [46, 47]. Yan et al. reported that a meta-analysis did not show a significant difference in mortality rates between HIPEC and control groups [47]. However, a

Table 34.2 HIPEC techniques in patients with gastric cancer, modified from table of the reference form Esquival J. Methodology of hyperthermic intraperitoneal chemotherapy in the United States, Europe, China, Japan, and Korea. Cancer J. 2009;15:249–254

Institution	Method	Drugs	Dosage	Outflow tempereture (°C)	Duration (minutes)
Regensburg Univerity	Closed	CDDP + Docetaxel	75 mg/m^2, 15 mg/m^2	41–42	60
Kusatsu general hospital, Kishiwada Tokushukai,	Open	MMC + CDDP	20 mg + 100 mg	42.5–43	60
		Docetaxel + CDDP	40 mg + 100 mg	43	40
Jilin Tumor Hospital	Closed	5FU	750 mg/L	41	60
		CDDP	20 mg/L		
		MMC	10 mg/L		
Zhongnan Hospital	Open	MMC	5 mg/ml	43	60–90
		CDDP	20 mg/ml		
Asian Medical Center	Semi-open	MMC	6 mg	42	60
		CDDP	60 mg		
		Etoposide	30 mg		
Kyungpook National University	Open	MMC	10 mg/m^2	42	60
Centro Hospitaier Lyon Sud	Closed	MMC	20–40 mg/m^2	42	90

significant increase was found in the incidence of abdominal abscess and neutropenia after HIPEC. Mizumoto et al. reported that morbidity rates after CRS with HIPEC were lower than those after CRS alone [46]. Accordingly, HIPEC is thought to be a safe and effective therapy even after aggressive cytoreductive surgery.

34.3.2.2 Colorectal Cancer

Table 34.3 shows the HIPEC methods for PM from colorectal cancer. MMC and oxaliplatin are used in almost all institutions. At the Washington Hospital Center, MMC and doxorubicin are introduced into the peritoneal fluid, and 5-FU is systematically injected during HIPEC. This method is called bidirectional intraperitoneal and systemic chemotherapy.

The open and closed abdominal techniques were performed in 9 and 5 institutions, respectively.

In the open-abdominal technique, 4 L of saline is dwelled in the abdominal cavity, and inlet and outlet tubes are placed on the pelvic and subdiaphragmatic spaces. Temperature probes are placed on the abdominal cavity. Roller pump forces the indwelled saline, which is warmed at the heat exchangers in the water bath. Surgeons gently but continuously manipulate all viscera to keep adherence of peritoneal surface to a minimum. After the temperature of all the temperature probes register an increases to 43 °C, chemotherapeutic drugs are introduced into the abdominal cavity. During HIPEC, the intraperitoneal fluid is maintained at 42.5 °C–43.5 °C. After perfusion is completed, the abdomen is suctioned dry of fluid and the volume of fluid is measured. Open abdominal techniques enable the even distribution of heat over the entire peritoneal surface and prevent excessive heating of normal tissue. Sampling tumors is also possible during HIPEC. However, the disadvantage of the open technique is the potential exposure of the operating room personnel to the chemotherapeutic agents. The heated drugs can aerosolize, creating the potential for exposure by inhalation. A smoke evacuator is used to pull air from beneath the plastic cover, preventing air contamination in the operating room.

The closed technique is popular in the United States. Its advantage is that the potential for environmental contamination by chemotherapeutic drugs is significantly lower than that of the open method. Because there is less opportunity for heat loss, the intraperitoneal temperature can be reached faster and maintained more easily than in the open technique. Additionally, intraabdominal pressure is higher than open technique, and the penetration distance of chemotherapeutic agents from the peritoneal surface is longer [48, 49]. The main disadvantage of the closed technique is that the heated intraperitoneal chemotherapy does not distribute uniformly [49]. Laparoscopic HIPEC I s a closed technique, and the even distribution of heat can be accomplished by stirring the saline with the outlet and inlet tubes.

CRS combined with HIPEC is now increasingly advocated for selected patients who suffer from colorectal cancer with PM. A large multi-institutional study, a randomized study, a national database, a systemic review, and a large number of

Table 34.3 HIPEC techniques in patients with colorectal cancer and appendiceal mucinous neoplasm, modified from table of the reference Methodology of hyperthermic intraperitoneal chemotherapy in the United States, Europe, China, Japan, and Korea. Cancer J. 2009;15:249–254

Institution	Method	Drugs	Dosage	Outflow tempereture (°C)	Duration (minutes)
Washington Hospital Center	Open	MMC Ip	15 mg/m^2	41	90
		Doxorubicin Ip	15 mg/m^2		
		5FU Iv	400 mg/m^2		
		Leucovorin Iv	20 mg/m^2		
Wake Forest University	Closed	MMC Ip	40 mg	40	120
St Agnes hospital	Closed	MMC Ip	40 mg	42	90
University California, San Diego	Closed	MMC Ip	20 mg/m^2	41–42	60
Regensburg University	Closed	MMC Ip	20 mg/m^2	41–42	60
		Doxorubicin Ip	15 mg/m^2		
MD Anderson Espania	Open	Oxaliplatin Ip	460 mg/m^2 (Female)	43	90
Uppsala University	Open	Oxaliplatin Ip 5FU Iv	460 mg/m^2	41	30
Kantonsspital St Gallen	Open	MMC Ip	25 mg/m^2	42	90
Basingstoke	Open	MMC Ip	15 mg/m^2	42	90
Kishiwada Tokushukai, Kusatsu, Kusatsu general hospital	Open	MMC Ip	20 mg	42.5–43.5	60
	Open	OHP + 5FUIp	100 mg, 500 mg	43	40
Institute Gustave Rossey	Open	OHPIp + 5FUIv	460 mg/m^2	43	30
Centro Hospitaier Lyon Sud	Closed	MMC + CDDP	20 mg/m^2, 20 mg/m^2	42	90
Winston Salem	Open	MMC	40 mg	38.5	120
University of Pittsburg	Closed	MMC	30–40 mg	40	100
Netherland Cancer Institute	Open	MMC	17.5 mg/m^2	40	90
St George hospital	Open	MMC	10–12.5 mg/m^2	42	90

single-institutional reports all shows the benefits of this method for the definitive management of PM. Mimezani et al. performed a meta-analysis of the results following CRS + HIPEC and systemic chemotherapy alone, focusing on survival outcomes. The analysis utilized an electronic literature search including 342 patients with colorectal cancer with PM [50]. Yonemura and Cao et al. studied prognostic factors related to colorectal cancer with PM [2, 37]. A significant improvement in survival was associated with treatment by CRS with HIPEC, complete cytoreduction, and PCI below cutoff level. The pooled data did not show a significant improvement in overall survival for patients treated by CRS with EPIC compared with those treated with CRS and systemic chemotherapy [37].

To date, two studies have reported the quality of life after CRS with HIPEC for PM from colorectal cancer. More than 90 % of patients had minimal to no limitation of activity and had functional assessments that compared favorably with national reference values for their respective age groups [51].

34.3.2.3 Ovarian Cancer

As shown in Table 34.*4*, CDDP and docetaxel are commonly used for ovarian cancer. Chua et al. performed a review analysis focusing on survival benefits of CRS + HIPEC in the ovarian cancer patients. CRS + HIPEC may be a feasible option with benefits that are comparable with the current standard of care [52].

Spiliotis et al. performed RCT consisting of CRS followed by HIPEC and then systemic chemotherapy versus CRS and systemic chemotherapy (without HIPEC). The survival duration for patients who received the former treatment was significantly better than for those who received the latter [53]. In the HIPEC group, survival did not differ between patients who had a platinum-resistant disease and those who had a platinum-sensitive disease (26.6 vs 26.8 months). In contrast, in the non-HIPEC group, there was a statistically significant difference in survival duration between platinum-sensitive and platinum-resistant disease (15.2 vs 10.2 months). Accordingly, HIPEC may become an alternative treatment method for conquering the platinum-resistant disease. Other prognostic factors in addition to HIPEC were reported to be completeness of cytoreduction, and a PCI cutoff level (=< 15) [53].

However, only one RCT has verified the effect of HIPEC in treatment of ovarian cancer [53]. Additional RCTs are required to establish the role of HIPEC in ovarian cancer.

34.3.2.4 Pseudomyxoma Peritonei

To date, no RCT has been conducted to verify the survival benefit of HIPEC in patents with pseudomyxoma peritonei (PMP). Drugs used for HIPEC in patients with PMP are the same as those for the treatment for colorectal cancer (Table 34.3). Chua et al. studied the effect of HIPEC on survival in 2298 patients with PMP. Multivariate analysis identified prior chemotherapy, major postoperative complication, histologic subtype of high grade mucinous neoplasm, high peritoneal cancer index, completeness of cytoreduction, and not using HIPEC as independent predictors of poorer progression-free survival [54]. In contrast, older age (younger than 53 years old vs older than 53 years old), postoperative complication, incomplete cytoreduction, prior chemotherapy treatment, and a histopathologic subtype of

Table 34.4 HIPEC techniques in patients with sarcoma, mesothelioma, pancreas cancer, ovarian cancer, and bile duct cancer

Diseases	Institution	Method	Drugs	Dosage	Outflow tempereture (°C)	Duration (minutes)
Sarcoma Liposarcoma	University of Chicago, Chicago	Closed	Doxorubicin	15 mg/L	42–43	90
			CDDP	45 mg/Lx3L		
Mesothelioma	National Milan Cancer Center, Milan	Closed	Doxorubicin	15 mg/L	42.5	90
			CDDP	45 mg/Lx3L		
	Washington Hospital Center, Washington DC	Open	Melphalan	50–70 mg/m^2	42	60–90
	Kishiwada Tokushukai, Kusatsu, Osaka, Shiga	Open	Doc + CDDP	40 mg + 100 mg	43	40
Pancreas cancer	Democritos University, Athen	Open	Gimcitabine	1000 mg/m^2	42.5–43.5	60
Ovarian cancer	Kishiwada Tokushukai, Kusatsu, Osaka, Shiga	open	Docetaxel + CDDP	40 mg + 100 mg	43	40
	Pietro Valdoni Hospital, Rome	Closed	CDDP	75 mg/m^2	42–43	60
	Hospital Alemän, Buenos Aires	Open	CDDP	100 mg/m^2	41–43	60
	Centre Jean Perrin, Toulouse	Open	OHP	460 mg/m^2	42–44	30
	Centre Hospitalo-Universitaire Lyon, Lyon	Closed	CDDP	20 $mg/m^2/L$ (4–6 L)	44–46 (inflow)	90
Bile duct cancer	Kishiwada Tokushukai, Kusatsu, Osaka, Shiga	Open	Doc + CDDP	40 mg + 100 mg	43	40

high-grade mucinous neoplasm were independent predictors of a poorer overall survival.

Figure 34.12 shows our data for the survival of patients with PMP, who underwent complete cytoreduction (CC or CC1). Patients treated with HIPEC survived significantly longer than those who were not. However, HIPEC did not improve the survival of patients who received CC2 or CC3 resection (Fig. 34.13).

34.3.2.5 Diffuse Malignant Peritoneal Mesothelioma (DMPM)

The latest treatment for patients with DMPM is CRS with HIPEC. Before comprehensive treatment was available, survival after systemic chemotherapy alone, palliative surgery, or total abdominal radiation therapy was very poor, and patients die within a year [55]. After introduction of CRS with HIPEC, survival has improved to 5-year survival rate of 42 % [56]. As shown in Table 34.3, cisplatin-based HIPEC is commonly performed. Helm performed a systemic review and meta-analysis of 1047 mesothelioma patients reported in 20 articles [56]. CRS with HIPEC was performed using both open and closed techniques, and the chemotherapy agents used were most frequently a combination on cisplatin, doxorubicin, and mitomycin. In the Washington Hospital Center, melphalan is used, because it has a

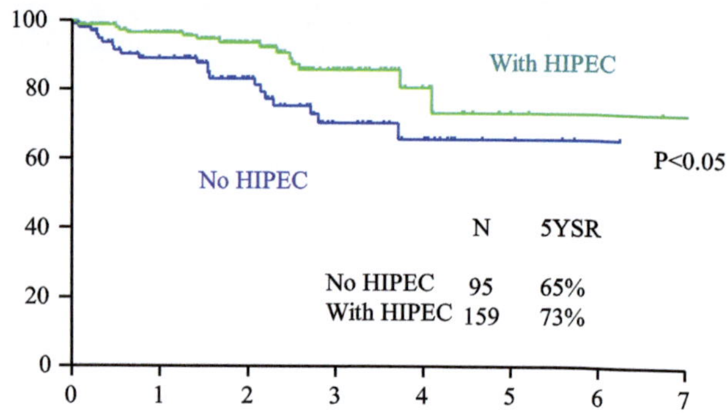

Fig. 34.12 Survival curves of 232 patients with pseudomyxoma peritonei, who underwent complete cytoreduction (CC0 or CC1)

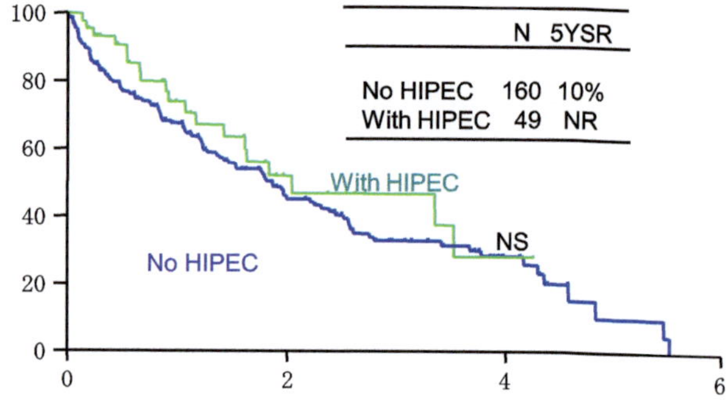

Fig. 34.13 Survival curves of 209 patients with pseudomyxoma peritonei, who underwent incomplete cytoreduction (CC2 or CC3)

strong synergistic effect when combined with hyperthermia [57]. However, long term survival after HIPEC with melphalan was not reported.

Schaub and Yan et al. reported that the histological subtype (epithelial type vs. biphasic or sarcomatoid type), PCI cutoff level (PCI > =20 versus PCI = <19, preoperative serum CA125, CC score (CC0/1 vs. CC2/3), lymph node metastasis (absent vs. present), nuclear size (30 μm > vs. 30 μm = <) and HIPEC had the greatest impact on overall survival [58, 59].

34.4 Postoperative Mortality/Morbidity and Learning Curve After CRS + HIPEC

There were no intraoperative complications or mortality after laparoscopic HIPEC. Four of 50 (8 %) patients developed mild Grade-II azotemia [22]. Accordingly, HIPEC alone is a very safe procedure.

In contrast, high morbidity and mortality rates have been reported after CRS + HIPEC. The comprehensive treatment is indeed a more aggressive approach compared with traditional palliative surgery, in that the goal of CRS is complete removal of all intraperitoneal tumors. Severe Grade-III or -IV morbidity (according to the common toxicity grading criteria) was experienced in 16–68 % of patients, and reoperation was needed in 7–11 % of cases [5, 51, 59, 60]. Grade-III or Grade-IV adverse events included visceral perforation, intraabdominal abscess, anastomotic leak, bleeding, line sepsis, urinary tract perforation, and infection [47, 59–62]. The most common cause of reoperations was fistula due to gastric or intestinal perforation. Anastomotic leak and postoperative massive bleeding necessitate reoperations [60].

The total 30-day or in-hospital mortality rates have been reported to be between from 0 %–9 % [46, 59, 60].

Causes of mortality include bleeding, sepsis, and multiple organ failure due to abdominal abscess, anastomotic leak, and pancreatic fistula. Pulmonary embolism is a lethal complication, and deep vein thrombosis of the lower extremities should be examined by ultrasound echography before surgery [47, 59–61].

Multivariate analyses revealed that PCI > =26, CEA > =6, operation time longer than 5 h, and a bleeding volume higher than 3 L are independent factors for the development of severe morbidity or mortality [46].

To perform the comprehensive treatment safely, surgeons should be experts in a lot of knowledge about anatomy, physiology, oncology, chemotherapy, postoperative care and nutrition. If postoperative complications occur, surgeons must treat patients promptly and adequately.

To reduce postoperative mortality and morbidity to an acceptable level, the beginning surgeons must train in operation techniques, HIPEC methods, and postoperative care under the guidance of experts. The number of cases over time it takes new doctors to become proficient at new treatment procedures is called learning curve. Smeenke et al. reported that the usual learning curve was about 130 cases [63]. This takes least 20 years after graduation from medical university. In contrast, well trained surgeons can shorten the learning curve to only 70 cases [59–61].

Peritoneal Surface Oncology International (POSGI) initiated training program in 2014 in Europe and Japan [60]. The European School of Peritoneal Surface Oncology (ESPSO) is a joint venture of the European Society of Surgical Oncology (ESSO) and PSOGI. Target audiences are chief residents, fellow surgeons, consultant surgeons, gynecologists, and medical oncologists. The duration of training is 2 years. The program consists of (1) Peritoneal carcinomatosis: overview and surgical techniques, (2) peritoneal carcinomatosis from colorectal cancer, ovarian cancer, gastric cancer, and rare peritoneal surface malignancies, (4) palliative care, (5) experimental and translational research, and (6) pseudomyxoma peritonei. Upon successful completion of this program, the candidates obtain the European Accreditation on Peritoneal Surface Oncology.

References

1. Sugarbaker PH. New standard of care for appendiceal epithelial malignancies and pseudomyxoma peritonei syndrome. Lancet Oncol. 2006;7:69–76.
2. Yonemura Y, Elnemr A, Endou Y, Ishibashi H, Mizumoto A, Miura M, et al. Surgical results of patients with peritoneal carcinomatosis treated with cytoreductive surgery using a new technique named aqua dissection. Gastroenterol Res Pract 2012; Volume 2012, Article ID 521487, 10 pages doi:10.1155/2012/521487.
3. Yonemura Y, Canbay E, Endou Y, Ishibashi H, Mizumoto A, Miura M, et al. Peritoneal cancer treatment. Expert Opin Pharmacother. 2014;15:623–36. doi:10.1517/14656566.2014.879571.
4. Verwaal VJ, van Ruth S, de Bree E, van Sloothen GW, van Tinteren H, Boot H, et al. Randomized trial of cytoreduction and hyperthermic intraperitoneal chemotherapy versus systemic chemotherapy and palliative surgery in patients with peritoneal carcinoamtosis of colorectal cancer. J Clin Oncol. 2003;21:3737–43.
5. National Institute of Health and Clinical Excellence (NICE). Cytoreduction surgery followed by hyperthermic intraoperative peritoneal chemotherapy for peritoneal carcinomatosis. IPG331. http://www.nice.org.uk/guidance/IPG331. Accessed 12 Aug 2013.
6. Jacquet P, Sugarbaker PH. Clinical research methodologies in diagnosis and staging of patients with peritoneal carcinomatosis. Cancer Treat Res. 1996;82:359–74.
7. Gomez Portilla A, Kusamura S, Baratti D, Deraco M. The intraoperative staging system in the management of peritoneal surface malignancy. J Surg Oncol. 2008;98:228–31.
8. Glehen O, Gilly FN, Arvieux C, Cotte E, Boutitie F, Mansvelt B, et al. Peritoneal carcinomatosis from gastric cancer A muti-institutional study of 159 patients treated by cytoreductive surgery combined with perioperative intraperitoneal chemotherapy. Ann Surg Oncol. 2010;17:2370–7. doi:10.1245/s10434-010-1039-7.
9. Yang XJ, Huang CQ, Suo T, Mei LJ, Yang GL, Cheng FL, et al. Cytoreductive surgery and hyperthermic intraperitoneal chemotherapy improves survival of patients with peritoneal carcinomatosis from gastric cancer: Final results of a phase III randomized clinical trial. Ann Surg Oncol. 2011;18:1575–81.
10. Coccolini F, Catena F, Glehen O, Yonemura Y, Sugarbaker PH, Piso P, et al. Complete versus incomplete cytoreduction in peritoneal carcinosis from gastric cancer, with consideration to PCI cut-off. Systemic review and meta-analysis. EJSO. 2015; doi.org/10.1016/j.ejso.2015.03.231.

11. Huang CQ, Yang XJ, Yu Y, Wu HT, Liu Y, Yonemura Y, et al. Cytoreductive surgery plus hypertyhermic intraperitoneal chemotherapy improves survival for patients with peritoneal carcinomatosis from colorectal cancer : A phase II study from a Chinese center. PLOS one, 2014 Sep 26;9(9):e108509. doi: 10.1371/journal.pone.0108509.
12. Huang Y, ALzahrani NA, Chua TC, Liauw W, Morris DL, et al. Impacts of low peritoneal cancer index on the survival outcomes of patients with peritoneal carcinomatosis of colorectal origin. Int J Surg. 2015;9(23):181–5. doi:10.1016/j.jsu.2015.08.078.
13. Yan TD, Black D, Savady R, Sugarbaker PH. Systematic review on the efficacy of cytoreductive surgery combined with perioperative intraperitoneal chemotherapy for peritoneal carcinomatosis from colorectal carcinoma. J Clin Oncol. 2006;24:4011–19.
14. Sugarbaker PH. Epithelial appendiceal neoplasms. Cancer J. 2009;15:225–35.
15. Cotte E, Glehen O, Mohamed F, Lamy F, Falandy C, Golfier F, et al. Cytoreductive surgery and intraperitoneal chemohyperthermia for chemoresistant and recurrent advanced epithelial ovarin cancer: Prospective study of 81 patients. World J Surg. 2007;31:1813–20.
16. González-Moreno S, Kusamura S, Baratti D, Deraco M. Postoperative residual disease evaluation in the locoregional treatment of peritoneal surface malignancy. J Surg Oncol. 2008;98:237–41. doi:10.1002/jso.21072.
17. Koh JL, Yan TD, Glenn D, Morris D. Evaluation of preoperative computed tomography in estimating peritoneal cancer index in colorectal peritoneal carcinomatosis. Ann Surg Oncol. 2009;16:327–33. doi:10.1245/s10434-008-0234-2.
18. Pôfannenberg C, Knigstainer I, Aschoff P, Oksűz MO, Zieker D, Beckers S, et al. (18)F-FDG-PET-CT to select patients with peritoneal carcinomatosis for cytoreductive surgery and hyperthermic intraperitoneal chemotherapy. Ann Surg Oncol. 2009;16:1295–303. doi:10.1245/s10434-009-0387-7.
19. Hong SH, Shin YR, Roh Y, Jeong EK, Song KY, Park CH, et al. Treatment outcomes of systemic chemotherapy for peritoneal carcinomatosis arising from gastric cancer with no measurable disease: Retrospective analysis from a single center. Gastric Cancer. 2013;16:290–300. doi:10.1007/s10120-012-0182-1.
20. Valle M, Federici O, Garofalo A. Patient selection for cytoreductive surgery and hyperthermic intraperitoneal chemotherapy, and role of laparascopy in diagnosis, staging and treatment. Surg Oncol Clin N Am. 2012;21(4):515–31. doi:10.1016/j.SOC.2012.07.005.
21. Kuramoto M, Shimada S, Ikeshima S, Matsuo A, Yagi Y, Matsuda M, et al. Etensive intraperitoneal peritoneal lavage as a standard prophylactic strategy for peritoneal recurrence in patients with gastric cancer. Ann Surg. 2009;250:242–6. doi:10.1097/SLA.0b013e3181b0c80e.
22. Yonemura Y, Canbay E, Sako S, Ishibashi H, Hirano M, Mizumoto A, et al. Management of peritoneal metastases developed from gastric cancer: Laparascopic hyperthermic intraperitoneal chemontherapy in neoadjuvant setting. Integr Oncol. 2014;3:1–4. doi:10.4172/2339-6771.1000117.
23. Baron MA. Structure of intestinal peritoneum in man. Am J Anat. 1941;69:439–97.
24. Franko J, Shi Q, Goldman CD, Pockaj BA, Nelson GD, Goldberg RM, et al. Treatment of colorectal peritoneal carcinomatosis with systemnic chemotherapy: A pooled analysis of north central cancer treatment group phase III trials N 9741 and N 9841. J Clin Oncol. 2012;30:263–7. doi:10.1200/JCO.2011.37.1039.
25. Kerscher AG, Chua TC, Gasser M, Maeder U, Kunzmann V, Isbert C, et al. Impact of peritoneal carcinomatosi in the disease history of colorectal cancer management: a longitudinal experience of 2406 patients over two decades. Brit J Cancer. 2013;108:1432–9. doi:10.1038/bjc.2013.82.
26. Los G, Mutsaers PHA, van der Vijgh WJ, Baldew GS, de Graaf PW, McVie JG. Direct diffusion of cis-diamminedichloroplatinum (II) in intraoperitoneal rat tumors after intraperitoenal chemotherapy: A comparison with systemic chemotherapy. Cancer Res. 1989;49(12):3380–4.

27. de Bree E, Tsiftsis. DD, "Experimental and pharmacologic studies in intraperitoneal chemotherapy from laboratory bench to bedside. In: Gonzalez-Moleno S editor. Advances in peritoneal surface oncology. Heidelberg/New York/Dordrecht/London: Springer; 2007. p. 53–73.
28. Yonemura Y, Endou Y, Shinbo M, Sasaki T, Hirano M, Mizumoto A, et al. Safety and efficacy of bidirectional chemotherapy for treatment of patients with peritoneal dissemination from gastric cancer: Selection for cytoreductive surgery. J Surg Oncol. 2009;15:311–16. doi:10.1002/jso.21324.
29. Verwaal VJ, Bruin A, Boot H, van Slooten G, van Tinteren H. 8-year follow-up of randomized trial:cytoreduction and hyperthermic intraperitoneal chemotherapy in patients with peritoneal carcinomatosis of colorectal cancer. Ann Surg Oncol. 2008;15:2633–5. doi:10.1245/s10434-008-9966-2.
30. Fujiwara K, Nagao S, Aotani E, Hasegawa K. Principle and evolving role of intraperitoneal chemotherapy in ovarian cancer. EXPERT OPINION. 2013;14:1797–806. doi:10.1517/14656566.2013.820705.
31. Ishigami H, Kitayama J, Kaisaki S, Hidemura A, Kato M, Otani K, et al. Phase II study of weekly intravenous and intraperitoneal paclitaxel (PTX) combined with S-1 for advanced gastric cancer with peritoneal metastasis. Ann Oncol. 2010;21:67–70. doi:10.1093/annonc/mdp260.
32. Yonemura Y, Canbay E, Endou Y, Ishibashi H, Mizumoto A, Li Y, et al. A new bidirectional intraperitoneal and systemic induction chemotherapy (BISIC) for the peritoneal metastasis from gastric cancer in neoadjuvant setting. Integrative Cancer Sci Therapeutics Integr Cancer Sci Therap. 2014;1(2):26–9.
33. Coccolini F, Cotte E, Glehen O, Lotti M, Poiasina E, Catena F, et al. Intraperitoneal chemotherapy in advanced gastric cancer. Meta-analysis of randomized trials. EJSO. 2013;40:12–26. doi:10.1016/j.ejso.2013.10.019.
34. Yu WS, Sugarbaker PH. Early postoperative intraperitoneal chemotherapy for gastric cancer. Cancer Treat Res. 1991;55:265–75.
35. Mahteme H, Hansson J, Berglund A, Pahlman L, Glimelius B, Nygren P, et al. Improved survival in patients with peritoneal metastases from colorectal cancer; a preliminary study. Brit J Cancer. 2004;26:403–7.
36. Cashin PH, Graf W, Nygren P, Mahteme H. Cytoreductive surgery and intraperitoneal chemotherapy for colorectal peritoneal carcinomatosis: Prognostic and treatment of recurrences in a cohort study. Eur J Surg Oncol. 2012;38:509–15. doi:10.1016/j.ejso.2012.03.001.
37. Cao C, Yan TD, Black D, Morris D. A systematic review and meta-analysis of CRS with perioperative intraperitoneal chemotherapy for peritoneal carcinomatosis from colorectal origin. Ann Surg Oncol. 2009;16:2152–65.
38. Lam JY, McConnell YJ, Rivard JD, Temple WJ, Mack LA. Hyperthermic intraperitoneal chemoterpay + early postoperative intraperitoneal chemotherapy versus hyperthermic intraperitoneal chemotherapy alone: assessment of survival outcomes for colorectal and high-grade appendiceal peritoneal carcinomatosis. Am J Surg. 2015;210:424–30. doi:10.1016/j.amjsurg.2015.03.008.
39. Yonemura Y, Endou Y, Bando E, Kawamura T, Tsukiyama G, Takahashi S, et al. The usefulness of oral TS-1 treatment for potentially curable gastric cancer patients with intraperitoneal free cancer cells. Cancer Therapy. 2006;4:135–42.
40. Yonemura Y, Canbay E, Sako S, Ishibashi H, Hirano M, Mizumoto A, et al. Phase II study of a comprehensive treatment using perioperative chemotherapy combined with cytoreductive surgery for curatively resected gastric cancer patients with positive peritoneal wash cytology. Global J Gastroenterol & Hepatol. 2014;2:108–13.
41. Glehen O, Kwiatkowski F, Sugarbaker PH, Elias D, Levine EA, De Simone M, et al. Cytoreductive surgery combined with perioperative intraperitoneal chemotherapy for the management of peritoneal carcinomatosis from colorectal; cancer: A multi-institutional study. J Clin Oncol. 2004;22:3284–92.

42. Sapareto SA, Dewey WC. Thermal dose determination in cancer therapy. Int J Radiation Oncol Biol Physic. 1984;10:787–800.
43. Yonemura Y, Canbay E. Intraperitoneal chemotherapy. In: Emel Canbay, Yutaka Yonemura, editors. Pertitoneal surface malignancy. Heidelberg/New York/Dordrecht/London: Springer, 2015, p. 45–55.
44. Shimizu T, Maeta M, Koga S. Influence of local huperthermia on the healing of small intestinal anastomosis in the rats. Br J Surg. 1991;78:57–9.
45. Yang XJ, Huang CQ, Suo T, Mei LJ, Yang GL, Cheng FL, et al. Cytoreductive surgery and hyperthermic intraperitoneal chemotherapy improves survival of patients with peritoneal carcinomatosis from gastric cancer: final results of a phase III randomized clinical trial. Ann Surg Oncol. 2011;18:1575–81. doi:10.1245/s10434-011-1631-5.
46. Mizumoto A, Canbay E, Hirano M, Takao N, Matsuda T, Ichinose M, et al. Morbidity and mortality outcomes of cytoreductive surgery and hyperthermic intraperitoneal chemotherapy at a single institution in Japan, Gastroenterol Res Pract, Volume 2012 (2012), Article ID 836425, 5 pages. doi:10.1155/2012/836425.
47. Yan TD, Black D, Sugarbaker PH, Zhu J, Yonemura Y, Petrou G, et al. A systematic review and meta-analysis of the randomized controlled trials on adjuvant intraperitoneal chemotherapy for resectable gastric cancer. Ann Surg Oncol. 2007;14:2702–13.
48. Thomas F, Ferron G, Gesson-Paute A, Hristova M, Lochon I, Chatelut E. Increased tissue diffusion of oxaliplatin during laparoscopically assisted versus open heated intraoperative intraperitoneal chemotherapy. Ann Surg Oncol. 2008;15:3623–4. doi:10.1245/s10434-008-0115-8.
49. Elias D, Antoun S, Goharin A, Otmany AE, Puizillout JM, Lasser PL. Research on best chemohyperthermia technique of treatment of peritoneal carcinomatosis after complete resection. Int J Surg Invest. 2000;1:434–9.
50. Mimezami R, Mehta AM, Chandrakumaran K, Cecil T, Moran BJ, Carr N, et al. Cytoreductive surgery in combination with hyperthermic intraperitoneal chemotherapy improves survival in patients with colorectal peritoneal metastases compared with systemic chemotherapy. Br J Cancer. 2014;111:1500–8. doi:10.1038/bjc.2014.419.
51. McQuellon RP, Loggie BW, Lehman AB, et al. Long-term survivorship and quality of life after cytoreductive surgery plus intraperitoneal hyperthermic chemotherapy for peritoneal carcinomatosis. Ann SUrg Oncol. 2003;10:155–62.
52. Chua TC, Robertson G, Liauw W, Farrell R, Yan TD, Morris DL. Intraoperative hyperthermic intraperitoneal chemotherapy after cytoreductive surgery in ovarian cancer peritoneal carcinomatosis: systematic review of current results. J Cancer Res Clin Oncol. 2009;135:1637–45.
53. Spiliotis J, Halkia E, Lianos E, Kalantzi N, Grivas A, Efstathiou E, et al. Cytoreductive surgery and HIPEC in current epithelial ovarian cancer: a prospective randomized phase III study. Ann Surg Oncol. 2015;22:1570–5. doi:10.1245/s10434-014-4157-9.
54. Chua TC, Moran BJ, Sugarbaker PH, Levine EA, Glehen O, Gilly FN, et al. Early- and long-term outcome data of patients with pseudomyxoma peritonei from appendiceal origine treated by a strategy of cytoreductive surgery and hyperthermic intraperitoneal chemotherapy. J Clin Oncol. 2012;30:2449–56.
55. Neumann V, Muller KM, Fischer M. Peritoneal mesothelioma: Incidence and etiology. Pathology. 1999;20:169–7.
56. Helm JH, Miura JT, Glenn JA, Marcus RK, Larrieux G, Jayakrishnan TT., et al. Cytoreductive surgery and hyperthermic intraperitoneal chemotherapy for malignant peritoneal mesothelioma: A systematic review and meta-analysis. Ann Surg Oncol. 2014; 15, doi 10.1245/s10434-014-3978-x.
57. Bijelic L, Sugarbaker PH, Stuart A. Hyperthermic intraperitoneal chemotherapy with melphalan: a summary of clinical pharamacological data in 34 patients. Gastroenterol Res Pract. 2012; Article ID 827534. doi:10.1155/2012/827534.
58. Schaube NP, Alimchandani M, Quezado M, Kalina P, Eberhardt JS, Hugher MS, et al. A novel nomogram for peritoneal mesothelioma predicts survival. Ann Surg Oncol. 2013;20:555–61.

59. Yan TD, Deraco M, Baratti D, Kusamura S, Elias DS, Glehen O, et al. Cytoreductive surgery and hyperthermic intraperitoneal chemotherapy for malignant peritoneal mesothelioma: Multi-institutional experiences. J Clin Oncol. 2009;20(27):6237–42. doi:10.1200/JCO.2009.23.9640.
60. Yonemura Y. Postoperative complication after CRS plus HIPEC. Comprehensive treatment for peritoneal surface malignancy with an intent of cure. In: Yutaka Y, editor. NPO organization to support peritoneal surface malignancy treatment. Oosaka: Kobe; 2015. p. 51–7. in Japanese.
61. Sugarbaker PH, Alderman R, Edwards G, Marquardt CE, Gushchin V, Esquival J, et al. Prospective mobidity and mortality assessment of cytoreductive surgery plus perioperative intraperitoneal chemotherapyto treat peritoneal dissemination of appendiceal mucinous malignancy. Ann Surg Oncol. 2005;13:635–44.
62. Yan TD, Links M, Fransi S, Jaques T, Black D, Saunders V, Morris DL. Learning curve for cytoreductive surgery and perioperative intraperitoneal chemotherapy for peritoneal surface malignancy- A journey to becoming a nationally funded peritonectomy center. Ann Surg Oncol. 2007;14:2270–80. doi:10.1245/s10434-007-9406-8.
63. Smeenk RM, Verwaal VJ, Zoetmulder FA. Learning curve of combined modality treatment in peritoneal surface disease. See comment in PubMed Commons belowBr J Surg. 2007;94:1408–14.

Chapter 35
Noninvasive Temperature Monitoring

Kagayaki Kuroda

Abstract Noninvasive thermometry of target tissue is one of the key issues for successful and safe thermal therapy. Although various techniques using X-ray, infrared, photoacoustics, radiometry, impedance, ultrasound, and magnetic resonance imaging (MRI) have been investigated, to date, infrared imaging for skin surface or MRI for cross-sectional temperature distribution is accepted as a tool for clinical use. Since cross-sectional temperature distribution deep inside a human body is important in thermal therapy in general, practically, MRI is the only practical modality applicable for the therapies. Among several intrinsic MR parameters including proton density, spin–lattice relaxation time, spin-spin relaxation time, diffusion coefficient and chemical shift, the chemical shift is known to be the most reliable for temperature imaging in aqueous tissues. Because this parameter can be measured only from the resonance frequency of water protons and hence independent from the other parameters, which are measured based on the intensity of the MR signals. Phase mapping or spectroscopic technique are applicable for measuring and/or imaging distribution of temperature change. For adipose tissue spin–lattice relaxation time of fat may be used.

Keywords Temperature • Thermometry • Hydrogen bonds • Shielding constant • pH • Susceptibility

35.1 Introduction

Temperature of tissue is one of the most important quantities in monitoring thermal therapy including hyperthermia. In general, temperature is indirectly measured by some physical property, whose value is a monotonous, univalent function of temperature. There are plenty of techniques for measuring temperature with this

---

K. Kuroda (✉)
Department of Human and Information Sciences, School of Information Science and Technology, Tokai University, 4-1-1 Kitakaname, Hiratsuka 259-1292, Japan

Center for Frontier Medical Engineering, Chiba University, 1-33 Yayoi-cho, Inage-ku, Chiba 263-8522, Japan
e-mail: kagayaki@keyaki.cc.u-tokai.ac.jp

S. Kokura et al. (eds.), *Hyperthermic Oncology from Bench to Bedside*,
DOI 10.1007/978-981-10-0719-4_35

prerequisite. In the field of heat transfer engineering, temperature measurement techniques are categorized into two groups, contact and non-contact. These categories correspond to the terminologies, “invasive” and “noninvasive” in the field f of thermal therapy.

Invasive thermometry measures temperature just around a probe. Liquid column, bimetal, thermocouples, resistance temperature detector and thermistors are widely used as probes in the variety of field. However, invasive thermometry in the human body is generally limited for a few probing points, and thus do not bring sufficient information on temperature distribution. Moreover, invasive thermometry is not comfortable for patients, especially for hyperthermia with repetitive weekly treatments lasting for 40–60 min. Therefore, noninvasive thermometry is strongly desirable. Noninvasive thermometry measures temperature by measuring various kinds of thermally induced physical phenomena remotely. When the phenomena are spatially distributing in the tissue under measurement, temperature distribution may be obtained. In other words, to image temperature distribution the measurement has to be noninvasive.

The noninvasive temperature imaging techniques include those using X-ray [1], infrared [2], photoacoustics [3], radiometry [4], impedance [5], ultrasound [6], and magnetic resonance imaging (MRI) [7]. Among them, only MRI is accepted to image cross sectional temperature distribution in a human body in the clinical medicine, while the others are still in the experimental stage. Thus noninvasive thermometry using MRI is focused here.

35.2 Temperature Dependent Mr Parameters

Most of the parameters observed in MRI are temperature dependent. For hydrogen nuclei (^{1}H) or proton MR, they include proton density or thermal equilibrium magnetization, M_0; [8, 9] spin–lattice or longitudinal relaxation time, T_1; [10–13] spin-spin or transverse relaxation time, T_2; [11, 14, 15] the diffusion coefficient, D; [16, 17] and resonance frequency or chemical shift [18, 19], δ.

35.2.1 *Proton Density, ρ*

Thermal Equilibrium Magnetization, M_0 is the strength of the macroscopic magnetization vector and is determined by the ratio of the average numbers of protons at the two Zeeman energy states. The ratio is subject to Boltzmann distribution and hence is a function of temperature as follows [20].

$$M_0 = \frac{N\gamma^2\hbar^2 B_0}{4kT} \tag{35.1}$$

where T is absolute temperature, N is the number of protons, γ is the magnetogyric ratio, B_0 is the magnetic flux density of a static magnetic field, $\hbar$ is the Plank constant divided by 2π, and k is the Boltzmann constant. In MR imaging, direct measurement of the exact thermal equilibrium is not possible. Instead, proton density ρ(T) giving the total intensity of the magnetization within a volume of a voxel is available [20]. Thus the more realistic model is as follows,

$$\rho(T) = \frac{N(T)\gamma^2 h^2 B_0}{4kT}, \tag{35.2}$$

where N(T) is the number of nuclear spins per a unit volume. The temperature coefficient of the proton density is expressed as follows.

$$\frac{1}{\rho(T)} = \frac{d\rho(T)}{dT} = \frac{1}{N(T)}\frac{dN(T)}{dT} - \frac{1}{T}. \tag{35.3}$$

The temperature coefficient of the proton density of pure water is known to be –0.36 %/°C at 40 °C (313 K).

Tanaka et al. [21] measured ρ(T) in the mouse thigh muscle by a 0.251 T (10.72 MHz proton resonant frequency) spectrometer and obtained the temperature coefficient of −1.2 % in 1981. They [22] made similar measurements in 1983 and obtained the temperature coefficient of −1.3 %/°C for both normal and pathological tissues in mouse thigh muscle. Kamimura [8] obtained the temperature coefficient of −0.44 %/°C in a 1.4 mM/l copper sulfate solution using a 0.0472 T (2.01 MHz proton resonant frequency) MR machine. In the other experiments [23], although the temperature dependence of ρ(T) was clear in the gel phantoms, that in the bovine tissues *in vitro* was changeable and quite different from that in pure water.

Temperature dependence of the thermal equilibrium magnetization M_0, which is obtained for the fixed number of protons, should not affected by the materials of the phantoms nor the kinds of organs. However that of the proton density $\rho(T)$ is affected by the materials of the phantoms and the kinds of organs, since the number of protons in a fixed volume varies. Chen et al. [9] demonstrated the difference in the temperature dependence of $\rho(T)$ in adipose and aqueous tissues. In contrast to that the proton density in muscle which had a clear hysteresis in its temperature dependence, the parameter in fat did not expose such phenomena.

35.2.2 *Spin–Lattice Relaxation Time,* T$_1$

Spin–lattice Relaxation time or longitudinal relaxation time, T_1, is a time constant of recovery of the longitudinal magnetization to the thermal equilibrium state accompanied by energy dissipation from a nuclei to the surrounding atoms and molecules, called lattice. Because the efficiency of energy dissipation changes with intensity of Brownian motion of molecules determined by temperature, T_1 becomes a function of correlation time of spatial position, and thus of viscosity and temperature of the molecular system as following.

A theoretical model in which the intensity of the molecular motion is quantified by the correlation length τ_c of the auto-correlation function of the proton position had been obtained [20]. According to the model, the following relationship is held, when ω_0 is the resonant frequency and $\eta(T)$ is the temperature dependent viscosity of the proton system;

(i) $\omega_0 \tau_c >> 1$ (solid state)

$$T_1 \propto \frac{{\omega_0}^2 \eta(T)}{T} \tag{35.4}$$

(ii) $\omega_0 \tau_c << 1$ (liquid state)

$$T_1 \propto \frac{T}{\eta(T)} \tag{35.5}$$

In the liquid state, T_1 becomes longer, as the energy dissipation takes longer. The temperature dependence of the proton density is neglected in these equations. From Eq. (35.6), the temperature coefficient in the liquid state is represented as follows

$$\frac{1}{T_1}\frac{dT_1}{dT} = \frac{1}{T} - \frac{1}{\eta(T)}\frac{d\eta(T)}{dT}. \tag{35.6}$$

The temperature coefficient is calculated as 2.2 %/°C at 40 °C (313 K) for the case of pure water [24].

Kato et al. [22] reported the temperature coefficient in a copper sulfate solution was 2 %/°C. Kamimura [8] obtained a temperature coefficient 1.9 %/°C for a copper sulfate solution of 1.4 mM/l. Parker et al. [12, 25] reported that the temperature coefficients are 2.4 % and 1.3 % for copper sulfate solution and human blood. Dickinson and Hall et al. [13] measured the changes in T_1 with temperature for the agar gel phantom, bovine muscle in vitro, rabbit muscle in vitro and human calf in vivo. The temperature dependence of T_1 varies remarkably among these objects. They [16] compared the temperature dependence of T_1 and the diffusion coefficient D for the agar gel phantom doped with copper sulfate and human calf in vivo. Our experiments [26] on T_1 showed that the temperature dependence of T_1 varies according to the kinds of organ tissues. Similar studies on the temperature dependence of T_1 have been made also by some other groups [14, 27, 28]. Lewa et al. [29] demonstrated hysteresis phenomena in the T_1 temperature dependence in tissues and attributed the phenomena to irreversible denaturation in protein.

The results described above say that the temperature dependence of T_1 differs significantly among the phantom materials and biological tissues although the temperature coefficients were consistently positive for aqueous objects. Based on the Zimmerman-Brittin model [30], water molecules are bound by macromolecules such as proteins with some different states, and those differently bound water molecules exchange to each other rapidly enough compared with their relaxation rates. A simplified model is called the FETS (First Exchange two states) model, in which only water molecules bound to macromolecules and those free from binding are considered. A single component of T_1 represented by the equation below is observed.

$$\frac{1}{T_1} = (1 - \mathrm{b})\left(\frac{1}{T_{1\mathrm{f}}}\right) + b\left(\frac{1}{T_{1\mathrm{b}}}\right). \tag{35.7}$$

In Eq. (35.7), b is the fraction of the bound water, T_{If} is the spin–lattice relaxation time of free water and T_{Ib} is that of bound water. As the fraction of the bound water varies with kinds and densities of macromolecules, the relaxation mechanism and hence the apparent T_1 differs with the kinds of organs. As T_1 also differs with ions and cell compartments in biological tissues. Therefore it is indispensable to calibrate T_1 with temperature in the target tissues.

One of the most significant application of using T_1 of proton is to measure temperature of fat, more specifically of the methylene chain or terminal methyl of fat [31]. Proton spectra of fat or oil comprise about nine components. Among these, the most pronounced peak is the signal at 2.1 ppm from protons of methylene chains $(-CH_2\text{-})_n$ of all the fatty acids in glyceride. The signal from the terminal methyl $(-CH_3)$ is the second prominent peak. Although the chemical shift of each of these signals is temperature insensitive, as the protons in glyceride have no hydrogen bonds, the spin–lattice relaxation time, T_1, changes linearly with temperature. Importantly, the values and temperature coefficients of these fat signal components are significantly different. The temperature coefficient of methylene chain protons in bovine fat is $1.79 \pm 0.07\ \%/°C$, while that of terminal methyl protons is $2.98 \pm 0.38\ \%/°C$. This means that spectroscopic parameter measurements are desired for fat temperature imaging.

35.2.3 Spin-Spin Relaxation Time $\mathbf{T_2}$

Spin-spin relaxation time, or transverse relaxation time T_2 is a time constant of elimination of the transverse component of the macroscopic magnetization due to the phase dispersion of the precession of isochromats of protons. The phase dispersion is due to a small difference in the resonant frequency caused by the microscopic local magnetic field created by nuclear magnetic moments. Since the influence of the local magnetic field becomes weak when molecular motion is intensified with temperature elevation, T_2 becomes longer. Thus T_2 becomes an increasing function of temperature as follows [20].

(i) $\omega_0 \tau_c >> 1$ (solid state)

$$T_2 = \frac{T}{\eta(T)} \tag{35.8}$$

(ii) $\omega_0 \tau_c << 1$ (liquid state)

$$T_2 = T_1 = \frac{T}{\eta(T)} \tag{35.9}$$

where the definitions of ω_0, τ_c and $\eta(T)$ are identical with those in Eq. (35.4). In the liquid state, T_2 is equal to T_1, and the temperature coefficient for pure water is 2.2 %/°C at 40 °C (313 K) [24].

Dickinson and Hall [13] obtained the temperature coefficients of T_2 as 0.76 %/°C for an agar gel phantom doped with copper sulfate and 0.51 %/°C for and human calf *in vivo*. Our experimental results [23] showed that the temperature dependence of T_2 is changeable according to the kinds of the organ tissues similarly with T_1.

Because a long echo time is generally used in the measurement of T_2, the signal to noise ratio in T_2 measurement is generally lower than that in T_1 measurement. Influence of the temporal fluctuation of the static magnetic field and the molecular diffusion is also stronger. Moreover the error in the flip angle of the π and $\pi/2$ pulse is considerable. Therefore measurement of T_2 is generally less accurate than that of T_1 [32]. However, some of the recent studies are conducted to use T_2 for fat temperature measurement [33, 34].

The transverse relaxation time including the effect of the static magnetic field inhomogeneity and susceptibility is an apparent transverse relaxation time and is often referred to as T_2^*. This parameter can be expressed by the following equation.

$$\frac{1}{T_2^*} = \frac{1}{T_2} + \frac{1}{T_{2e}}, \tag{35.10}$$

where T_2 is the above-mentioned transverse relaxation time determined by the spin-spin interaction, while T_{2e} is the transverse relaxation time determined by the contribution from the environmental factor. Since exact modeling of the second term of the above equation is not available as is subject to change with tissue environments, temperature dependence of T_2^* is not simply modeled. Nevertheless, one of the realistic applications of this parameter to thermometry for monitoring cryosurgery; T_2^* becomes a liner function of temperature in the frozen tissue in the temperature range under 0 °C [35].

35.2.4 *Diffusion Coefficient* D

Diffusion coefficient D is the average area of proton diffusion in an unit time. The diffusion coefficient depends on temperature like the model function below [17, 24].

$$D \propto \frac{T}{\eta(T)} \propto \exp\left(\frac{E_a}{kT}\right), \tag{35.11}$$

where $\eta(T)$ is the viscosity coefficient, E_a is the activation energy for translational diffusion and k is Boltzmann coefficient. The temperature coefficient of the diffusion coefficient in pure water is calculated as 2.2 %/°C at 43 °C (313 K), which is similar to the case of T_1.

Le Bihan et al. [17] reported that the temperature coefficient of the diffusion coefficient in a Polyacrylamide phantom of 7.5 %w was 2 %/°C. They claimed that the diffusion coefficient is superior to T_1 by the following three reasons; first, the diffusion coefficient is more sensitive to temperature than T_1, since D is the quantification of translation of the molecules, which is more sensitive to temperature than rotation, the primary contribution to T_1. Second, the relationship between D and temperature is direct, in contrast to that between T_1 and temperature. T_1 is affected by the resonant frequency and other factors. Third, the diffusion coefficient is obtained from the ratio of two echo signals and hence is less affected by the static magnetic field inhomogeneity than T_1. Hall and Dickinson [16] obtained the temperature coefficients of D for the agar gel phantom doped with copper sulfate, and the human calf in vivo were 2.4 %/°C and 2.25 %/°C, respectively. The uncertainty in estimated temperature by D for these objects were ±6.7 °C and ±10 °C. These figures are inferior to those of T_1; they are ±3.2 °C and ±5.32 °C for those materials. They explained this inferiority in D to T_1 as D was measured with a long echo time and was more sensitive to the noise, temporal fluctuation of a static magnetic field and body motion. They conclude, from these points of view, that T_1 is superior to D for thermometry. The comparison of D and T_1 has not yet reached the absolute answer.

Ventricular cerebrospinal fluid (CSF) temperature measurement based on D have been reported recently [36, 37] using the temperature-diffusion relationships obtained by Mills as follows [38];

$$T = \frac{2256.74}{\ln\left(\frac{4.39221}{D}\right)} - 273.15. \tag{35.12}$$

35.2.5 *Chemical Shift δ*

Chemical Shift δ is the normalized expression of the resonance frequency. Compared with the above mentioned parameters, the proton chemical shift is the only parameter determined solely from the frequency of the resonant signal; in contrast, the other parameters are determined from the amplitude. Hence, it is possible to detect the proton chemical shift independently from other parameters.

Resonance frequencies of protons of the hydroxyl group, such as in methyl alcohol [39], ethyl alcohol [40], and water [41], linearly depend on temperature. This is because these protons are associated with hydrogen bonds between the protons in hydroxyl groups and other more negatively electrified atoms [42]. Since the temperature dependence is linear and reproducible within a specified temperature range, researchers have commonly used measurement of the internally referenced chemical shift between the hydroxyl (−OH) and methyl ($-CH_3$) or ethyl ($-C_2H_5$) group in methanol or glycol to precisely measure temperature inside a sample in a tube [43, 44] in the field of NMR for analytical chemistry. For monitoring body temperature in vivo, water protons' chemical shift is accepted as the most reasonable and practical choice. Because water is the simplest and most abundant specimen of this type of proton system in biological tissues, although the amide proton also has hydrogen bond–induced temperature dependency, and thus represents our focus here.

35.2.5.1 Temperature Dependence of Water Protons' Chemical Shift

Among several factors determining the resonance frequency of water protons, the shielding due to the electric field effects of the charge distributions in the neighboring molecules is related to hydrogen bonds [18, 41]. Briefly, the electrical currents caused by the electron motion in the electron cloud, apparently in the direction opposite to the external field, shield the magnetic flux of the external field [42, 45]. A schematic diagram of the shielding effect for the protons in the hydroxyl group under temperature change in relation to chemical shift is depicted in Fig. 35.1 [18]. At a low temperature, the electron motion in the electron cloud is restricted by the electrostatic motive of hydrogen bonding, and thus, the shielding effect of the electron cloud is weak, giving a high resonance frequency. Since the proton exchange between the broken and unbroken bonds is rapid enough compared with the time scale of our observation, the observed resonance frequency reaches an average value for all possible configurations in the proton system [41]. When we warm up the proton system, the motion of water molecules intensifies, thereby distorting, extending, and/or breaking the hydrogen bonds of the molecules. Thus, an electron cloud at higher temperatures is freer from the restraint of the electrical bonding force than at lower temperatures, thereby strengthening the shielding effect. As a result, the resonance frequency of the water proton decreases. The apparent change of the proton resonance frequency in pure water is known to be

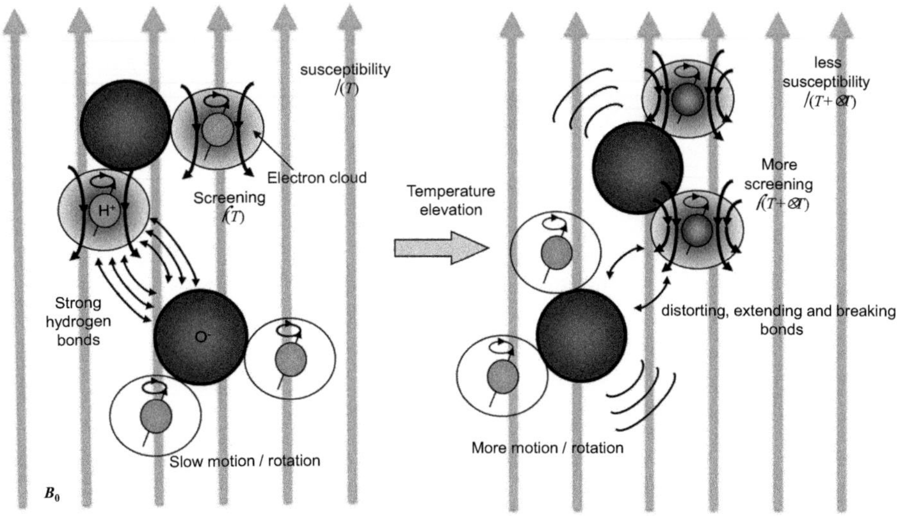

Fig. 35.1 Principle of temperature dependence of proton magnetic resonance frequency in water molecules

−0.0108 ppm/°C at a temperature range of 0–50 °C and −0.00091 ppm/°C at a temperature range of 50–100 °C [41]. There is a slight change in the temperature coefficient at 50 °C.

In biological tissues, factors including electrolytes, macromolecules, pH, and susceptibility, in addition to fluid motion and elements, contribute to the absolute value and the temperature change of the proton resonance frequency of water. Researchers have not been able to model the temperature dependence of the resonance frequency in a strict way that includes of all these elements to date. Instead, we need to model this dependence in a simplified form, such as that shown in the following [18]:

$$\omega_W \cong \gamma(1 - \sigma_W(T) - c\chi(T))B_0, \tag{35.13}$$

where ω is the resonance frequency [rad/s], γ is the magnetogyric ratio [rad/T], and c is the shape factor for volume susceptibility, and $\chi(T)$ is the magnetic susceptibility. The subscript W denotes water protons.

When there is a proton component within the same macroscopic volume (ie, a voxel, or a volume of a surface coil sensitivity), whose screening constant is insensitive to temperature and the volume susceptibility is common with the water proton, this component can be used as an internal reference for measuring the water proton resonance frequency, as follows:

$$\delta_{\text{W-R}}[\text{ppm}] = \frac{\omega_{\text{W}}(T) - \omega_{\text{R}}(T)}{\omega_{\text{R}}(T)} \cdot 10^6 \\ \cong \frac{\omega_{\text{W}}(T) - \omega_{\text{R}}(T)}{\omega_{\text{RF}}} \cdot 10^6 \\ \cong (-\sigma_W(T) + \sigma_{\text{R}}).10^6, \tag{35.14}$$

$$\omega_R(T) = \gamma B_0(1 - \sigma_{\text{R}} - c\chi(T)), \tag{35.15}$$

where subscript R denotes the reference and RF denotes the radio frequency of the transceiver/receiver. Note that the susceptibility term in Eqs. (35.13) and (35.15) is canceled out. When the hypothetical temperature coefficient is a [ppm/°C] and the chemical shift at 0 [°C] is b [ppm], the relationship between temperature and the chemical shift is denoted as:

$$T = \frac{\delta_{\text{W-R}} - b}{a}. \tag{35.16}$$

We can obtain the temperature difference, ΔT [°C], using the following equation:

$$\Delta T = \frac{\Delta\delta_{\text{W}-\text{R}}}{a}, \tag{35.17}$$

where $\Delta\delta_{\text{W-R}}$ denotes the change in $\delta_{\text{W-R}}$. Temperature measurement based on Eq. (35.16) an (35.17) is available when spectroscopic techniques [18] described later in this chapter are used.

When no internal reference is available, a relative chemical shift change from the value at a baseline temperature, T_{R}, may be detected as:

$$\Delta\delta_{\text{w}}[\text{ppm}] = \frac{\omega_{\text{W}}(T) - \omega_{\text{W}}(T_R)}{\omega_R(T)} \cdot 10^6 \\ \cong \frac{\omega_{\text{W}}(T) - \omega_{\text{W}}(T_R)}{\omega_{RF}} \cdot 10^6 \\ \cong (-\sigma_W(T) + \sigma_W(T_R) - c(\chi(T) - \chi(T_R)))10^6. \tag{35.18}$$

In Eq. (35.18), the susceptibility term remains. Depending on the paramagnetic and diamagnetic contribution from the tissue of interest, this susceptibility term acts in both positive and negative ways [46]. Thus, we can obtain only the relative temperature change, ΔT, using the temperature coefficient a [ppm/°C], as follows:

$$\Delta T = \frac{\Delta\delta_W}{a}. \tag{35.19}$$

Temperature change observation based on Eq. (35.19) is available for spectroscopic technique as well as phase mapping technique [47] with components other than water sufficiently suppressed.

35.2.5.2 Temperature Dependence in Biological Tissues

A number of representative results of temperature dependence of the proton chemical shifts in tissues obtained with internal references are summarized in Table 35.1 Arus et al. [48] focused on the chemical shift of the amide proton in NAA and obtained a temperature coefficient of −0.00789 ppm/°C and an estimation error of temperature of ±0.4 °C. Moreover, Lutz et al. [49] examined the feasibility of measuring temperature in a cell culture perfusion using the proton chemical shift of water, pyruvate, acetate and lactate relative to the acetone shift, and found that the chemical shift of the water proton has a temperature coefficient of −0.0107 ppm/°C, while the shifts of other compounds are temperature-insensitive. In our experiments using fat proton as the internal reference for skeletal muscle extracted from mice, the coefficient was −0.00853 ppm/°C [50]. For rabbit brain in vivo with NAA as the internal reference, temperature coefficient was −0.0097 [51].

Cady et al. [52, 53] reported that the correlation between tympanic brain temperature and the chemical shift of water (H_2O) from NAA in newborn piglets was linear with a temperature coefficient of −0.0106 ppm/°C. According to Corbett et al., [54, 55] the coefficient of the similar chemical shift difference in porcine brains varied −0.0102 to −0.0139 ppm/°C based on direct calibration with an implanted temperature probe. They also found that the chemical shift of water from NAA and trimethylamine (TMA) in infarct and noninfarct adult canine brains had temperature coefficients of −0.0122 and −0.0143 ppm/°C, respectively [56]. In their examination, the reason for the difference in the temperature coefficients was attributed to minor contributions from other compounds such as N-acetylaspartylglutamate, glutamate, or acetate to the NAA signal, and glycerophosphocholine, phosphocholine, or choline to the TMA signal. Similar results were obtained for brain samples by the other groups [57, 58]. Most studies have reported that pH might not change the temperature coefficient, but that it does affect hydrogen bonding and water proton resonance frequency, making absolute temperature measurement difficult [52, 59].

35.3 Acquisition Techniques

For water protons' chemical shift, research has focused on the investigation of two major techniques—phase mapping and spectroscopic methods.

Table 35.1 Representative results on Temperature dependence of proton chemical shift in tissue

	Temperature range (°C)	δ (ppm) at 0 °C	$\partial\delta/\partial T$ (ppm/°C)	Correlation coefficient	Internal reference	Sample
Arus [48]	20–40	8.127 (amide proton in NAA)	−0.00789 (amide proton in NAA)	–	N-CH_3 proton in NAA	Brain cytosol in mammalian Ringer solution buffered with KH_2PO_4 and $NaHCO_3$ (pH = 6.05–8.31)
Lutz [49]	31.4–50.2	5.050 ± 0.0031	−0.0107 ± 0.00008	–	Methyl group of acetone 2.225 ppm	HAMs F-12 cell culture medium (pH = 7.8)
Kuroda [50]	27–50	2.929 (+0.08/−0.07)	−0.00853 (+0.00057/−0.00034)	−0.998 ± 0.001	CH_2 of fat	Mouse muscle in vitro immediately after sacrifice, calibrated with sample tube temperature
Kuroda [51]	32–51	3.011	−0.0097	−0.932	NAA	Rabbit brain in vivo calibrated with fiber-optic probe–measured temperature
Cady [52, 53]	30–40	3.059	−0.0108	−0.985	NAA	Swine brain in vivo calibrated with tympanic temperature
		1.876	−0.0113	−0.980	Cr[c]	
		2.010	−0.101	−0.990	Cho[c]	
Corbet [54]	20–44	3.013	−0.0102	–	NAA[a]	Swine brain in vivo under whole body hyperthermia (calibrated in model solutions, pH = 5.5–7.6)
Corbett [55]	28–40	3.161	−0.0139	–	NAX[b]	Swine brain in vivo and postmortem with fiber-optic probe–measured temperature
		1.923	−0.0129		TMA[b]	
Corbett [56]	34–44	3.109	−0.0122	–	NAA	Canine brain in vivo with fiber-optic probe–measured temperature
		1.997	−0.0143		TMA	
Zhu [57]	31–38	3.022	−0.00963	−0.99	NAA	Rat brain in vivo with fiber-optic probe–measured temperature
			−0.00983	−0.99	Cr	
			−0.00943	−0.99	Cho	

Weis [58]	33–38	3.123	−0.0121	−0.893	NAA	Swine brain in vivo with fiber-optic probe–measured temperature
		1.829	−0.00949	−0.934	Cr	
		2.123	−0.0124	−0.860	Cho	

[a] *NAA* N-acetylaspartate
[b] *NAX* N-acetylaspartate and related compounds, *TMA* Trimethylamine
[c] *Cr* total creatine (phosphorylated plus unphosphorylated), *Cho* choline

35.3.1 Phase Mapping

The temperature imaging technique most widely accepted today is the phase mapping technique [47, 60] generally in conjunction with a group of gradient echo sequences. Although this type of technique is often referred to as "proton resonance frequency (PRF) Shift" [61] method, this terminology should include spectroscopy or spectroscopic imaging techniques as they are all based on the same principle of water PRF shift or water protons' chemical shift. Thus a term "phase mapping" is used for distinguishing this specific technique from the spectroscopic technique.

In the phase mapping technique, the change in the resonance frequency is detected as the change in the phase of the complex resonant signal of water proton, $\Delta\phi_W$, during an echo time, *TE*, as depicted in Fig. 35.2 and the following equations.

$$\begin{aligned}\Delta\phi_W &= \phi_W(T+\Delta T) - \phi_W(T) \\ &= [\omega_w(T+\Delta T) - \omega_w(T)]\cdot TE \\ &= \Delta\delta(T)\cdot\omega_{RF}\cdot TE \\ &= a\cdot\Delta T\cdot\omega_{RF}\cdot TE,\end{aligned} \qquad (35.20)$$

where $\phi_W(T)$ is the phase at temperature T and the other variables are identical with those used in Eq. (35.18) and (35.19). The phase difference in Eq. (35.20) is obtained basically by a voxel-by-voxel complex subtraction between the baseline image taken before temperature change and the image after temperature change.

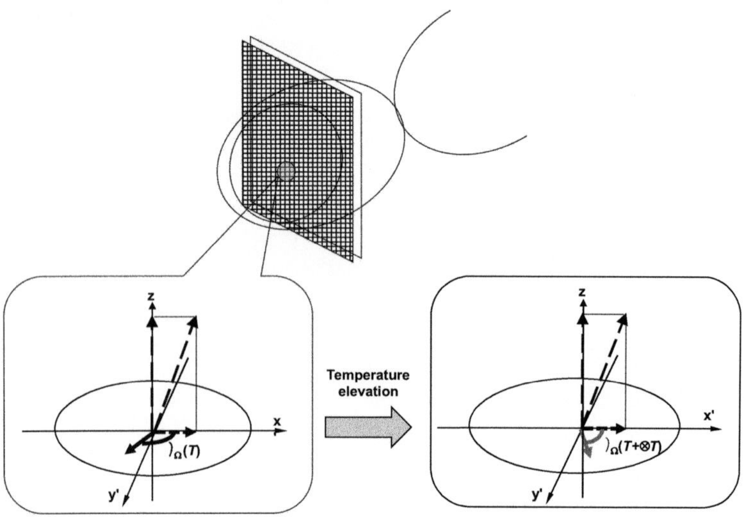

Fig. 35.2 Principle of phase mapping technique to measure change in the proton magnetic resonance frequency

$$\begin{aligned}\Delta\phi_W &= \arg\left(S_1^* \cdot S_2\right)\\ &= \tan^{-1}\left(\frac{S_{R1}\cdot S_{I2} - S_{R2}\cdot S_{I1}}{S_{R2}\cdot S_{R1} + S_{I2}\cdot S_{I1}}\right)\end{aligned} \tag{35.21}$$

where S_1 is a complex baseline image, S_2 is an objective complex image, and * denotes a complex conjugate. Index R means real part, while I means imaginary part. From Eq. (35.20) and (35.21), temperature change is obtained as follows.

$$\Delta T = \frac{\Delta\phi_W}{a \cdot \omega_{RF} \cdot TE}. \tag{35.22}$$

In this technique, the temperature coefficient a is often set as −0.01 ppm/°C for simplicity. In order to compensate for the change in spatial distribution of the static magnetic field, external phantoms containing oil are placed and the phase change in those phantoms are used [61]. As gradient echo sequences are available on most of the clinical scanners, this temperature imaging technique is readily applicable. This technique has been used to monitor tissue temperature changes under thermal therapies [62–65], following initial feasibility reports [66, 67]. A typical signal acquisition time for a single slice temperature image is 2 s with TE of 3 ms, a repetition time (TR) of 7.8 ms, and an acquisition matrix of 256 by 256, which leads to a temperature sensitivity of phase of 0.00384 rad/T at a 3-Tesla scanner.

The primary drawback of this technique is the necessity of the image subtraction process, which is affected significantly by tissue's susceptibility to change [46, 68] and inter-scan motion [60, 69, 70]. Moreover, any non-water component contained in the signal, such as lipid, needs to be suppressed in order to avoid erroneous phase calculation. [71, 72] Among these factors, the most significant is inter-scan motion, body or organ movement between a pair of scans at two different time points of temperature observation. In contrast, intra-scan motion, or movement during image acquisition is relatively insignificant because it can be avoided by using rapid scan techniques [73] combined with breath-holding. Inter-scan motion involves two primary effects. First, tissue motion and/or deformation causes the protons in a voxel to move away to another voxel(s). Second, it induces distortion of the magnetic flux of the external magnetic field. A few techniques for compensating for these motion effects in temperature imaging have been previously proposed [69, 74]. However, these methods are effective only when the motion is translational, periodic, in-plane, and/or without deformation.

When the size of the heated region is relatively small compared to the size of a target tissue as often encountered in localized heating with microwaves, radio frequencies (RF), lasers, or focused ultrasound (FUS), the phase change induced by thermal effect can be detected as a localized phase change. In such a case, a technique called "Referenceless PRF thermometry" [60], or the "Self-reference method" [75], can be applied. As is shown in Fig. 35.3b, a region of interest (ROI) covering the heated area is set, and then the phase distribution in the ROI is estimated in the surrounding regions for estimation (RFE). The "estimated"

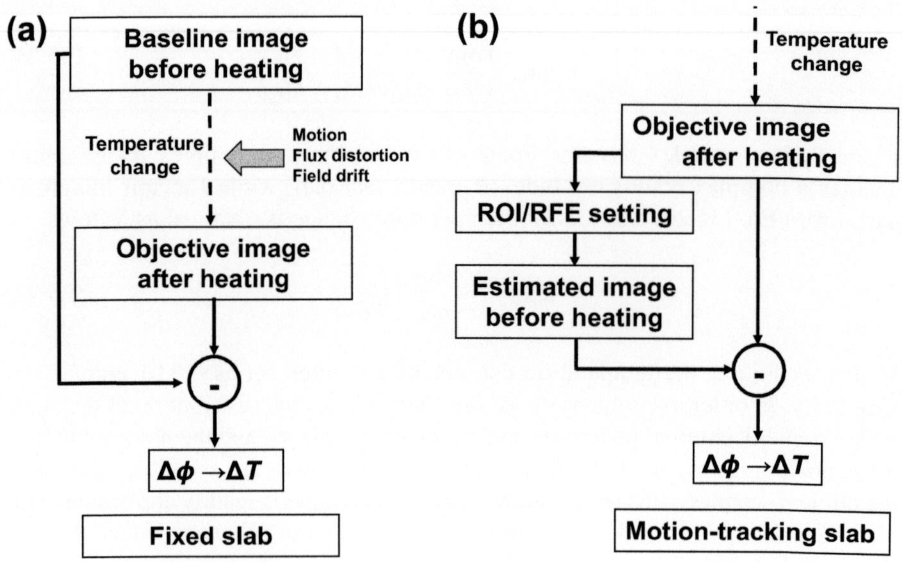

Fig. 35.3 Comparison between the conventional phase mapping technique (**a**) and the referenceless or self-reference technique (**b**). In contract to the conventional technique with the image subtraction between the baseline and objective images, the referenceless technique does not require the baseline image, and hence motion, flux distortion, and/or field drift do not affect

reference phase image is then deduced from the original phase image, and thus the temperature change distribution in the ROI is obtained. Unlike the conventional method shown in Fig. 35.3a, this technique does not require any external baseline images to be taken before heating; therefore, it reduces the effects of inter-scan motion. As is shown in Figs. 35.4a and b, the resultant temperature elevation obtained by the conventional and referenceless or self-reference techniques agrees well. Note that the new technique is effective only when the phase distribution is smooth and the heated area is sufficiently small compared with the size of target tissue [75].

35.3.2 Spectroscopy and Spectroscopic Imaging

A simple and reasonable approach to avoiding the errors induced in the phase mapping methods is to use a spectroscopic technique that detects a temperature-insensitive proton component as an internal reference as was shown in Eq. (35.14) [18] for the sake of a relatively longer acquisition time in relation to that of phase mapping techniques. The simplest way to obtain temperature information from tissue is to place a local surface coil and observe the signal [52, 54, 55, 76]. This approach may be useful when a localization process should be avoided in animal experiments.

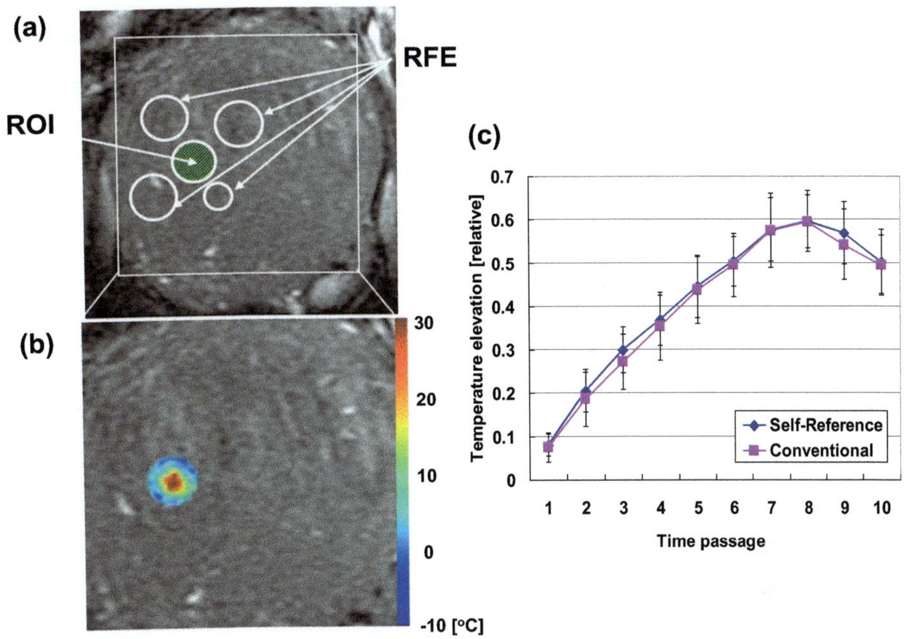

Fig. 35.4 Settings of ROI (region of interest) and RFE (region for estimation) surrounding ROI for the self-reference technique (**a**) in the uterine fibroid. Resultant temperature elevation image obtained by the self-reference technique (**b**). Comparison of the temperature elevation values obtained by the conventional and self-reference techniques (**c**)

Single voxel spectroscopy has been intensively examined to investigate the temperature dependence of the water proton chemical shift, as described above, and to measure temperature at spatial points in tissue [52, 54–57]. A magnetic field higher than 1.5 T with high temporal stability and spatial homogeneity may be necessary to detect the small metabolite signal(s) at concentrations around 10 mM for internal referencing.

2D (two-dimensional) or 3D (three-dimensional) SI (spectroscopic imaging) has been explored for imaging temperature in various animal tissues in vitro and in vivo [58, 77–79]. The long acquisition time for SI increases the possibility of encountering motion and temperature change during scanning. Spectral estimation techniques to maintain a sufficient temperature resolution (0.01 ppm) are indispensable. Overall, conventional MRS imaging would be appropriate only for a slow and distributed temperature change, for which temporal and spatial resolutions are not crucial.

To overcome the time limitation of the conventional SI approach, researchers [80] have proposed a fast method using EPSI (echo planar spectroscopic imaging) [81]. The primary feature of this method is that one of the spatial directions and the chemical shift direction of the signal space are encoded by a series of gradient reversals. As a practical example, the authors examined temperature imaging of

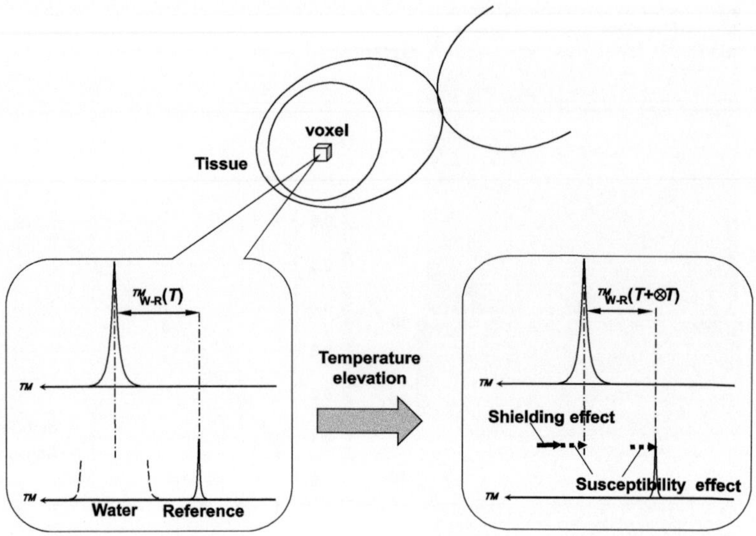

Fig. 35.5 Basic principle of thermometry using magnetic resonance spectroscopy. A single voxel technique is depicted for simplicity. Since the ^{1}H spectra of water and an internal reference signals moves together with susceptibility effect, the frequency difference or the chemical shift between water and the reference may give the water resonance frequency change by the thermally induced screening constant change.

porcine liver in vitro at 1.5 T. They executed two separate scans, with and without water suppression, to obtain the chemical shift difference ($\delta_{H2O\text{-}CH2}$) between the water and lipid signals. The acquisition time per temperature image was 3 min. The temperature error induced by inter-scan sample movements was 30 % of that using the phase mapping method [80].

Researchers have adopted the LSEPSI (line scan echo planar spectroscopic imaging) method, [79, 82] a combination of the echo planar technique with the concept of column scanning, to accelerate multi-voxel spectroscopy and to avoid the spectral degradation due to field instability by eliminating the phase encoding process. They applied the technique to breast imaging with the lipid signal as a reference [83] and brain imaging with the NAA signal as a reference [51]. In experiments using a 3.0 T scanner, temperature distributions in a rabbit brain under laser heating was visualized [51]. The signal acquisition time per column in the brain experiment was 66 s with 2-s TR and 32 spatial encoding for a pair of nonsuppressed and water-suppressed scans. The acquisition time is fairly acceptable for hyperthermia, which requires temperature, spatial, and temporal resolutions of 1 °C / 1 cm^3 / 1 min. The EPSI and LSEPSI approaches are substantially similar to the multiple gradient echo phase mapping technique. [84]

After selecting an acquisition method, the effects of signal-to-noise ratio, length of signal observation time, intravoxel inhomogeneity of the static magnetic field, and eddy current magnetic field are thought to be significant error factors on the instrumentation side in thermometry, as is common in the MRS parametry. We may reduce most of the effects by using appropriate spectral estimation techniques [53].

35.4 Use of Extrinsic Indicators

35.4.1 The Lanthanide Complex

Researchers have conducted many works on thermometry using lanthanide complex agents such as $TmDOTP^{5-}$, $TmDOTMA^{-}$, Pr[MOE-DO3A], and so on. Readers can consult the well-featured review by Hekmatyar et al. [85] for the details related to this category. Briefly, the observed chemical shifts of some specific protons in the lanthanide complexes of Pr^{3+}, Yb^{3+}, Tm^{3+}, Dy^{3+}, or Tb^{3+} consist of three terms, as follows:

$$\delta_{obs} = \delta_{dia} + \delta_{con} + \delta_{dip}, \tag{35.23}$$

where δ_{dia} is the diamagnetic contribution; δ_{con} is a scalar paramagnetic, or the "contact" contribution; and δ_{dip} is a vector paramagnetic, or the "dipole" contribution from the complex. Since the latter two terms have a linear temperature dependence, we can express the relationship between the chemical shift δ_{obs} and temperature T as follows:

$$T = \frac{\delta_{\mathrm{obs}} - b}{a}, \tag{35.24}$$

where the definitions of a and b are similar to those in Eq. (35.16).

The most significant advantage of using lanthanide complexes is that the temperature coefficient is much larger than that of the endogenous chemical shift of water protons. For PrMOE-DO3A, the coefficient of methoxy (CH_3O-) protons is -0.13 ppm/°C, which is more than 10 times higher than that of water. For $TmDOTP^{5-}$, the most pronounced and narrowest proton signal, called H_6—one of the side chain methylene protons at around -140 ppm—exhibited temperature sensitivity of 0.83 ppm/°C in the abdominal part of rats in vivo and 0.89 ppm/°C in 20 mM $TmDOTP^{5-}$ solution [86]. These coefficients are more than 80 times higher than that of water. This complex also had temperature dependence in its ^{31}P signal, and the coefficient was 2.18 ppm/°C in vitro. Because this complex also has pH sensitivity in its proton chemical shifts, we can determine both temperature and pH simultaneously using two different proton signals, such as H_6, as mentioned above, and H_2 at around +80 ppm [87, 88].

Although these lanthanide complexes are capable of measuring absolute temperature in vivo with markedly higher sensitivity than water, their toxicity makes their application to humans difficult. A technique should be developed to carry out in vivo measurements under a dosage of 1–2 mmol/kg to translate this approach into clinical use [85].

35.4.2 The CEST Agent

Recent studies have revealed that agents for PARACEST (paramagnetic chemical exchange saturation transfer) can be used to monitor temperature in addition to pH. In the CEST spectrum, or z-spectrum, of a lanthanide complex agent with Eu^{3+} (eg, Eu^{3+}-DOTAM-Gly (Glycine)-Phenylalanine (Phe)), the frequency of the CEST peak depends linearly on temperature in the physiological pH range (6–8 pH) [89]. The temperature coefficient of the frequency change is as large as −0.302 ppm/°C (or −3.31 °C/ppm). Because the proton exchange rate between bound water and the complex and free bulk water depends on temperature, and the effect of the exchange appears asymmetrically on the bound water peak, the frequency of the bound water peak gets closer (downward) to the bulk water peak, with a nonlinear change in its amplitude. Since this peak shift rate does not depend on pH, this technique is applicable to temperature change monitoring. This temperature sensitivity was good for imaging temperature in a 10 mM agent solution. Other PARACEST agents with Tm^{+3}, such as Tm^{3+}-DOTAM-Gly-Lys (Lysine) have a similar temperature dependence to Eu^{3+} in its z-spectrum frequency of the bound water peak, in addition to its spectrum bandwidth dependency on pH [90]. Once we know the pH from the bandwidth of the bound water peak, we can obtain the absolute temperature with the equations below [90] for:

$$CS = 0.268T + b, \tag{35.25}$$

$$b = 0.3738pH^2 - 4.782pH - 40.679, \tag{35.26}$$

in the pH range of 6–8 and temperature range of 35–39 °C. This technique determines both the temperature and pH of tissue simultaneously. McVicar [90] demonstrated temperature maps of mice in vivo. To our best knowledge, however, there has been no report on the endogenous DIACEST (diamagnetic CEST) temperature application to date.

35.4.3 Non-proton Nuclei

A few nuclei other than those from protons, such as ^{13}C [91], ^{31}P, [92, 93] or ^{58}Co [94], demonstrate feasibility for thermometry, but it is difficult to find a specific reason to use them, because they are much less abundant than ^{1}H. The other possibility is to use hyperpolarized 129Xe [95], whose temperature dependence of the chemical shift in a cryptophane-A cage was reported to be 0.29 ppm/°C.

Although the temperature coefficient is more than 10 times larger than that of water protons, and this phenomenon is interesting as basic science, the laborious hyperpolarization process makes its clinical translation difficult.

35.5 Absolute Temperature Measurement

The ultimate goal of MR thermometry is to monitor the absolute temperature in the human body. It should be noted that, although the internal reference technique makes image subtraction unnecessary and yields temperature change images, this does not mean that the technique enables direct and accurate temperature mapping in biological tissues. The absolute value of the chemical shift difference between water and the reference may vary with tissue location and inherent conditions because of the dependence of the water proton resonance frequency on other parameters such as pH, magnetic ion concentration, Blood Oxygenation Level Dependent (BOLD)-effect, and so on, as mentioned above. The feasibility of absolute temperature imaging has not yet been established, and further investigation is required [96–98].

35.6 Conclusion

Noninvasive thermometry techniques were discussed briefly. At this moment, MRI techniques using water protons' chemical shift is the only practical modality, and actually used in clinical thermal treatments. The features may be summarized as follows: First, the temperature dependence is simply linear, with a high correlation in a temperature range of biological systems; Second, the temperature coefficient is mostly the same for different tissues and is around -0.01 ppm/°C, which is close to that for pure water. The slight variations of the temperature coefficient may be attributed to the susceptibility effect and the presence of ions; and Third, the water proton chemical shift allows measurement of absolute temperature when an appropriate internal reference is used in sample solutions. In biological tissues, however, the absolute value of the shift depends on pH, susceptibility, and the presence of ions and macromolecules. Thus, temperature errors in the order of a few degrees are conjectured. Moreover, MRI is not very handy and thus not applicable for bedside use. Efforts to develop alternative noninvasive thermometry techniques based on ultrasound and other less-sumptuous techniques should be continued.

Acknowledgments The author thanks Dr. Daisuke Kokuryo and Prof. Etsuko Kumamoto, Kobe University, and all other colleagues for their collaborations.

References

1. Fani F, Schena E, Saccomandi P, Silvestri S. Int J Hyperth. 2014;30:219.
2. Bakhsheshi M, Diop M, Lawrence KS, Lee T. J Biomed Opt. 2014;19:057005.
3. Pramanik M, Wang L. J Biomed Opt. 2009;14:054024.
4. Rodrigues D, Maccarini P, Salahi S, Oliveira T, Pereira P, Limao-Vieira P, Snow B, Reudink D, Stauffer P. IEEE Trans Biomed Eng. 2014;61:2154.
5. Conway J, Hawley M, Mangnall Y, Amasha H, van Rhoon G. Clin Phys Physiol Meas. 1992;13:185.
6. Maraghechia B, Koliosa M, Tavakkolia J. Int J Hyperth. 2015;31:666.
7. Lüdemann L, Wlodarczyk W, Nadobny J, Weihrauch M, Gellermann J, Wust P. Int J Hyperth. 2010;26:273.
8. Kamimura Y. Nagoya University, PhD Thesis. 1985.
9. Chen J, Daniel B, Pauly K. J Magn Reson Imaging. 2006;23:430.
10. Lewa C, Baczkowski A. Bull Cancer. 1977;64:37.
11. Bottomley P, Foster T, Argersinger R, Pfeifer L. Med Phys. 1984;11:425.
12. Parker D. IEEE Trans Biomed Eng. 1984;BME-31:161.
13. Dickinson R, Hall A, Hind A, Young I. J Comput Assist Tomogr. 1986;10:468.
14. Nelson T, Tung S. Magn Reson Imaging. 1987;5:189.
15. Kuroda K, Tsutsumi S, Saitoh A. Trans IEICE. 1990;J73-A:1431.
16. Hall A, Prior M, Hand J, Young I, Dickinson R. J Comput Assist Tomogr. 1990;14:430.
17. Le Bihan D, Delannoy J, Levin R. Radiology. 1989;171:853.
18. Kuroda K. Int J Hyperth. 2005;21:547.
19. Rieke V, Butts P. J Magn Reson Imaging. 2008;27:376.
20. Bloembergen N, Purcell EM, Pound RV. Phys Rev. 1948;73:679.
21. Tanaka H, Kato H, Ishida T, Eno K. Nippon Acta Radiol: J Jpn Radiol Soc. 1981;41:897.
22. Kato H, Kano E, Sugahara T, Ujeno Y, Nishida T, Ishida T. International conference cancer therapy by hyperthermia. 1983;33.
23. Kuroda K, Tsutsumi S, Saito K. Trans IEICE (in Japanese). 1990;J73-A:1431.
24. Simpson J, Carr H. Phys Rev. 1958;111:1201.
25. Parker DL, Smith V, Sheldon P, Crooks LE Fussel L. Med Phys. 1983;11:321.
26. Kuroda K, Miki Y, Tsutsumi S. JJME (in Japanese). 1990;28:253.
27. Finch ED, Homer LD. Biophys J. 1974;14:907.
28. Ling CR, Poster MA, Hutchison JMS. Phys Med Biol. 1980;25:748.
29. Lewa CzJ, Majewska. J. Bull Cancer. 1980;67:525.
30. Zimmerman J, Brittin W. J Phys Chem. 1957;61:1328.
31. Kuroda K, Iwabuchi T, Obara M, Honda M, Saito K, Imai Y. Magn Reson Med Sci. 2011;10:177.
32. Farrar T, Becker E. Pulse and fourier transform NMR: introduction to theory and methods. 1st ed. New York: Elsevier; 1971.
33. Baron P, Deckers R, Knuttel FM, Bartels LW. NMR Biomed. 2015;28:1463.
34. Eriksson M, Parmala M, Rytioja M, Tanttu J, Köhler M. J Magn Reson Imaging. 2015. doi:10.1002/jmri.25064.
35. Butts K, Sinclair J, Daniel BL, Wansapura J, Pauly JM. JMRI. 2001;13:99.
36. Yamada K, Sakai K, Akazawa K, Yuen S, Sugimoto N, Sasajima H, Mineura K, Nishimura T. Neuroreport. 2010;21:851.
37. Sakai K, Yamada K, Mori S, Sugimoto N, Nishimura T. NMR Biomed. 2011;24:1063.
38. Mills R. J Phy Chem. 1973;77:685.
39. Arnold J, Packard M. J Chem Phys. 1951;19:1608.
40. Liddel U, Ramsey N. J Chem Phys. 1951;19:1608.
41. Hindman J. J Chem Phys. 1966;44:4582.
42. Muller N, Reiter R. J Chem Phys. 1965;42:3265.
43. Van Geet A. Anal Chem. 1968;40:2227.

44. Van Geet A. Anal Chem. 1970;42:679.
45. Kuroda K, Tsutsumi S. Chapter C: NMR tomography II: Temperature imaging by proton chemical shift. In: Miyakawa BJ, Boca M, editors. Non-invasive thermometry of the human body. Raton: CRC Press Inc.; 1996.
46. Poorter J. Magn Reson Med. 1995;34:359.
47. Ishihara Y, Calderon A, Watanabe H, Okamoto K, Suzuki Y, Kuroda K, Suzuki Y. Magn Reson Med. 1995;34:814.
48. Arus C, Chang Y, Baramy M. J Magn Reson. 1985;63:376.
49. Lutz K, Kuesel A, Hull W. Magn Reson Med. 1993;29:113.
50. Kuroda K, Abe K, Tsutsumi S, Ishihara Y, Suzuki Y, Satoh K. Biomed Thermol. 1994;13:43.
51. Kuroda K, Takei N, Mulkern R, Oshio K, Nakai T, Okada T, Matsumura A, Yanaka K, Hynynen K, Jolesz F. Magn Reson Med Sci. 2003;2:17.
52. Cady E, D'souga P, Penrice J, Ann L. Magn Reson Med. 1995;33:862.
53. Cady E, Penrice J, Robertson N. NMR Biomed. 2011;24:865.
54. Corbett R, Laptook A, Weatherall P. J Cereb Blood Flow Metab. 1997;17:363.
55. Corbett R, Laptook A, Tollefsbol G, Kim B. Neurochem J. 1995;64:1224.
56. Corbett R, Purdy P, Laptook A, Chaney C, Garcia D. Am J Neuroradiol. 1999;20:1851.
57. Zhu M, Bashir A, Ackerman J, Yablonskiy D. Magn Reson Med. 2008;60:536.
58. Weis J, Covaciu L, Rubertsson S, Allers M, Lunderquist A, Ortiz-Nieto F, Ahlstrom H. Magn Reson Imaging. 2009;30:1505.
59. Kuroda K, Itoh Y, Chung A, Hynynen K, Jolesz F. 4th Meeting ISMRM. 1996;1:43.
60. Rieke V, Vigen K, Sommer G, Daniel B, Pauly J, Butts K. Magn Reson Med. 2004;51:1223.
61. Poorter J, Wagter C, Deene D, Thomsen C, Stahlberg F, Achten E. Magn Reson Med. 1995;33:74.
62. Hynynen K, Freund W, Cline H, Chung A, Watkins R, Vetro J, Jolesz F. Radiographics. 1996;16:185.
63. Moriaty J, Chen J, Purcell C, Ang L, Hinks R, Peter R, Henkelman R, Plewes D, Bronskill M, Kucharczyk W. J Magn Reson Imaging. 1998;8:128.
64. Kettenbach J, Silverman S, Hata N, Kuroda K, Saiviroonporn P, Zientara G, Morrison P, Hushek S, Black P, Kikinis R, Jolesz F. J Magn Reson Imaging. 1998;8:933.
65. Kahn T, Harth T, Kiwit J, Schawarzmaier H-J, Wald C, Mödder U. J Magn Reson Imaging. 1998;8:160.
66. Cline H, Hynynen K, Schneider E, Hardy C, Maier S, Watkins R, Jolesz F. Magn Reson Med. 1996;35:309.
67. Peters R, Hinks R, Henkelman R. Magn Reson Med. 1998;40:454.
68. Young I, Hajnal J, Roberts I, Ling J, Hill-Cottingham R, Oatridge A, Wilson J. Magn Reson Med. 1996;36:366.
69. Vigen K, Daniel B, Pauly J, Butts K. Magn Reson Med. 2003;50:1003.
70. de Senneville BD, Ries M, Maclair G, Moonen C. IEEE Trans Med Imaging. 2011;30:1987.
71. Kuroda K, Oshio K, Chung A, Hynynen K, Jolesz F. Magn Reson Med. 1997;38:845.
72. de Zwart JA, Vimeux FC, Delalande C, Canioni P, Moonen CT. Magn Reson Med. 1999;42:53.
73. Weidensteiner C, Kerioui N, Quesson B, de Senneville BD, Trillaud H, Moonen CT. J Magn Reson Imaging. 2004;19:438.
74. Vogel M, Moelker A, Ouhlous M, Pattynama P. 11th ISMRM meeting. 2002;2214.
75. Kuroda K, Kokuryo D, Kumamoto E, Suzuki K, Matsuoka Y, Keserci B. Magn Reson Med. 2006;56:835.
76. Farrant R, Lindon J, Nicholson J. NMR Biomed. 1994;7:243.
77. Hall L, Talagala S. J Magn Reson. 1985;65:501.
78. Kuroda K, Suzuki Y, Ishihara Y, Okamoto K, Suzuk Y. Magn Reson Med. 1996;35:20.
79. Mulkern R, Chung A, Jolesz F, Hynynen K. Med Phys. 1997;24:1899.
80. Kuroda K, Oshio K, Mulkern R, Panych L, Nakai T, Moriya T, Okuda S, Hynynen K, Jolesz F. Magn Reson Med. 2000;43:220.

81. Posse S, DeCarli C, LeBihan D. Radiology. 1994;192:733.
82. Oshio K, Kyriakos W, Mulkern R. Magn Reson Med. 2000;44:521.
83. McDannold N, Hynynen K, Oshio K, Mulkern R. Med Phys. 2001;28:346.
84. Mulkern R, Panych L, McDannold N, Jolesz F, Hynynen K. J Magn Reson Imaging. 1998;8:493.
85. Hekmatyar S, Poptani H, Babsky A, Leeper D, Bansal N. Int J Hyperth. 2002;18:165.
86. Zuo C, Bowers J, Metz K, Nosaka T, Sherry A, Clouse M. Magn Reson Med. 1996;36:955.
87. Zuo C, Metz K, Sun Y, Sherry A. J Magn Reson. 1998;133:53.
88. Sun Y, Sugawara M, Mulkern R, Hynynen K, Mochizuki S, Albert M, Zuo C. NMR Biomed. 2000;13:460.
89. Li A, Wojciechowski F, Suchy M, Jones C, Hudson R, Menon R, Bartha R. Magn Reson Med. 2008;59:374.
90. McVicar N, Li A, Suchy M, Hudson R, Menon R, Bartha R. Magn Reson Med. 2013;70:1016.
91. Bemsk G, Leon V, Manzo F. Acta Cient Venez. 1980;31:125.
92. Temussi P, Guidoni L, Ramoni C, Podo F. Physiol Chem Phys. 1979;11:445.
93. Gorenstein D, Findlay J, Momii R, Luxon B, Kar D. Biochemistry. 1976;15:3796.
94. Smith N, Webb A, Ellis D, Wilmes L, Obrien W. IEEE Trans Ultrason Ferroelectr Freq Control. 1995;42:489.
95. Schilling F, Schröder L, Palaniappan K, Zapf S, Wemmer D, Pines A. Chem Phys Chem. 2010;11:3529.
96. Jenista E, Branca R, Warren W. Int J Hyperth. 2010;26:725.
97. Hentschel M, Findeisen M, Schmidt W, Frenzel T, Wlodarczyk W, Wust P, Felix R. MAGMA. 2000;10:52.
98. Sprinkhuizen S, Konings M, van der Bom M, Viergever M, Bakker C, Bartels L. Magn Reson Med. 2010;64:360.

Chapter 36
Development of Heating Device

Masahiro Kuroda, Kazuyuki Saito, Koichi Ito, and Shin-ichiro Umemura

Abstract Hyperthermia has a long history as a treatment modality for tumors, with the rapid development of heating devices for hyperthermia starting more than a century ago. Ideal heating devices would enable the targeted area to be heated in accordance with the depth and width of the tumor. They would also be integrated with image-guided, four-dimensional, real-time temperature monitoring in the body and a simulation function for the prediction of temperature fluctuations according to treatment and changes in blood flow in order to heat tumors selectively and avoid damage to healthy tissue. However, economic viability is also an important factor, in that the ideal heating device must be able to be covered under national medical health insurance systems if it hopes to gain widespread use in individual medical economies. The development of a sophisticated yet cost-effective device with automatic control functions that is safe for patients and easy for medical staff to use would be expected to contribute to more effective treatment, a reduction in labor costs, and savings to national health care systems.

Keywords Heating technology • Heating device • Selective heating • Electromagnetic heating • RF capacitive heating • Microwave heating • HIFU • Focused transducer • Transient thermal response

---

M. Kuroda (✉)
Radiological Technology, Graduate School of Health Sciences, Okayama University, 2-5-1 Shikata-cho, Kita-ku, Okayama 700-8558, Japan
e-mail: kurodamd@cc.okayama-u.ac.jp

K. Saito • K. Ito
Center for Frontier Medical Engineering, Chiba University, Chiba, Japan

S.-i. Umemura
Graduate School of Biomedical Engineering, Tohoku University, Sendai, Japan

S. Kokura et al. (eds.), *Hyperthermic Oncology from Bench to Bedside*,
DOI 10.1007/978-981-10-0719-4_36

36.1 Overview of Heating Technology

36.1.1 History of Heating Technology

The first known use of hyperthermia, a procedure that increases the temperature of tumor-loaded tissue to 40–43 °C, was to treat breast cancer more than 3500 years ago. Around 400 BC, Hippocrates, in reference to hyperthermia, stated the following: “Those who cannot be cured by medicine can be cured by surgery. Those who cannot be cured by surgery can be cured by fire (hyperthermia). Those who cannot be cured by fire, they are indeed incurable.” [1]. In the era of Hippocrates, breast cancer lesions were treated by fire or hot pokers. Fast forward to 1866, when the disappearance of a facial sarcoma after high fever due to infection of erysipelas was reported by Busch [1, 2]. Soon after 1910, Coley reported treating tumors using fever induced by mixed bacterial toxin, which would later be known as “Coley’s toxin” [3]. Based on the subsequent development of biological research on hyperthermia, which clarified its efficacy and antitumor mechanism, the development of hyperthermia equipment for clinical use advanced rapidly starting sometime around 1900 [4]. However, in order to create a device capable of selectively heating a tumor to around 43 °C without harming other tissues in the body, a high level of design and engineering skill is necessary.

There are three major obstacles that must be overcome before such a device can be realized. The first is the presence of plentiful vessels in normal tissue. Blood flow helps cool elevated temperatures in tumors and normal tissue back down to body temperature. Local heating induces a physiological response to dilate blood vessels and increase blood flow, which cools the tissue by carrying heat away. When the skin surface is heated with infrared radiation or during bathing, cooling due to an increase in blood flow limits the penetration depth of thermal conduction and constitutes barriers for deep-seated tumors to be heated to the target temperature. To date, a number of advanced heating devices that overcome these traditional limitations by utilizing modalities such as electromagnetic waves, ultrasound, and lasers have been developed and used in the clinical setting to treat various types of tumors [5–7].

The second obstacle is the specific limitation of each modality. Regarding devices using electromagnetic waves, the penetration depth of electromagnetic waves into the body is inversely related to the localization of the heating area [5]. In other words, when a deep-seated tumor is heated using electromagnetic waves, in principle, it is difficult to limit the heated area to the tumor site alone. Regarding devices using ultrasound and lasers, the width of the heating area is limited, and thus the movement of the tumor due to factors such as respiration during heating becomes problematic for the uniform heating of large tumors.

The third obstacle is the difficulty of four-dimensional (4D) real-time temperature measurement in the body during heating, which is an essential technology for the development of improved heating devices. By the 1990s, temperature measurement was limited to the use of temperature probes such as thermistors and optical fibers at adequate insertion points in the tumor. Thereafter, image-guided non-invasive 3D temperature measurement using magnetic resonance imaging was developed, which marked a great improvement and was closer to an ideal technology, as described in detail in the former chapter. At present, some commercially available heating devices are equipped with image-guided thermometers. The use of these devices has provided useful data regarding temperature distribution during heating, and led to a greater understanding into what is necessary for improving their heating properties. From the late 1990s, technology that allows the simulation of specific energy absorption rates and temperature distribution in the body during heating has also been developed, and some commercially available heating devices have already been retrofitted with these simulation systems. Although simulation technology should be developed further to predict 4D temperature distribution, including an exact calculation of changes in blood flow both inside and outside of tumors during heating, it is already possible in clinical practice to select patients before treatment whose tumor can be heated uniformly and to predict temperature distribution in the body and make corresponding adjustments during heating.

Over the past two decades, heating devices using radio waves, microwaves, radio-frequency waves, ultrasound (such as high-intensity focused ultrasound [HIFU]) and laser, have become commercially available and enjoyed widespread clinical use, as described in detail later in Sect. 36.2 and 36.3. Other heating devices expected to be available for clinical use in the near future include those using technology such as inductive heating, heating using magnetic materials, and interstitial microwave heating. Several other promising types of devices such as cylindrical cavity resonators and nanoparticles excited by laser for photothermal therapy are currently under development.

In relation to the difficulty in developing heating technology for hyperthermia, Hall stated "the biology is with us, the physics is against us." [8]. Although the development of such technology remains difficult, cutting-edge heating devices that truly allow the selective heating of tumors alone are getting closer to the ideal and can be realistically expected in the near future.

36.1.2 Ideal Heating System from Clinical View

Although the basic biological research for hyperthermia to date consistently supports its clinical utility, unfortunately, hyperthermia has yet to be universally accepted as a standard treatment for tumors in many countries. To promote the widespread use of heating devices for hyperthermia in the clinical setting, two issues need to be addressed. First, a device that can heat large tumors selectively is necessary. Second, to gain widespread use in individual medical economies, such a device must be economically viable and able to be covered under national medical health insurance systems.

Ideally, devices should have the capability to deliver heat selectively at a temperature adequate for the depth and width of tumors in the body using 4D

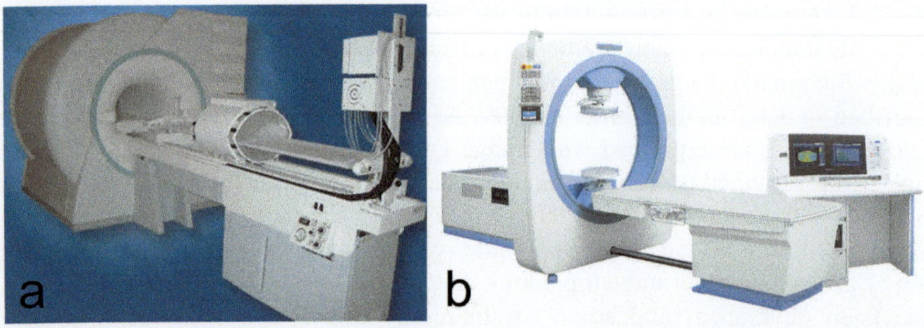

Fig. 36.1 Currently available commercial heating devices (**a**) the BSD-2000 3D and 3D/MRI system (image taken from http://www.medicalexpo.com/prod/bsd-medical/product-84451-540429.html) [10]. (**b**) the Thermotron RF8 EX Edition (image taken from http://www.vinita.co.jp/medical_div/medical/product/) [11]

real-time temperature measurement both after the start of heating and during treatment. Simulating the specific energy absorption rate and temperature distribution during heating is of clinical importance and necessary in order to make positioning adjustments to the target area in the shortest amount of time. Finally, to promote the widespread use of such devices in clinical practice, an integrated function that automatically repeats each step of heating—including 3D temperature analysis and heat-control simulation until the ideal temperature distribution is achieved—is indispensable. Although the development of new heating devices around the world is ongoing, an ideal device has yet to be realized. Some 3D-temperature-image-guided heating devices (Fig. 36.1a) have recently become commercially available, and as a result, have stimulated clinical research. Several clinical trials on hyperthermia using these devices are currently underway in the US and EU. As previously mentioned, the economic viability of these new devices will also be paramount.

One of the reasons that hyperthermia has not been universally accepted as a standard clinical treatment is its temporal inefficiency; the number of patients treated with hyperthermia is relatively low compared to the number treated with surgery and radiotherapy. Furthermore, each hyperthermia session takes at least 1 h, and is usually repeated several times for each patient. This temporal inefficiency results in the need for a larger number of medical staff and is less cost-effective in terms of both labor and technology compared with surgery and radiotherapy for the same number of patients. Newly developed commercially available 3D-temperature-image-guided heating devices are rather expensive, and in many countries, the revenue generated from hyperthermia treatment for healthcare services provided by national insurance systems does not cover its expenses. This economic efficiency is a crucial factor in terms of the widespread use of hyperthermia in the future. Therefore, the development of sophisticated yet cost- and labor-effective devices with automatic control functions that are safe for patients and easy for medical staff to use is expected.

36.1.3 Present Status of Heating Technology

The types of heating treatments for tumors currently being used in clinical practice are summarized in Table 36.1.

Thermal treatment that utilizes heat higher than body temperature includes hyperthermia, which is typically conducted in a range between 40 and 43 °C, and ablation, which is typically conducted at temperatures over 50 °C. Types of ablation that use an interstitial applicator, such as microwave and radiofrequency ablation, have recently become popular in clinical practice, as have other more specific types of ablation, including laser ablation (interstitial laser therapy), ultrasound ablation (HIFU), and magnetic resonance-guided focused ultrasound.

Hyperthermia can be divided into the following three categories based on the width of the heating area (from small to large): local, regional/part-body, and whole-body hyperthermia, respectively [6, 7]. To heat the tumor selectively, the clinical treatment is selected according to the size and depth of the tumor inside of the body.

Local hyperthermia uses a wide variety of applicators (interstitial, intraluminal, and intracavital) and electrodes for superficial tumors that transmit microwaves typically between 433 and 2450 MHz, radio waves between 100 kHz to 150 MHz, and lasers. Oncothermia is a type of capacitive heating that uses specially arranged radio waves for impedance instead of plane-wave coupling.

Regional/part-body hyperthermia typically uses two types of heating devices. In the EU and US, the use of a heating device (Fig. 36.1a) with applicators consisting of coherent arrays of dipole antenna pairs positioned in a ring pattern to adjust the heating area in the body by regulating the phase and amplitude of electromagnetic waves from 70 to 150 MHz has been rapidly increasing. In Japan and some other Asian countries, whose populations tend to have thinner layers of subcutaneous fat, capacitive heating devices (Fig. 36.1b) using 8 MHz radio waves have been predominate [5, 9]. Other special heating techniques being utilized include regional thermal perfusion using extracorporeal heat-exchange passing between a supplying artery and a draining vein or hot water blankets, and continuous hyperthermic peritoneal perfusion/intraperitoneal chemotherapy, which is performed intra-operatively for patients such as those with peritoneal carcinomatosis or mesothelioma.

Whole-body hyperthermia utilizes thermal chambers, hot water blankets, and infrared radiators to treat carcinoma patients with distant metastases.

In contrast, some thermal treatments use cooling below body temperature. One example is cryotherapy, in which tumors are frozen using cryogens such as liquid nitrogen or argon gas.

Table 36.1 Types of thermal treatments for tumors currently used in clinical practice

Type of treatment	Temperature range (°C)	Equipment/applicators	Energy source	Tumor sites or tumor	Tumor size
Ablation	>50 °C	Interstitial applicator	Microwaves, radio waves, or lasers	Lung, liver, breast, kidney, bone, pancreas, bile duct, etc.	Commonly, <3 cm
		HIFU or MRgFUS	Ultrasound	Prostate, breast, bone, brain, liver, pancreas, rectum, kidney, etc.	–
Local hyperthermia	40–48 °C (Typically 40–43 °C)	Superficial applicator (waveguide, horn, spiral, current sheet, compact applicators, etc.) or capacitive heating electrodes	Typically microwaves, radio waves, or lasers	Superficial tumor (skin, head and neck, breast, etc.)	Commonly, <6 cm
		Interstitial, intraluminal, and intracavital applicator (antenna, electrode, ferromagnetic implants, resistive wire implants, etc.)		Breast, brain, prostate, rectum, vagina, esophagus, etc.	
		Magnetic nanoparticles such as iron oxide	Magnetic field	Brain, prostate, uterus, etc.	
Regional/part-body hyperthermia	Typically 41–42 °C	Applicators consist of coherent arrays of dipole antenna pairs positioned in a ring pattern, capacitive heating electrodes, inductive heating applicators, etc.	Microwaves or radio waves	Prostate, bladder, colon, rectum, uterus, ovary, soft tissue, lung, pancreas, etc.	–
	39–40 °C	Regional thermal perfusion using extracorporeal heat-exchange passing between a supplying artery and a draining vein or hot water blankets	Warming perfusion or warming blanket	Soft tissue, melanoma, liver, lung, etc.	–
	41.0–42.5 °C	CHPP, HIPEC	Warmed fluid	Peritoneal carcinomatosis of stomach, colon, rectum, ovary, unknown origin or mesothelioma, etc.	–

Whole-body hyperthermia (WBH)	≤42.0 °C (extreme WBH), 39.5–41.0 °C (moderate WBH)	Thermal chambers, hot water blankets, infrared radiators	Warmed fluid, infrared	Carcinomas with distant metastases	–
Cryotherapy	< −40–0 °C	Cryoprobe	Liquid nitrogen or argon gas	Prostate, liver, retina, skin, uterus, kidney, breast, colon etc.	–

HIFU high-intensity focused ultrasound, *MRgFUS* magnetic resonance-guided focused ultrasound, *WBH* whole-body hyperthermia, *CHPP* continuous hyperthermic peritoneal perfusion, *HIPEC* hyperthermic intraperitoneal chemotherapy

36.2 Electromagnetic Heating

36.2.1 Heating by Electromagnetic Energy

Although several energy sources can be considered for hyperthermia heating systems, at this moment, electromagnetic energy devices are mainly used. However, characteristics of electromagnetic wave strongly depend on their frequency. Therefore, performance of typical heating devices is explained according to their operating frequency. Moreover, some new approaches are also introduced.

36.2.2 Classification by Frequency

36.2.2.1 RF Capacitive Heating

Today, large numbers of capacitive heating devices, which utilize electric current of 8 MHz, are employed in Japan. Figure 36.2 illustrates main unit of the device. This device was approved as a medical device of thermal treatment for cancer by the Japanese Ministry of Health, Labour and Welfare. The main part of this device consists of two electrodes with water boluses, oscillation unit, and controller. The maximum output power is 1.5 kW [12]. Electric current of 8 MHz is applied to a patient via pair of electrodes. Temperature of target tumor(s) inside the patient and their periphery increased by Joule heating of the current. In this scheme, the maximum temperature occurred at the surface of the patient body where the electrodes contact with bolus. Therefore, large output power and strong body surface cooling by low temperature water bolus are utilized to heat the deep region of the body. Moreover, the heating region can be adjusted based on the arrangement of several sizes of electrodes (Fig. 36.3). The temperature in and around the target can be measured by some thermocouples equipped with the device. However, the temperatures are not measured in many cases because of invasiveness.

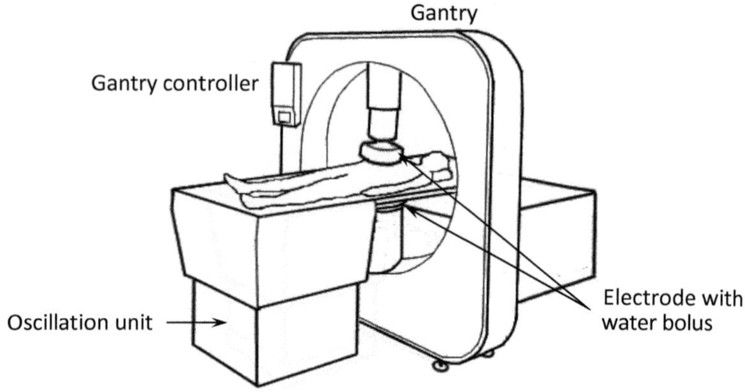

Fig. 36.2 Conventional capacitive heating device

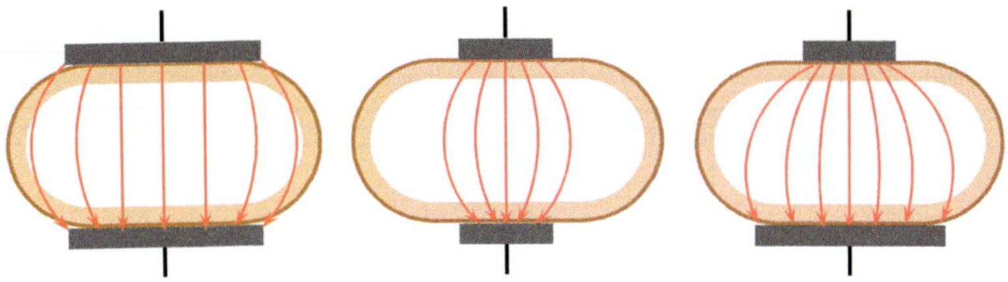

Fig. 36.3 Heating regions by different size of electrodes

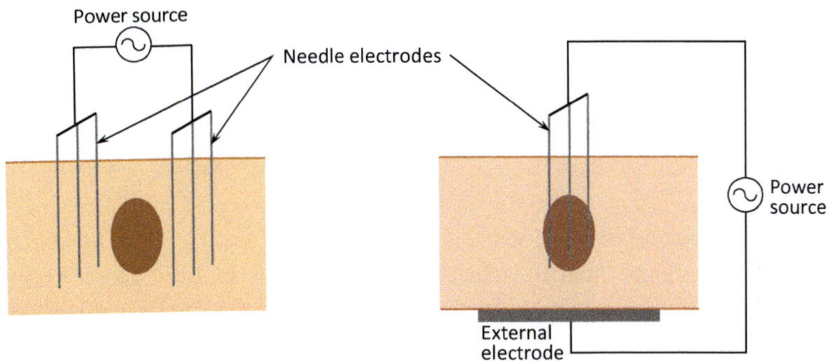

Fig. 36.4 Interstitial capacitive heating

In the capacitive heating, there are some other schemes. Figure 36.4 shows the capacitive heating using interstitial technique. Subsequently, longitudinal heating region can be adjusted easily by changing the length of the exposed metallic part of needle electrodes. However, it is difficult to cover the larger heating region in perpendicular to the needle electrode. Moreover, tissue coagulation system based on the interstitial capacitive heating is used for the treatment of hepatocellular carcinoma [13]. In addition, intracavitary capacitive heating system can be realized using the same scheme which was sold in Japan [14].

36.2.2.2 Microwave Heating

Microwave energy is one of the heating sources used for localized hyperthermia. Electromagnetic waves are classified by their frequencies. The frequencies which include in microwave range from several hundred to several thousand MHz (i.e. up to several GHz). For hyperthermia treatments, heating devices using 430 MHz, 915 MHz, and 2.45 GHz have been developed.

There are some characteristics which are unique in microwave heating. One of the most unique characteristics of microwave heating compared to RF capacitive heating is the number of antennas (or electrodes in the case of RF capacitive heating) needed for a treatment. As shown in Fig. 36.2, for RF heating, the target tumor is heated by a “current” running from one electrode to the other. Therefore,

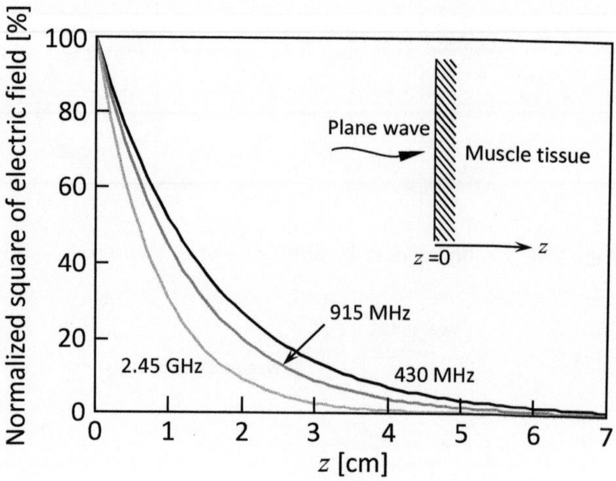

Fig. 36.5 Attenuation of electric fields in muscle tissue

more than two electrodes are needed for this type of heating [15]. However, with microwave heating, the heating energy is propagated in biological tissue as an "electromagnetic wave" generated from the antenna, so only a single antenna can heat the target tumor.

The characteristics of heating devices strongly depend on the frequency even in the microwave frequency band. Figure 36.5 shows the attenuation of electric fields in various frequencies, which are commonly used for hyperthermia therapy. In Fig. 36.5, a planar wave with various frequencies is incident on the surface of a half-infinite muscle tissue. The penetration depth of this microwave energy is only several cm, and decreases with the radiation frequency. Therefore, there are difficulties in heating a deep-seated tumor using microwave radiation without any artifices.

The size and the shape of target tumors also vary. Therefore, it is impossible to treat many different types of tumors with a single type of antenna. Figure 36.6 shows several types of microwave antennas for hyperthermia treatment. The "Radiator type" and the "Contact type" are classified into "External antennas," and the "Intracavitary type" and the "Interstitial type" are called "Internal antennas". Here, interstitial microwave heating is introduced.

Interstitial microwave hyperthermia can be used with a localized tumor. In this method, thin microwave antennas are inserted into the tumor radiating the microwave energy directly into the target [16]. This method allows effective heating of the tumor. Figure 36.7 shows a coaxial-slot antenna [17] which is one type of thin microwave antenna used for interstitial heating. Although this technique is an invasive technique, a large volume of heated region can be generated using a small number of antennas by choosing an appropriate array structure.

Figure 36.8 shows a photograph during the treatment. In this treatment, three coaxial-slot antennas were inserted for heating. Moreover, three fiber optic thermosensors were also utilized for temperature measurement. In this case, enough temperature rising around the target tumor was obtained.

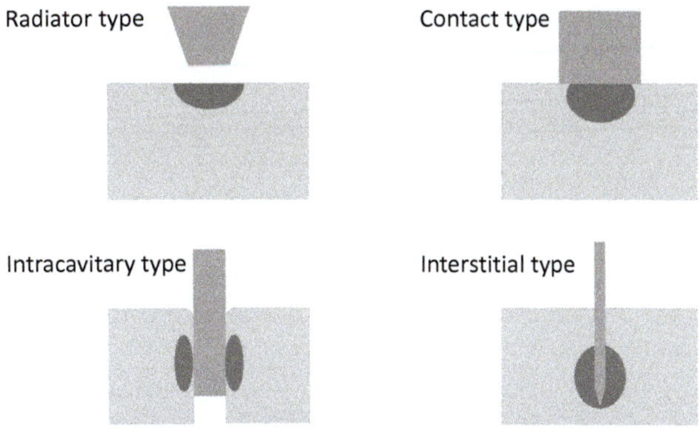

Fig. 36.6 Various types of antennas used for microwave heating

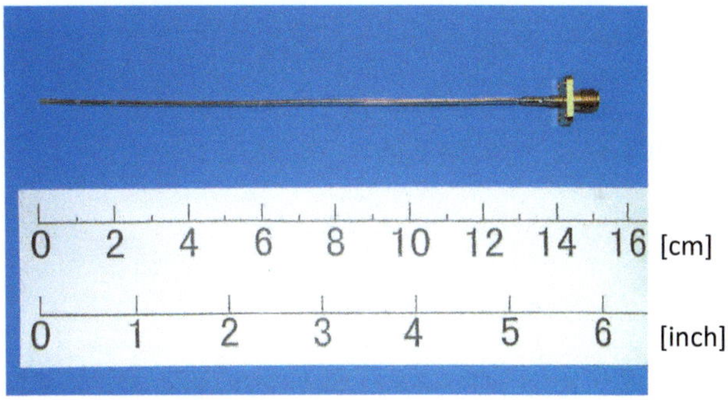

Fig. 36.7 Coaxial-slot antenna for interstitial microwave hyperthermia

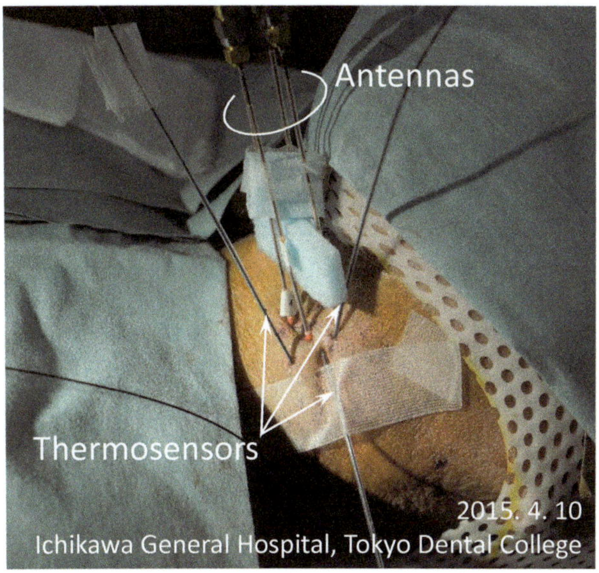

Fig. 36.8 Photograph during the treatment

36.2.2.3 New Topics

Treatment planning is also important in the hyperthermia. For example, the radiation dose can be calculated in short period for the radiation oncology. However, it is difficult to realize such planning for hyperthermia because of high computational load of electromagnetics and temperature analysis. Even so, the treatment planning for the hyperthermia is tried based on the high speed computing system. In [18], temperature distributions around a neck by microwave antenna array are calculated for the treatment planning. It is desirable that this kind of system can be employed in other parts.

Another type of capacitive heating system is developed in [19]. The device outputs modulated high frequency current, whose dominant frequency is 13.56 MHz, and the maximum output power is 150 W. The output power is smaller than the conventional capacitive heating device. Therefore, it is considered that the treatment effect is based on the "mild hyperthermia", because the target tumor cannot be heated until the hyperthermic temperature. However, detailed biological mechanisms of the treatment are still not clear. Therefore, researches on the clinical mechanisms are desirable.

36.3 Ultrasound Heating

36.3.1 Devices for High Intensity Focused Ultrasound Treatment

Therapeutic ultrasound treatment can be categorized into those employing either focused or unfocused ultrasound. Focused ultrasound mechanical treatment, such as lithotripsy and histotripsy, uses pulsed ultrasound with a wide relative bandwidth close to that for ultrasonic imaging. The reason is that cavitation can be induced in an uncontrolled manner and unwanted adverse effects might be caused if a long burst ultrasound is reflected by the bubbles created by the earlier part of the burst. Unlike these applications, however, CW or long burst ultrasound is used for thermal ultrasonic treatment. Such treatment inducing thermal coagulation of target tissue by high intensity focused ultrasound is widely known as HIFU treatment.

HIFU treatment was first introduced by Fry's group in 1950s [20] but its clinical application was very limited. In 1990s, its clinical application to transrectal treatment of prostate, as shown in Fig. 36.9, started, and is gradually gaining momentum [21]. A typical focal field of an ultrasonic transducer for such application is shown in Fig. 36.10. The width and length of the focal spot less than 1 and 5 mm, respectively, are achieved with a F number of 1 at an ultrasonic frequency of 3 MHz. The tissue temperature elevation, which should be achieved by this HIFU transducer, was numerically simulated and is shown in Fig. 36.11.

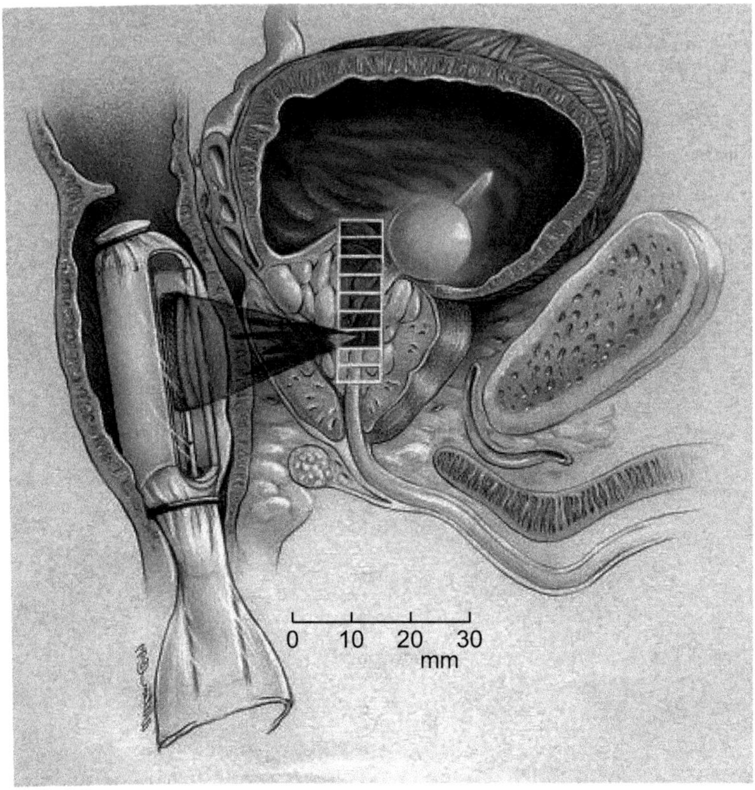

Fig. 36.9 Transrectal HIFU treatment of prostate

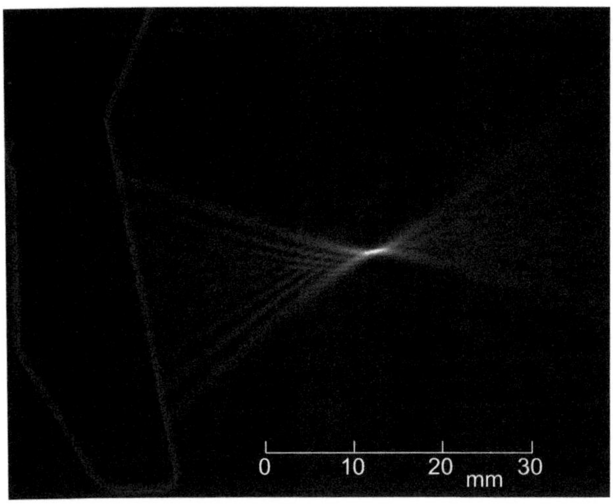

Fig. 36.10 Optical image of typical HIFU field

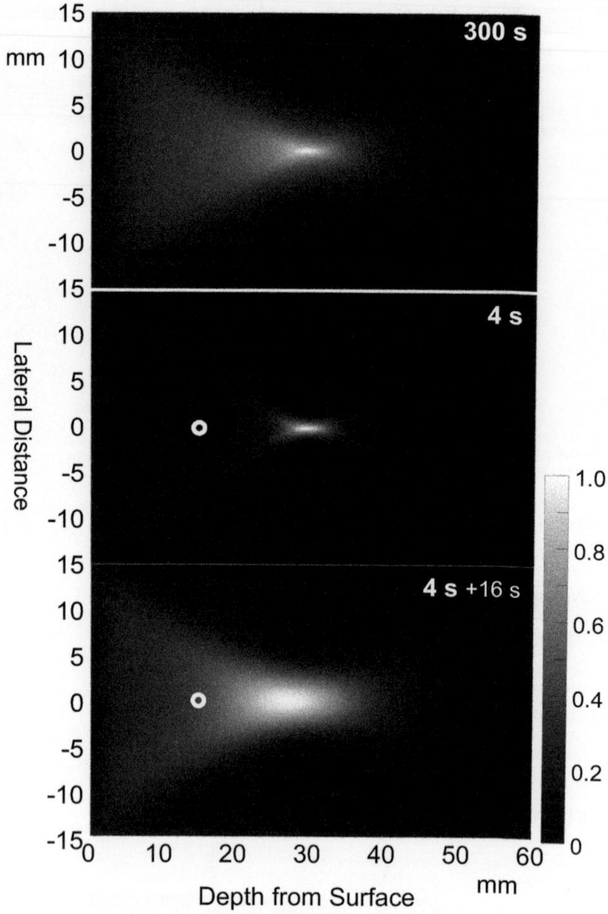

Fig. 36.11 Tissue temperature rise created by HIFU

Unlike a typical form of hyperthermia, HIFU treatment uses a transient thermal response. A single HIFU exposure of prostate is normally continued less than 5 s. The basic concept is to induce thermal coagulation before the ultrasonically generated heat diffuses away from the tissue at the focal spot. Figure 36.11 shows that the heated volume stays in the same order of magnitude as the focal spot when the HIFU exposure is continued for 4 s. In contrast, the geometric selectivity of HIFU heating will be lost if the exposure is too long. The generated heat will diffuse to the region around the focal spot forming a heated volume larger than the focal spot by orders of magnitude if the HIFU exposure is continued for 300 s as shown in Fig. 36.11.

When the target tissue to be treated is larger than the focal spot, the focal spot must be scanned to paint throughout the tissue either mechanically or electronically. Between each consecutive HIFU exposure, a certain period of time to cool the tissue, intervening between the focal spot and the HIFU transducer, is needed to protect the intervening tissue from irreversible change, because the cooling rate of the intervening tissue is much lower than the tissue at the focal spot by orders of magnitude as shown in Fig. 36.12. The tissue temperature distribution 16 s after the

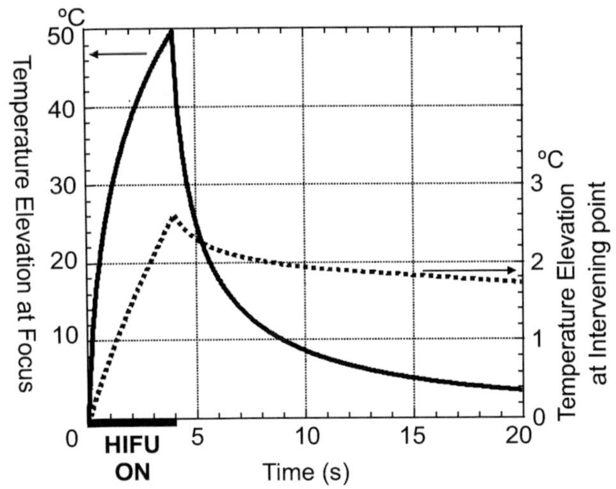

Fig. 36.12 Tissue temperature change induced by HIFU

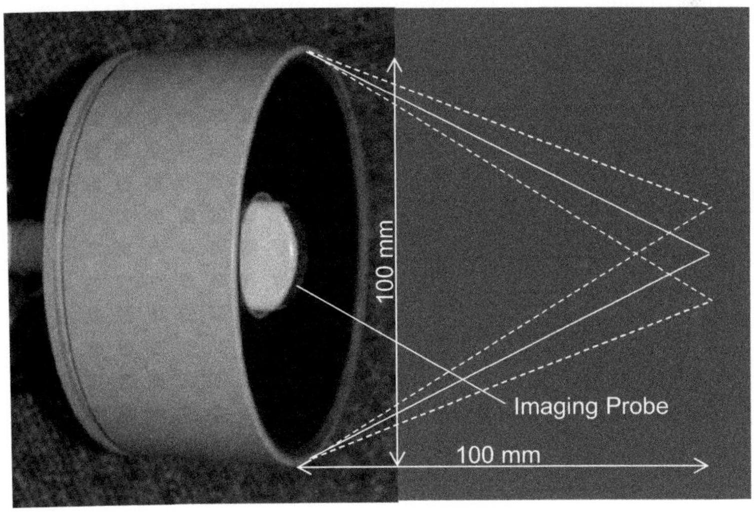

Fig. 36.13 Focused array transducer for extracorporeal HIFU treatment

cease of 4 s HIFU exposure as well as the sampling intervening point in Fig. 36.12 are also shown in Fig. 36.11. The reason behind is that the heat diffuses away from the intervening point almost only by blood perfusion while in case of the focal spot largely by heat conduction.

After the beginning of this century, clinical application of HIFU expanded to extracorporeal treatment of non-superficial target tissue [22]. A typical focused array transducer at 1 MHz for this purpose is shown in Fig. 36.13. The HIFU frequency is chosen to maximize the heat generated in the tissue at the focus.

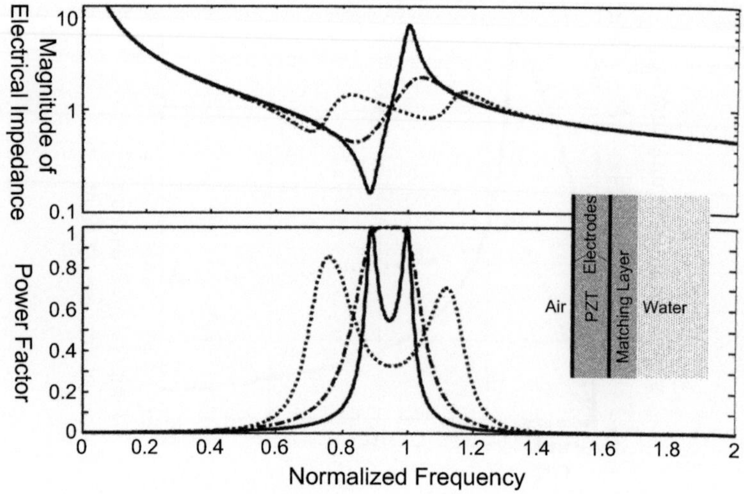

Fig. 36.14 Electrical impedance of air-backed transducer with matching layer

The resulting frequency is approximately inversely proportional to the depth of the target tissue [23] because the ultrasonic attenuation as well as absorption of biological tissue is approximately proportional to the frequency.

A typical HIFU transducer for heating tissue in either intracavitary or extracorporeal application has an air backed structure to prevent overheating of the transducer itself. When a matching layer is on the transducer, its optimum acoustic impedance to maximize the power factor is significantly less than that of the conventional design, which is the geometric mean of the acoustic impedances of the piezoelectric material and water to maximize the bandwidth, as shown in Fig. 36.14. Here, the solid curves denote the values without a matching layer, the dotted curves with the conventional matching layer, and the chained curves with the optimum matching layer with acoustic impedance 0.4 times the conventional [24].

Noninvasive treatment such as HIFU treatment must be guided by also noninvasive imaging, instead of naked eye for open surgery. Ultrasonic echography and MRI are used for such guidance. The advantage of MRI is thermometry while ultrasonic echography has advantages in self-consistency, speed, and cost. The self-consistency is in that both therapeutic and imaging ultrasound deflects in a similar way if there is nonuniformity in sound speed in the intervening tissue.

References

1. Hall EJ. Hyperthermia. In: Giaccia AJ, Hall EJ, editors. Radiobiology for the radiologist. 6th ed. Philadelphia: Lippincott Williams & Wilkins; 2006. p. 469–70.
2. Wei MQ, Mengesha A, Good D, Anne J. Bacterial targeted tumour therapy-dawn of a new era. Cancer Lett. 2008;259:16–27. doi:10.1016/j.canlet.2007.10.034.

3. Coley WB. The Treatment of inoperable sarcoma by bacterial toxins (the mixed toxins of the *Streptococcus erysipelas* and the *Bacillus prodigiosus*). Proc Roy Soc Med. 1910;3:1–48.
4. Seegenschmiedt MH, Vernon CC. A historical perspective on hyperthermia in oncology. In: Seegenschmiedt MH, Fessenden P, Vernon CC, editors. Thermoradiotherapy and thermo-chemotherapy. Berlin: Springer; 1995. p. 3–44.
5. Kikuchi M. [History of technology development and challenges for the future (in Japanese)]. In: Japanese Society for Thermal Medicine, editor. [Hyperthermia (in Japanese)]. Kobe: Shinryobunko; 2008. p. 154–155.
6. Wust P, Hildebrandt B, Sreenivasa G, Rau B, Gellermann J, Riess H, et al. Hyperthermia in combined treatment of cancer. LANCET Oncol. 2002;3:487–97.
7. Chicheł A, Skowronek J, Kubaszewska M, Kanikowski M. Hyperthermia – description of a method and a review of clinical applications. Rep Pract Oncol Radiother. 2007;12:267–75.
8. Dewhirst MW. Future directions in hyperthermia biology. Int J Hyperthermia. 1994;10: 339–45.
9. Kato H, Kondo M, Imada H, Kuroda M, Kamimura Y, Saito K, et al. Quality assurance: recommended guidelines for safe heating by capacitive-type heating technique to treat patients with metallic implants. Int J Hyperthermia. 2013;29:194–205.
10. BSD-2000/3D/MR. http://www.medicalexpo.com/prod/bsd-medical/product-84451-540429.html. Accessed 15 Oct 2015.
11. Thermotron RF8 EX Edition. http://www.vinita.co.jp/medical_div/medical/product/. Accessed 15 Oct 2015.
12. Hiraoka M, Mitsumori M, Hiroi N, Ohno S, Tanaka Y, Kotsuka Y, Sugimachi K. Development of RF and microwave heating equipment and clinical applications to cancer treatment in Japan. IEEE Trans Microw Theory Tech. 2000;48:1789–99.
13. Wright AS, Mahvi DM, Haemmerich DG, Lee FTJ. Minimally invasive approaches in management of hepatic tumors. Surg Technol Int. 2003;11:144–53.
14. Sugimachi K, Inokuchi K, Kai H, Sogawa A, Kawai Y. Endotract antenna for application of hyperthermia to malignant lesions. Jpn J Cancer Res. 1983;74:622–4.
15. Takahashi H, Tanaka R, Uzuka T, Grinev I. Stereotactic technique for radio-frequency antenna implantation for brain tumor treatment -technical note and report of complications-. Jpn J Hyperthermic Oncol. 2005;21:221–30.
16. Lin JC, Wang Y-J. Interstitial microwave antennas for thermal therapy. Int J Hyperthermia. 1987;3:37–47.
17. Ito K, Ueno K, Hyodo M, Kasai H. Interstitial applicator composed of coaxial ring slots for microwave hyperthermia. Proc Int Antennas Propagation Symp 1989;253–56.
18. Van Rhoon GC, Paulides MM, Togni P, Canters RAM, Rijnen Z, Van de Velde-verduijn G, Levendag PC. Challenges of the clinical application of hyperthermia for head and neck tumors. Proc European Conf Antennas Propagation.2013;635–6.
19. Szasz A. Challenges and solutions in oncological hyperthermia. Thermal Med. 2013;29:123.
20. Wall PD, Tucker D, Fry FJ, Mosberg WH. The use of high intensity ultrasound in experimental neurology. J Acoust Soc Am. 1953;25:281–5. doi:10.1121/1.1907032.
21. Uchida T, Sanghvi NT, Gardner TA, Koch MO, Ishii D, Minei S, Satoh T, Hyodo T, Irie A, Baba S. Transrectal high-intensity focused ultrasound for treatment of patients with stage T1b-2NOMO localized prostate cancer: a preliminary report. Urology. 2002;59:394–8. doi:10.1016/S0090-4295(01)01624-7.
22. Funaki K, Fukunishi H, Funaki T, Kawakami C. Mid-term outcome of magnetic resonance-guided focused ultrasound surgery for uterine myomas: from six to twelve months after volume reduction. J Minimally Invasive Gynecology. 2007;14:616–21. doi:10.1016/j.jmig.2007.04.009.
23. Hill CR. Optimum acoustic frequency for focused ultrasound surgery. Ultrasound Med Biol. 1994;20:271–7. doi:10.1016/0301-5629(94)90067-1.
24. Umemura S. Focused ultrasound transducer for thermal treatment. Int J Hyperthermia. 2015;31:216–21. doi:10.3109/02656736.2015.1008059.

Chapter 37
Epilogue

Takeo Ohnishi

Abstract This chapter includes a summary of some of the publications cited in this book, and this work focuses on the scientific basis for interest in thermal medicine. In particular, research topics in thermal medicine and hyperthermic oncology are described which appear to have significance for thermal medicine based on basic research. Much of this work has not been discussed in this book.

Keywords Apoptosis • DSB • Arrhenius plots • Adaptive response • Psychological stress • NO radical

37.1 History of Hyperthermic Oncology in Japan and the Publication of This Book

In 1985, the Japanese Society for Hyperthermic Oncology (JSHO), the former name of the Japanese Society for Thermal Medicine (JSTM) was established, and the 1st Japanese Congress of Hyperthermic Oncology (JCHO) was held at the same time with an annual Japanese meeting in Kyoto, and Dr. Tsutomu Sugahara and Dr. Mitsuyuki Abe helped found the JSHO. Initially, about 800 participants attended the JSHO meetings. When JSHO was established, the scientific fields covered in the JSHO meetings ranged from studies of *Escherichia coli* bacteria, Micrococcus, yeast, and other organisms to human cultured cells and clinical field trials. At that time, I presented work on heat sensitivity in cellular slime molds, a low eukaryotic organism. We have published many papers concerning heat sensitivity, heat shock proteins and thermo-tolerance in systems ranging from microorganisms to human cultured cancer cells. However, there were few reports concerned with clinical hyperthermic cancer therapy at that time. Several Japanese companies have designed and manufactured individual types of machines for applying hyperthermic cancer therapy, and consequently, many physical scientists

T. Ohnishi (✉)
Department of Radiation Oncology, Nara Medical University, 840 Shijo-cho, Kashihara, Nara 634-8521, Japan
e-mail: tohnishi@ares.eonet.ne.jp

S. Kokura et al. (eds.), *Hyperthermic Oncology from Bench to Bedside*,
DOI 10.1007/978-981-10-0719-4_37

have joined these meetings to describe methods used to measure physical temperatures in mammalian tumors and at the whole body level.

The 1st International Congress of Hyperthermic Oncology (ICHO) was held in 1970 in Washington DC, USA (the president was J. Robinson). The 5th ICHO was held in Kyoto, Japan on August 29–September 3, 1988. The president was Dr. T. Sugahara. The 15th ICHO in 2012 was again held in Kyoto on August 28–31, 2012, and the president was Dr. Toshikazu Yoshikawa. In that congress, many current advanced topics concerning hyperthermic oncology were reported on, and the subjects ranged from basic science to clinical efforts in cancer therapy. Since we would like most participants to understand progress in this field, scientific reports will be summarized here, and the opportunity to publish this work will hopefully enable workers in this field to understand and appreciate progress in this field. In addition, we would like to acknowledge the contributions from many Japanese scientists over the years. I have been the president of the JSTM and JSHO for about 13 years, and this September, we welcomed a new president, Dr. T. Yoshikawa. In view of our recent history, this book is being published to recognize the history and accomplishments of the JSTM and the 15th ICHO.

37.2 Specific Critical Temperatures in Arrhenius Plot Analysis

In the fields of thermal medicine and hyperthermic oncology, there are many relevant scientific topics (Table 37.1), and it is hoped that we will continue to progress and advance in these fields. Basically, it is accepted that tumor cells are more sensitive to heat than normal cells, because the blood vessels are very

Table 37.1 Scientifically relevant topics in thermal medicine and hyperthermic oncology

Scientific topics of interest in thermal medicine and hyperthermic oncology	
1	Tumor and normal organization
2	Cancer and normal cells
3	Critical temperature at 42.5 °C in Arrhenius plot
4	Target of heat-sensitivity
5	Heat-induced signal transduction
6	Thermal tolerance
7	Heat shock proteins
8	Cross talk or communication between different stress response pathways
9	Combination with other anti-tumor therapies
	(1) Radiation
	(2) Anti-tumor drugs
	(3) Immune therapy
	(4) Surgery
10	Progress in developing heating methods

different in normal and tumor tissues. The blood vessels in tumor tissues are not well developed when compared with normal tissues. The differences are very important in considering heating tissues with hyperthermic machines and physical techniques. Normal tissues can easily disperse or transmit higher temperatures due to well-developed blood vessels and vascular systems. Therefore, when heat is delivered to tumor tissues, temperatures increase more readily than in normal tissues. Recent molecular biology studies have analyzed heat sensitivity differences between cancer cells and normal cells by considering their genetic backgrounds and the presence of oncogenes and tumor suppressor genes. In this book, the focus is on the *p53* tumor suppressor gene, and the role that this gene plays in heat-induced signal transduction pathways affecting cell death, the cell cycle, and cell survival. Earlier work with heated human cultured cancer cells showed that there was a critical point for the temperature at 42.5 °C in Arrhenius plot analysis (Fig. 37.1). When the survival levels were graphed against heat energy delivered to human cultured cells, the resulting linear graph had a sudden change in the slope at 42.5 °C (Fig. 37.1). This inflection or break point at 42.5 °C where the linear slope changed is the critical temperature. The question is what does this critical temperature mean, and can it be understood from a molecular biology viewpoint. Is this critical temperature associated with proteins, DNA lesions and/or structures in the membranes or in the chromosomes? One possible proposed mechanism suggested that this phenomena involved DNA DSBs induced by a heat treatment. It was observed that the frequency of DSB formation was associated with the critical point temperature [1]. These important observations were described in this book. In addition, various other types of environmental stresses and chemicals were also found to induce DSBs (Fig. 37.2). The experimental reports concerning this work were reported over the last two decades.

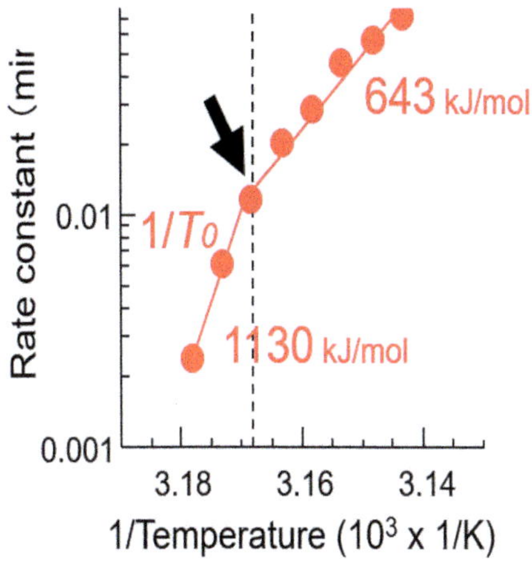

Fig. 37.1 Relationship between cell killing and temperature in human cultured lung cancer H1299 cells. A critical point was detected at 42.5 °C (*black arrow*)

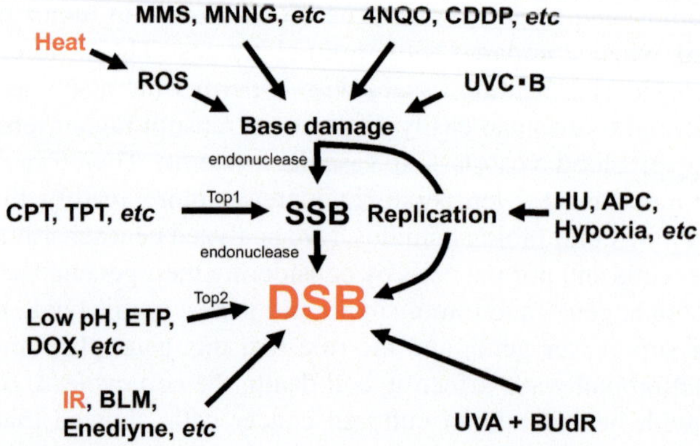

Fig. 37.2 DSB induction by various environments and types of chemicals in cells. DNA-damaging agents and anti-tumor drugs produced many types of DNA lesions in DNA. MMS, methylmethanesulfonate; MNNG, N-methyl-N'-nitro-N-nitrosoguanidine; 4NQO, 4-nitroquinoline-1-oxide; CDDP, cisplatin; ROS, reactive oxygen species; UV, ultraviolet; BUdR, bromodeoxyuridine; CPT, camptothecin; SSB, single strand break; HU, hydroxyurea; DOX, doxorubicin; IR, ionizing radiation; BLM, bleomycin; Top, topoisomerase

37.3 Adaptive Responses to Small Conditioning Stresses Protects Cells against Challenging Physiological Stresses

Many organisms can recognize and respond to various environmental stresses, even at very low levels, and these low level exposures can serve as a low dose conditioning stress against future larger challenging stresses. It was reported that exposures to low levels of stress could lead to more effective responses to a subsequent larger challenging physiological stress. This form of the adaptive response can function in the case of stomach damage in intact mice which is induced by physiological stresses [2]. Mice are usually maintained at 23–24 °C, and treatment with RWIS ("Restraint plus Water Immersion Stress" which generates a physiological stress) clearly induces ulcerative stomach damage (Fig. 37.3Ab) when compared to growth without any RWIS exposure (Fig. 37.3Aa). However, when mice were pre-exposed to 37 °C for 30 min at 12 h before experiencing RWIS, such stomach damage was almost completely suppressed (Fig. 37.3Ac). These results suggest that pre- or conditioning treatments may introduce tolerance to physiological stresses through the induction of HSPs. When mice were pre-irradiated with 0.45 Gy 2 weeks before an RWIS treatment, the grades of subsequent stomach ulcers were classified by using three grades of carmine and red colors observed when compared with normal stomach tissue colors observed. RWIS treatments were for periods of 1 h and 3 h. A depression in the appearance of stomach ulcers was observed after the two RWIS treatment periods of 1 h and 3 h (Fig. 37.3B). Severe damage indicated by a carmine color was largely depressed by the 3 h pre-exposure

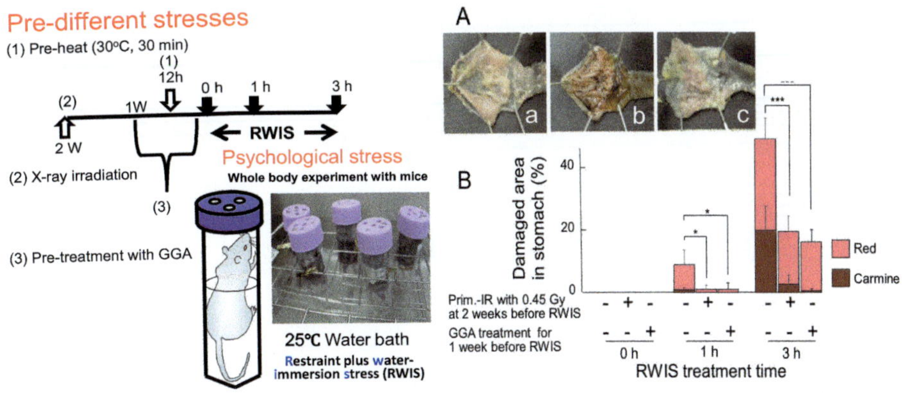

Fig. 37.3 Cross-talk or communication in adaptive responses after a pre-treatment with physical and chemical stresses against physiological stress in mice. (1) Pre-heat treatment at 37 °C room temperature for 30 min was performed at 12 h before physiological RWIS stresses were delivered. (2) X-rays were delivered to mice 2 weeks before RWIS. (3) An HSP-inducer of GGA was given to mice 1 week before RWIS. **A**, stomach ulcer induced by RWIS. **a**, no-treatment (control experiment); **b**, RWIS treatment for 3 h without pre-treatment; **c**, RWIS treatment for 3 h with pre-heat treatment at 30 °C for 30 min. **B**, stomach ulcer induced by RWIS. The grades of stomach ulcer were classified from the affected areas into two grades. Carmine, heavy damage; red, light damage

to RWIS. In addition, HSP inducer, geranylgeranylacetone (GGA), was given to mice in drinking water at a concentration of 200 mg/kg twice per day by pipette for 1 week before an RWIS treatment. In this case, a depression of ulcer damage was observed after a 3 h RWIS treatment (Fig. 37.3B). Stomach damage was also measured by analysis of apoptosis using immuno-histochemical staining. All of the pre-treatments depressed RWIS induced apoptosis. It was concluded that pre-irradiation with X-rays as well as pre-treatments with GGA induced Hsp32 (to about 2 times normal levels), Hsp70, and Hsp90 in the stomach wall. From these results, it appears that physiological stress-induced stomach ulcers may be depressed by HSPs, and this phenomenon is associated with different pre-treatments or with mild challenging stresses.

37.4 Synergistic Enhancement between Heat and Radiation Therapies

Evidence for the efficacy of hyperthermic therapy is very important. Scientific evidence provided by research in the basic sciences should provide evidence for the effects of hyperthermia. Human cultured cancer cell lines with a different *p53* gene status (wt*p53* and m*p53*) were established. The genetic background of the two cell lines is identical except for their *p53* gene status. These cells were transplanted into nude mice, and cancer therapies were delivered which consisted of a heat treatment at 42 °C for 20 min, or X-irradiation with a dose of 2 Gy [3]. Each of these

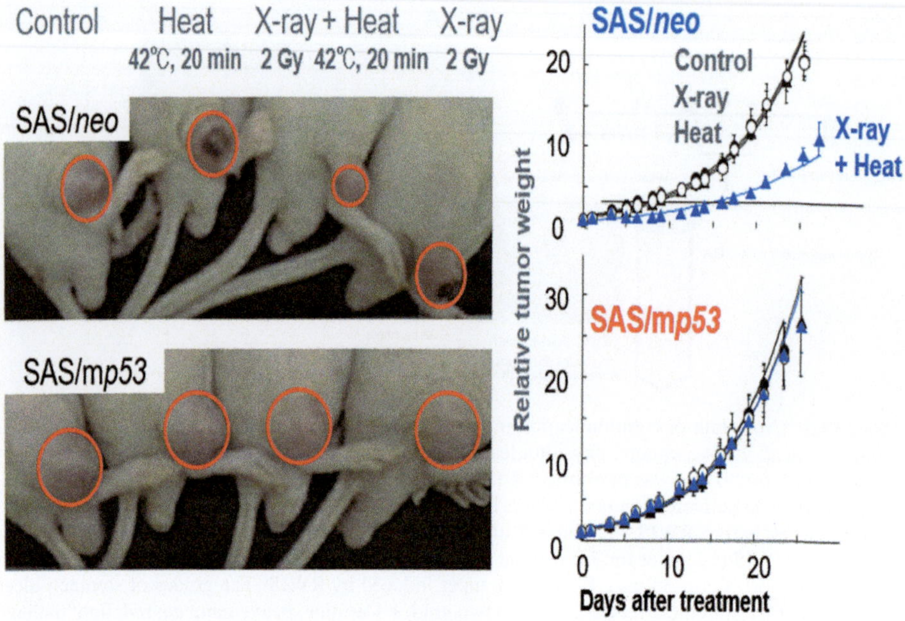

Fig. 37.4 Two human cultured cancer cell lines derived from human squamous cell carcinoma (SAS) cells bearing wt*p53* (SAS/neo) and m*p53* (SAS/m*p53*) genes were transplanted to the legs of nude mice. Treatment with heat (42 °C, 20 min) or X-ray (2 Gy) alone was delivered to the transplanted mouse tumors. Tumor growth was measured after treatment with heat and/or radiation

individual treatments was insufficient to obtain effective therapeutic results. Tumor growth of the wt*p53* cell line was not depressed by either treatment. However, these combined treatments apparently depressed tumor growth. These results indicated the existence of viable synergistic effects using hyperthermic therapy and radiation therapy. With m*p53* cancer cells, all treatments, especially the combined treatment, was unable to depress the tumor growth at all. These results suggest that a combination therapy using hyperthermia and radiation are affected by *p53* status and are *p53*-dependent (Fig. 37.4). Consequently, the genetic characteristics of cancer cells are clearly very important and must be considered when developing effective cancer therapies.

References

1. Takahashi A, et al. Evidence for the involvement of double-strand breaks in heat-induced cell killing. Cancer Res. 2004;64:8839–45.
2. Yamagata R, et al. Restraint plus water-immersion stress (RWIS)-induced stomach ulcers in mice are depressed by a conditioning heat treatment. Thermal Med. 2009;25:81–7.
3. Asakawa I, et al. Radiation-induced growth inhibition in transplanted human tongue carcinomas with different *p53* gene status. Anticancer Res. 2002;22:2037–43.